THE SCIENCE OF MEASUREMENT

THE SCIENCE OF MEASUREMENT

A Historical Survey

Herbert Arthur Klein

DOVER PUBLICATIONS, INC. *New York*

Published in Canada by General Publishing Company, Ltd., 30 Lesmill Road, Don Mills, Toronto, Ontario.
Published in the United Kingdom by Constable and Company, Ltd.

This Dover edition, first published in 1988, is an unabridged, corrected republication of the work originally published by Simon & Schuster, Inc., New York, in 1974, and by George Allen & Unwin Ltd., London, in 1975 under the title *The World of Measurements: Masterpieces, Mysteries and Muddles of Metrology.*

Manufactured in the United States of America
Dover Publications, Inc., 31 East 2nd Street, Mineola, N.Y. 11501

Library of Congress Cataloging-in-Publication Data
Klein, H. Arthur.
      [World of measurements]
      The science of measurement : a historical survey / Herbert Arthur Klein.
            p.   cm.
      Reprint. Originally published: The world of measurements. New York : Simon & Schuster, 1974.
      Includes index.
      ISBN 0-486-25839-4 (pbk.)
      1. Weights and measures.   2. Weights and measures—History. I. Title.
QC88.K57   1988   88-25858
530.8′09—dc19   CIP

Introduction

Expression of hail and farewell is the welcome prerogative of an author at the penultimate point, when the text is crystallized, for better or worse, and before the start of the production processes that are to transform the word into substance. This recalls the insight of Blaise Pascal, one of the great men of measurement, to whom the following pages seek to do justice; for he noted that "The last thing we learn when writing a book is what to put first."

First place appears, beyond doubt, to belong to two outcomes of the long labors that produced the following fifty-four chapters: one a small confession, the other a multiple statement of sincere thanks. The confession can be conveyed in a phrase from another Introduction, written nearly a score of years ago by James R. Newman for his esteemed four-volume anthology, *The World of Mathematics:* "what was envisaged as a volume of moderate size has assumed dimensions which even a self-indulgent author must acknowledge to be extended." First thanks are accordingly offered to Peter Schwed of Simon and Schuster for permitting the self-indulgence. And to the readers of every age, kind and condition, the author confesses that he found the subject matter so infernally fascinating that briefer treatment seemed undesirable, even when attainable.

Though the present work can be called "extended," the author feels the effects of the fisherman syndrome. This is the recollection expressed by "Yes, but you should have seen the ones that got away!" Or better, by "the ones that were tossed back and not brought home." With a mixture of resignation and regret the author recalls a variety of aspects of modern metrology that remain

7

unmentioned, some because of considerations of length, others because of their peculiar characteristics.

In particular, there is here no exploration of the realm of biochemical and pharmaceutical metrology, with its units of potency or efficacy, so basic in endocrinology, enzyme studies and other active areas. To deal properly with all this, a separate work obviously is required.

The efforts, spread over nearly ten years, which culminate in the present book were aided by many acts of assistance and encouragement from a number of patient, generous and informed persons. Some of that help came in correspondence, some in conversations. Much of it was in response to inquiries initiated by the author, and brought either correction or validation of conclusions that were as yet but tentative. And an appreciable as well as appreciated part of the help amounted mainly to a bolstering of the author's belief that a work as wide-ranging as this one could prove useful to laymen and specialists alike.

Quite possibly some of those named in the following thank-you list may by this time not recall the occasion or nature of the help for which they are thanked; but the author assuredly has not forgotten. There is no rational way to assign a rank-order of thanks here, so the accidental sequence of the alphabet is followed. Doctoral and professorial titles are omitted, but individuals are identified by positions held at the time their help was given:

James A. Barnes, Chief, Time and Frequency Division, Institute for Basic Standards, National Bureau of Standards; J. E. Barrell, Director, Customs Cooperation Council, Brussels, Belgium; Bruce B. Barrow, Technical Activities Board, Institute of Electrical and Electronics Engineers; J. de Boer, Instituut voor Theoretische Fysica, Universiteit van Amsterdam, Holland; Stephen Dresner, London, author of the valuable *Units of Measurement: An Encyclopedic Dictionary* (Aylesbury, England, 1971).

Also: Stanley Feuer, Feuer Corporation, air conditioning contractors, Los Angeles; L. E. Howlett, Division of Applied Physics, National Research Council of Canada, Ottawa; H. S. Hvistendahl, Newcastle upon Tyne, author of *Engineering Units and Physical Quantities* (London, 1964); Daniel Kleppner, Department of

Physics, Massachusetts Institute of Technology; Joel L. Lebowitz, Physics Department, Yeshiva University, Belfer Graduate School of Science, New York City.

Also: A. G. McNish, Chief, Metrology Division, Institute for Basic Standards, National Bureau of Standards; H. Moreau, Bureau International des Poids et Mesures, Sèvres, France; Sheldon Novick, Editor, *Environment,* St. Louis, Mo.; Ronald F. Peierls, Acting Leader, High Energy Theory Group, Brookhaven National Laboratory, L. I., N.Y.; H. Preston-Thomas, Division of Physics, National Research Council of Canada, Ottawa; Theodore T. Puck, Director, Institute for Cancer Research, and Professor of Biophysics and Genetics, University of Colorado Medical Center, Denver.

Also: Arthur Schawlow, Department of Physics, Stanford University; Arnold H. Silver, Physics Department, Scientific Laboratory, Ford Motor Company, Dearborn, Mich.; Louis F. Sokol, President, Metric Association, Inc.; Martin V. Sussman, Department of Chemical Engineering, Tufts University, Medford, Mass.; M. P. Thekaekara, NASA Goddard Space Flight Center, Greenbelt, Md.; Alvin Wald, University Medical Center, University Hospital, New York City; George P. Yost, Department of Physics, University of California at Berkeley.

A special category of thanks also goes to several individuals for courtesies and attentions that once again justify a truism that the present author put into doggerel in the Foreword to a quite different book, about a dozen years ago:

> Could books like these be finished, sans
> The work of fine librarians?
> By words alone can't be repaid
> Helpful librarians' great aid!

Such thanks are due the units of the great library complex of the University of California at Los Angeles, as represented by Wally Pegram, Physics Library; John Hill, Geology Library; Rosalie Wright, Engineering and Mathematical Sciences Library; and also through University Librarian Page Ackerman, to each and every staff member who made the author's efforts more pleasant and productive.

Similar thanks are offered to Evelyne Hamilton, Librarian, Malibu Branch of the Los Angeles County Library, and her staff.

In an era when libraries great and small suffer under short-sighted budgetary cuts and enforced reductions of services, the continuing helpfulness of all these staffs merits more than perfunctory mention.

In the "private sector," thanks are given for access to the reference facilities of the Hughes Research Laboratory of Malibu, through the good offices of Toby Ann Mandel, Librarian, and Ed Reese, Technical Information Manager. Good neighbors indeed!

Needless to emphasize, the author alone bears the blame for any errors, whether of commission or omission, that may mar the pages that follow this true confession.

Finally, the essential assistance given by Mina C. Klein at every stage of the long road is inadequately suggested by the dedication.

HERBERT ARTHUR KLEIN

Malibu, California, 1974

To the memory of

MINA C. KLEIN (1906–1979)

for more than can be measured

Galileo Galilei
(1564–1642)

Evangelista Torricelli
(1608–1647)

Blaise Pascal
(1623–1662)

Isaac Newton
(1642–1727)

René Antoine Ferchault de Réaumur
(1683–1757)

Anders Celsius
(1701–1744)

Contents

IV / This Thermal Universe

V / The Body Electric

VI / Problems of Pressures, Densities, Strong Drinks, Sound, and Flow

VII / Nuclear Disintegrations and Some Other Pressing Problems

Antoine Baumé
(1728–1804)

Charles Augustin de Coulomb
(1736–1806)

James Watt
(1736–1819)

Alessandro Volta
(1745–1827)

André Marie Ampère
(1775–1836)

Hans Christian Oersted
(1777–1851)

The eminence of measurements among the major attainments of human-kind is stressed in the great engraving known as *Temperance,* after a 1560 drawing by Peter Bruegel the Elder. (The engraving itself probably was done by Philippe Galle, who has been called "the most subtle and resourceful of the engravers who worked from Bruegel's originals" during the lifetime of the great Flemish master.)

This part of the engraving shows typical measurements on Earth. Using a divider, one measurer checks the dimensions of a pillar, while seated precariously on a support suspended from its top. Above, another measurer ascertains the pillar's angular position, using a plumb line. The angular separation of distant objects is measured by a man sighting along what looks like a stick with a wheel at its far end—a suggestion of trigonometric determinations of lengths. Behind the latter, an engraver measures angles around a work of art at his feet.

At the other side of the pillar, cannons and crossbows suggest the measurement of trajectories.

These practical metrological activities, incorporating the most advanced geometry of the time, contrast sharply with the efforts of the five disputants in the foreground: they are theologians, using the arts of rhetoric to set forth the relative merits of the Catholic, Protestant, and Jewish faiths.

Extraterrestrial or cosmic measurements are suggested also in Bruegel's picture. At right here, a robed geo-metrologist uses a divider to measure distances on the great globe of Earth. Above, perched as if on the north pole, the astro-metrologist boldly measures the heavens themselves. The Moon, the Sun, and the stars are all subject to his quantitative determinations. The face in the Sun indeed appears to regard with interest all this insatiable measuring activity.

Here is Temperantia herself, presented by Bruegel as the queen and patron of the useful, applied, and liberal arts, including those of mensuration and metrology.

Five significant symbols reveal what she meant to Bruegel. On her head is supported the most advanced, sophisticated machine of his era: the clock, which measures passage of time. Under her feet is the blade of a windmill, an important power source in the Lowlands, whose life Bruegel knew so intimately. Her left hand (to our right) holds a pair of spectacles, a rather recent as well as useful aid when Bruegel worked. Her right hand grasps the reins to the bit in her own mouth. This suggested firm self-control and direction, or purpose.

Her girdle is a knotted snake, probably suggesting mastery over the crude physical desires or impulses surging from below. This Temperance to Bruegel was no exponent of ascetic or withdrawn ways of life. Rather she represents the opposite of sloth, of backward ignorance, and of primitive helplessness. Confident and competent she stands, surrounded by evidences of the traditional liberal arts and also of the proliferating technologies of the artist's time.

Both the older arts and the newer technologies intersect and interact in the work of measurement. That work was expanding and diversifying in Bruegel's land as it has continued to do worldwide during the four centuries since the first prints appeared from this instructive engraving.

Karl Friedrich Gauss
(1777–1855)

Georg Simon Ohm
(1787–1854)

Michael Faraday
(1791–1867)

Joseph Henry
(1797–1878)

Wilhelm Eduard Weber
(1804–1891)

James Prescott Joule
(1818–1889)

I / The Babel Behind Us: Profusion of Confusions and Contradictions

> . . . *learning, the sciences, all shall be here,*
> *None shall be slighted, none but shall here be*
> *honor'd, help'd, exampled.*
> —WALT WHITMAN,
> *Song of the Exposition* (1881)

> . . . *we must remember that measures were made for man*
> *and not man for measures.*
> —ISAAC ASIMOV,
> *Of Time and Space and Other Things* (1965)

Rudolf Julius Emanuel Clausius
(1822–1888)

William Thomson, Baron Kelvin
(1824–1907)

Alexander Graham Bell
(1847–1922)

Nikola Tesla
(1856–1943)

Heinrich Rudolf Hertz
(1857–1894)

Ernest Rutherford
(1871–1937)

1 WHY IN THE WORLD MAKE MUCH OF MEASUREMENT?

> . . . *thought,*
> *Which is the measure of the universe.*
> —SHELLEY,
> *Prometheus Unbound* (1820)

> *Why, who makes much of a miracle?*
> *As to me I know of nothing else but miracles* . . .
> —WALT WHITMAN,
> *Miracles* (1856)

> *Science is a continuum, and any attempt to fragment it into its component parts does violence to the whole.*
> *As physical measurements have steadily become more elegant, definitive, and profound, it has become obvious that all scientists are dealing with the same basic laws. Scientists are increasingly being required to learn one another's language in order to remain effective* . . .
> —Announcement, New England Institute,
> Ridgefield, Conn., 1966

"Man is the measure of all things," declared Protagoras, the great Sophist philosopher of the fifth century B.C.E. That pithy, cryptic statement is, at any rate, credited to him by Plato in *Theaetetus,* written during the following century.

In this era of accelerating technology and science, some twenty-five centuries later, it may be more suitable to modify the Protagorean epigram as follows: "Man is the measurer of all things."

This, though obviously an overstatement, points in the direction that distinguishes our era. Accurate, repeatable numerical measurement or quantification is a constant objective, even in areas where it is not yet fully achieved. Measurement is a massive, many-sided activity in all branches of production of the necessities, conveniences and luxuries of life. Moreover, the tools and techniques of measurement provide the most useful bridge between the everyday worlds of the layman and of the specialists in science.

Molière's "bourgeois gentleman" was gratified to learn that all his life he had been speaking *prose,* a category of rhetoric. Nonscientists may be similarly impressed to discover that units of measurement—for length, area, volume, time duration, weight, and all the rest—are essentials of science.

The author, during research and writing of several books on recent advances in science, became impressed, and finally almost possessed, by the conviction that modern measurement methods provide the broadest, most natural, least restricted road on which nonscientists can approach modern science.

This work is the result of that conviction. It should prove serviceable to professionals in science, but its main purpose is to make outsiders realize that in their daily lives and concerns they too are involved in the activities and ideas classified as *metrology,* the science of measurement—a subdivision of science that underlies and assists all others.

The histories of weights and measures are rich, quaint, and—for most of us—abundantly fascinating. This book is not primarily, however, a historical survey of the manifold measuring units, standards and methods of the past. It is oriented toward the present and the future.

The framework of the following chapters is provided by the units and definitions of the most complete, consistent, and serviceable approved system evolved by the world's scientists—the International System of Units, commonly designated SI, the initials of the first two words of its name in French:

Système International des Unités—The SI is the logical expansion of the metric system of units, whose origins date back 175 years. The General Conference on Weights and Measures, which met in Paris in October 1960, gave the name International System of Units to the metric system based on the meter, unit of length; the kilogram, unit of mass; the second, unit of time; the ampere, unit of electric current; the kelvin, degree of temperature; and the candela, unit of luminous (light) intensity. Associated with these six precisely defined units are a variety of derived and supplementary units, which will all be presented in the following pages.

The International System has not remained rigid and static since it was formulated and recommended for use in all scientific work. Improvements and extensions are recommended after careful study by the appropriate bodies of the International Committee on Weights and Measures, and approved by an international general conference, which meets every six years.

The international agreement known as the Treaty of the Meter is the foundation for the continuing International Committee on Weights and Measures. From that treaty flows the program that brings together the periodic Conference on Weights and Measures.

In 1960, when the SI came of age, thirty-six nations had declared their adherence to the Treaty of the Meter. Today the number is larger.

Approval of the International System is nearly worldwide; and its application, though not complete, is enormous and is increasing with impressive rapidity.

It is, beyond all doubt, *the* international system of units for science and technology. The changes in unit usage that are taking place throughout the world today are, with insignificant exceptions, changes *from* units superseded by the SI *to* units contained in the SI.

The breadth and flexibility of the SI makes it more than a mere "extension" of the metric system of units. In its use of the meter and kilogram as basic units of length and mass, it has rendered obsolescent the CGS (centimeter-gram-second) system, which was general in scientific work not many years ago.

The SI is not only an MKS (meter-kilogram-second) system of units, but an MKSA system, the A (for ampere) giving it marked advantages over the CGS system, which contained no basic electrical unit.

Woven into the fabric of the International System are a variety of other *single-named units*. They include two so-called supplementary units: the radian, to measure plane angles, and the steradian, to measure solid angles; also a variety of derived units, named (like the ampere) for great scientists, mathematicians and inventors, such as:

The newton, unit of force.

The joule, unit of energy, work or quantity of heat.

The watt, unit for measuring power, the rate at which energy is developed or transferred from one form to another.

The coulomb, unit of electric charge, or "quantity of electricity."

The volt, unit of electromotive force, electric potential, or voltage.

The ohm, unit of resistance to the flow of electricity.

The farad, unit of electric capacitance.

The weber, unit of magnetic flux (or flow).

The henry, unit of electric inductance.

The tesla, unit of magnetic flux density.

In the realm of light measurement the SI provides, besides the basic candela of luminous intensity, the lux, unit of illumination.

In addition to the second, basic unit of time, the SI provides a special unit, the hertz, for measuring frequency of cyclical or repeating processes, 1 hertz meaning 1 cycle or completed event per second.

The single-named units of the SI are combined readily and variously to provide *compound units* for the measurement of many important physical relationships. All of us are familiar with such compounds, for we speak of automobile speeds in *miles per hour,* or *kilometers per hour.* The equivalent, in basic SI units, is the compound *meters per second.*

There are also *triple-unit combinations* in the SI, such as the *newton-second per square meter,* a suitable measure of the physical effect called kinematic viscosity. It multiples *force* by *time,* then divides by *area.*

Let us turn back from the contemporaneity and futurity of the International System of Units to some historical highlights that will exhibit the importance of units of measurement.

In the United States, though we are reputed to be the most scientifically and technically advanced among nations, ancient and illogical units are retained in everyday use. We have the familiar inch, foot and mile of length; the pint, quart and gallon of capacity; the pound and ton of gravitational force (weight), all derived from ancient British units. Let us dip into history for a few significant records and relics from the complex, crazy, mixed-up, sometimes even mad metrology of the past.

2 VARIETIES OF VOLUMES

. . . that orbit of the restless soul
Whose circle grazes the confines of space,
Bending within the limits of its race
Utmost extremes . . .
     —GEORGE H. BOKER,
      Sonnet

Measurement demands some one-one relations between the
numbers and magnitudes in question—a relation which
may be direct or indirect, important or trivial, according to
circumstances.
     —BERTRAND RUSSELL,
      The Principles of Mathematics (1937)

So long as there was an England there were persistent problems
of units of measurement—recognized and reproducible standards
of weight and capacity, or volume. The establishment and maintenance of standards of weights and measures have long caused
concern and created complications.

After the Roman conquerors of Britain had departed, the Anglo-
Saxon invaders made use of the abandoned Roman camps and
colonies. There the Angles and Saxons established their tribal
strongpoints and fortified defense centers called *burhs,* later to be
known as *boroughs* and to give rise to a variety of place names
ending with -bury or -borough. The burh was the seat of the local
chieftain or regional ruler. Commerce flowed to and from the
burh; courts and offices for levying taxes and tolls were established
there. The law of the land—to the extent that legislation of such a
scope existed—provided the burh with authorized scales and
standards for weights and measures.

The reeve, a royal officer responsible to the king, collected fees
from the services provided by the burh, including supervision of

the weights and measures used in commercial transactions. Annually a traditional burh meeting took place involving the members of the reeve's council, the witan. At this yearly burh-wara problems of weights and measures might be considered and, if not settled, passed along up to the king.

During the reign of Ethelred the Unready, king of England from 968 to 1016 C.E., a statute appeared, approved by him and his witan. In bland and pious rhetoric not unlike that discernible in legislative texts for nearly a thousand years since, it urged that "hateful illegalities be earnestly shunned" throughout the realm. First among the illegalities mentioned were "false weights and wrongful measures."

In 1066 William the Bastard of Normandy crossed the Channel with his forces and in the Battle of Hastings became William the Conqueror. Overleaping some four generations of Norman rule and reorganization, during which conflicts and contradictions of units of measurement continued in new contexts, we arrive at another much-memorized date and site: mid-June 1215, C.E., at the lovely Runnymede meadow. There a reluctant King John, under heavy pressure from the barons, set his seal to the famous Great Charter, or Magna Charta. Among the reforms it promised was an important metrological policy: henceforth there should be but "one weight, one measure" for the realm—in other words, uniform units, recognized, reproducible, and reproduced, on which buyers as well as sellers could rely.

Unfortunately, chicanery, evasion and trickery tend to make history more colorful than do routine observances of law and order. The record of units of weight and measure used in England is, notwithstanding the Magna Charta, contradictory, complex and racy, replete with deviations, devices and dodges, most of them tending to enrich cheating merchants, local officers and money-hungry monarchs at the expense of those more honest but less influential.

Revisions and reforms appeared and reappeared—on parchment, if not in reality. Henry III, coming to the throne in 1216—in the very shadow of Magna Charta, so to speak—proclaimed that, in keeping with the great document, there should be in his realm but one standard of weight and but one measure of volume or capacity. Let us look first at capacity.

Standard measures of capacity, area, and weight were promised by King John of England when he unwillingly set his seal to the famed Magna Charta, June 15, 1215, at Runnimede. Arrows indicate the beginning and the ending of the "measurements" pledge. It was the thirty-fifth of the sixty-three clauses or pledges to his barons. Above, the scribe's long script lines cross two book pages. The original is in the British Museum (Cotton MS. Augustus ii 106).

Translated from its medieval Latin into modern English, this clause stipulates: "Throughout the kingdom there shall be standard measures of wine, ale, and corn. Also there shall be a standard width of dyed cloth, russet, and

haberject; namely [a width of] two ells within the selvedges. Weights [also] are to be standardized similarly."

Uniformity and specificity of units of measurement were desired even by the rude barons of the early thirteenth century, as they forced an unpopular monarch to promise to meet their most urgent demands. In our era, more than 750 years later, the benefits of universal uniformity and specificity of standards are still incompletely attained; however, the International System (SI) is widely and increasingly used and seems certain to be in use worldwide before the year 2000 arrives.

Which barrel do you mean?—Conflicting measures of capacity were, by established practice, used to measure amounts of different kinds of substance. An ale barrel was recognized; likewise a beer barrel; each of these contained a specified number of gallons. The ale barrel was based on a so-called corn gallon of just under 269 cubic inches capacity. The beer barrel was based on a wine gallon of 231 cubic inches capacity.

In 1493, during the reign of King Henry VII, a statute whose original was written in Old French was translated into English. It declared that since the capacity standards for "Wine, Oyle, and Honie" ought to contain a "certain" (i.e., sure, or definite) measure, "then every Pipe" should contain "six score and six gallons," that is, 126 gallons.

Now, a pipe was equal to half a tun and also to 2 hogsheads and to 4 barrels. Each barrel, then, was to contain 31½ gallons—meaning, in this case, the wine gallon of 231 cubic inches capacity. Thus the statutory barrel for "Wine, Oyle, and Honie" and some other liquids represented a volume of $31½ \times 231 = 7,276½$ cubic inches.

Under Henry VIII, who reigned from 1509 to 1547, appeared a statute decreeing that the beer barrel should contain 36 gallons, as against only 32 gallons in the ale barrel. However, it is the opinion of Edward Nicholson, in his *Men and Measures* (London, 1912), that each of the 32 gallons in the ale barrel was a still larger gallon of 282 cubic inches capacity, or about 22 percent greater capacity than the contemporary wine gallon. Thus the ale barrel would contain 9,024 cubic inches as against 36×231, or 8,316 cubic inches, in the beer barrel.

This Tudor beer-barrel polka gave the ale-barrel buyer an advantage of some 8 percent in capacity. However, since there was presumably no hiding the difference, it can be argued that there was no deception of the wholesalers and publicans who bought their ale and beer by the barrel.

The metrology of Elizabethan England—The reign (1558–1603) of the first Queen Elizabeth of England brought important changes in units of measurement and weight. The so-called corn gallon was restored to its former size of 268.8 cubic inches, while the wine gallon remained at 231 cubic inches. The old magnitude of the corn gallon, 282 cubic inches, became the magnitude of the new ale gallon, which was used as the measure for malt liquor. The use of three different gallons—of 231, 268.8 and 282 cubic inches capacity—certainly violated the principle of "one weight, one measure" supposedly ensured by Magna Charta.

Yet, one of the important reforms accomplished under Good Queen Bess, in the 1590s, was a general standardization of measures of capacity, weight and length for use in her realm. There were complications, as indicated, but also clarity prevailed—and standards were provided to help judge working measures suspected of deficiencies.

During the first few decades of the seventeenth century, it appears, the ordinary man or woman of England could feel reasonably certain of receiving a known quantity when making a purchase of food or drink, unless the vendor was a downright rascal.

With this new security in measures and weights of commodities in common use came an improvement in the standards of coinage. Debasement of British coinage had been something of a scandal. Edward Nicholson has said that following "the conquest of England by Henry Tudor [who, as Henry VII, reigned from 1485 to 1509] a cloud of deceit came over the coinage." Elizabeth, however, established the coinage on "an honest basis."

By this time there had arisen a remarkable doubling design in the common units of capacity and volume. So far we have singled out only two—the gallon (or the various gallons), and the barrel, or barrels. They were but two rungs on a long ladder built on the "doubling my predecessor" pattern. Let us run through it from smallest to largest, simplifying somewhat as we go.

The poor of necessity lived literally from hand to mouth, much as now. Their purchases of honey or oil might be made in quantities as small as the *mouthful,* a unit of not quite 1 cubic inch capacity. (A mouthful of a liquid whose density is similar to that of water weighs about ½ ounce, or $\frac{1}{32}$ pound, in the nonscientific weight units still in use in the United States.)

Twice the size of a mouthful was the *jigger,* or *handful.* Next came the *jack,* or *jackpot,* or *double jigger.* Then *gill, jill,* or *double jack.* Twice the jill brings us to the *cup,* and twice a cup presents us with the familiar *pint.*

Thus a pint equaled 32 mouthful measures, 16 handful or jigger measures, 8 jack measures, 4 gill measures, or 2 cup measures. The liquid pint, as used in the United States today, has a capacity of 28⅞ cubic inches or 473.179 cubic centimeter. The British liquid pint is a shade over 20 percent greater in capacity than the U.S. liquid pint, which holds about 1.042 pounds of water at ordinary room temperature.

The supposedly international truism "A pint's a pound the world around" should be altered for Britain to "A pint's a pound plus a fifth, or a pound plus 3 ounces, all around Britain." Soon, however, full British application of the metric system will make the pint and all related measures obsolete.

A sixteenth of a U.S. liquid pint today equals 1 U.S. fluid ounce, or a little over 1.8 cubic inches in volume. If water is the liquid measured, a U.S. fluid ounce weighs close to 1 ounce (avoirdupois) or $\frac{1}{16}$ pound (avoirdupois).

One imperial (British-Canadian) pint of water weighs 1¼ pounds, or 20 ounces. The imperial jigger—if we may suppose that it exists today—represents $\frac{1}{16}$ of 20 ounces, or 1¼ ounces of water.

The common jigger measure used by bartenders in the United States contains not 1 ounce, but about 1½ ounces of water, and somewhat less weight of vodka or other alcohol-rich distilled spirits, which are substantially less dense than water.

A fairly clear line of descent has thus been traced from the jigger, or handful, of Elizabethan England to the customary unit for dispensing the "firewater" that is the most prevalent drug used in our own time and culture, nearly four centuries later. In the United States the half jigger, sometimes called a *pony,* is half again the Elizabethan mouthful.

Twice a pint is a *quart;* twice a quart is a *pottle;* and twice a pottle is a *gallon.* Today the U.S. gallon contains 231 cubic inches —the same as the old British wine gallon. Today's imperial gallon, which is also used in Canada, contains 277.42 cubic inches— that is, 46.42 cubic inches more than the U.S. gallon. A gallon plus 1½ pints U.S. measure almost equals 1 British-Canadian gallon.

This review of the double-my-predecessor procession has

reached the gallon in eight steps. Beyond that lie many other units, beginning with the double gallon, or *peck;* the double peck, or *half bushel;* and the quadruple peck, or *bushel,* corresponding to 8 gallons.

Next in ascending order we come to a volumetric unit with three intriguing names: the *cask,* the *strike,* and the *coomb.* At certain times and places in Britain's past these may have occupied independent places in the long double-my-predecessor procession. By now, however, they have become merely names for measures containing about 16 British gallons, or 8 pecks, or 2 bushels.

The terms *strike* and *coomb* are so obsolete that they are not listed in the extensive *Random House Dictionary of the English Language* (1966). The cask is listed there merely as an undefined volume—in other words, as the name for whatever capacity an actual cask may contain.

Next in doubling up we arrive at the important *barrel,* sometimes called the chaldron, which holds very nearly 32 gallons. The U.S. liquid barrel contains 31.5 U.S. gallons. However, the U.S. oil or petroleum barrel contains 42 U.S. gallons, and today's British dry barrel contains 36 imperial gallons.

In the old English doubling procession, however, the barrel unquestionably occupied the place of 32 gallons and—if we wish to go all the way back to the bottom of the great ladder—of 4,096 jiggers or of 8,192 mouthfuls!

Doubling the barrel brings us to the *hogshead,* of close to 64-gallon capacity.

The standard hogshead, according to the *Oxford English Dictionary,* once contained 63 "old wine gallons," equal to 52½ imperial or Canadian gallons. However, some hogsheads actually held as much as 140 such gallons, and in the 20th century the British hogshead has been stabilized at 63 imperial gallons.

The earliest prescription for hogshead capacity appears in a statute of 1423. Hogsheads, like gallons, long varied widely in capacity according to the liquid they were intended to contain. The ale hogshead once contained only 48 gallons; the beer hogshead outside London, 51 gallons; and the more generous London beer hogshead, 54 gallons.

The next doubling brings us to the double hogshead, or *pipe,* sometimes known also as the *butt.* (This last name is not to be confused with the *butt* that refers to various flatfishes, especially the turbot, flounder and halibut. The -bot in turbot and the -but in halibut are derived from this butt.)

The pipe, or butt, contained 126 old wine gallons. It was a measure well known to Shakespeare's Sir John Falstaff, that prodigious consumer of sack (sherry). It has been described as a measure equivalent to 2 hogsheads or 4 barrels, containing usually 105 imperial (British) gallons. This corresponds very nearly to the seventh power of 2, for $2^7 = 128$. Thus the pipe, or butt, is seven steps or rungs up the ladder from the old wine gallon.

We must add as a safety clause amid this historically accumulated confusion of measures that the actual capacity of the pipe varied "for different commodities."

The wine gallon's volume was about five sixths of the British imperial gallon's.

At the top of the doubling ladder we come to the double pipe, or *tun,* which contained 252 wine gallons. This volume is equivalent to about 2,120 pounds of water; somewhat less of wine; and still less of hard liquor.

The names of a measure of capacity (tun) and of weight (ton) look and sound much alike. The resemblance may be accounted for by strands extracted from the tangled skein of historical measurements. While the tun is 2,120 pounds of water, the long ton (or tun) in Great Britain is set legally at 2,240 pounds avoirdupois; the short ton in Britain and the United States contains 2,000 pounds avoirdupois.

In *Pandosto* (1588), by Robert Greene, one may read that "a pound of goold is worth a tunne [sic] of lead." In Bradley's *Dictionary* (1725) the tun is described as a "measure in averdupois [sic] consisting of twenty hundredweight," each of which is 112 pounds. Hence a total of 2,240 pounds for the tun. (Incidentally, averdupois is the original and more defensible spelling of the word now corrected, or corrupted, to avoirdupois, and abbreviated as av., avdp., or even avoir.)

Varying tons have been used to measure different substances. Even when the legal ton was set at 2,240 pounds, Cornwall had a miner's ton of 21 hundredweight, or 2,352 pounds, and a long weight ton of 2,400 pounds.

From tun to tunning to the English Revolution and regicide—The term tun and its variations appeared and reappeared in important connections during the 14th, 15th and 16th centuries. Tunning itself meant brewing, because the end product was poured into tuns. About 1510, that racy, raucous, rough-hewn poet and humanist John Skelton, who tutored Henry VIII when Henry was a boy, wrote his *Tunning of Eleanor Rumming,* well described as a "monstrous caricature of an alewife and . . . those who throng to buy her brew."

Of deeper concern to English monarchs and their masters of the exchequer was the institution of tunnage (or tonnage)—taxation (by the tun) levied on wine imported in tuns or casks. Tunnage and poundage were closely linked, the former being a levy by volume, the latter a levy by weight, of imported liquid and solid goods respectively.

The proceeds of tunnage and poundage, first levied in the 14th century, were granted for life to a number of rulers, beginning with King Edward IV (reigned 1461–83). But the first taking of tunnage actually began about a century earlier, in 1371, during the reign of Edward III. It was instituted as a source of funds to provide for the "protection [at sea] of the merchant navy."

The *Rolls of Parliament* for 1422 mentions "a subsidie of Tonage and poundage . . . that is to sey of every Tunne iii s. [3 shillings], and xii d. [12 pence] of every Pounde."

In 1629, questions of "tunnage" (or "tonage") and poundage were still agitating members of Parliament. King Charles I had been crowned in 1625, and within a few years he had come into bitter conflict with most of his subjects and with the squires who made up much of the House of Commons.

Influenced by George Villiers, Duke of Buckingham, Charles had married the strongly Roman Catholic Henrietta Maria of France and had indulged in a series of unpopular and unsuccessful overseas military expeditions against the France of Richelieu and against Spain. The news of Buckingham's assassination by a fanatical Puritan was greeted with scarcely concealed satisfaction by the common people of England.

King Charles tried to rule without Parliament, which he resented and feared. In particular, he was outraged by a violent event in the House of Commons in 1629, when members forcibly held the

Speaker in his chair so that they might pass resolutions against "Popery" on the one hand and illegal tonnage and poundage on the other. That was the last Parliament the King permitted to meet until 1640.

Meanwhile, three members of Parliament were imprisoned for what they had done there, and one of them, Sir John Eliot, was executed in the Tower when he refused to make submission to the King's illegalities.

Tax-hungry and tyrannical, the King ruled without Parliament. He and his inner circle devised one method after another to extract money from his subjects. It has been said that he sought to exercise "rigid economy" in this interval, but to growing numbers of his subjects it seemed that his court was shameful in its splendor and rotten with extravagance.

More and more Englishmen became convinced that the units of measurement meant to serve their daily needs were being manipulated to defraud and rob them. Besides the enforced payments of tonnage and poundage, there were the monopolies. The King granted such monopolies to corporations, since it was forbidden by statute that they be granted to individuals. A commentator of this time—the incubation period of the British Revolution—complained of these royally invested corporate monopolists: "They sip in our cup, they dip in our dish, they sit by our fire; we find them in the dye-vat, wash-bowl, and powdering tub . . . they have marked us and sealed us from head to foot."

A tax was levied on the *jackpot,* the measure halfway between the jigger and the gill. Not only was the jack-sized pot taxed, but in order to increase the amount of tax, the capacity of the jackpot was reduced. A kind of inflation from above oppressed the poor man. His penny bought less, because the capacity measure contained less. With the jackpot reduced in content, its double, the gill, also shrank.

Truly, arrogant royalty was "sipping in our cup" and "dipping in our dish." But how could the urban poor protest? They had no representation. They had only the recurrent recourses and resources of the impotently indignant: the jeer, the gibe, the jest, whispered perhaps behind the hand, or sung out like a folk ditty when the King's informers were least likely to be near.

We still repeat one of the rhymed satires from those teasing times, assuming it to be only a children's jingle, devoid of real significance:

Jack and *Jill* went up the hill
To fetch a *pail* of water.
Jack fell down and broke his crown,
And *Jill* came tumbling after.

Three units of capacity measurement appear in this venerable Mother Goose verse: the jack, the jill (or gill) and the pail (then a capacity measure approximating the gill). The third line of the verse is prophetic: it was the crown of the elegant divine-righted Charles that broke or was lost. The consummation of the bourgeois revolution in Britain, almost a century and a half before the French Revolution, not only broke the crown of the King, but took his head off. This decapitation was performed in 1649, not many years after the degradation of the jackpot.

Gamblers talk of "hitting the jackpot," unaware that they are referring to a unit of capacity. Charles I can be said to have gambled and lost at least in part because of the royal mishandling of the jackpot.

When the Long Parliament came into session in 1640, one of its first acts was to declare fully and finally that there must be no taking of tonnage and poundage or of ship money without the sanction of Parliament. Nearly a century and a half later, in 1787, when George III was king, tonnage and poundage were finally abolished.

Chronicles without consistency—Confusions and contradictions aplenty appear in the annals of the units of capacity, area, length and weight that have been used or misused in Britain and in its colonies since the reign of Charles I.

The fate of the Jill of the 17th-century nursery rhyme has been quite ambiguous. The gill today may refer to (1) a U.S. liquid unit of 7.218 cubic inches capacity, just ¼ U.S. pint or 1/32 U.S. gallon; or (2) a British unit of 8.669 cubic inches capacity, ¼ imperial pint or 1/32 imperial gallon; or even (3) the "dialect English" gill, which is ½ pint, not ¼ pint.

3 CURIOSITIES OF CAPACITY

*Like doth quit like, and measure still
 for measure."*
 —SHAKESPEARE,
 Measure for Measure

*. . . his mind was such that whatever was little seemed to
him great, and whatever was great seemed to him little.*
 —THOMAS BABINGTON MACAULAY,
 Horace Walpole

*. . . the pint measure and quart measure. . . .
The brewery, brewing, the malt, the vats,
everything that is done by brewers,
wine-makers, vinegar-makers. . . .*
 —WALT WHITMAN,
 A Song for Occupations (1855)

The Mow and the measures—Folk verse and song contain traces of many quaint and otherwise almost forgotten units of measure from the past. One of the most amusingly informative of such folk souvenirs is a drinking song known as "Here's Good Luck to the Barley Mow!" The "mow" here means a pile of sheaves of grain stored for protection in a barn.

As usual with folksongs, different versions arose in different districts. The author has found two, quite similar. One comes from the upper Thames region, and goes like this:

> We'll drink it out of the nipperkin,
>   My brave boys,
> Here's a health to the barley mow,
> The nipperkin and the jolly brown bowl.
> Here's a health . . . etc.

The nipperkin was a small fluid measure, about $\frac{1}{8}$ pint in size, containing 2 to $2\frac{1}{2}$ ounces. It was still a moderately familiar word at the turn of the century. Thomas Hardy, in a brief but memorable antiwar poem, "The Man He Killed," wrote, in part:

> Had he and I but met
> By some old ancient inn
> We should have sat us down to wet
> Right many a nipperkin!
>
> But ranged as infantry,
> And staring face to face,
> I shot at him as he at me,
> And killed him in his place.

The pledges of good luck to the barley mow begin but do not end with the nipperkin. They go on, verse after verse, each using a unit of measure double that before. But in local versions the units do not always match those given in the preceding chapter.

The second round calls for a drink of a quarter pint, then a half pint, then a pint. This is followed by the quart, the pottle, the gallon, the peck, the half bushel, and the bushel. We tend to assume that the peck and bushel are and always were dry measures. But the tavern roisterers drinking their way verse by verse through the Barley Mow knew them as liquid measures—or at least as containers capable of holding quantities of beer and ale to be downed by a drinker of fabulous capacity.

W. Chappell, in *Popular Music of the Olden Time* (London, n.d.), notes that in the Barley Mow "the size of the drinking measure is doubled at each verse."

So, beyond the bushel came the half sack, the sack, the half hogshead, and the hogshead. Then the stout drinkers of the upper Thames taverns abandoned all contact with their local metrology and plunged head first into beery, cheery fantasy. The measure after the hogshead was literally "the river," and after that, "the ocean."

If we shift to the records of the Dorset folk songs, we find a variant, recorded in the 1940s by J. Caddy, of Melplash, in which each in the ascending order of drinking measures is presented as a kind of challenge, a question spoken, not sung. The Dorset words are these:

SUNG:
Here's a good health to the barley mow,
The nipperkin and the brown bowl,
And merrily pass it around, brave boys.
Here's a good health to the barley mow.

SPOKEN:
Can you drink out of the quarter-pint?

When the next verse is sung, it substitutes "quarter pint" in its second line for "nipperkin." And so on. A sort of recapitulation line is added to each verse which passes in review all the measures from the smallest (nipperkin) to the largest. The full cycle as given in *A Dorset Book of Folk Songs,* edited by J. Brocklebank and B. Kindersley (London, 1948), gives this sequence: nipperkin, quarter pint, half pint, pint, quart, (no pottle or half gallon), gallon, flagon (instead of peck). It proceeds to the firkin, then jumps to the barrel, the hogshead, the butt, and finally the tun. In the recapitulation or review refrain at the end of each verse, the names of all the measures previously mentioned are sung in a monotone, on a single note.

The Barley Mow songs, whether from the upper-Thames or Dorset region, began moderately with the smallest units of liquid measures, then went on up, as long as the local topers could still swill. Drinking songs of England provided the tune for our difficult national anthem, "The Star-Spangled Banner." They also lend substance to the lore of metrology.

Folk song and folk say offer many evidences of regional variations in units. For example, the gill, or jill, in southern England was officially ¼ pint, as in the Jack and Jill nursery rhyme. In Lancashire and the north of England the same name meant half a pint. The jack, or double gill, was not always called so. When used to measure brewed or spirituous liquors, it was sometimes given the name of jug.

A descending series of nicknamed measures for alcoholic beverages was linked with the jug, or pint, of 20 ounces: ½ pint, called gill, or jill (10 ounces); ¼ pint, called jack, or noggin (5 ounces); ⅛ pint, called jock (2½ ounces); and ¹⁄₁₆ pint, called joey (1¼ ounces).

A jolly quartet they seem to be: the joey, the jock, the jack (or jug) and the jill. Jollity, however, cannot mask the lack of simplicity and consistency in historical measures which varied, sometimes abruptly, from county to county, district to district, and from era to era.

Other commonly used containers for alcoholic drinks introduce new complexities. The ordinary English wine bottle contained 26½ ounces. Half a dozen bottles held 159 ounces, or an ounce less than 10 pounds. This was very close to an imperial gallon, which holds 10 pounds avoirdupois of water. Wine then, if bottled, was purchased by the "sixth," not by the "fifth" (of a gallon) as distilled spirits—whiskey, vodka, gin, rum, and so forth—are commonly sold in the United States. A "fifth," in this country, represents ⅕ of 231, or 46.2 cubic inches of volume. A "fifth" of the imperial British or Canadian gallon, however, is ⅕ of 277.42, or 55.484 cubic inches. A bottle holding a "fourth" (of a gallon) in the U.S. today would contain exactly 1 quart, or 57.75 cubic inches of liquid. Thus, the British-Canadian "fifth" almost equals the U.S. quart.

It is all curiously complicated and confusing, not only to casual consumers (of measures and spirits) but often to those in the trade.

The wine gallon of 231 cubic inches capacity, which was made official in an order signed by the first Queen Elizabeth, was for a long time identified with her name. It later came to be called Queen Anne's gallon. As such it became the standard fluid measure of the United States, of Ceylon—and also of Canada (until the 20 percent larger "imperial" gallon was adopted there).

When we scan historical records of larger containers constructed as cask measures, we find discrepancies between those built to transport wine and those intended for beer. The traditional beer barrel came to be built large enough to contain 36 gallons, though it probably arose as an offshoot from what had been the half hogshead of 32 gallons or even 31½ gallons capacity. Thus the beer barrel deviated by 12½ percent from the pure "doubling" pattern: 1:2:4:8:16:32:64:128 . . . and so forth.

A half barrel for beer in our time would contain 18 gallons. Centuries ago the half barrel was called the *kilderkin,* probably a corruption of the Flemish *kinderkin,* which in turn seems linked with the German diminutive *Kinderchen,* or "little children."

The kilderkin corresponded to a measure known in 1483 as the runlet, and also to a German measure, the eimer. In modern German, *Eimer* means pail or bucket, though a bucket commonly contains far less liquid than did the half barrel of old. The measure called kilderkin, runlet, or eimer had still another alias, the double anker. In German today *Anker* means anchor (of a ship), or armature (of an electric motor). The kinship between the shape of an armature and a small cask suggests the linguistic influences that operated here.

A step down the capacity ladder we find the quarter barrel of beer, called the *firkin*. In the 15th century this was commonly spelled ferdekyn. It had nothing to do with a "little horse" (*Pferdchen*, in German). It came from *vierde-kin* or "little fourth." The closest analogy in English would be *fourth-let* as a small brook is a brooklet or a small book a booklet.

The firkin, insofar as it may be used in the United States today, is a measure of 9 gallons, or $9 \times 231 = 2,079$ cubic inches. The British firkin contains 9 imperial gallons or $9 \times 277.42 = 2,496.78$ cubic inches. The kilderkin, by 1960, was nonexistent in United States usage, but still appeared occasionally in Britain, with a capacity of 2 firkins, equal to 18 imperial gallons, or 4,993.56 cubic inches. The British hogshead had by then been "standardized" at 63 imperial gallons, equal to 7 firkins or 17,477.46 cubic inches. The 20th-century tun, of 252 imperial gallons capacity, corresponds to 28 firkins, or 14 kilderkins, or 69,909.84 cubic inches. The modern imperial gallon has about 1.2 times the capacity of the old wine gallon. Hence the modern tun corresponds to 302.4 old wine gallons. And such a tun of pure water weighs 2,526 pounds avoirdupois, or roughly $1\frac{1}{4}$ tons of 2,000 pounds each.

The Flemish-Dutch array of units of capacity included several others of interest, especially to tipplers and topers. They have survived, somewhat modified by metric influences on their magnitudes, but still recognizable by name, in the Netherlands today. There one may find traces of the *kan* (can), now of 1 liter capacity; the *spint*, 5 liter; and the *stoop,* a measure for Pils beer, of 6 liter.

That stoop, or stoup, has a long background. It was a familiar term in Elizabethan England, though better known in northern England and Scotland than in the south. A stoup was not only a holy-water basin to be seen near the church entrance, but also a drinking vessel or tankard. The name derived from Middle English *stowp,* was related to Icelandic *staup* (cup) and Old English *steap* (flagon). In the final scene of *Hamlet,* as the swordplay between the Prince of Denmark and Laertes is about to begin, the guilty uncle, King Claudius, orders his servitor: "Set me the stoups of wine upon that table . . ."

Later in the same scene the King calls the wine container "the poisoned cup." In the excellent German translation of the play by August Schlegel and Ludwig Tieck, King Claudius's order becomes: *"Setzt mir die Flaschen Wein auf diesem Tisch." Flaschen* means "bottles." Thus, in literature as well as in commerce and the common culture, one finds amid manifold confusions of measure the additional confusion between the open drinking vessel and the semiclosed containing vessel, or bottle.

Netherlands units of liquid capacity included their own barrel of 220 liter size, their *okshoofd* of 232.8 liter size, and several larger units, each used to measure and hold a particular kind of wine. For sherry there was the *boot* of 530 liter; for port the *pijp,* or pipe, of 530 liter; and for Rhine wine the *halbstück* (half piece) of 600 liter, and the *fuder* of 1,000 liter capacity.

Though the Dutch pijp and the British pipe are akin in sound and derivation, they differed in magnitude as measures so often do when they wander geographically. The pipe in England came to represent commonly—though by no means always—a measure of 126 wine gallons, equal to 105 British imperial gallons. But this equals 477.33 liter, about 63 liter less than the Dutch pijp in its 20th-century form.

And it still goes on . . . —This does not, by any means, exhaust our sometimes bewildering display of variety and disparity in units of capacity.

Even past the middle of the 20th century, a tolerably complete guide to volumetric measures that may be encountered in the United States or Britain has to include the following units, which we list in order of increasing magnitude, in cubic inches. One cubic inch equals 16.387 162 cubic centimeters (cc)—or, for purposes of approximation, 16.4 cc.

UNIT	CUBIC INCHES
drachm (fluid, British)	0.023 281
drachm (fluid, U.S.)	0.024 233
noggin (British)	8.669
quartern = ½ imperial gallon	138.71
bucket = 4 imperial gallons	1,109.7
bushel (dry)	2,219.3
strike = 2 bushels (dry)	4,438.6
bag, or sock = 3 bushels (dry)	6,657.9
coomb = 4 bushels (dry)	8,877.2
barrel (dry) = 36 imperial gallons	9,987.017 04
seam, or quarter = 8 bushels (dry)	17,754.4
butt = 126 imperial gallons	34,954.9
chaldron = 32 bushels (dry)	71,017.6
wey	c. 89,000
puncheon = 70 imperial gallons	155,351
register ton = 100 cubic feet	172,800
last (British) = c. 80 bushels (dry)	177,544
rod (British) = 1,000 cubic feet	1,728,000

Of the foregoing miscellany of volume measurements, two—the wey and the last—have still not settled down to one exclusive magnitude. The magnitude assigned in the list is simply one of the more common in use.

A number of the units listed are included in the group called "the excellent octonary series of measurements" by Edward Nicholson in his *Men and Measures:*

UNIT	CUBIC INCHES
noggin	8.669
gill	17.342
pint	34.675
quart	69.35
pottle	138.71
gallon	277.42
peck	554.8
tuffet (or bucket)	1,109.7
bushel	2,219.3
strike	4,438.6
coomb	8,877.2
quarter (or seam)	17,754.4

Greater than the quarter was the chaldron, or corn ton, or quadruple quarter, with a volume of 71,017.6 cubic inches. The units measure both dry and fluid volumes. Those below the pint—the gill and noggin—have been more commonly employed to measure fluids, however.

Distinctions between dry and fluid capacity units have survived to this day in the extravagantly unsystematic "system" prevalent in the United States of America, whose citizens commonly regard it as the country most advanced, technologically speaking, in all the world. There is, on the one hand, the U.S. dry quart, with capacity of 0.038 889 cubic foot, or 67.200 192 cubic inches; and there is the U.S. liquid quart, with capacity of only 0.033 421 cubic foot, or 57.751 488 cubic inches. The discrepancy is substantial—9.5 cubic inches. A proportional difference is found between the U.S. dry pint of 33.6 cubic inches and the U.S. liquid pint of 28.875 cubic inches capacity—4.7 cubic inches difference.

Such complications, confusions and contradictions in traditional capacity measures are paralleled, even surpassed, by the "good old" measures of length, area, and weight. To eliminate such inefficiencies, the metric system was developed and then extended step by step into the International System (SI).

The metric system and its SI expansion are decimal-based, like our currency and our number system itself. Alteration of U.S. practices seemingly must follow the lead of almost all the rest of the world, adopting metric and SI units with their exact decimal multiples and fractions.

A few voices have been raised in favor of another sort of change —toward the past. Harold F. Larson, an engineer, has urged a return to a modernized binary system of volumetric units, on the old "double-my-predecessor" principle of which only the pint-quart pair remain in active use in the United States.

Larson's elaborate proposal was submitted as a paper in November 1967 to the Committee on Commerce of the United States Senate, 90th Congress, then holding hearings on two bills, S. 441 and S. 2356.

Larson's binary system would begin with the tablespoon, of precisely 1 cubic inch capacity. Seventeen units would form the rungs in the ascending ladder of volumes or capacities (the number in parentheses gives capacity in cubic inches): tablespoon (1), handful (2), jack (4), gill (8), cup (16), pint (32), quart (64), pottle (128), gallon (256), pail or peck (512), keg (1,024), bushel (2,048), cask (4,096), barrel (8,192), hogshead (16,384), pipe or butt (32,768), tun (65,536).

The handful (2 cubic inches) would provide his "universal ounce," since 2 cubic inches of water weigh something like 1 ounce avoirdupois; and the pint (32 cubic inches) would provide his "universal pound."

In this system the tun, which equals 2,048 pints, is thus 2,048 "universal pounds."

The metric system, whose adoption Mr. Larson opposed, provides but one basic unit volume—the cubic meter (equivalent to 61,023 cubic inches) and one subsidiary or derived unit, the liter, now defined as one thousandth, or 0.001, cubic meter (61.023 cubic inches).

Since, as we know from computers that use a binary-number system, any decimal number can be represented by a binary number (which the computer finds less unwieldy than we do), it may be amusing to note that the cubic meter can be expressed exactly as the sum of twelve Larson units: 1 pipe plus 1 hogshead plus 1 barrel plus 1 bushel plus 1 keg plus 1 pail plus 1 quart plus 1 cup plus 1 gill plus 1 jack plus 1 handful plus 1 tablespoon. In tabular form:

UNIT	CUBIC INCHES
1 pipe	32,768
1 hogshead	16,384
1 barrel	8,192
1 bushel	2,048
1 keg	1,024
1 pail	512
1 quart	64
1 cup	16
1 gill	8
1 jack	4
1 handful	2
1 tablespoon	1
1 cubic meter	61,023

In the binary system, our 61,023 becomes 1110111001011111; the zeros represent the units *not* listed in the table.

Larson's system not only offers a clumsy, hard-to-remember 17 units in a number system not our own, but we would have to replace the U.S. standard pint, quart and gallon with Larson units that bear the same name, but do not have the same capacity (again in cubic inches):

	U.S. LIQUID	LARSON BINARY	DIFFERENCE
pint	28.875	32	3.125
quart	57.749	64	6.251
gallon	231	256	25

No great prescience is needed to predict that future officially recognized measures and weights in the United States will not include an "excellent octonary" or "beautiful binary" system of volumetric measure.

Today, in almost all science, in pharmacy, and in much of technology, volumes and capacities are already expressed metrically: in cubic meter (m^3), in cubic centimeter (cc or cm^3), or in cubic millimeter (mm^3) units. No table or slide rule need be used to compute the basic relationships of metric magnitudes. Not even a pencil is necessary to recall that 1 m^3 = 1,000,000 cc = 1,000,000,000 mm^3. By using the "powers pattern" for writing multiples of 10 (described simply in Chapter 11), we may state the same relationship this way:

$$1 \ m^3 = 10^6 \ cm^3 = 10^9 \ mm^3$$

As soon as one metric equivalent of a given binary unit is known, it is pleasantly easy to find others as needed. For example, to convert cubic centimeter to cubic meter, move the decimal point six places to the left:

1 U.S. gallon = 3,785.4 cc = 0.003 785 4 m^3

1 U.S. liquid pint = 473.179 cc = 0.000 473 179 m^3

Among the Greeks and Hebrews—Long before Anglo-Saxon and Norman influences created the traditional British measures of capacity, the Greeks used two capacity measures of their own. The *medimnos* contained about 32.9 cubic inches (a little more than the U.S. pint of 28.875 cubic inches). The *metretes* contained 2,439.4 to 2,441.3 cubic inches, or 10.57 U.S. gallons, or about 40 liter in the metric system. The metretes held about 2 percent less than the British firkin, which had a capacity of 9 imperial gallons = 2,496.78 cubic inches.

In the Gospel of John, 2, Jesus and his disciples were guests at a wedding at Cana in Galilee when the wine gave out. Jesus took water which had been placed in six stone water jars and changed it to wine. Each of these jars, according to the New English Bible translation (2:6), held "from twenty to thirty gallons." The King James version gives "two or three firkins." The very early Greek version gave "two or three metretes," and the term "metretes" was retained in Wycliff's English Bible translation (1388) and in the modern Dutch Bible translation. The reader will readily observe that the three measures given are roughly consistent.

In the Old Testament, I Kings 7:38, we read of the ten mobile bronze lavers, or purification-water carts, made for the Temple of Solomon on Mount Moriah, Jerusalem, nearly a millennium before the time of Jesus. The capacity of a laver is given as "forty baths" both in the 1917 translation of the Jewish Publications Society and in the King James version. In James Moffett's *The Bible: A New Translation* (1922) the capacity is given as 326 gallons. Accordingly, the Hebrew *bath* appears to have contained 8.15 U.S. gallons. However, the *Random House Dictionary of the English Language* (1966) defines *bath* as a Hebrew unit of liquid measure equal to "between 10 and 11 U.S. gallons."

One looks in vain today in the Middle East for a unit of capacity that corresponds closely in magnitude to the ancient bath. The nearest is the Syrian *mudd,* which in its Sonada and Salkhad version is equivalent to about 7.26 U.S. gallons, and in its Chahba version, to about 6.9 U.S. gallons.

Today, in Israel, Jordan, Syria, Lebanon and Egypt, the metric system has long prevailed.

Return to the present, via a pub crawl—To end our review of the curiosa of capacity units, the author offers some results of researches willingly undertaken in various pleasant pubs of contemporary Britain. Posted in these public houses was a modest but rather reassuring notice, pursuant to the Weights and Measures Act, 1963: "The standard measure of whiskey, gin, rum, and vodka used in this [hotel, public house] is $\frac{1}{6}$ of a gill or multiple thereof." This refers to the British gill, whose capacity is 8.669 cubic inches, or 142.07 cc. Thus $\frac{1}{6}$ gill is only 1.44 cubic inches, or 23.68 cubic centimeter. It is smaller by one-third than the traditional British jigger, which equaled $\frac{1}{16}$ British pint, $\frac{1}{128}$ British gallon, or 35.52 cubic centimeter.

In the United States some well-known chains of motels and hotels ask their bartenders to use $1\frac{1}{4}$ ounce shot glasses in dispensing distilled spirits. Such a glass holds a shade under 40 cubic centimeter of liquid. Hence it provides about $20\frac{1}{2}$ drinks from a familiar four-fifths quart bottle and about $25\frac{1}{2}$ from a full quart bottle. Even more precisely, $20\frac{1}{2}$ drinks per "fifth" means 36.93 cubic centimeter per drink, and this is 56 percent more liquor than the British $\frac{1}{6}$ gill standard.

That British standard happens to represent almost exactly $\frac{1}{32}$ of the contents of a U.S. fifth, and $\frac{1}{160}$ of a U.S. gallon. It is, however, $\frac{1}{192}$ of a British imperial gallon and $\frac{1}{48}$ of a British imperial quart.

Britain is committed to complete metrification, more popularly known as metrication. What are the prospects for patrons of pubs and hotel bars when the process gets that far? If "whiskey, gin, rum, and vodka" are supplied in liter-size bottles, and if the standard measure provides 40 drinks per bottle, that would mean 25 cubic centimeter per drink—an increase of 5.6 percent over the present $\frac{1}{6}$ gill. A standard of 24 cubic centimeter per drink would take slightly less than 42 drinks from a liter bottle. A standard of 20 cubic centimeter per drink would provide 50 drinks per liter bottle, but represent a decrease of $15\frac{1}{2}$ percent from the present $\frac{1}{6}$ gill amount.

One liter (1,000 cc) is 5.7 percent larger than one U.S. quart. It equals 0.264 2 U.S. gallon, but only 0.22 British imperial gallon.

If metrication reaches American bars and taverns in the form of liter-size bottles, bartenders should have no need for extensive postgraduate schooling. A skilled operator who today can pour $20\frac{1}{2}$ drinks from a fifth bottle need only dispense the same amount per drink 27 times to empty a liter bottle.

Problems of spirituous metrication can readily be liquidated, so far as concerns volume dispensed and imbibed. The inevitable conversion from our present confusion to the rationality of metric and SI measures need create no crisis of conviviality.

Demonstration of the historical hodgepodge among units of capacity has been the principal purpose of this and the preceding chapter. Following chapters turn to the mixed-up character of leading measures of length with which humans have managed—albeit inefficiently—to make do in the past.

4 HOW FAR FROM HERE
TO THERE?

The search for suitable and reproducible units of length has gone on for many millennia. Several approaches can be recognized, and each is identified by a different origin.

Most quaint and colorful, though not necessarily always oldest, are anatomical units, based directly or by derivation on dimensions of parts of the human body.

Other length units, which may be called botanical, are derived from plants—in particular, from seeds.

Later we have terrestrial or geophysical units based on the dimensions of the Earth. Closely related are proposed gravitational units derived from lengths of pendulums whose swing has a certain duration when the pendulum is at a spot on Earth with a given gravitational intensity.

Finally, there are important length units that must be called arbitrary, because they cannot be equated with known models or theoretical lengths.

Assorted anatomical lengths—Here we find the clearest, if crudest, illustration of the Protagorean assertion, "Man is the measure of all things." Many parts of the human body have supplied standards of length. Even the earliest users of such units must have been aware that one man's foot or finger differed in length from that of his neighbor or fellow tribesman, but individual variations did not seem important, considering the low degree of precision required for most measurements in those days.

Various precautions were incorporated into the earliest definitions of anatomical length units. For example, a pithy definition of the Scottish inch is attributed to King David I of Scotland, about 1150 c.e.: "The thowmys [thumbs] of iii [3] men, that is to say a mekill [big] man, and a man of messurabel [moderate] statur, and of a lytell man. The thoums [another spelling for thumbs] are to be mesouret at [meaning across] the rut [root] of the nayll."

King David's definition provided a kind of automatic averaging. The combined width of the thumbs at the base of the thumbnail on a small, a medium, and a large man was divided by 3 to arrive at a working length for the Scottish inch. Presumably the right thumbs of right-handed men and the left thumbs of left-handed men were measured.

The author, a pronounced right-hander, finds his thumb breadth measures very nearly $15/16$ "modern" inch, if the thumb is not pressing against a hard surface. When such pressure is applied, the breadth unmistakably expands to a full inch or perhaps $1\frac{1}{32}$ inches. The author's son, large-boned and standing 6 feet 2 inches tall, has a greater thumb breadth, appropriate to a "mekill man."

Thumbnails in particular and fingernails in general have played special roles in length measurement. The Greek word onych or onyx designated the fingernail of a human or the claw of an animal. Though today onyx to most English speakers means a kind of stone (chalcedony), onyx as a medical term still refers to fingernail or toenail; onyxis is the pathological condition known to the layman as the ingrowing of the nail of a toe or finger—a peculiarly painful affliction!

Some words whose meanings correspond to Greek onych are unguis and unglia in Italian, unglo in Provençal, and ongle in French. Un ongle may be a human fingernail, an animal claw, or the hoof of a horse. And *un ongle in carné* is a nail ingrown into the flesh or "meat."

Our inch is derived etymologically from the Old English *unce* or *ynche*, which in turn derives from the Latin *uncia*, meaning one-twelfth. The same derivation led by different routes to another unit named the ounce. The Latin word *uncia* came from *unus*, "one."

Old English *ynche—ynch—unce*, or inch, apparently always was used to mean the thumb breadth, as described by Scot King David I. But the *nail*, which here meant the breadth of a thumbnail, began as a smaller length unit, roughly three quarters of a modern inch. The nail (0.75 inch) was the English counterpart of a most important Roman length unit, the *digit* (*digitus*), based on finger breadth, usually close to ¾ inch or 1.9 centimeter in modern terms. The digit was widely used throughout the ancient world.

Roman units of length continued to influence the metrological measurements of Continental Europe and Britain long after the Roman Empire declined and fell.

Measuring lengths as the Romans did—The digitus, in and around Rome, was about 1.854 cm or 0.73 inch long. It was related to the larger Roman length units by an interesting sequence of ratios. (Approximate metric equivalents are given in parentheses.)

$$4 \text{ digiti} = 1 \text{ palmus} (7.4 \text{ cm})$$
$$4 \text{ palmi} = 1 \text{ pes} (29.5 \text{ cm})$$
$$5 \text{ pes} = 1 \text{ passus} (1.48 \text{ m})$$
$$125 \text{ passus} = 1 \text{ stadium} (184.5 \text{ m})$$
$$8 \text{ stadia} = 1 \text{ milliare} (1,476 \text{ m})$$

The *pes*, equal to 16 digiti, was the Roman "foot," measuring on the average 11.62 inches. Roman colonies in Africa and Britain appear to have used a slightly larger pes, the difference being about 1.5 mm, or less than one half of 1 percent.

Besides the six Roman units listed above, there was the twelfth part of the pes, called the *uncia* (2.46 cm or 0.97 inch—just 3 percent less than the present inch). There was also the Roman *palimpes*, a name that fused palmi and pes. It equaled 5 palmi or 1.25 pes or 20 digiti. The *cubitus* equaled 24 digiti.

The nine basic Roman length units in order of increasing size were digitus, uncia, palmus, pes, palimpes, cubitus, passus, stadium, and milliare.

From the length of the digitus came the English nail. From the uncia came the inch. The counterpart of the pes is the foot, in all its worldwide metamorphoses. The pes of 11.6 inches implies a human foot of generous size from heel to toe, but is a little more credible than our 12-inch unit.

The passus originally meant a step, or pace, such as that of a Roman legionnaire on a long march. Its length, nearly 1.5 meter or 58-plus inches, may seem impossibly long. It was, however, doubtless the full cycle left-right-left or right-left-right, hence twice a single pace of about 29 inches or 0.75 meter.

The milliare, just 1,000 times the passus in length, came to be called, with a sort of wry affection, the *mille passuum,* or "thousand paces." Thus was born the name of the present mile, which, with its length of 1,609.35 meter, is about 9 percent longer than the Roman milliare.

Nailing down metrological illogicalities—The metamorphoses of measurements form a special chapter in the annals of illogical human behavior. The fate of the nail was strange. It parallels the way in which the Latin *unced* came to stand for one twelfth. In England the nail ($\frac{3}{4}$ inch) was $\frac{1}{16}$ foot, just as the digit was $\frac{1}{16}$ Roman foot. But the nail, as $\frac{1}{16}$ foot, came to mean by analogy the sixteenth part of other measures also. A sixteenth part of a yard was called a nail. This nail-of-a-yard measured $\frac{36}{16} = 2\frac{1}{4}$ inches. The British nail is still defined in listings of measures of length as a unit of 2.25 inches or 5.715 centimeter.

Then there were the nail-of-a-bushel, or sixteenth part of a bushel, a volume measure, and the nail-of-a-hundredweight, or sixteenth part of a hundredweight. (However, these nails were units not concerned with length.)

The 1 nail length ($2\frac{1}{4}$ inches) was also defined as the half finger, that is, the length from the tip of the middle finger to the center of the second joint from the tip.

The *finger* was double the nail. It was the length of the whole middle finger to the knuckle, the same that women habitually used to measure lengths of linen fabric. Its length was ⅛ yard, or 4½ inches.

The author, testing tradition against his own anatomy, finds his half finger to be only slightly over 2¼ inches in length. His full finger, when extended straight, measures 4½ inches.

Here is the hand—Another "manual" unit survives today in restricted use: the *hand,* sometimes in the past called the palm, but different in magnitude from the old Roman palmus (2.9 inches). The hand was the transverse length (breadth) of the four extended fingers. By a statute of the reign of Henry VIII, the hand was defined as a length of 4 inches. It is used today in measuring the height of a horse from ground to top of shoulder. The horse-height hand of 4 inches seems substantially to exceed the widths of the four fingers of most human hands—even of most palms measured across the knuckles.

To repeat in ascending order, 1 nail = 2¼ inches; 1 hand = 4 inches; 1 finger = 4½ inches.

The nail enjoyed widespread, if confusing, use as a cloth measure. It was multiplied into the ell, sometimes spelled el, a family of units especially devised for measuring lengths of cloth.

The shortest recorded ell was the Dutch, 22 inches or slightly less than 10 nails. Then came the Danish ell, 24.7 inches or very nearly 11 nails. The well-known Flemish ell measured 27 inches, exactly 12 nails.

The Scottish ell was 36 inches (16 nails), even sometimes 37 inches. The English ell, 45 inches or 20 nails, was exceeded by the French ell, 54 inches and 24 nails. The longest ell (French) was more than double the shortest (Dutch).

Almost all these ells were dutifully listed in standard works on Arithmetick, the subject that encompassed all common commercial measurements in the 18th century.

The ell survives today in at least two forms. In the Netherlands the meter itself (39.37 inches) is sometimes called an ell. And a cautionary proverb in English warns against the greedy man who if you "give him an inch, will take an ell."

The anatomical relevance of the foot unit seems self-explanatory. A large man wearing shoes and walking heel-to-toe steps off distances approximately in feet. The author, a man of only medium height but supported by two of the flattest feet possible, finds that his heel-toe stepping measures off distances of about 11¾ inches.

Even a largish man, striding freely while walking, is likely to find he has a single stride less than 3 feet (1 yard) long. The yard did not derive from the single stride. It originated in another anatomical dimension: the length of a string or tape reaching from a man's nose to the farthest fingertip of his arm extended straight out to his side horizontally. Today's yard (0.914 401 83 meter) is almost identical with the cloth ell once used in Scotland.

The foot of 12 inches may seem to be a stable standard amid the incessant drifting and shifting of other length units, but at different times and places the foot has measured anywhere from 11 to 14 present inches; thus, from 17 percent greater to 8 percent less than the U.S. foot.

The U.S. foot and the British foot are now defined in terms of the meter. The British legal definition, however, gives 30.479 97 centimeter, which is $\frac{9}{100,000}$ of a centimeter less than the 30.480 06 of the U.S. definition. The difference is, to be sure, minute, but it does underscore that nonstandard measurement units must lead to unnecessary complications.

The case of the cubit—Between the foot and the yard in length is found a historically famous length unit—the *cubit*. Actually there was a whole family of cubits associated with different regions and eras of the ancient world. The cubit was originally the distance from a man's elbow joint to the farthest fingertip of his extended hand.

Eight different cubits have been identified; they average slightly more than half the length of the modern meter (that is, 50 centimeter). The longest cubit unit, however, was 14 centimeter greater than the average length; the shortest, nearly 6 centimeter less. The range of known cubits was thus 20 centimeter, or 40 percent of average cubit length.

The table shows the approximate lengths, in U.S. inches and meter, of the eight ancient cubit standards as given by A. E. Berriman in *Historical Metrology* (London, 1953) and by other authorities.

CUBIT	INCHES	METER
Roman	17.48	0.444
Egyptian "short"	17.72	0.450
Greek	18.23	0.463
Assyrian	19.45	0.494
Sumerian	19.76	0.502
Egyptian "royal"	20.62	0.524
Talmudist	21.85	0.555
Palestinian	25.24	0.641

It has been said that when cubits originating in western Asia, Greece and Africa are compared, the variations in length are surprisingly small. It seems the cubit had been more or less standardized by the time of the Hellenistic era. Establishment of larger settlements and towns permitted the measurement and averaging of the lengths of sufficient forearms to approximate the length standards of other communities of the area.

One might guess that travel and trade carried cubit standards from region to region. But transportation problems made it difficult to maintain at one center a single standard cubit for ready comparison by the whole Hellenistic world. Length standards remained local. Multiple measurements and averaging operations were still used to establish local length standards 1,500 years later.

Fate of a family of measures—What has become of the cubit family? Prevalent opinion assigns it to the limbo of obsolete length units. However, its disappearance may not be complete. Despite the worldwide advance of metric measurement, descendants of the old cubits were found occasionally in the second and third decades of this century. In India (then under British rule) the Bombay region offered a cubit, or *covid,* of 18 inches; in the Madras region, like-named units ranged from 18 to 21 inches (0.457 2 to 0.533 4 meter). In Siam the cubit, or covid, was 18 inches long; in Arabia the *corido* had about the same length. Arabia also occasionally used the *great corido,* about 27 inches long, or 1½ times the ordinary corido.

In the mid-1960s, United Nations specialists, compiling for statisticians a useful handbook *World Weights and Measures,* reported only the cubit or *taim* (18 inches) in Burma, the *cubito* (22 inches) in Somalia, and the *côvado* (14.8 inches) in Macao. (Côvado was also used in Macao as a unit of area, slightly less than 200 square inches.)

To such remnants had shrunk the scope of the cubit, once known and used throughout the "civilized world." In Somalia and Macao the meter was the official unit of length; the cubit-related units were among the obsolescent "also used" units. (In Burma since 1920 both the meter and the British inch have been standard lengths.)

The far-flung family of the foot—In many regions the forearm-based cubit was supplemented by a linear subunit whose length was 60 percent of the cubit. This subunit matches today's foot within 1 or 1½ inches.

Foot was not its name. Names varied: the Persians had the *vitasti* (10.7 inches), the Romans had the pes (11.6 inches), and so on.

Archeologists have tried to approximate the bygone "foot" of each region and era. They are by no means all in agreement.

Hellenistic influences are shown in three different "Greek feet" —short, medium, and long. There are two Roman feet; they differ by about ⅘ inch.

Geographical scattering of units meant numerical scattering of their magnitudes until the worldwide spread of the metric system in the 19th and 20th centuries.

O. J. Lee, in *Measuring Our Universe* (New York, 1950), places the Roman "foot" at 12.44 inches (31.6 centimeter). A. E. Berriman, in *Historical Metrology,* gives 11.654 inches (29.6 centimeter)—2 centimeter less. They agree in finding that the common Roman foot was either equal to or slightly shorter than its most common Greek counterpart. By the Middle Ages the foot had lengthened to 12.47 inches (31.7 centimeter).

Actually units based on the human foot antedated the units (cubits) based on the forearm. A new foot of about ⅗ cubit apparently was adopted as the cubit became the basic measure.

To sum up, the average or typical "short" pes (foot) of Rome measured 29.6 cm. The "long" Roman foot ranged from 30 to 32 cm, or 2 to 8 percent longer.

The Parthenon foot (29.7 cm), deduced from painstaking analyses of the dimensions of that glorious Greek structure, was a short foot much like that of the Romans. The later foot of Athens was 31.6 cm, that of Aegina 31.4 cm, of Etruria 31.6 cm, of Miletus 31.8 cm. These discrepancies have led to the grouping of the foot-like measures of Greek and Hellenistic communities into three categories—the short Greek foot of 30.9 cm, the medium of about 31.8 cm, and the long of 33.9 cm. (The latter was close to units long prevalent in Egypt and Asia Minor.)

The foot of Athens (31.6 cm) was the best-known Greek foot. On it was based the *arkana* (3.16 meter), equal to 10 Athenian feet; 10 arkanas equaled 1 *plethron* (31.6 meter). The next step departed from decimal ratio for the Mesopotamian sextuple: 6 plethrons equaled 1 foot-based *stadion* (189.6 meter). The Greek stadion was thus slightly longer than the Roman *stadium* (184.5 meter).

The Hellenistic world evolved another sequence of length units with more consistently decimal ratios. The Greek digit (1.84 cm) was close to the Roman digitus (1.845 cm). Then, 100 digits = 1 *orguia* (1.84 m); 10 orguias = 1 *amma* (18.4 m); 10 ammas = 1 digit-based stadion (184 m).

Still another stadion—a sort of superstadion—of 1,896-meter length was recognized in the ancient world. It was based upon the foot, and it was very nearly ten times the length of the digit-based Greek stadion and the later Roman stadium.

Variations in footlike units continued through the centuries as the course of empire wended its way north and west from declining Rome. The Belgic foot, which lasted until the third or fourth century C.E., measured between 33.2 and 33.6 cm; the old French foot, about 32.5 to 33.1 cm. The old foot of medieval England has been placed at 31.8 cm by some metrologists, at 33.5 or 33.6 cm by others. All of these are longer than the modern English foot (30.48 cm), abandoned when Britain went metric.

Two rather conjectural early British foot equivalents are known. From the sizes and positions of the stones at Stonehenge, that end-lessly intriguing relic on Salisbury Plain in England, it has been calculated that its builders used a footlike unit of 0.296 7 meter or 11.68 inches. Multiple measurements made in Britain on hill figures and stone circles other than Stonehenge indicate use of a "foot" of 0.295 meter or 11.6 inches, virtually identical with that already mentioned. The stones speak, and in this respect they do not quite agree, but they do not differ much either.

In the 20th century there survived, as hand-me-downs, some local units of length close to the 12-inch Anglo-American foot. In the Shi'zhi' (market) system of mainland China, a unit called the *che* was 9.4 percent longer than the foot. The Hong Kong *cheh* or "Chinese foot" was even longer (14.625 inches). The traditional *shaku* of Japan (11.93 inches) closely approximated the U.S. foot. Many other Japanese length units were barred from use after March 1966, 45 years after national adoption of the metric system.

Korea's *chok* is 11.93 inches long, like the Japanese shaku. In Malaysia, the *check* (14.75 inches) might be considered a footlike unit. A subunit, $\frac{1}{10}$ check, was called the *chum* (1.475 inches).

Also in Malaysia a unit called *tokhoi* or *chi* was shorter than the check, being 5 percent longer than the U.S. foot. It retained a limited use into this century.

Latin American regions, on becoming metric, discarded a number of footlike units; some traditional units have continued in occasional, though diminishing use.

The *tercia* of Costa Rica is $\frac{3}{100}$ inch less than 11 inches; the *pie* of Mexico, where the metric system has been mandatory since 1862, is $\frac{2}{100}$ inch less than 11 inches. One twelfth of a pie is a *pulgada* (0.917 inch).

In Ecuador and El Salvador a unit known as tercia or pie was exactly 12 inches. The tercia of Nicaragua was 11 inches long; a subunit, the pulgada, was $\frac{11}{12}$ inch.

The Paraguayan pie was 11.37 inches; the pulgada was again $\frac{1}{12}$ as long (0.947 5 inch).

The Philippine *piye,* 1 foot in length, was 12 times the pulgada, an inch equivalent, and one third the *yarda* (1 yard). The telltale name *milya* referred to 5,280 piye lengths. These units were used even after the Philippine government adopted metric measures.

In Haiti two vestigal foot units survived the adoption of metric measures and weights in 1921. The *pied* (1.066 feet) reflected French heritage. The *pied anglais* equaled the U.S. foot. In Honduras the pie (10.95 inches) persisted after metric adoption in 1912.

Iran went metric in 1933, but there was still some use of the *charac* (10.24 inches). Inhabitants of the Ryukyu Islands, though officially metric, remained familiar in the 1960s with the traditional *shaku* (11.93 inches). In the Seychelles, likewise metric, a *pied* (1.016 feet) was encountered occasionally.

In the German Federal Republic (West Germany) the old *Fuss* (12.36 inches) is used occasionally, and so is the *Zoll* (1.03 inches). In Amsterdam, though the Netherlands has been metric since 1820, the *voet* (0.928 foot) survived obscurely and unofficially, past the mid-1920s.

In Finland the traditional *jalka,* or *fot* (12 inches), was used to measure logs. Its twelfth part, the *tuuma,* or *tum,* was 1 inch exactly.

In the South African provinces of Cape of Good Hope, Orange Free State and Transvaal, and in South-West Africa, a number of Cape units were officially tolerated for use in land measurements into the mid-1960s, along with metric and standard British units. The *Cape foot* (1.033 British feet) was 12 times the Cape inch; the *Cape rood* was 12 times the Cape foot.

Botanical measures, and how they grew—After footing it around the world we arrive at old botanical length units. The most striking botanical unit was the *barleycorn,* $\frac{1}{3}$ inch, or $\frac{1}{36}$ foot, or $\frac{1}{108}$ yard.

In early 14th-century England, during the reign of King Edward II, the inch was defined as "three grains of barley, dry and round, placed end to end, lengthwise." To provide a shorter unit, the barleycorn was split into four equal parts, each called a *line* (2.116 7 millimeter). Thus there were 12 lines to the inch, 12 inches to the foot.

An old folk saying in the class of "A pint's a pound the world around" declared that "three barleycorns make an inch." This was a crude approach. Three especially large barley seeds may noticeably exceed an inch, and three small seeds can total a good bit less. However, repeated samples of three randomly chosen barley seeds —especially seeds of the barley prevalent in early 14th-century England—should average close to 1 inch.

Other plants too supplied units of weight. In the Plantagenet period, the *pennyweight* was described as the weight of 32 (average) wheat corns (seeds). (The pennyweight today is defined as 24 grains troy weight, equal to $\frac{1}{20}$ ounce, or 1.555 gram.) The average wheat seed in Plantagenet days probably weighed about 48.6 milligram; hence 20,600 such seeds would weigh 1 kilogram. For comparison, 1 carat = 200 milligram, or roughly 4 wheat corns.

The inch has also been defined as the combined lengths of 12 poppyseeds, a barley seed being on the average four times as long as a poppyseed.

A rule from the England of 1566 declared that "foure graines of barley make a finger; foure fingers a hande; four handes a foote." This bridges botanical and (human) anatomical standards of length. The rule indicates that 64 barley grains made a foot; later, 36 barleycorns made a foot.

Is the average barley seed longer than it was four centuries ago? Or was the average foot longer then? The latter supposition is not plausible, since men are larger now, not smaller. One need not assume, either, that the barleycorn is four times as long now. The confusions and contradictions of historical unit usage defy the most ingenious present-day attempts to harmonize them or to explain them away.

5 MORE LORE OF LENGTH

Which of you by mental effort can add a span to his height?
—MATTHEW 6:27,
trans. H. J. Schonfield

Have you reckon'd . . .
. . . the attraction of gravity, and the great
laws and harmonious combinations
and the fluids of the air,
as subjects for the savans?
Or the brown land and the blue sea for maps
and charts?
Or the stars to be put in constellations
and named fancy names?
—WALT WHITMAN,
A Song for Occupations (1855)

The rood, mentioned in the preceding chapter, proves to be another nest of metrological diversity. The word can be traced back to Germanic *Rute,* thence to Old English *rod.*

Rood meant a large crucifix like those that stood beside entrances of medieval churches. By extension to the past, the rood came to represent the cross on which Jesus was crucified. As a unit of measurement, rood has been used to denote a land area of approximately ¼ acre or 40 square rods, equal also to 10,890 square feet, or 1,210 square yards. Less frequently rood has meant 1 square rod, containing 272.25 square feet.

The rood we are interested in here is a British linear unit, or rather a family of units, for from place to place it has ranged from 16.5 to 24 feet in length. At 16.5 feet it was identical with the rod, a surveyor's measure. At 24 feet the rood was 45 percent longer than the rod.

The German rute is an old linear land measure. Despite the common origin of rute, rood and rod, the rute was considerably the shortest of them all. It was 12.36 to 12.47 feet (3.77 to 3.8 meter). The German fuss is $\frac{1}{12}$ rute, or 12.36 to 12.47 inches, and the zoll is $\frac{1}{12}$ fuss, or 1.03 to 1.04 inches.

Denmark's related rode (3.762 meter, or 12.34 U.S. feet) is obsolete.

The road to an honest rood in the 16th century—There is an interesting prescription for producing a "right and lawful" length standard for a rute or rood in the 16th-century how-to-do-it compendium of surveying written by Master Koebel.

The surveyor should station himself by a church door on Sunday. When the service ends he should "bid sixteen men to stop, tall ones and short ones, as they happen to come out . . ." Note the early use of random selection. It would never do to choose sixteen short men or sixteen men of basketball-player stature.

The chosen sixteen should be made to stand in a line with "their left feet one behind the other." The resulting sum of sixteen actual left feet constituted the length of "the right and lawful rood," and the sixteenth part of it constituted "the right and lawful foot."

If sixteen churchgoers laid off a length equal to today's German rute, their average left-foot length would be 9.27 inches, which seems short, even for men of four centuries ago. It seems likely that the rood Koebel was seeking was closer to today's rod (16.5 feet) than to a unit of 12.36 feet. Thus, on the Continent as in Britain, unit magnitudes could vary irregularly even when their venerable names remained the same.

This lining-up of sixteen churchgoers garbed in their Sunday best is pictured in an old engraving. It shows that the search for recognizable, reproducible standards was a matter of communal concern and good citizenship.

The author has found no records of comparable procedures for adding and averaging 16 forearms to determine a "right and lawful" cubit. (Note that 16 is a power of 2. If a string is cut to match the combined lengths of 16 measurements, it can then be folded end to end and cut in half. Three more such cuts and an average length is obtained, just one sixteenth of the original string.)

Nosing out the original yard—The cubit was longer than the foot, but was not the longest unit based on human anatomy. A string or tape stretched from the end of a man's nose to the farthest fingertip of an arm extended straight out to the side, horizontally, measured a yard (today 3 feet, or 36 inches, or 0.914 401 83 meter).

The word "yard" comes from the Anglo-Saxon *yard,* which meant a rod or stick used for measuring. Beyond that lies *gyrdan,* signifying a person's girth or waist (circumnavigated by a like-sounding girdle). Since *gYRDan* contains the root of *YaRD,* another anatomical basis is suggested: a stout fellow, after years of loyal service to his lord at the mead bench, might measure a "perfect 36" or 1 yard around his middle. But in Middle English a *yerde* was the spar attached to the mast from which a sail was hung

Queuing-up to arrive at a "right and lawful" rood in the 16th century. This authentic old depiction of the process laid down by Master Koebel does indeed place 16 assorted individuals toe-to-heel, old and young, obscure and important, just as they happen to come out of the door of the church at right. The three observing personages in the background are very likely the local commissioners for weights and measures, overseeing this averaging-out operation. (COURTESY OF ZEITLIN AND VER BRUGGE, BOOKS AND PRINTS, OF LOS ANGELES)

(as in yardarm). It has been suggested that the old practice of measuring a yard of cloth by holding it between nose and outstretched fingertip made a picture something like a sail hanging down from a yard.

The longest anatomically derived unit is the fathom (2 yards), which originated as the distance from middle fingertip to middle fingertip of a good-sized man holding his arms fully extended. The fathom has long been a nautical unit. In *The Tempest,* Ariel sings, "Full fathom five thy father lies"—that is, 30 feet. Even in our 20th-century age of sonar, one widely used apparatus for finding the distance to the bottom is called a fathometer.

Human anatomy and metrology have been linked since antiquity. Leonardo da Vinci (1452–1519), in considering the proportions of the human body, wrote approvingly of the views of the Roman architect-engineer Vitruvius Pollio (1st century B.C.E.):

> . . . Vitruvius declares that Nature has thus arranged the measurements of a man: four fingers make one palm. and four palms make one foot; six palms make one cubit; four cubits make once a man's height; four cubits make a pace, and twenty-four palms make a man's height . . .

For a 6-foot man we have:

UNIT	INCHES
finger	0.75
palm	3
foot	12
cubit	18
man's height	72
man's pace	72

By "pace" Leonardo meant not a single stride, but rather the complete cycle left-right-left or right-left-right. If a 6-footer has a 3-foot stride, his pace is 6 feet.

Leonardo might have added another ratio of 4:1. His finger unit (0.75 inch) was a finger breadth. The cloth measurer's whole finger length, already described, was 4.5 inches. And 1 cubit (18 inches) = 4 cloth measurer's fingers.

Going up in length units—We have reached the maximum in anatomically based standards of length. One cannot in practice lay human beings end to end to measure distances on a geographic scale. Long ago rough motion-based estimates such as "a day's journey" and "an hour's walk" were used, along with equestrian estimates such as "a day's ride" and "half an hour's canter."

One of the earliest, perhaps *the* earliest of all tables of English linear measures, Arnold's *Customs of London,* appeared about 1500. It contains the following sequence (in which we have substituted Arabic for Roman numerals):

> The length of a barley corn 3 times make an ynche and 12 ynches make a fote and 3 fote make a yerde and 5 quaters [quarters] of the yerde make an elle.
> 5 fote make a pace.
> 125 pace make a furlong and 8 furlong make an English myle.

Thus in 1500, 1 ell equaled 3.75 feet; 1 furlong equaled $125 \times 5 = 625$ feet; and 1 mile equaled $625 \times 8 = 5,000$ feet. If 1500's foot was the same as ours, their mile was 280 feet shorter than ours.

During the reign of Queen Elizabeth I, a statute increased the length of the furlong from 625 to 660 feet (40 rods). The mile thus gained the missing 280 feet.

The furlong is a familiar distance in U.S. horse racing. The "220" in our track meets is 1 furlong (220 yards). Since 1 furlong equals 201.168 meter, a runner who has trained to sprint the 220 is in good shape also for the 200-meter run. The "440" is only about 2 meter longer than a 400-meter race; the "880" only about 5 meter longer than an 800-meter race.

The mile run would be only about 9 meter longer than a 1,600-meter race, but in fact the nearest Olympic race is only 1,500 meter (109 meter or 120 yards less than 1 mile). The 5,000-meter race is about 3 miles plus 188 yards; the 10,000-meter race, about 6 miles plus 376 yards. The 50,000-meter walk is about 31 miles plus 120 yards—almost 5 miles longer than the Olympic marathon, which measures 26 miles 385 yards.

Many a mile ago—An elaborate hypothesis offered by H. J. Chaney (1842–1906) suggests that a primal or original mile was brought to Britain by Belgic tribes who migrated across the Channel. With them came the Belgic foot, between 33.2 and 33.6 cm long (13 to 13.2 inches). Chaney pictured the resulting measurement systems as persisting until the ninth or tenth century of this era. The full Belgic-German unit system is as follows:

B.G. UNIT	INCHES
foot	13.22
yard	39.66
fathom	79.32
chain	793.2
furlong	7,932
mile	79,320

The B.G. mile, of 6,610 U.S. feet, was 1,330 feet longer than the modern statute mile. More remarkable was the decimal relationship: 10 fathoms = 1 chain; 10 chains = 1 furlong; 10 furlongs = 1 mile.

The B.G. mile was exactly 6,000 B.G. feet. The B.G. yard, of 39.66 present inches, was very close to the later meter (39.37 inches).

Chaney's evidence for the existence of the B.G. system leans on statutes and regulations issued to suppress that system and substitute the present foot. These laws he traces back to 950 C.E., a century before the Battle of Hastings.

The new system substituted shorter units from a foot up. The new mile, Chaney believed, grew out of an "old London mile," first of 5,000 feet, then of 5,280. Elsewhere in the British Isles long mile units were used, including an Irish mile of 6,721 feet, and a Scottish mile of 5,951 feet.

Across the Channel, milelike measures varied widely. Behind this assortment lay the Roman milliare (4,840 feet).

Half a league onward—Changes in the league have been no less complicated than changes in the mile. The Gallic unit called *leuga,* or league, was not a direct outgrowth of the suggested B.G. mile. Being 8,395 feet, it was 1,785 feet longer than the B.G. mile and 3,115 feet longer than today's statute mile.

The old English league was even longer (about 3 statute miles, or 4,830 meter). It was only 950 feet less than 2 Gallic leugas.

Many other linear units have been used in the British Isles, including the *link* (7.92 inches), the *span* and the *quarter* (9 inches, or a quarter yard), the *pace* (30 inches), the *perch* (1 rod, or 16½ feet), and even the *skein* (360 feet).

Besides these varied land units we find such special seagoing units as the *nautical mile* (6,080 feet) and the *nautical league* (2,400 feet longer than the land league). The extended nautical mile of 6,080 feet is the distance used in the speed unit called the *knot,* identical with 1 nautical mile per hour, which is very nearly 1.15 statute miles per hour. A ship traveling at 30 knots is moving as fast as an auto at about 34.5 miles per hour.

The knot has also been a length unit. Log lines in the past were marked off by knots tied at intervals, each interval, or "knot," being 47.29 feet long.

Marine metrology need no longer be a special subdivision of general metrology. Much needless labor is now saved by the metric system, which uses the same units on land and at sea.

Inconsistencies everywhere—Not only British long-distance linear units are marred by deviations or discrepancies that cannot be left out of account. The ancient Arabian "mile" of 7,093 feet was much longer than our statute mile. The stadium of antiquity, much shorter than the mile or even the kilometer, did not have a single, invariable length. The Greek stadium, associated with the ascendancy of Periclean Athens, was about 607 feet or 185 meter long. Another stadium was 622 feet or 190 meter. The Roman stadium at 606 feet was very close to the Athenian stadium.

The Greek *stade* (stadium) has been equated with 600 Greek "feet." According to this, the Athenian foot would be 1 percent longer than the U.S. foot. The Roman foot has been equated with 96 percent of the Greek foot, making the Roman foot about 3 percent shorter than the U.S. foot.

The Roman foot included 16 digits, each digit being about 1.85 centimeter, or just less than ¾ inch. Twenty Egyptian digits, each equal to a Roman digit, made 1 *remen* (15 inches).

Lengths in Europe of old—In France we can find reasons enough for the dissatisfactions that led to the birth of the metric system at the end of the 18th century.

The shortest measure in common use was the *ligne* (2.255 83 millimeter), called the Paris line in Britain. It was somewhat longer than the old British line (2.116 7 millimeter), which was $\frac{1}{12}$ British inch. Twelve Paris lines made a *pouce,* or Paris inch (about 2.71 centimeter), and 12 ponces made a *pied,* or Paris foot (about 32.5 centimeter), 2 centimeter longer than our foot. Then 6 pieds equaled a *toise* (1.95 meter), and 3,000 toises equaled a French league (5,850 meter), which was about 1,000 meter longer than the British statute league and not far from 300 meter longer than even the British nautical league.

Also, the Paris foot, or pied, times 22 equaled 1 *pole* (7.15 meter). A pole equaled 6 *annes*—the anne (1.15 meter) being the old French unit closest in length to the meter.

The Paris foot equals about 12.785 inches or 32.5 centimeter. Times 5,000 the Paris foot becomes a "mile" of 5,327 U.S. feet.

Measurements in Paris feet played an important role in 17th-century British science. Isaac Newton used the Paris foot in his calculation of the force of gravity. Newton employed this foreign pied, because French surveyors and expeditions had made the most precise determinations of the size of the Earth. In his early twenties Newton had attacked the problem of gravitation with the best available estimates of the Earth's dimensions, which were, as we know now, about 15 percent short. He could not fit the falling toward Earth of the orbiting Moon into the same equations as the falling toward Earth of apples and such, until revised and increased measurements of Earth segments (in Paris feet) were made.

The old Russian unit of distance, the verst, has been dealt with thus by incorrigible punsters:

> SHE: Come see me sometime; I live just a few steppes away.
> HE: Yes, that's the verst of it!

The verst, which we see used to better purpose in translations of Dostoevski and Tolstoi, was a little longer than a kilometer (3,500 U.S. feet compared to 3,281). It included 500 *sajons* of 7 feet each; 1 sajon equaled 6 *fusses* (14 inches each); and 1 fuss equaled 8 *versocks* (1.75 inches each). Thus 1 verst = 500 sajons = 3,000 fusses = 24,000 versocks. It is no coincidence that a verst is exactly equal to 3,500 feet. Peter the Great (1682–1725), in his relentless drive to westernize Holy Russia, not only ordered forcible reductions of the lengths of the abundant beards of his boyars (nobles), but decreed a reduction of about 1 percent in the scale of Russian linear units to make the correspondence to English measures exact. This was a minor but useful and symbolically significant adaptation of the unit system in a backward country to that of an advanced country.

Today, adoption of the metric system has brought many underdeveloped countries into the mainstream of the most advanced metrology.

The land of Ivan the Terrible, Peter and Catherine has been completely metric since January 1927, less than ten years after the Russian Revolution. By mid-1960, UN statisticians could report that among nonmetric units only the *pood* or *poud* (36.11 pounds avoirdupois or 16.38 kilogram) was "also used" in the Ukraine and elsewhere in the U.S.S.R.

Whether in "backward" lands like tsarist Russia or industrially "advanced" lands like Britain, units of measurement remained inconsistent and subject to arbitrary change until adoption of the metric system and its expansion, the International System of Units.

In Greek legend, Procrustes, whose name meant "the stretcher," was an innkeeper-robber who lived near Eleusis until slain by Theseus. Procrustes was a metrologist of sorts, his unit of length being a bed—or, according to one variant of the legend, two beds, one very long, one very short. In any case, when a traveler came to seek shelter, Procrustes insisted that he must precisely fit the bed, rather than vice versa. If a tall traveler's feet projected beyond the bed, Procrustes cut them off. If a short traveler had headroom, Procrustes pulled his limbs, as on a rack. Either way the unlucky guest died before dawn.

Humankind's inconstant and inconsistent units of measurements have provided a kind of Procrustean bed. Not until the metric system and the SI have the wayfarers of technology, science and industry been able to move in safety and comfort.

6 COVERING THE SUBJECT
(AREA AND SURFACE)

. . . every Space that a Man views around his dwelling
    place
Standing on his own roof or in his garden on a mount
Of twenty-five cubits in height, such space is his Universe:
And on its verge the Sun rises and sets, the Clouds bow
To meet the flat Earth & the Sea in such an order'd Space:
The Starry heavens reach no further, but here bend and set
On all sides . . .
Such are the Spaces called Earth & such its dimension.

. . . that false appearance which appears to the reasoner
As of a Globe rolling through Voidness, it is a delusion . . .
The microscope knows not of this nor the Telescope: they
    alter
The ratio of the Spectator's Organs, but leave Objects
    untouch'd.
        —WILLIAM BLAKE,
        Milton (1804)

. . . every Space larger than a red Globule of Man's blood
Is visionary . . .
And every space smaller than a Globule of Man's blood
    opens
Into Eternity of which this vegetable Earth is but a shadow.
        —WILLIAM BLAKE,
        Milton

Plant seeds—barleycorns—supplied standards for tiny units of weight such as the carat. And seeds were part of the origin of some primitive measures of land area for agricultural use. Italian *moggio* and French *boisselée, strée,* and *seterée* were all measures of seed grain (corn). Norway had the *tunn-land,* and Germany the *scheffel,* the name of several measures of capacity. Its etymological kinship to English shovel is apparent. The *scheffel* as a measure of capacity came to equal 14.56 times the *metz* (3.44 liter). Thus the scheffel totaled 50 liter in the metric system. The scheffel of land—a measure long obsolete— was the amount of land that could be sown with a scheffel (measure of capacity) of seed.

Capacity measures of grain with corresponding area measures of similar names developed in Provence, where civilization first flowered in medieval Europe. Four such measures are listed in Edward Nicholson's *Men and Measures* (1912):

CAPACITY UNIT	LITER	AREA UNIT	SQUARE METER	RATIO (m²/l)
quartiero	5	quarteirado	202	40.4
elmmo	20	eiminado	810	40.5
sestie	40	sesteirado	1,619	40.5
saunnado	160	saunnado	6,475	40.5

Other area units were based on labor time. The Old English *daieswork* (day's work) equaled about 10 modern square rods (2,700 square feet, or 250 square meter). The German *tagwerk* ("day's work"), equal to 3,407 square meter, or 0.84 acre, was occasionally still used in Bavaria in the mid-1960s, though use of metric weights and measures had been mandatory there since 1872.

In viniculture, the French many centuries ago developed area unit called *ouvrée hommée* (about 180 square feet, or 16.7 square meter). Obviously, the vineyard was cultivated far more intensively than the grain fields.

Other area units were based on how much land a man could plow with oxen drawing the crude wooden plows of the time: the

Latin *jugerun,* the Italian *giornata* (a day's plowing area), and the French *journal* and *arpent.*

The German *joch* (cf. English *yoke,* as in "a yoke of oxen") and *acker,* also derived from typical areas covered by plowmen, were about equal to the later British acre (4,047 square meter).

Even the "standard" nonmetric measures are sometimes not quite consistent from country to country. British and American acres, as legally defined, differ very slightly, the U.S. acre being larger by 0.024 m².

The common German measure for arable land was the *morgen* (literally, "morning"). It varied from 2,500 to 3,600 square meter, depending on the region of Germany. In Hesse it was 2,500 square meter; in Hannover 2,621; in Bavaria 2,726; in Württemberg 3,152; and in Baden 3,600.

When the Bavarian morgen (2,726 m²) is subtracted from the Bavarian tagwerk (3,407 m²), not much is left for the afternoon. Perhaps a heavy lunch of beer and white sausage was to blame.

The morgen unit of Denmark and Norway, with 2,553.6 square meter, was near the lower limit of the morgen units of Germany, but the morgen of the Netherlands was far larger, with 8,124.4 square meter. The morgen, as a land measure, entered South Africa with the first Boer settlers; there it reached values ranging all the way from 2,500 m² to 8,550 m², a backbreaking area for the labor of 1 morning.

Links between linear and area measures—Relations that once existed between important units of area and length measurements are revealed in a manuscript in Oxford's Bodleian Library. The key passage, with Arabic numerals substituted for Roman, reads:

> Sixteyne foote and a half maky a perche; and in sum cuntre [some countries] a perche ys 13 foot. Fourty perchys in lengyth makyth a rode [rod] of lande; put 4 thereto in brede [breadth], and that makyth an acre.

Thus, 40 perches, of $16\frac{1}{2}$ feet each, form a length of 660 feet. Multiplying by a breadth of 4 perches, or 66 feet, we obtain 43,560 square feet = 160 square perches = 1 acre. But these must be the $16\frac{1}{2}$-foot perches of England, not the 13-foot foreign perches.

Caution: The 660-foot "rode of lande" was not the modern 16½-foot rod. The old rode equaled today's furlong, or "220," and it was the perch that equaled today's rod of 16½ feet (5½ yards). The old acre, equal in area to that of today, was formed, in the example of the Bodleian document, by a length of 660 feet times a breadth of 66 feet, equal to ⅛ mile by 1/80 mile. Thus 1 square mile = 640 acres.

The Bodleian manuscript continued:

> . . . 14 acrys makyth a yerde of lande and 5 yerdis makyth a hyde of lande, which is 70 acrys. And 8 hydis makyth a knyghtes fee, which is 560 acrys of lande.

Summing up:

UNIT	SQUARE FEET	ACRES
square perch	272.25	
(today's square rod)		
acre	43,560	
square rode	435,600	10
(today's square furlong)		
yerde	609,840	14
hide	3,049,200	70
knight's fee	24,393,600	560
square mile	27,878,400	640

The "hyde," or hide, was not always "70 acrys." It varied from place to place, from 60 to 120 acres.

"Hide" is not, as one might think, derived from a skin or hide cut into long strips and fastened together to form a cordlike boundary around an area of land. It is derived from the Old English *hīgid*, which signifies the area of land appropriate to a family or household (*hīgan*). A hide, then, is a household-size area of arable land.

The larger unit with the glamorous name of "knyghtes fee" more often corresponded to 5 than to 8 hides; however, when a large 120-acre hide was used, this canceled out most of the difference. In regions with poor soil and scanty arable portions, a knight's fee might be 800 to 900 modern acres, but perhaps only half as much in more fertile areas.

Or again one may think of the feudal knight's fee as being about

1 square mile of land, some arable, some not. In feudal England, as elsewhere, units of measurement reflected economic and class conditions of the society in which they were applied. Surviving fragments of metrological history reveal the economic facts of life rather than romances of King Arthur and the Table Round. The knight errant was not on the road to search for distressed damsels to be rescued from dragons, but because he lacked land and serfs. He had not found his "fee," or perhaps had been ousted from it by force. His knightly exploits were designed to demonstrate his worth to some landed feudal lord. His fealty would flow to that lord in exchange for the fee—the land, which, with its hard-working peasants, would support him in the style to which he wished to be accustomed.

His square mile or so of land would keep half a dozen peasant households busy in his behalf and—after that was taken care of—their own.

Across the English Channel existed a counterpart of the square perch—the French *perche,* equal to 220 square *pieds de roi.* (As a measure of length, the *pied de roi,* or "king's foot," is equal to 0.324 8 meter—a little longer than the U.S. foot of 0.304 8 meter.) The perche was equal to about 250 square feet, whereas the square perch of England was equal to 272.25 square feet.

Another familiar French area unit, the *arpent* (51.07 are), covered about 55,000 square feet, or 1.26 acres. (The are is a metric measure of area; 1 are = 100 square meter = 1,076.4 square feet.)

The French-Canadian arpent, a unit of length, equaled 180 French feet (191.9 U.S. feet); the Canadian square arpent (about 36,825 square feet) equaled 0.85 acre, considerably less than the French arpent. Given enough diffusion in time and space, a nonmetric measure will give rise to two or more different measures of the same name.

The French-Canadian perche being 10 percent of the Canadian arpent, the corresponding square perche was 1 percent of its square arpent, or 368 square feet, considerably *more* than the French perche.

Lands in which Romance languages are spoken have measures of area clearly related to the French *are.* Thus Italy has its *ara* and

Spain its *área*. That word, like our own "area," can be traced back to the Latin *area*, which designated an open space, courtyard or "empty lot." It began in Latin as *arere*, signifying "to be dry," hence, a dry space extensive enough for games and recreations; and a cousin to the ancestor of another familiar modern word—*arid*. The best-known metric unit of area is the hectare $= 100$ are $= 10,000$ square meter $= 2.471$ U.S. acres. In the Netherlands the "old hectare," called the *bunder*, also equaled 2.471 acres.

Other measures of roughly acre size around the world have included the *dessiatine* (2.7 acres) of old Russia; the *faltche* (23 are $= 0.568$ acre) of Moldavia; the *feddan* (1.038 to 1.127 acres, varying locally) of Egypt; the *fanega* (1.615 acres, but varying locally) of premetric Peru; and the *cawnie* (1.332 acres) of Madras, India.

Among many units of land area used in China the *mou* was about ⅙ acre, or 7,260 square feet (in Shanghai, however, only 6,690 square feet). Later, in a treaty with a Western power the mou equaled 55,225 square feet. The *ch'ing* (72,600 square feet) was not to be confused with the *ching* (only 121 square feet).

Measuring midget areas—Tiny areas are often measured in the United States by the *square mil*, which is one thousandth of an inch on a side, and thus one millionth of a square inch, or 6.45×10^{-6} square centimeter.

Surveyors in the past have used *square link* (0.040 5 square meter) and *square chain* (404.69 square meter). At the top of the nonmetric scale, the states west of Ohio use the *section* (1 square mile); a total of 36 sections, duly numbered, make up the *township*, with an area of 1,003,622,400 square feet or 23,040 acres. Thus:

$$1,000,000 \text{ sq mil} = 1 \text{ sq in.}$$
$$144 \text{ sq in.} = 1 \text{ sq ft}$$
$$9 \text{ sq ft} = 1 \text{ sq yd}$$
$$4,840 \text{ sq yd} = 1 \text{ acre}$$
$$640 \text{ acres} = 1 \text{ sq mi}$$
$$36 \text{ sq mi} = 1 \text{ township}$$

These varied numerical relationships—1,000,000: 144: 9: 4,840: 640: and 36—constitute ratios without a rationale!

Perhaps the most complex table of area measurements is the one that defines standard paper sizes in premetric Britain and America. Britain devised some forty paper sizes, each with its own peculiar designation. The following list presents only a dozen rather typical examples, arranged in order of increasing area:

NAME	AREA (in square inches)	DIMENSIONS (in inches)
Foolscap octavo	28.687 5	4.25 × 6.75
Foolscap quarto	57.375	6.75 × 8.5
Typewriter quarto	85	8.5 × 10
Foolscap folio	114.75	8.5 × 13.5
Foolscap	229.5	13.5 × 17
Crown	300	15 × 20
Ledger Demy	310	15.5 × 20
Demy	393.75	17.5 × 22.5
Royal	500	20 × 25
Elephant	644	23 × 28
Imperial	660	22 × 30
Double Elephant	1,080	27 × 40

The United States has used about thirty paper sizes, some identical with British sizes in name and dimensions, and some identical in name only. Following are some samples:

NAME	AREA (in square inches)	DIMENSIONS (in inches)
Typewriter	93.5	8.5 × 11
Legal	119	8.5 × 14
Flat Letter	160	10 × 16
Typewriter Cap	208	13 × 16
Crown	285	15 × 19
Folio	374	17 × 22

Typewriter, or letter, size is by far the most common size, familiar to most students, writers and secretaries. It is the size most often used for typescripts such as the one from which this book was set in type for printing.

Will all these traditional paper sizes survive the inevitable metrication, or metricization, as it is sometimes called? On the Continent and in Britain a new basic size, 21 × 30 cm (8.27 × 11.8 in.; called A4), is rapidly replacing such sizes as typewriter quarto and foolscap folio; it contains 97.6 square inches, compared with 93.5 square inches for U.S. typewriter size.

The writer estimates that over the past forty years he has used a couple of million sheets of 8½ × 11 inch paper for note taking, rough drafts, direct typing and carbon copies. Although he has acquired a sort of sentimental attachment to the letter-size sheet, with its width 77.3 percent of its depth, he finds himself willing, even eager, to eventually switch to A4, despite its austerely impersonal name. A4's width is just 70 percent of its depth; its surface area exceeds that of letter size by 4 to 5 percent.

The metric A4 paper size will be supplemented by multiples and submultiples. For instance, the U.S. folio sheet (17 × 22 = 374 sq in.) will probably be replaced by a quadruple A4 sheet with doubled dimensions of 42 × 60 cm (390 sq in.), and the British demy will probably be replaced by the same quadruple A4 sheet.

The excessive proliferation of "standard" paper sizes has been matched by the units used to count paper quantities. There is not only a *quire* of 24 sheets, but a *printer's quire*—a sort of baker's two dozen—of 25. The *ream* has 480 sheets; the *printer's ream* 516. The *large ream* may be a printer's ream of 516 sheets, or may contain only 510 sheets, or even 500.

In the western United States, moreover, a ream (not "large ream") of 500 sheets has long been prevalent in stationery stores. The writer would, indeed, feel put upon if he bought a ream which contained only 480 sheets.

Beyond the ream is the *bundle* (a double ream, of usually 1,000 sheets) and the *bale* (5 reams, usually 2,500 sheets). One is tempted here to suggest that the illogic of paper measures is the result of a baleful influence.

We leave the more venerable paper sizes—quarto, octavo, duodecimo, and the rest—to the antiquarian and collector of bibliographical and metrological curiosa.

7 WAYWARD WEIGHTS

The first units of weight, like those of length, were often based on botanical objects, particularly seeds, the least variable identifiable parts of plants.

The word *carat* is derived from the Arabic *qirat,* the seed of the so-called coral tree. Such seeds have been used for thousands of years in determining the weights of precious stones. Today's carat is 0.2 gram; it marches in step with the metric system. It is one five-thousandth of the metric kilogram of mass. Later we shall make clear the important distinction between a kilogram of mass (an amount of matter) and a kilogram of weight (a measure of force).

For many centuries the carat varied from country to country. In 1887, leading jewelers agreed on a 205-milligram carat. In 1913 this was reduced to 200 milligram, the present value.

Diamonds are measured by the carat or by the *point,* which is equal to 2 milligram, or one hundredth of the carat.

There are several other units of measurement called *point.* In printing in the United States, the point is a unit of measurement of type size and is equal to $\frac{1}{72}$ in., or 0.013 888 in. Thus, 12-point type is $\frac{1}{6}$ in. high; 6-point type is $\frac{1}{12}$ in. high; and so on. Another *point* is a unit of measurement of thickness of card or paper stock and equals one thousandth of an inch (0.001 in.).

The carat too, although it is universally standardized, is subject to confusion with another unit, one that sounds identical but is spelled differently—the *karat*. The latter is used to measure the relative purity of gold; it is not a measure of weight or of mass, but an index of proportion. It is a twenty-fourth part, or 4.166 7 percent, of the whole. Unalloyed gold—that is, chemically pure Au— would be designated 24-karat gold (24 twenty-fourths gold). An alloy of 3 parts gold to 1 part other metal would be rated 18-karat (18 k) gold.

A jeweler offering a 3-carat diamond in a setting of 18-karat gold is confusing neither his units nor—it is to be hoped—his customers.

The carat is not the oldest unit based on botanical items. Under England's Plantagenet kings (12th to 15th centuries), it was ordered that 32 wheat corns should equal 1 *pennyweight.* The pennyweight (1.555 17 grams), abbreviated *dwt,* survived into the 20th century as part of the troy system of weights. It equaled one twentieth, or 5 percent, of a troy or apothecary *ounce,* and just a bit less than 5½ percent of the more commonly used avoirdupois ounce.

The average wheat corn in Plantagenet times must have weighed 0.048 6 gram, or about one fourth as much as the average coral-tree seed, which gave rise to the modern carat.

The *grain,* being $\frac{1}{24}$ pennyweight, is larger than a wheat corn. It is an average of the weight of grains of wheat taken from the middle of the ear. A troy pound weighs less than an avoirdupois pound, and a troy ounce weighs more than an avoirdupois ounce, but 1 troy grain = 1 avoirdupois grain = 0.064 798 918 gram.

Note that the decimals running to a billionth of a gram in this metric equivalent are not merely theoretical. Modern scientific instruments can compare weights to within about one six-billionth of a gram.

For very small measurements, instead of calculating in kilogram, gram or milligram, the microgram (10^{-6} gram) is often convenient. Its symbol is μg, the Greek letter being *mu;* 1 grain = about 65,000 μg.

A nongarlic clove—Another old English weight unit is the *clove* (7 pounds). It has nothing to do with a clove of garlic, or cloves (the spice) with baked ham, or any botanical object. Like the French *clou,* the clove is derived from Latin *clavus,* nail. It is, next to the British imperial (avoirdupois) pound, the smallest unit in the stone series of weights:

$$7 \text{ imperial pounds} = 1 \text{ clove}$$
$$2 \text{ cloves} = 1 \text{ stone}$$
$$4 \text{ stones} = 1 \text{ half hundredweight}$$
$$2 \text{ half hundredweights} = 1 \text{ hundredweight}$$
$$20 \text{ hundredweights} = 1 \text{ ton}$$

Thus a clove is 7 pounds, a stone is 14 pounds, a half hundredweight and hundredweight are 56 and 112 pounds respectively, and a ton (a long ton) is 2,240 pounds.

This ungainly series has survived because, in Britain, the stone is the habitual unit for expressing a person's weight. An American may be puzzled when a Britain says he weighs "10 stone 10 pound," meaning 150 pounds. With some effort one can learn to state human weights in terms of stones plus pounds. The British metrological historian Edward Nicholson recalls that he worked for years at tasks that obliged him to use not only the stone but the clove and hundredweight as well. Applying "this inconvenient set" of units, a man weighing 152 pounds would be recorded as weighing 1 hundredweight, 2 stones, 1 clove, 5 pounds (because 112 plus 28 plus 7 plus 5 equals 152). Small wonder that "errors were necessarily frequent." By the 1940s and '50s at least the hundredweight and the clove had gone the way of so many superfluous units before them.

Among hundreds of weight systems outside the United States and Britain, we may note the binary or doubling relationship in a sequence used in Burma. It begins with the *ywegale* or *ruay* (0.025-51 gram): 1 ywegale × 2 = 1 *ywegi;* 1 ywegi × 2 = 1 *pai;* 1 pai × 2 = 1 *moo;* 1 moo × 2 = 1 *mat;* 1 mat × 2 = 1 *ngamu;* 1 ngamu × 2 = 1 *tical* or *kyat.* Thus 1 tical = 64 × 1 ywegale, or 1.633 gram.

In Afghanistan was found the *nakhod* (19.17 milligram): 1 nakhod × 24 = 1 *misqal;* 1 misqal × 24 = 1 *khord;* 1 khord × 4 = 1 *pow;* 1 pow × 4 = 1 *charak;* 1 charak × 4 = 1 *seer.* Thus 1 nakhod × 36,864 = 1 seer (about 707 gram). Finally, 1 seer × 80 = 1 *kharwar* (56.53 kilogram), an increase in magnitude of 2,949,-120 times in seven steps.

The convenient ratio of 10 appears in several regional systems of weight units. In Japan one sequence began with the *mo* (3.75 milligram): 1 mo × 10 = 1 *rin;* 1 rin × 10 = 1 *fun;* 1 fun × 10 = 1 *manne.* Then the 10 ratio is dropped, for 1 manne × 160 = 1 *kin* (600 gram). Finally, 1 kin × 6.25 = 1 *kan* (3.75 kilogram).

In China the old Shi' zhi' system of market weights began with the *si* (0.5 milligram): 1 si × 10 = 1 *hao;* 1 hao × 10 = 1 *li;* 1 li × 10 = 1 *fen;* 1 fen × 10 = 1 *gian;* 1 gian × 10 = 1 *hang;* 1 hang × 10 = 1 *jin.* Finally, 1 jin × 100 = 1 *dam* (50 kilogram). Thus the dam is 10^8, or 100 million, times greater than the si.

In a very real sense, the basic decimal principle of the metric system and the SI was used in China long ago.

The Hong Kong region also developed a system with "times ten" ratios: 1 *fun* (378 milligram) × 10 = 1 *chin;* 1 chin × 10 = 1 *leung* (37.8 gram).

Cambodia had a decimal sequence that began with the *bin* or *li* (37.5 milligram): 1 bin or li × 10 = 1 *hun;* 1 hun × 10 = 1 *chin* or *chi;* 1 chin or chi × 10 = 1 *tael* or *damleng* (37.5 gram).

The well-known *maund* of India weighed on the average 37.32 kilogram and was 40 times a smaller unit, the *seer* (about 933 gram). Bombay had its own typical maund of 12.7 kilogram weight; Madras, one of about 12.3 kilogram; and other deviations no less striking could be found in different districts.

Aden and South Arabia also had seer units of 933 gram, but in Afghanistan a seer more than 7 times as heavy could be found. In Aden, too, a maund of 12.7 kilogram was prevalent, but in nearby Bahrein, a maund double that size, or about 25.4 kilogram, was in use.

And so, far and wide, identical or similar unit names labeled weight standards that were confusingly unlike or disparate in magnitude.

In Iran as recently as the mid-1960s more than twenty nonmetric units of weight still were in use. They ranged from tiny units with intriguing names, such as the *una,* the *gandon* and the *dung* to the *rattel* (30 gram) and the *kharvar* (nearly 270 kilogram).

In Arabia, however, the rattel, or rattle, at 7.83 kilogram, had about 260 times the weight of the Iranian rattel. In Egypt the *ratl,* or *rattolo,* had 450 gram weight in Cairo, while the "great" rattolo of Alexandria had 964 gram. Also in use was a "rattolo mina" of 780 gram.

Derivations of drachma and dram

The *drachma* of the Netherlands (3.906 gram) was about equal to the apothecaries' dram (3.887 gram). Like the drachma of Turkey and the royal *drachmé* of Greece (both 1 g), it was used for weighing gold. Constantinople (now Istanbul) had its own drachmé (0.375 g) as well as the regular Turkish drachma.

The ducat, a gold coin mentioned in Shakespeare's plays with Italian background, appeared as a weight unit (53.9 g) in Vienna. The old Austro-Hungarian Empire also used the *loth* (270 g) and the *marc* (280.673 g); the latter recalls the *marco* of Spain (230.1 g) and of Portugal (229.5 g). In France, the *poide de marc* weighed 245 g, and the Swedish marc, or *mark,* weighed only a mere 211 g.

The best-known Anglo-American *ounce* is the *ounce avoirdupois,* whose metric equivalent is 28.349 527 g. Sixteen such ounces make 1 avoirdupois pound (453.592 4 g). The "other" ounce is the troy or apothecaries' ounce, with a weight of 31.103 481 g, or 2.753 954 g greater than the avoirdupois ounce.

Chapters 2 and 3, on measures of liquid capacity, referred to the U.S. fluid ounce (29.573 7 cubic centimeter) and the slightly smaller British fluid ounce (28.413 cc).

We can tabulate these four ounces and their symbols:

OUNCE	ABBREVIATION	SYMBOL
avoirdupois	oz. av.	℥ av.
troy or apothecaries'	oz. t. or ap.	℥ t. or ap.
fluid (U.S.)	oz. fl.	℥ fl.
fluid (British)	oz. fl.	℥ fl.

The ounces' cousins in other lands include an old Netherlands *onze,* 10 of which made 1 *pond.* After the metric system was legalized in the Netherlands in 1810, there appeared the *ons* (exactly 100 g) and the five-times-larger pond (500 g).

The old French *once* weighed 30.6 g, while that of Portugal weighed 28.7 g. The old Roman *oncia* was much heavier, with 436.2 gram. Sound-alike units all too often are not weigh-alike units!

Commodity-based units of weight—There are many special units used only to weigh particular commodities.

The German Federal Republic, though fully metricated, still uses two units redolent of herring: the *kauje* (keg, 74 kg) and the *barrel* (100 kg).

In Indonesia, so-called "opium weights" included the *timbang* or *matt* or *hoon* (386.01 mg), the *tiji* (3.860 1 g) and the *thail* (38.601 g)—another "times 10" sequence.

Special coffee weights in Latin America have not been completely eliminated by metric units. There are, for example, the *saco de café,* or coffee sack, which is 69 kg in Costa Rica, 90.2 kg in Cuba, 75.1 kg in the Dominican Republic, 27.27 kg in El Salvador, and 60.1 kg in Haiti. The *farnega de maiz* (corn) of Costa Rica weighed 350 kg, that of Ecuador only 92.2 kg. Costa Rica also had the *farnega de tabaco* (46 kg) and the *carga de papa* (828 kg).

Among the few nonmetric units with some restricted use in the Netherlands is a special "ton" measure (725 kg) for oats, another (1,006 kg) for rye, and a third (1,015 kg) for wheat and maize.

At least, readers in every country can get a rough idea of the magnitude of all these odd measures because there is a standard metric scale of gram and kilogram to translate them into.

During the centuries in which the books of the Jewish Bible took form, two units of weight were dominant near the Mediterranean and the Fertile Crescent: the shekel (smaller) and the mina (larger). Generally 1 mina equaled either 50 or 60 shekels—sometimes, only 45 shekels.

In Persia the ratio was 50 or 60 to 1; in Greece and Rome, 50 to 1.

The shekel, apparently of Babylonian origin, weighed ¼ to ½ ounce (7 to 14 gram). A mina, then, weighed from $50 \times 7 = 350$ gram to $60 \times 14 = 840$ gram, that is, 0.8 to 1.9 avoirdupois pounds.

As the mina overshadowed the shekel, so the talent overshadowed the mina, with weight ratios of 50:1 up to 120:1. In Palestine and Syria, the lands of the Old Testament, 1 talent equaled 50 or 60 minas as well as about 3,000 shekels; elsewhere there were talents equal to 120 minas and to 5,400 shekels.

The Greek talent (23.23 kg, or 51.21 pounds avoirdupois) equaled 6,000 drachms, one drachma being about equal to the later troy or apothecaries' dram (3.888 gram). The old British drachm also weighed 3.888 gram. In view of the British propensity for eliding syllables, it is not hard to imagine how drachm evolved into dram.

Some authorities say the Greek drachm was heavier than a dram —namely, 4.3 or 4.4 gram. Assuming a ratio of 100 drachms to the mina, the mina would weigh 430 to 440 gram; the Greek talent (25.8 to 26.4 kg), about 57 to 58 pounds avoirdupois.

The 1970 edition of the *Encyclopaedia Britannica* offers a rounded weight of 57 pounds (25.8 kg) for the Greek talent, and a weight of 66 pounds (30 kg) for the Hebrew talent (or kikkor) in the "holy" or "sacred" system of weights. Holy units were, in fact, essentially secular; they were used in weighing all objects not made of gold or silver. In the sacred system 1 talent equaled 60 minas; in the profane or nonsacred system, used to weigh gold and silver, 1 talent equaled 50 minahs or minas. ("Minah" may well be a better spelling for the Hebrew measure.)

The Hebrew sacred system sometimes used a "heavy" talent of double weight.

Representative values for the Hebrew and Greek weights are compared here:

UNIT	WEIGHT IN GRAM
Hebrew talent or kikkor = 60 minahs	29,900
Greek talent = 60 minas	25,900
Hebrew minah = 60 shekels	499
Greek mina = 100 drachmas	431
Hebrew shekel = 2 bekahs	8.33
Greek drachma	4.3
Hebrew bekah = 2 rebahs	4.17
Hebrew rebah = 5 gerahs	2.08
Hebrew gerah	0.416

Thus the Hebrew bekah or half shekel weighed about the same as the Greek drachma.

The basic Roman weight was the *libra,* or libbra (327.45 gram). This is 0.722 lb, considerably less than our pound avoirdupois, or troy, whose abbreviation "lb" is derived from libra. It is also somewhat less than the Greek or Hebrew mina(h). Here are the fractions of the libra:

UNIT	FRACTION OF LIBRA	WEIGHT IN GRAM
libra	1	327.45
deunx	11/12	300.17
dextans	5/6	272.88
dodians	3/4	245.59
bes	2/3	218.29
setunx	7/12	191.01
semis	1/2	163.73
trieme	1/3	109.14
quadrans	1/4	81.86
sextans	1/6	54.57
uncia	1/12	27.29
semiuncia	1/24	13.64
quarter uncia	1/48	6.82
eighth uncia	1/96	3.41
sixteenth uncia	1/192	1.71

Latin *uncia,* one-twelfth, is the ancestor of our word ounce. The troy or apothecaries' ounce is still $\frac{1}{12}$ pound, but the common avoirdupois ounce is $\frac{1}{16}$ pound. Both are roughly equal to the Roman uncia.

In recording important weights, the Romans used submultiples. Thus a Roman clerk probably would have noted a 100-lb (avoirdupois) bale or box as having a weight of:

CXXXVIII LIBRAE ATQUE SEMIS.

That is, 138 libras plus 1 half-libra unit. The many Roman submultiples were provided, presumably, to avoid the need to use more than two unit names for one weighing.

Metric units should last as long as there are quantities to be measured, but the nonmetric units of the past were constantly fading away and being replaced—or, if they survived, they did so in bafflingly modified or muddled forms.

Vanished weight units—In the late 1960s a United Nations survey, *World Weights and Measures,* listed many nonmetric units of weight, length, area or volume still in use. None of them bore a name resembling the mina, shekel or talent. Of the name drachma, only the British apothecaries' dram remained.

More than a dozen Spanish-speaking nations showed some use of units named *libra:* Chile (460 g); Colombia (500 g); Costa Rica (460 g); Cuba (460 g); Dominican Republic and El Salvador (453.592 g); Ecuador (460 g); Guatemala (460 g); Honduras, Mexico, Nicaragua, Peru and Spain (460 g); and Paraguay (459 g).

These records suggest how the Roman libra migrated, via Spain, to the New World, and even how it gained about 40 percent in transit. Brief inspection shows that most of the New World libras weigh about the same as the U.S. avoirdupois pound (453.592 gram). Dictionaries and historical works have sometimes called the Roman libra the Roman pound. Indeed, in the Dominican Republic and El Salvador, the libra has become identical with the U.S. pound weight. The modification of the libra can be traced to a sort of "gravitational" influence of the pound of the giant power of North America.

Beside the *maund* units of various Middle Eastern and South Asian countries, there may be mentioned the maud or maund of Basra, Iraq (42.28 kilogram). Such measures may have arrived with the Macedonian Greek, Alexander the Great, his generals, and their dynastic successors. Perhaps they are derived from the Greek talent through a lengthy process of transmission.

Fairness versus fraud in weighty matters—Uniform and recognized weights and measures do not suffice to produce fair trade. There must be also freedom from fraud in measurement.

For thousands of years the balance, or two-pan scale, was used to compare the weight of the object to be bought or sold and the standard weights that should just balance it. A scale or weighing machine may be defective, however. If the distance from the center fulcrum to the suspension of the left-hand pan is 5 percent greater than that from the fulcrum to the suspension of the right-hand pan, then 5 percent more gold or silver—say 1.05 minas—will have to be placed in that right-hand pan to balance 1 mina in the left-hand pan.

To deal with this danger, the payer can insist that the contents of the two pans be interchanged. Only if the scale is true will it continue to balance after such a reversal. There are also methods for getting a true result even with a balance which is not true, but we shall not examine them here.

A second kind of fraud is harder to prevent: false weight standards. No matter how true the scale itself, if a standard weight marked 1 mina weighs in fact only 0.85 mina, then there will be a 15 percent disadvantage to one party and a corresponding advantage to the other. Besides true weights, both parties must understand and agree on the standards that they use. A purchaser who agreed to pay 3 minas of gold, meaning the Greek mina of 431 gram, would suffer fraud if the weighing were done with Hebrew minahs of 499 gram each.

The Bible contains a number of powerful denunciations of the practice called, in English translation, diverse weights. This means carrying two sets of weights—one actually lighter than marked, the other heavier. The light weights are for selling, for they result in less merchandise for the money. The heavy weights are for buying. The secret substitution of one set or the other is done by the same kind of sleight-of-hand today used to work a crooked pair of dice into or out of a craps game.

Naturally, the false weights could be used also when paying customs duties and other levies—or when collecting them.

Leviticus 19:35–36 commanded flatly: "Ye shall do no unrighteousness in judgment, in meteyard, in weight, or in measure.

"Just balances, just weights, a just ephah, and a just hin, shall ye have: I am the Lord your God . . ."

The *ephah* was a Hebrew dry measure containing 0.182 to 0.273 hectoliter, or 0.5 to 0.75 U.S. bushel. (1 hectoliter = 100 liter.) The *hin* was a liquid measure of about 4.7 liter, or 1.25 U.S. gallons capacity. The *mete-* in meteyard meant the act of apportioning out, as in the use of weights and measures. Figuratively, to mete out rewards and punishments is to measure them out.

The most legalistic and prescriptive part of the Pentateuch, Deuteronomy, issues this mandate (25:13, 15): "Thou shalt not have in thy bag divers weights, a great and a small. . . . But thou shalt have a perfect and just weight, a perfect and just measure shalt thou have . . ."

Similarly in the book of Proverbs several condemnations appear: "A false balance is abomination to the Lord; but a just weight is his delight"; also "Divers weights and divers measures, both of them are alike abominations unto the Lord"; and "A just weight and balance are the Lord's: all the weights of the bag are his work."

The prophetic book of Ezekiel (45:10) says: "Ye shall have just balances, and a just ephah, and a just bath." The bath was a Hebrew liquid measure of from about 18 to nearly 25 liter capacity (4.75 to 6.5 U.S. gallons).

Hosea (12:7) makes a bitter comment on the businessman of the time: "He is a merchant, the balances of deceit are in his hand . . ."

And Micah (6:11) asks indignantly: "Shall I count them pure with the wicked balances, and with the bag of deceitful weights?"

The most remarkable metrological story in the Bible is in Daniel. The time is 539 B.C.E.; the place is the great feast given by Belshazzar, ruler of Babylon, attended by thousands of his lords, captains, their wives, concubines, serving women, and men-at-arms. The monarch orders his guests to drink from golden vessels that had been taken from the Temple of Solomon in Jerusalem when the Babylonians (Chaldeans) destroyed it in 587 B.C.E. They drink wine from the sacred vessels, then praise their own "gods of gold, and of silver, of brass, of iron, of wood, and of stone."

As they do so, the fingers of a man's hand appear and write on the palace wall. Seeing "the writing on the wall," Belshazzar is terrified. "The joints of his loins were loosed, and his knees smote one against another." He calls his astrologers and soothsayers, but they cannot interpret what has been written. At last the captive Jew, Daniel, is brought. He dares to decode for the king the message on the wall:

MENE, MENE, TEKEL, UPHARSIN

In translation, a metrological metaphor appears. In somewhat modernized language: MENE: God has measured your kingdom and reduced it or terminated it. TEKEL: You yourself have been weighed in the balance and found wanting. UPHARSIN (or PERES): Your kingdom [shall be] divided and given to the Medes and the Persians.

That very night the Medes and Persians, under their dynamic leader Cyrus, were lying in wait outside Babylon. Within hours they struck and "in that battle Babylonian resistance was swiftly broken, Belshazzar was slain, and the Babylonian-Chaldean people became subjects of the new Persian empire." (Mina C. Klein and H. Arthur Klein, *Temple Beyond Time*, New York, 1970, page 72.)

Biblical scholars now generally agree that MENE stands for the mina, unit of weight; TEKEL is a variant or dialectal form of shekel, another weight unit; and U-PHARSIN may be a weight unit of about two half shekels, symbolizing the twin forces of the Medes and Persians united under Cyrus—or it may be only a reference to the Persians or Parsis (pharsin).

The symbolism is undoubtedly, however, one of weights that have been found wanting, thus drawing down the wrath and retribution of the Lord God.

II / Times in Turmoil, and the Origins of the Metric System

. . . every Moment has a Couch of gold for soft repose,
(A Moment equals a pulsation of the artery),

. . . every Minute has an azure Tent with silken Veils:
And every Hour has a bright golden Gate carved with skill:
And every Day & Night has walls of brass & Gates of
*   adamant,*
Shining like precious Stones & ornamented with appropriate
*   signs:*
And every Month a silver Terrāce builded high:
And every Year invulnerable Barriers with high Towers:
And every Age is Moated deep with Bridges of Silver &
*   Gold:*
And every Seven Ages is Incircled with a Flaming Fire.

Now Seven Ages is mounting to Two Hundred Years.
Each has its Guard, each Moment, Minute, Hour, Day,
*   Month & Year.*
All are the work of Fairy Hands of the Four Elements . . .
  —WILLIAM BLAKE,
   Milton (1804)

8 KEEPING TRACK OF TIME

*There is an appointed time for everything and a time for
every affair under the heavens.*
—ECCLESIASTES 3:1
(New American Bible trans.)

Time is but the stream I go a-fishing in.
—HENRY DAVID THOREAU,
Walden (1854)

*To me every hour of the light and dark is a miracle,
Every cubic inch of space is a miracle,
Every square yard of the surface of the earth
   is spread with the same,
Every foot of the interior swarms with
   the same.*
—WALT WHITMAN,
Miracles (1856)

Among the physical variables that men have found means to measure, time and its twin, duration, form a particular and peculiar
pair. In many respects they appear to be the most baffling and ambiguous of variables. Only temperature may offer so many uncertainties as does time to the metrologist who seeks to go beyond
conventions to fundamentals.

But from the standpoint of practical rather than theoretical metrology, the greatest triumphs of consistency and precision have
been attained in the realm of time or duration. Time is more precisely measurable than the other two basic variables, mass and
length. Also, the units in which time and duration are measured
are more nearly uniform throughout the world, and became so
earlier, than the units of mass (weight) or length. Even the diminishing number who advocate retaining American foot and pound
units would not support abandoning the second, the minute and

the hour for some other, special, "American" system of keeping track of time.

The worldwide system of time measurement predates, in fact, the metric system. We shall review the effort made in the last years of the 18th century to introduce a decimal and metric system of time units. The failure of that effort reflects the priority and the firm acceptance of the world's rather peculiar time units.

Today's metric and SI systems use only one basic time unit—the second, abbreviated s. So standardized and widely accepted is this unit that we present it first among the basic SI units.

The second is commonly regarded as a tiny amount of time. "Just a second!" "Can't you be still for a second?" and other every-day phrases testify to this. Yet 1 second is a duration of enormous magnitude compared with many events now measured with great precision in scientific and technical work.

Decimalized submultiples of the second are the millisecond (0.001 s), the microsecond (0.000 001 s) and the picosecond (0.000 000 001 s), abbreviated respectively as ms, μs and ps. The Greek letter mu is used for *micro-*, because m is already preempted for *milli-*, as in mm (millimeter) and ms (millisecond).

If man is still conducting scientific and technical researches in 2000 A.D. or C.E., important measurements will doubtless be reported in terms of the femtosecond (fs), which is the million-millionth (or trillionth) part of a second. To recapitulate, 10^{-3}, 10^{-6}, 10^{-9}, and 10^{-12} are represented by m, μ, p and f.

Thus, the second is rather a "middling" measure of time than a "small" one. In the other direction *kilosecond* (10^3s), *megasecond* (10^6s), *gigasecond* (10^9s) and *terasecond* (10^{12}s) are abbreviated as ks, Ms, Gs and Ts, respectively. A kilosecond is a little more than a quarter of an hour. Conversely, a year is about 31.6 mega-second; a millennium is about 31.6 gigasecond; and the 4.5 billion years that our Earth has existed are about 140,000 terasecond. The last number is a little unwieldly, but we still have notation based on powers of 10 to express any number without using a lot of zeros. Thus a billion years can be expressed as 3.16×10^{16} second and the age of the Earth as 1.4×10^{17} s. About the largest number used in considering cosmology is 15 billion years, or 4.74×10^{17} s.

To get back to familiar measures of time—we have other common units larger than the second: 60 s = 1 minute; 60 min = 1 hour; 24 hr = 1 day; 365¼ d = (just about) 1 year.

Between the day and the year is a nonastronomical unit, the week, of 7 days (= 168 hr = 10,080 min = 604,800 s).

The month (the word is derived from Old English *mona*, "moon") is irregular; in the Christian calendar it may have 31, 30, 29 or 28 days; the non-leap-year February is the only month that has a whole number of weeks.

Each year begins just past midnight on the day designated as the first of January and ends at midnight of the day designated as December 31. A calendar year with 365 days of 24 hours each contains 8,760 hours, but the Earth completes a full cycle around the Sun in about 8,766 hours. To be more exact, the sun-measured year has a duration of 365.242 2 d, or 8,765.812 8 hr, or 31,556,- 926.08 s. The day is also an astronomical unit, the duration of the rotary motion of the Earth around its imaginary axis extending from the North to the South Pole.

The scientific name of the sun-determined year is "tropical mean solar year." For rough calculation its duration is often rounded to 365¼ days (average of three "regular" years and a leap year). This is too large by 0.007 8 days, or 673.92 s—in short, about 11 minutes.

Calendar makers have always been plagued by the absence of a whole-number ratio between the solar year and the day. Some of the devices and fictions they have used to overcome that difficulty will be discussed later.

Units of duration greater than the year are the decade, the century and the millennium, each ten times the magnitude of the one before. Occasionally one encounters also the aeon of 1 million years. Then why do we have ratios of 60:1 and 24:1 connecting the day, hour, minute and second?

These relationships were born 5,000 to 6,000 years ago in the flat alluvial plains of Mesopotamia (now Iraq), where civilization and city-states first appeared on Earth. Here, in the valley of the Tigris and Euphrates rivers, at the eastern end of the Fertile Crescent, sky watching and measurement of stellar positions first became precise and continuous. Here, as a result, the movements of heavenly bodies across the sky were converted into clocks.

It was in Mesopotamia, not in Egypt, that systematic measurement and comparison of angles was first developed—angles formed by structures on the flat surface of the Earth and angles formed between stellar objects and the Earth.

The Mesopotamian number system was based on the number 60, just as ours is based on 10, and that of most digital computers is based on 2. We do not know why the Mesopotamian sky watchers and Earth measurers rejected the common base of 10 founded on the ten fingers, but it seems likely that they were influenced by the "360" days in a year and the "30" days in a month, based on the time from one full moon to the next. (The actual time is about 29.25 days.)

In dealing with such numbers before the development of decimal fractions or a comprehensive system of calculating with common fractions, a sexagesimal or base-60 system has substantial advantages. Our base 10 is divisible evenly only by 2 and 5, while the Mesopotamian base 60 is divisible evenly by ten numbers: 2, 3, 4, 5, 6, 10, 12, 15, 20 and 30. Religious and magical considerations may have magnified these advantages in the eyes of the Mesopotamians.

They may well have divided the year into 360 equal units, with a few "extra" days; and they did divide the circle (which, like the year, is a full rotation) into 360 equal parts or angles. Thus arose our present angular measure of 360 degrees in a circle with angles and arcs represented by symbols such as 30° and 60°.

Modern mathematics, however, has replaced the degree as the basic unit of angular measure. The radian is the new measure, and is the only approved unit of plane angle in the International System. One radian is the angle which, with vertex at the center of a circle, cuts off on its circumference an arc equal in length to the circle's radius. That arc is a little shorter than $\frac{1}{6}$ the circumference of that circle.

A radian (1 rad) $= 180°/\pi = 57.295\ 8°$. Inversely, 2π radian $= 360$ degrees, or a full circle, sometimes known also as 1 turn.

Attention-attracting angles—Certain angles imposed themselves on the attention of sky watchers, however early they began their observations and record-keeping. Thus from the zenith, the sky point directly overhead, to the horizon on the flat mud-baked plain the angle was just a quarter of a full circle, or 90 degrees (90°).

The angular separation represented by just 1° was too coarse for the needs of the sky measurers. Stars could be distinguished though separated by less than 1°. Hence the Mesopotamians subdivided each degree by their basic number, 60. Today each sixtieth of a degree is called 1 minute of angle and is symbolized by 1'. Accordingly, 9 degrees plus $\frac{13}{60}$ degree is written 9° 13'.

Later the minute of angle again was split 60 times, into the second of angle, symbolized by 1''. Accordingly, 9 degrees plus $\frac{13}{60}$ degree plus ½ (that is, $\frac{30}{60}$) minute is written 9° 13' 30''. A full circle contains 360 degrees, or $360 \times 60 = 21{,}600$ minutes, or $21{,}600 \times 60 = 1{,}296{,}000$ seconds of angle.

The relation between angular measurement and units of time duration goes beyond the fact that they both use minutes. Every nondigital clock relates an angular position on a circle to a point in the unceasing continuum of time. The hands of the clock compactly perform rotational motions somewhat like those of the heavenly bodies and the Earth—the face of the clock or watch is a kind of replica of the sky.

The longer (minute) hand of a clock completes a full circle in just one hour—that is, it moves 360° in 60 minutes, or 6° per minute, or 6' per second. The shorter (hour) hand is geared to rotate 360° in 12 hours—that is, it moves 30° in 60 minutes, or 30' per minute, or 30'' per second. Its rate of angular motion is $\frac{1}{12}$ that of the longer hand.

Many watches and clocks have a third hand—the so-called "second hand"—which makes a full turn in 60 seconds, thus moving at a rate of 6° per second. It rotates 60 times as rapidly as the minute hand and 720 times as rapidly as the hour hand.

In clocks that have 24 hour divisions, the hour hand completes 1 revolution in 24 hours. The second hand still rotates 60 times as rapidly as the minute hand, but 1,440 times as rapidly as the hour hand.

In World War II, crews of warplanes instantly announced the relative location of enemy aircraft in terms of hour-hand positions on an imaginary clock face, with the 12 dead ahead, the 6 aft, the 3 on the right, the 9 on the left. The warning "Three coming in, two o'clock high" told the gunner to look for three airplanes 60° to the right of the plane's axis, well up toward the zenith, and so on. A generation that may grow up accustomed to digital rather than "hand" clocks will be unable to make such connections between time and angle measurement.

Are the Mesopotamian ratios of 360:1 and 60:1 acceptable to all? By no means. The British astronomer and science writer Fred Hoyle, in a useful work, *Astronomy* (London, 1962), suggests that it would be far better to divide the circle or turn into 1,000 equal parts, each called a *milliturn,* and the milliturn into 1,000 *microturns.* The milliturn would correspond to an angle of 21′ 36″, or less than half a degree; the microturn to just 1.296″. A right angle would be 0.25 t (where 1 t means full turn) or 250 mt (milliturn).

To decimalize our timepieces and time units, we might begin not with the arbitrarily defined second, but with the astronomically determined day (d). Dividing by 10 we get 1 deciday (dd) = 2.4 hours. Then 1 centiday (cd) = 0.24 hours, or 14 minute and 24 second. Then 1 milliday (md) = 86.4 second. Thus the deciday would afford a generous lunch hour; a centiday would be nearly a quarter of an hour; and a milliday about a minute and a half. For scientific purposes there would be also the microday (μd), or 10^{-6}d, of 0.086 4 s.

Clocks with new looks—A decimal clock has three hands. The slowest is the deciday hand, rotating once, not twice, each full day; the next, the centiday hand, rotating 10 times daily and arriving at a new mark at intervals of 14 minutes 24 seconds; and finally the speediest hand, rotating in 86.4 seconds. If it made 100 forward jumps, or ticks, each revolution, it would mark off time intervals of 0.864 s.

Only historical mischance, according to astronomer Hoyle, saddled us with Mesopotamianized ratios and relations in our timekeeping and our angular measurements. He holds it to be high time for changes, and he warns, "Man's inability to rid himself of inconvenient conventions is a trait that could lead to his undoing."

Hoyle's own country, Britain, long labeled tradition-bound and conservative in formal matters, caught up with the United States by decimalizing its ancient currency system, and then leaped out ahead by adopting the metric system.

Another informed critic of usage with regard to time and angle measurement is Professor Maurice Danloux-Dumesnils, a French engineer-scientist. His book *Esprit et bon usage du système métrique* has been translated into English as *The Metric System: A Critical Study of Its Principles and Practices* (London, 1969).

Danloux-Dumesnils accepts the second as a convenient time unit, but objects to durational use of the names second and minute, which belong to angular measurement. That is, the 3,600th part of an hour should not have the same name as the 1,296,000th part of a circle, and so on. (He also remarks that use of the word "degree" in angular measurement is confusing, since there are also temperature degrees—for example, degrees Fahrenheit—and degrees of viscosity and specific gravity, Baumé degrees, Engler degrees, and so on.) The second, he suggests, might be renamed the *dur* (from duration) as a basic unit of time.

Danloux-Dumesnils also favors use of decimal multiples of the second: the *hectosecond* (hs) of 100 s; the *kilosecond* (ks) of 1,000 s; and the *megasecond* (Ms) of 1 million s.

He does not propose, however, to decimally subdivide the day, since this would call for two basic units of time. Hectosecond is an awkward name, so he would call it a prime. As we shall see, the minute itself was originally called a prime minute.

For thousands of years the smallest duration unit in common use was the hour. At first it varied, being the twelfth part of one daytime or one nighttime. In the northern hemisphere the daytime hour increased, and the nighttime hour decreased, in summer. Equalizing daytime hours and nighttime hours, and summer and winter hours, had to await the development of accurate time-measuring machines, or clocks.

Durations of less than 1 hour were at first expressed as fractions —half an hour, quarter of an hour, and so on, as still used today. All smaller segments of time were called a minute, meaning a "little one," as we might say, "Wait a minute." As a need arose for an exactly defined small unit, a few timepieces actually did divide the hour into 100 parts, but the influence of the Mesopotamian pattern of angle measurement prevailed. Just as the degree of angle had been split into 60 minutes of arc, so the hour was split into 60 minutes of time.

Originally $\frac{1}{60}$ hour had been called (in Latin) *minuta primam,* or "first small one." In the late 16th century, around the time when Galileo was measuring how long it took balls to roll down inclined planes, the *minuta secundam,* or "second small one," came into use also. Its name was later shortened, fortunately, to the one word: *second.* It ceased then to be called the "second minute."

Danloux-Dumesnils takes the "prima" from *minuta prima* to name his 100-second minute, or prime. Note that 1 day = 864 primes. Since the day is not a decimal multiple of the second, one could expect to use only a few simple subdivisions: for instance, quarter day (216 primes), half day (432 primes = 43,200 second), and so on. Even $\frac{1}{32}$ day can be expressed as 27 primes.

9 REVOLUTION AND MEASUREMENTS

> . . . *others of the Sons of Los build Moments & Minutes*
> *& Hours*
> *And Days & Months & Years & Ages & Periods, wondrous*
> *Buildings* . . .
> —WILLIAM BLAKE,
> *Milton* (1804)

> *The struggle and the daring, rage divine for liberty,*
> *Of aspirations toward the far ideal, enthusiasts dream of*
> *brotherhood,*
> *Of terror to the tyrant and the priest.*
> *O star! O ship of France, beat back and baffled long!*
> *bear up O smitten orb! O ship continue on!*
> —WALT WHITMAN,
> *O Star of France* (1871)

The French Revolution gave birth to three significant plans for reform in measurements. The first, applied to the basic variables of length and mass (and, so, weight), launched the metric system, recently expanded into the International System (SI). The second, reform of the calendar, remained in effect in France for thirteen years, from 1792 until terminated by the dictator-emperor Napoleon in 1805. The third was a far-reaching reform of the clock face and of units of time. It was virtually stillborn, but is nonetheless interesting.

No history of the French Revolution should scant any of these three efforts. In general, the aims and achievements of the French Revolution, which ended in the Napoleonic *coup d'état* of the 18th Brumaire (1799), have been distorted and besmirched, especially for those whose concepts are derived mainly from two classics of English letters, *A Tale of Two Cities,* which reveals more about the genius of Charles Dickens than about social forces in France, and

105

Thomas Carlyle's *The French Revolution,* a crabbed and tortured piece of special pleading.

The clatter of the tumbrils carrying the condemned to the guillotine has been allowed to drown out the main themes of a period of overdue and irreversible change. The Terror was only the blown spray on the great wave that washed away feudal remnants. It came relatively late in the Revolution, was largely localized in Paris, and was confined to the period of late 1791–1794. The summary condemnations and executions of that overemphasized period were matched, probably exceeded, by those that took place under the reaction after the coup of Thermidor (July 1794), and those perpetrated still later as Napoleon organized his police state in France.

From 1789 until late 1791, as H. G. Wells has noted, the Revolution remained an orderly, ongoing process. And again, after 1794, the republic that the Revolution had made possible was "an orderly and victorious state," despite its defects and disappointments. Without understanding this, the victorious thrusts of the French armies from the early 1790s to Napoleon's enormous error of invading Russia are incomprehensible, and there is no way to properly appreciate the forces of good sense and consistency that gave birth to the metric system and other metrological reforms.

Revolutionary rapidities—The preliminaries to the Revolution took very little time. The bankruptcy of the monarchy of Louis XVI led to convoking the French National Assembly in May 1789. Accumulated pressure for change plus the king's erratic actions precipitated a series of decisive events culminating in the storming of the Bastille on July 14. Like wildfire a new awareness spread through Paris and into the provinces: the jig was up, at last, for the feudal *ancien régime.* Peasants seized land, put the torch to chateaux, destroyed the hated documents that listed their feudal obligations and penalties. The lords and their families, stewards, agents and hangers-on fled in fear for their lives.

Almost at once the Assembly seemed to take fire. A progressive member of the nobility, Count de Noailles, who served under Lafayette in the American Revolution, took the floor the night of August 4 and moved to end the ancient feudal services, dues and tithes. Before morning the Assembly had acted to destroy the underpinnings of feudalism in France.

An enormous throwing-out of objectionable baggage from the past began in that late summer of '89. Its consequences were felt throughout France and later, spread by French arms and administrators, in many another land of Europe. It was in this period that the first steps were taken to reform the measurement of time, length and weight in France.

One of the most astute among the militant Assembly members was the Bishop of Autun, Charles Maurice de Talleyrand-Perigord (1754–1838), referred to in history as Talleyrand. He is known to us as a nimble turncoat and careerist, the reactionary leader under the monarchy after Waterloo. Talleyrand had entered the political arena as a representative of the clergy in the States General; but in 1789 he opposed inclusion of the clergy in the National Assembly. His skill in trimming his sails to the most favorable winds already was notable. Increasingly he came forward as an advocate of changes likely to prove acceptable to the majority of his fellow Assembly members.

In July 1790, to celebrate the first anniversary of the storming of the Bastille, Bishop Talleyrand officiated at a so-called mass in the great Parisian square of Champ de Mars. The radical bishop headed a staff of one hundred priests, and the congregation was estimated at 300,000, including the unhappy king, Queen Marie Antoinette, General Lafayette, and the National Guard. Highlight of the religiopolitical spectacle was a civic oath of allegiance to the French fatherland, one and indivisible. The ceremony was in the fashionable style of the old Romans as distilled from Plutarch, but with much late-18th-century seasoning added.

That was Talleyrand's last mass. In surefooted, feline fashion he moved further in the anticlerical direction. By October 1790 he was proposing that the Assembly confiscate the great holdings of the Church in France and apply them to the general welfare. After renouncing his bishopric, he was excommunicated by the pope. By this time he had been appointed by the Assembly to the government of the Department that included the city of Paris.

As the first Constituent Assembly approached its end in 1791, Talleyrand spoke in favor of free education all the way to the university level. He urged that religion—but not any particular dogma or creed—be taught in the new public schools.

It was this "radical" Talleyrand who in the Assembly initiated a proposal to the French Academy of Sciences that it work to reform weights and measures. The Academy, which, with the Royal Society of England, was preeminent among the world's institutions of science, responded to the Assembly's proposal by appointing a "blue-ribbon" committee to formulate a practical program for fundamental reform of weights and measures for all France.

The problem that Talleyrand had singled out was neither remote nor trivial. Metrologically as well as politically, France was still absurdly divided, confused and complicated. A given unit of length recognized in Paris, for example, was about 4 percent longer than that in Bordeaux, 2 percent longer than that in Marseilles, and 2 percent shorter than that in Lille. And so it went, from region to region, department to department.

Metric reforms were pressed mainly by moderate, even somewhat conservative, reformers—Girondins rather than Jacobins, at least in the early phases.

Talleyrand was not, after all, the first French churchman interested in metrological reform. More than a century before, the Abbé Gabriel Mouton, a vicar from Lyons, had proposed that all distances be measured by a decimal system of units based on dimensions of the Earth itself. His ideas anticipated much that later became part of the metric system.

In his 1670 work, *Observationes diametrorum solis et lunae apparentum,* Mouton suggested a primary length standard equal to 1 minute of arc on a great circle of the Earth. Our planet is not a perfect sphere, but if it were, its volume would fit into great circles everywhere close to 40,075,510 meter in circumference. Then 1 minute of arc, being $\frac{1}{21,600}$ of such a full circle, would equal 1,855.3 meter, or 1.855 3 kilometer.

The crude geodesic measurements available to Mouton indicated a greater length for this 1 minute of arc, but here we take advantage of the modern measurements.

For this basic length Mouton offered a name: the *milliare.* He followed it by seven subunits, each $\frac{1}{10}$ the length of the one preceding: the *centuria* (185.53 m), the *decuria* (18.553 m), the *virga* (1.855 3 m), the *virgula* (18.55 cm), the *decima* (1.855

cm), the *centesima* (1.855 mm), and the *millesima* (0.185 5mm). The first syllables of most of these names suggest numbers: *milli* (1,000) and *mille* ($\frac{1}{1,000}$); *centu* (100) and *cente* ($\frac{1}{100}$); *decu* (10) and *deci* ($\frac{1}{10}$). The milliare as largest unit and, so to speak, the parent of the others, was 10^7 (10 million) times its last, least descendant, the millesima.

Mouton's proposed virga, conveniently in the middle, was close to a familiar old French unit of length, the *toise* (1.949 m in Paris, varying elsewhere).

Abbé Mouton knew well enough that accurate measurement of 1 minute of arc of a meridian of Earth was a difficult task. Accordingly, he discussed an alternative way to arrive at a standard length, also using physical characteristics of the Earth: pendulums with precisely timed swings.

By this time the general relation between the length of a true pendulum and the period of its swing was familiar. A true pendulum is formed by a very light rod or wire with a low-friction fulcrum at the upper end and all, or almost all, of its mass concentrated at the lower end. When such a pendulum swings freely back and forth, the longer it is the more time its swing will take, but the time, or period, of its swing increases in proportion to the square root of the length from fulcrum to center of mass.

That time, or period, is also inversely proportional to the square root of the gravitational intensity—more correctly called the gravitational acceleration—where the pendulum is swinging.

At sea level in latitude 45°, halfway between pole and equator, the standard acceleration due to gravity is 9.806 21 meter per second per second (m/s^2). Under those conditions a pure pendulum just 0.993 577 meter long will complete its cycle of swinging (to and fro) in just 2 seconds: 1 s for the swing one way, 1 s for the swing back.

Such a pendulum is called a "seconds pendulum." A pendulum just twice as long would require 1.414 times 2 seconds, or 2.818 s for the full cycle; and a pendulum just four times as long would require 2 times 2 (or 4) s.

Mouton pointed out that a standard seconds pendulum could also define a basic length—its own. In fact, 1,867.3 times the standard seconds pendulum length would also provide a milliare of length 1.855 3 kilometer. The virga, being just $\frac{1}{1,000}$ milliare, would be matched by 1.867 3 times the length of the standard seconds pendulum (provided that sea-level gravitational intensity prevailed).

Somewhat similar proposals were made also by an eminent French astronomer and geodesist, Jean Picard (1620–82). In 1669 he even showed that the rates of clock pendulums changed as their lengths were altered by expansion caused by increased temperature, or by contraction from decreased temperature.

Mouton and Picard each measured the length of seconds pendulums at Lyons. Their results differed by about 5 parts in 1,000, Mouton being closer to the figure now recognized as correct. Even under sea-level conditions, seconds pendulums are found to vary in length by somewhat less than 3 parts in 1,000, since gravitational acceleration at sea level is not everywhere identical.

Neither of Mouton's proposals—the second of arc or the seconds pendulum standard—took hold. However, they served as seeds and stimulated many thoughtful people who were discontented with the contradictions and confusions in measuring length and other important physical variables in France.

The founding fathers of the metric system were true polymaths, men of encyclopedic learning. Surely, at least those who had contributed to or studied the 36-volume *Encyclopédie* deserved this title.

Nominal head of the Academy's weights and measures committee was the mathematician and nautical astronomer Jean Charles Borda (1733–99). Other members were also eminent scientists and scholars: Joseph Louis Lagrange (1736–1813); M. J. A. N. de Caritat, Marquis de Condorcet (1743–94), known as Condorcet, a Girondist who died in prison during the Terror; Gaspard Monge (1746–1818); and Pierre Simon de Laplace (1749–1827). In close touch with them until guillotined by order of the Convention was the great chemist Antoine Lavoisier (1743–94).

The caliber of this committee was high, even though the Academy of Sciences had been shackled by the *ancien régime* and aristocracy. Like the Royal Society in England, the Academy existed by grace of the king and under the sponsorship of the nobility. Its presiding officers were nobles, not professional scientists.

Twelve peers and eighteen pensionaries formed the Academy's

membership and executive body, selecting new members and making policy. There were also twelve associates, twelve adjunct members, some free associate members, and foreign or corresponding members outside France. In its rigidity, the Academy of Sciences seemed typical of a society that was on its way out rather than one that was being raised to new importance. The metrological committee members were regarded as salaried state officials.

Jean Charles Borda, the chairman of the committee, had been a naval officer. He was only twenty-three when he presented a paper on the movement of projectiles, and his presentation led to his selection as a member of the Academy. He had entered the French navy and there helped map the Azores and the Canary Islands. When his frigate was captured by the British in 1782, he was imprisoned in England, but was soon paroled and restored to France. His contributions to the Academy included highly valued papers on hydrodynamics, and he devised improvements in the repeating and reflecting circles used in celestial and secular surveying. As one of the leaders of the actual survey of the arc of the meridian (Dunkirk to Barcelona), Borda designed most of the equipment used in that historic enterprise.

Joseph Louis Lagrange was one of the preeminent mathematicians of all time. His great work on analytic mechanics so impressed another mathematician, William Rowan Hamilton, that the latter called it "a scientific poem." As a teacher and exemplar of other mathematicians, Lagrange helped to develop, among others, the celebrated astronomer and mathematician Laplace.

Condorcet was described by James R. Newman, in the well-known anthology *The World of Mathematics* (New York, 1956), as an "eminent French philosopher, mathematician, and revolutionist." Condorcet's philosophy influenced Thomas Jefferson as much as, or more than, that of John Locke. Condorcet's remarkably prescient predictions of the future course of society may establish him as the first "futurist." Francis Galton, in the 19th century, listed Condorcet as one of several great mathematicians who were more than "mere" mathematicians.

In 1794 Condorcet became a fugitive from the Jacobins, who had seized power in Paris. He was recognized and jailed, and the next morning he was found dead.

Gaspard Monge is best known as the developer of descriptive

geometry. At sixteen he was appointed professor of physics in Lyons. His career was aided by Condorcet. Monge rose to high office under Napoleon, but died in poverty during the Bourbon restoration that followed Napoleon's defeat at Waterloo.

Pierre Simon de Laplace, youngest of the group, was the most successful careerist among them—and a mathematician of great ability. He was an examiner for the Royal Artillery at fifteen and professor of mathematics in the Ecole Militaire of Paris at twenty. Later he and Lagrange taught mathematics at the newly established Ecole Normale, one of the great institutions founded under the First Republic. James Newman credits Laplace with having "suggested, in keeping with the reform spirit of the Revolution, the adoption of a new calendar . . ."

Laplace, often called the "Newton of France," seemed able to ingratiate himself with whatever regime came into power. He held important posts under the Republic, and later, under Napoleon, he was for a time minister of the interior. Still later, after the defeat at Waterloo, as a senator, he voted to depose Napoleon. Having thrown himself at the feet of the restored monarchy, he became a member of the French Academy in 1816 and its president in 1817.

The first edition (1796) of Laplace's great work on celestial mechanics was inscribed to the Council of Five Hundred; the third edition (1802) contained a worshipful paean to Napoleon, who had dispersed the Council. The 1812 edition of Laplace's work on the theory of probability was dedicated to "Napoleon the Great"; in the 1814 edition that dedication was replaced by a comment criticizing "empires that aspire to universal dominion."

We have mentioned Lavoisier as, so to speak, an ex-officio member of the committee. He was one of chemistry's immortals, yet by modern standards also as much a physicist as chemist. He had been head tax collector for the French monarchy. A few years later this connection proved fatal, for during the Terror he was condemned as an enemy of the Republic and guillotined.

As an Anglophile, Talleyrand had seen to it that an English representative was invited to sit on the committee on weights and measures, but the British foreign secretary sent back a chilling negative, noting that reform of weights and measures was considered "almost impracticable." A similar invitation sent to the young

United States of America was similarly rejected. Danloux-Dumesnils calls these rejections the "beginning of the opposition of the English-speaking countries to the metric system—when one has refused to collaborate in a scheme, one cannot forgive its subsequent success."

Choosing a numerical base—From its first deliberations, the committee inclined toward using the basic multiple 10 not only for measurements but also for the monetary system of the new France. However, it did consider also the base-12, or duodecimal, system. One member expressed regret that the base 10 could be divided evenly only by the factors 2 and 5, whereas the base 12 could be divided by 2, 3, 4 and 6.

Lagrange, indeed, surprised his colleagues by favoring an 11-base system, not evenly divisible save by itself and 1.

The committee, once having decided on a base of 10, agreed that the unit of length, when chosen, would be used to derive the units of area and volume. And the basic weight would be that of a designated volume of pure water.

The length unit chosen by the committee was one ten-millionth (10^{-7}) of a quadrant, or quarter, of an arc of meridian. Thus was born the unit later named the meter. It was not, after all, to be based on the length of a seconds pendulum, as Talleyrand and others had anticipated. A great surveying operation would be required to determine the length of the quadrant of meridian.

The unit of weight (mass), the committee decided, would be the amount of water in a cube whose side was one-hundredth of the still unmeasured meter. Thus the weight of 1 cubic centimeter of pure water was defined as 1 gram. The weight of a given volume of water was known to vary depending on its temperature. At first the decision was to use water at the temperature of melting ice, or $0°$ C by the modern scale. Later it was found necessary to change that to the temperature at which water attains its maximum density —that is, $4°$ C. Oddly enough water is denser at that temperature than when it is chilled still further toward $0°$ C.

The committee's proposals, when presented to the Constituent Assembly in 1791, were accepted.

With the rapidity typical of a time of swift change, the Assembly proceeded to implement the recommendations of the Academy committee in order to put an end to "the shameful diversity in our measures."

Two astronomer-geodesists were appointed to work with Borda on measuring the length of a quadrant of meridian: J. B. J. Delambre (1749–1822) and P. F. A. Méchain (1744–1804). There was no need to triangulate a complete quadrant from the North Pole to the Equator; they chose a sector of about one-ninth of a quadrant (10 degrees of latitude) extending from Dunkirk, on the English Channel, to a site near Barcelona, on the Mediterranean coast of Spain. One of the advantages of the choice was that both ends of the sector were at sea level.

The task took from 1792 to 1799. The surveyors encountered obstacles both comic and tragic. More than once they were accused of being foreign spies or, because of their white marker flags, royalist agents. Detentions and even formal arrests took place. Their strange exercise in painstaking precision amidst vast social transformations acquires, through the lenses of history, a strangely symbolic stature. Metrology too has its martyrs and heroes.

These premetric meridian measurers used special "repeating circles" that had been designed by Borda and were calibrated not in "Mesopotamian" degrees, minutes and seconds of arc, but in terms of the *grade,* the hundredth part of a right angle and hence one four-hundredth part of a full circle. One grade is thus equivalent to 0.9 degree, or 54 minutes, of arc.

The *grade,* then, rather than the meter or the gram, was the first decimalized unit arrived at in the development of the metric system. Professor Danloux-Dumesnils accordingly claims for the grade the title of "oldest unit of the metric system." Ironically, the official metric system established by a law of 1837 did not include the grade of angle as a mandatory measure, and today the grade is only rarely used by any but French geodesists.

The grade has two submultiples: the *centesimal minute,* the hundredth part of a grade; and the *centesimal second,* the thousandth part of a grade. Today these submultiples are called, respectively, the *centigrade* and the *milligrade.* Now that the former "centigrade" thermometric scale is officially called the Celsius scale, the name "centigrade" may safely be used again to designate the hundredth part of the grade of angle.

Danloux-Dumesnils would like to see students taught no units of angle save the grade and the right angle (100 grade). The new trigonometric tables giving sine and cosine, tangent and cotangent, for each centigrade would have 5,000 entries from 0 to 50 grades. Or, since conventional trig tables have 2,700 entries from $0°$ to $45°$, the grade tables could be cut to a corresponding 2,500 entries, one for each 2 centigrades—for example, 10.00 grade, 10.02, 10.04, 10.06, 10.08 . . . 10.96, 10.98, 11.00 grade, and so on. The massive measurement trek from Dunkirk to Barcelona changed the estimated standard meter by less than 0.3 millimeter—less than 3 parts in 10,000. This shows that projections from early, limited surveys had been fairly accurate, even when compared with the more-than-six-year survey covering over 1.1 million meter. Furthermore, the meter of 1799 was itself inaccurate by a similar factor. We now know that the length of a standard quadrant of meridian is 10,002,288.3 meter, a little over 2 parts in 10,000 greater than the quadrant length established by the Dunkirk-Barcelona expedition.

   Basis of the Metric System, a detailed account of the survey, was published in three volumes by Delambre (1806–10).

10 CHANGING TIMES IN FRANCE

Come what come may,
Time and the hour runs through the roughest day.
　　—SHAKESPEARE,
　　Macbeth

A fine wind is blowing the new direction of Time.
　　—D. H. LAWRENCE,
　　Song of a Man Who Has Come Through (1917)

King Louis XVI and his family were in prison. Revolutionary France, at war with Austria and Prussia, had won an impressive victory at Valmy, September 20, 1792. The next day the National Convention, newly elected by universal male suffrage to replace the Legislative Assembly, met for the first time. There were no monarchists in the Convention, which promptly deposed Louis XVI and proclaimed France a republic.

On September 22 the Convention's first order of business was a new constitution and metrological reform. That day was the autumnal equinox. As an expression of revolutionary fervor and national self-confidence the delegates decreed that it should become the first day of the Year 1 (An I) of the Republic of France. A new calendar was to follow, with no ties to tradition, Roman emperors or Pope Gregory XIII.

Side glance across the Atlantic—On June 12, 1782, a decade earlier, the Congress of the new United States of America adopted a Great Seal for the nation. It can be found displayed on the back of the nearest dollar bill. On the reverse, or verso, of the Great Seal is pictured a pyramid whose base bears the words *Novus Ordo Seclorum* ("A new order of the ages") and the year 1776.

116

However, Congress did not go so far as to renumber the years, with a Year 1 replacing 1776, to inaugurate this new order of the ages.

In France, the National Convention named a committee to work out details of the new calendar and to reform the units for measuring time. This committee was headed by G. C. Romme (1750–95), who, as chairman of the Convention's committee of public instruction, was in effect the minister of education. Other members were Lagrange and Monge of the previous weights and measures committee, and the poet-dramatist Fabre d'Eglantine (1750–94). He became the only man in modern history to be given the privilege of renaming the months of the year.

Members of this committee were active also in other great events of the period. Monge, for example, was minister of marine and colonies, and ran second only to his friend Danton in balloting for the post of minister of justice. In 1793–94 Monge was occupied with metallurgy and the manufacture of ordnance; his writings on these subjects were circulated among the busy arsenals of the embattled Republic.

After almost a year, in September 1793, the committee presented its proposals to the Convention, which adopted them. October 5, 1793, became the 14th Vendémiaire, An II; and retroactively the former September 22, 1792, was designated as the 1st Vendémiaire, An 1.

In the new Republican calendar each year was to begin with a 1st Vendémiaire, which would fall on September 22 or 23 in the Gregorian calendar in use elsewhere. The twelve months of thirty days each were all renamed to suggest prevailing weather or agricultural conditions:

AUTUMN
1. Vendémiaire (wine or vintage month: Sept. 23–Oct. 21)
2. Brumaire (foggy month: Oct. 22–Nov. 20)
3. Frimaire (frosty month: Nov. 21–Dec. 20)

WINTER
4. Nivôse (snowy month: Dec. 21–Jan. 19)
5. Pluviôse (rainy month: Jan. 20–Feb. 18)
6. Ventôse (windy month: Feb. 19–March 20)

SPRING
7. Germinal (bud month: March 21–April 19)
8. Floréal (flower month: April 20–May 19)
9. Prairial (meadow month: May 20–June 18)

SUMMER
10. Messidor (reaping month: June 19–July 18)
11. Thermidor (heat month: July 19–Aug. 17)
12. Fructidor (fruit month: Aug. 18–Sept. 17)

The final syllable of the name of each month revealed the season of the year in which it occurred. Autumn months ended in -*aire,* winter months in -*ôse,* spring months in -*al,* and summer months in -*dor.* The syllable -*dor* was derived from the Greek word for gift. (Fructidor, for instance, signified nature's gift of fruits.)

Fabre d'Eglantine, with elegance and ingenuity, had forged a chain of twelve lovely links. But his chain comprised only 360 days; it lacked five days of the conventional calendar year. Those five days formed an extra-monthly period in the Republican Calendar—a festival period—and were called *jours sans-culottides* (literally, "days of those without knee breeches"). This term represented the defiant acceptance of the French aristocrats' contemptuous epithet *sans-culottes,* for republicans, who wore pantaloons rather than the knee breeches of the aristocrats and their servants.

As for the leap-year day, every four years the *jours sans-culottides* were increased to six.

The *sans-culottides* sequence was: (1) Festival of Genius; (2) Festival of Labor; (3) Festival of Actions; (4) Festival of Rewards; (5) Festival of Opinions; and (6), every fourth year, Festival of the Revolution. Leap years occurred in An III, An VII and An XI. The four-year period was called a Franciade.

The Republican calendar continued through the period of the Directory in France but was abolished January 1, 1806, by Napoleon, who was moving toward traditionalism, the Church, and personal aggrandizement. By that time, the calendar had been in effect 12¼ years, plus 1 year retroactively.

Today perhaps only a few of the names of the revolutionary months are remembered outside France: Thermidor, for the end of Jacobin rule; Germinal, for Emile Zola's novel of that name; and

Brumaire, as in the title of the work by Marx on the 18th Brumaire of Louis Napoleon.

The Republican calendar abolished the familiar seven-day week. Instead, each 30-day month was divided into three décades, meaning periods of ten days, not ten years.

The days in each décade were renamed after the numbers 1 to 10, instead of for gods and goddesses. In succession their names were: Premedi, Duodi . . . up to Decadi, which was a day of rest but definitely not a religious sabbath.

Each year contained 36 Decadis plus an average of $5\frac{1}{4}$ *sans-culottide* days of festival. This total of $41\frac{1}{4}$ did not equal the Gregorian calendar's 52-plus Sundays and its other holidays.

The Thermidorean coup that overthrew Robespierre and the Jacobins put men of money and property in power. These men of Thermidor, as it turned out, also favored radical reform of weights and measures—from motives of profit. Reform was needed by the Directory to rationalize commerce, trade, finance, and technology, just as Napoleon later needed a rational measuring system for efficient solution of the supply problems of his armies. Whatever was to be the future of France, there could be no return to feudal stagnation, metrological or otherwise.

11 CLOCK CORRECTIONS OF
LIMITED LIFE

Brief as the lightning in the collied [blackened] night,
That, in a spleen, unfolds both heaven and earth,
And ere a man hath power to say 'Behold!'
The jaws of darkness do devour it up.
—SHAKESPEARE,
A Midsummer Night's Dream

To carve out dials, quaintly point by point,
Thereby to see the minutes how they run,
How many make the hour full complete;
How many hours bring about the day;
How many days will finish up the year;
How many years a mortal man may live.
—SHAKESPEARE,
Henry VI, Part 3

The National Convention's committee also proposed a clock reform (like that mentioned first in chap. 8), but it failed almost immediately. It divided the day into 10 equal units (each equivalent to 2.4 present hours). One-tenth percent of a day equaled 86.4 second, and 1 percent of a thousandth day equaled 0.864 second. Using the standard metric prefixes:

$$1 \text{ deciday} = 2.4 \text{ hours}$$
$$1 \text{ milliday} = 86.4 \text{ second}$$
$$10 \text{ microdays} = 0.864 \text{ second}$$

The latter measure, 10 microdays, might be called 1 dekamicroday, but there is no standard metric prefix for $\frac{1}{100,000}$. At any rate, full decimalization extended from the décade (10 days) to a millionth of a décade.

An inescapable problem lies beyond the top of this range. The year and the day are both essential time units; they are determined astronomically and, of course, are not in a decimal ratio. As an intermediate, 30 days = about $\frac{1}{12}$ year is probably more useful than either a unit of 100 days or one of $\frac{1}{10}$ year = 36.525 days.

The decimalized clock units were introduced in France in the autumn of 1793 (An III) and immediately met with stiff resistance. It was relatively easy to print and use new Republican almanacs and calendars, but scrapping a nation's valuable clocks and portable timepieces was a far different matter. The clockmakers and watchmakers of France opposed the innovation. This was before the era of planned obsolescence; furthermore, units of time probably are used more often than other units and so are more stubbornly retained.

Unlike reform of the calendar, clock reform was not pushed by the Directory after the overthrow of the Jacobins. The actual life of this reform—which was never truly adopted in practice—was less than a year. In 1795 a law "tabled" clock reform. It is still tabled.

Professor Danloux-Dumesnils states that people in and out of France "would have been justified in rejecting the new system . . . thus leading to its complete collapse," even if the regime had tried to push it through. He correctly stresses that while length and weight units replaced by the metric system were numerous, contradictory and chaotic, the conventional subdivisions of the day are uniform, consistent, and clearly understandable—and, at the end of the 18th century, universal, or at least international.

Having defined a new unit of length by the actual survey from Dunkirk to Spain, the Academy of Sciences considered the question of a name for that unit, instead of the old length designations, which were monosyllables such as *toise, pouce, ligne,* and so on. The first proposal was that the name of the new basic length unit also be a monosyllable, and that other monosyllabic names be given to its approved multiples and submultiples (fractional parts). However, the Academy came to realize that the need was for linguistic barriers, not ties, to the past. Also it became clear that a set of prefixes for multiples and submultiples of length could be applied to the names of every other basic unit, such as the unit of weight. This would greatly reduce the number of new names required.

The first names of basic units were, for length, the *meter* (*"metre"* in France and Britain); for area or land surface, the *are;* for volume or capacity, the *liter* (*"litre"* in France and Britain); and for weight, the *gram* (*"gramme"* in France and England), which was subsequently to be designated as the unit of mass. The word "meter" came from the Greek *metron,* meaning "measure."

The prefixes for multiples were derived from Greek roots: deka- (da), 10; hecto- (h), 100; kilo- (k), 1,000; and myria-, 10,000. Recent decisions by international bodies supervising the metric and SI systems have called "deka-," "hecto-" and "myria-" unnecessary, leaving only "kilo-" (and additional multiples by 1,000: "mega-," "giga-," "tera-"). "Hecto-" survives in metric countries, where the hectare, a contraction of *hecto-are* (100 are), is more commonly used than the are, in nonscientific work.

Latin roots were used for submultiple prefixes: originally "dec-" (d), $\frac{1}{10}$; "centi-" (c), $\frac{1}{100}$; and "milli-" (m), $\frac{1}{1,000}$.

Milli-, again the only one still in good standing, was introduced in a law of the 30th Nivôse, An II (January 19, 1794). The other two appeared in a law of the 18th Germinal, An III (April 2, 1795). This 18th Germinal, as we shall see, can be regarded as the birthday of the metric system.

Further modern submultiple prefixes—again by steps of 10^3—are "micro-" ($\frac{1}{1,000,000}$), "nano-" ($\frac{1}{1,000,000,000}$), "pico-" (10^{-12}), "femto-" (10^{-15}), and "atto-" (10^{-18}). The "atto-" prefix represents one quintillionth, or a millionth of a millionth of a millionth!

More such "tiny" prefixes are in use than "giant" prefixes because of the enormously accelerated modern researches into the submicroscopic worlds of atomic and nuclear and even subnuclear physics.

Increasingly today, only the basic unit is used, and the count of units is expressed by a few significant figures of a number between 1 and 10, times a power of 10. Thus 54 gigameter is 5.4×10^{10} meter, and 0.54 femtometer is 5.4×10^{-16} meter. The simplicity and elegance of this approach are evident. Beyond all else, it enormously simplifies estimations and computations.

The law of the 18th Germinal, An III, in its fifth article defined six basic units of the metric system. Those definitions, somewhat simplified, are the following:

1. The meter (length), the ten-millionth part of the arc of meridian from Pole to Equator.

2. The are (area), the area of a square with sides of 10 meter. (Devised as a measure of arable and residential land.)

3. The stere (bulk), the volume of a cube 1 meter on edge. (Devised primarily to measure cut wood for fuel, like the American cord, which equals 3.625 stere or cubic meter.)

4. The liter (volume or capacity), the volume of a cube 0.1 meter on edge, thus 0.001 cubic meter. (Intended as a measure of dry and liquid substances, or gases for that matter.) Both stere and liter are measures of volume: 1 stere = 1,000 liter.

5. The gram (weight), the absolute weight of pure water at the temperature of melting ice, contained in a cube 0.01 meter (1 centimeter) on edge, with volume of one-millionth cubic meter (a thousandth the volume of a liter).

6. A special unit, the franc, a monetary unit replacing the livre of the monarchy. (Like "pound," "livre" was a unit of weight as well as of currency.)

The new law made no mention of any new time units. Clock reform had been abandoned.

The legislation of An III (1795) was a kind of declaration of intentions rather than the complete codification of the metric system. Detailed labors of the French metrologists were still under way. Not until 1799 (Ans VII and VIII) were actual prototypes of the principal units made and placed in the archives for reference or comparison.

Intriguing travels of Talleyrand—Talleyrand had managed to be sent to England in 1792. Instead of being guillotined for not being radical enough, he was merely denounced in France as an émigré. And when he seemed too radical for the English, he only had to spend two and a half uncomfortable years among the unsophisticated citizens of the young United States of America. Talleyrand soon realized that the events of Thermidor opened many an opportunity in France for a clever man with an eye to his own interest. After a transitional sojourn in the free city of Hamburg, he felt it safe to return to France. By 1797 (Thermidor, An V) he had become foreign minister for the Directory, which was the executive branch of the French government.

Talleyrand had not wavered, however, in his support of the metric system. Indeed, he urged that this system must "belong to the entire world." In 1798, European scientists were invited to France to take part in perfecting the system. Among those who arrived in response to that summons was the Danish astronomer Thomas Bugge, who later wrote a book about his visit to France. Such works were popular; conservative and literate Europe looked toward regicide Paris with a mixture of curiosity and apprehension like that directed nowadays toward the People's Republic of China or Castro's Cuba.

In all Paris, Bugge noted, he had found only two clocks whose dials were divided "according to the new time." All the rest followed the ancient Mesopotamian pattern. Bugge did not regret this; he thought of the burden of recalculating vast masses of astronomical data using new measures of angle and duration. Although a delegate to the international commission for the metric system, Bugge believed the scientific world could manage well enough without that system.

Other scientists from abroad were more responsive. Eight nations besides Denmark sent representatives, who participated in committee work, in validating computations, and in setting up the prototype *étalons* (the French word for "standards," as of weights or measures). The head of the committee on the standard of length was J. H. van Swinden, of the Netherlands, and the head of the committee on the standard of weight was M. Tralles, of Switzerland.

In 1799, the national legislature of France ratified the new standards, and they were deposited in the archives. The significance of this was that the law now established certain *objects,* rather than the concepts which had led to them, as *the* meter and *the* kilogram. The end-to-end length of a carefully made bar of platinum defined the distance of 1 meter. The kilogram mass was a no less carefully made cylinder. With these standards safely under guard, the Earth might alter its circumference if it so chose, but the meter would remain the same.

Such prototypes remained as standards until very recently, when more exact—and less singular—standards could be defined.

The legislation of 1799 made the metric system the sole legal pattern for measuring lengths, areas, volumes and weights in France. Actually, however, it left a good many loopholes, and these loopholes were expanded after the *coup d'état* of the 18th Brumaire, An VII (November 9, 1799).

General Napoleon Bonaparte had relatively little enthusiasm for the metric reforms, as such. From the start of his rule as First Consul until 1812, the year of his mistaken attempt to become dictator of Russia also, the metric system occupied a somewhat ambiguous role in France. It was used by scientists, savants and bureaucrats, and thanks to the unified and closely controlled national educational system, a new generation of schoolchildren became thoroughly familiar with its units and concepts. But common people, in everyday market transactions, continued to employ the old familiar units—pieds, pouces, livres, bosseaux, and others.

In 1812, in fact, this mixed state of affairs was legalized. What one Frenchman preferred to call a pied, another called the third of a meter. The livre (weight) of one citizen was called the five tenths (or one half) kilogram by another. The boisseau, an old measure of bulk, was reset at 12½ liter, and so on. This dualism was reflected in the new regulation requiring that measures and scales calibrated in the old measures be marked also with the metric equivalents. Only in 1837 did France abolish the nonmetric units and impose fines for their use.

By then, too, the metric system had made headway in other countries. It had entered Switzerland, Italy and parts of Germany (then still a collection of separate states). The Netherlands had previously gone metric when occupied by Napoleon, and Belgium, for a time reunited with the Netherlands, did likewise.

Metrication marches on—In the 1870s, as science, technology, and industrial production expanded in western Europe, the French government once again invited representatives of other nations to consider current problems of weights and measures and of metrication in general. Fifteen nations sent representatives, including the United States and Great Britain. The ensuing Metric Convention of 1875 established an International Bureau of Weights and Measures, housed in the Pavillon de Breteuil, Sèvres, France. Though expanding Paris is about to swallow up that former suburb, the bureau remains in its rather idyllic original site. It is in France, but it enjoys extraterritorial status; it serves as a metric embassy for the many nations that have signed the Treaty of the Meter. The bureau has close connections with such standardizing agencies as the Bureau of Standards in the United States and the National Physical Laboratory at Teddington in Britain. It has become a symbol of metrological internationalism.

Almost a century has gone by since the bureau at Sèvres was founded—a century full of wars and rampant nationalism. Born in an era of revolution, the metric system has survived, adapted, and become a demonstration that humans can choose to cooperate rather than conflict.

12 CALENDRIC CURIOSITIES

Canst thou number the months . . . ?
—JOB 39:2

Ah, but my Computations, People say,
Reduced the Year to better reckoning? —Nay,
Twas only striking from the Calendar
Unborn To-morrow, and dead Yesterday.
—EDWARD FITZGERALD,
Rubáiyát of Omar Khayyám (1859)

The neat regularities of the French Revolutionary calendar reveal, by contrast, the compromises and makeshifts of the calendars in widespread use today. Best known of these is the Gregorian calendar of the Christian world. Pope Gregory XIII superseded the Julian calendar by decreeing that the day after October 4, 1582, would be October 15, 1582. Thus ten days in that October never existed—at least in the Roman Catholic countries. More than a century passed before some Protestant lands followed. It was 1752, for example, before Britain switched over, with September 2, 1752 (O.S.—for Old Style), being followed by September 14, 1752 (N.S.—for New Style). By that time, eleven days had to be deleted from the old calendar. In America, George Washington was born February 11, 1732 (O.S.), but by the time he had attained his legal majority his birthday had been retroactively shifted to February 22, 1732 (N.S.). In France the Gregorian calendar was reimposed by Napoleon in 1806. The Gregorian calendar was adopted in 1873 by Japan, in 1912 by Greece, and in 1927 by Turkey.

The Julian calendar, which dated back to Julius Caesar in 46 B.C.E., made every fourth year a leap year, without exception. The mean, or average, year was $\frac{1}{4}(365 + 365 + 365 + 366)$, or

127

365.25, days. However, there are actually 365.242 2 days in the solar year (365 days, 5 hours, 48 minutes, 45.5 second), so the Julian calendar slowly fell behind.

The Gregorian reform removes three leap days every 400 years by excepting from the quadrennial leap-year rule those century years that are not evenly divisible by 400. Thus, 1600 was a leap year; 1700, 1800 and 1900 were not, but 2000 will be; 2100, 2200 and 2300 will not be, but 2400 will be; and so on. In 2,000 Gregorian years there are 500 − 15, or 485, leap days plus 730,000 regular days, or 730,485, which is 365.242 5 days per year. The error is only $\frac{3}{10,000}$ day per year. Since 1582 the accumulated error has been only a small fraction of a day. (The Mayan calendar was even more accurate.)

Each of the two calendars—Gregorian and Julian—is made up of months that are not of uniform length. Our calendar contains seven 31-day months (January, March, May, July, August, October and December); four 30-day months (April, June, September and November); and one 28-day month (February), which adds a 29th day in leap years. Leap day, by the way, comes at the end of February because the Roman year began in March.

Two other widely used calendar systems are the Muslim and the Jewish. The Muslim calendar is based entirely on the phases of the Moon as it revolves about the Earth. Its disregard for the mean solar year is shown in a year with a length of 354.366 7 mean solar days, a whopping 10.875 5 days (10 days, 21 hours) short of the solar year. Thus the Muslim New Year moves backward by ten to twelve days each year in relation to the Gregorian calendar. For example, the Muslim New Year 1391 fell on February 16, 1972; and the Muslim New Year 1392 fell on February 4, 1973. In about $32\frac{1}{2}$ years, the Muslim New Year moves backward through our four seasons and is just about where it started.

From our point of view the Muslim year is 3 percent too short. But the traditional purpose of the Muslim calendar is to keep track of lunar changes, and it is better suited to its purpose than the Gregorian calendar.

The Muslim calendar is based on a cycle of thirty lunar years, each containing twelve lunar months that start at or very near the new moon. In a rather complex pattern, nineteen of the thirty are ordinary years, with 354 days, and the other eleven are "leap" years, with 355 days. The 30-year cycle includes 30 × 354, or 10,620, days plus 11 leap days, a total of 10,631 days, or an average of 354.366 7 solar days per Muslim year.

The Muslim year 1 began on Friday, July 16, 622 c.e. It commemorates the flight of the Prophet from Mecca to Medina, started the day before. By February 4, 1973, or 1,351 years later by the Gregorian calendar, the Muslim year 1392 had been reached. And 1,392 is 3 percent greater than 1,351.

The average Muslim month in the 30-year cycle is 354.366 7 divided by 12, or 29.530 6 days; and the lunar synodic or phase month, as measured by astronomers, is the same 29.530 6 mean solar days. Truly, a good fit!

The twelve months of the Muslim calendar alternate from 30 to 29 days, with no Gregorian "moonstrosities" of 31 or 28 days: Muharram (30), Safar (29), Rabi I (30), Rabi II (29), Jumada I (30), Jumada II (29), Rajab (30), Sha'ban (29), Ramadan (30), Shawwal (20), Dhu'l-Qa'dah (30), Dhu'l-Hijja (29 or 30).

In eleven years of the 30-year cycle the twelfth month contains an extra day. The positions of those years in the cycle are: 2, 5, 7, 10, 13, 16, 18, 21, 24, 26, 29.

Muslim special days, other than the New Year, include the eleventh of Rabi I (birth of Muhammad); the twentieth of Jumada I (capture of Constantinople by Muslim Turks); the fifteenth of Rajab (Day of Victory); the twentieth of Rajab (Day of Exaltation of Muhammad); and the fifteenth of Sha'ban (Night of al-Burak).

Al-Burak ("lightning") is the name of the steed that carried the Prophet from Mecca to Jerusalem, where Muhammad climbed on a ladder of light to the presence of Allah and then flew back to Mecca, all in a single night. (Details and an imagined picture are in Mina C. Klein and H. Arthur Klein, *Temple Beyond Time,* New York, 1970.)

The Gregorian year 1976 includes portions of three Muslim years: 1395, 1396 and 1397.

The Jewish calendar, replete with leaps—The Jewish calendar uses "lunar" years of 353, 354 and 355 days, all comparable in length to the Muslim year. However, by inserting also some giant leap years of 383, 384 and 385 days, the average year length is held close to the mean solar year of 365.242 2 days.

The shortest Jewish year, that of 353 days, is called defective, the middle year (354 days) is called regular, while the largest (355) is called abundant. The extra one or two days are provided by adding a thirtieth day to certain 29-day months. As in the case of the Muslim calendar, all Jewish months have 29 or 30 days.

The first Jewish month is Tishri, and the first day of Tishri is the holiday of Rosh Hashanah, the New Year, which usually falls between September 10 and October 5 (Gregorian). Following months are: Heshvan, Kislev, Tebet, Shebat, Adar, Nisan, Iyar, Sivan, Tammuz, Ab and Elul. In leap years, an extra, or thirteenth, month is added. Its name, Veadar, means "month after Adar," which is indeed its position. Thus while the Gregorian calendric system has leap days, the Jewish has leap months.

The Jewish year 1, according to tradition of that religion, began with the creation of the world, as described in the Book of Genesis, placed at 3761 B.C.E. in the common calendar. Adding 3,761 to September 27, 1973, brings us to the 1st of Tishri, 5734. While the Gregorian calendar struggles through its third millennium, in 2239 C.E. the Jewish calendar will reach its year 6000.

In the Jewish decade 5731–40 every possible year length is represented. Of those 10 years, 4 have 355 days each, 2 have 354 days each, while the remaining 4 contain, respectively, 385, 384, 383 and 353 days.

Astronomical motions provide us with three great "natural" time units: the day, the month and the year, in order of increasing length. The day is determined by the Earth's spin, the year by its orbital path around the Sun, and the month by the Moon's orbital movement around the Earth.

In our Gregorian calendar the months are each stretched by about a day, so that 12 of them together fill out a 365-day solar year. Twelve truly lunar months, however, would make a 354-day lunar "year." The relation of month and moon remains, nonetheless, more than merely etymological.

Between the three astronomical time measures, day, month and year, the connections are complex, the more so because of the irregular and irrational patterns of adjustment typified by the three calendars just discussed.

13 CELESTIAL CLOCKS THAT
CAN'T KEEP STEP

*On all-important time . . . the man is yet unborn who
duly weighs an hour . . .*
  —EDWARD YOUNG,
    Night Thoughts (1742–45)

*I wake and feel the fell of dark, not day.
What hours, O what black hoürs we have spent
This night! what sights you, heart, saw; ways you went!
And more must, in yet longer light's delay.*

*With witness I speak this. But where I say
Hours I mean years, mean life. . . .*
  —GERARD MANLEY HOPKINS,
    "Sonnet" (c. 1888)

We have seen that the synodic, or lunar, month has a duration of
29 days, 12 hours, 44 minutes and 2.8 second (or 2,551,442.8 s).
Thus, 12 synodic months = 354.367 days (or 30,617,313.6 s),
about 3 percent shorter than the mean solar year of 365.242 2 days
(or 31,556,926 s).

There is also the so-called *sidereal* year, equal to the Earth's
full circuit around the Sun, as it would appear to an observer on a
distant star. The sidereal year is longer than the mean solar year by
a mere 20 minutes 20.5 second. Hence 1 sidereal year = 365 days,
6 hours, 9 minutes, and 9.54 second—or less than 4 parts in 100,-
000 longer than the mean solar year.

Stated most simply, the situation is this: The lunar month is very
nearly 29½ days long—not 30. The solar year is very nearly 365¼
days long—not 360. And the solar year itself is 12⁴⁄₁₀ lunar months
long—not 12. The motions of the Earth and Moon thus provide no
whole-number ratios between the various natural time periods
(day, month and year).

131

If the number of planets circling suns in the universe is as great as some astronomers believe, there may be distant planets on which numerologists, astrologers and other manipulators of the mystical can delight in neat whole-number ratios of motion cycles. We, however, remain passengers aboard our space ship Earth.

Possibly whole-number ratios between the great natural time divisions will develop even upon Earth in the far future. The Earth, it is now obvious, does not spin steadily and uniformly on its axis. It rotates with slight but significant irregularities, and with a marked trend toward slowing down. The length of the average solar day, in fact, grows by about 20 microsecond (2×10^{-5}) per year, 2 millisecond per century, 20 s per million years, and 200 s (3 min 20 s) per 10 million years.

Some 500 to 600 million years ago the day was only 21 hours long. About 500 to 600 million years in the future it will be 27 hours long, by our current time measures. Some recent analysts, in fact, envisage a nearly 30-hour day 500 million years from now.

This remorseless deceleration of Earth's rotational rate results from the incessant friction of the tides on and within the Earth. They are generated by the gravitational attractions of the Moon and Sun, and act like a brake on the spinning globe. The tides *on* Earth are those of the oceans and seas, forced to flow through narrow channels and straits. Those *within* Earth are the slow deformations of its rocky but not totally rigid interior, which also responds in its own way to the powerful gravitational and tidal attractions of the Moon and, to a lesser extent, the Sun.

Very nearly 31,556,900 second make up 1 mean solar (or "tropical") year at present. When the mean solar day increases to a length of 87,658 second, just 360 such days will fit into the present year's duration. Each of these days will contain 1,258 second more than our present day. Lengthening of this extent can be looked for in some 60 to 70 million years. Further refinements in astronomical observation and theory may alter such forecasts somewhat, but the basic direction is not in doubt.

Whole-number ratios between major natural time divisions may arise also in another combination. The Moon is slowly spiraling in closer to Earth. As it does so, its orbital period and the resulting synodic, or lunar-phase, month grows shorter. Meanwhile, the Sun-determined day is lengthening. If the shortened month reaches the same time span as the lengthened day, then days one month long and months one day long will become possible in the calendars of the inhabitants, if any remain then on this small but basically beautiful planet.

The second (s), basic unit of time in the International System, is not an astronomical measurement. It began, however, as an arbitrary fraction of such—that is, $\frac{1}{86,400}$ of a mean solar day (24 × 60 × 60 = 86,400).

The trouble with astronomically based time measurements is that they do not remain constant. As just noted, the mean solar day is lengthening; the phase month is shortening. For centuries the day, and then its subdivision, the second, served as foundations for time measurement. Then tiny but intolerable irregularities were revealed, and in 1956 a new second was defined—a second that was a complicated fraction ($\frac{1}{31,556,926}$) of one particular and carefully measured tropical or mean solar year, that of 1900. This was the so-called "ephemeris" second. But it was not long in effect. By 1967 extraordinary technical advances in atomic devices had created clocks which provided far more exact, uniform and reproducible seconds. Hence, in 1967 the second was redefined again, in terms of a specified number of oscillations of radiation generated in such atomic clocks.

Today's "atomic" second is thus cut loose from its astronomical origins, though astronomical time scales and atomic time scales are constantly compared, to serve the needs of humanity for navigation and timekeeping of many other kinds.

Following pages will explore more fully some of the complexities of timekeeping resulting from the many motion patterns exhibited by the Earth and by its great satellite, the Moon. The continuing relationships between the second, as time unit, and the great time scales by which we all live and control our activities, can be clarified only by facing the observed facts about the Earth's spin, its angular movements in space, and the many irregularities that cannot be disregarded if the time-measuring needs of our era are to be met.

Inconstancies and inconsistencies—Juliet spoke of "the inconstant moon," because of the way it changes phases throughout the months. The Sun too appears far from constant. While this book was written, the Sun was above the horizon sometimes less than 10 hours in 24; sometimes as much as 14 hours.

An old metrological joke says the expansive power of heat is shown because daylight expands (grows longer) in summer, but contracts in winter. In Southern California the Sun is above the horizon about 45 percent longer in midsummer than in midwinter. Toward the North Pole the Sun for six months never disappears completely, then for six months more becomes a mere memory.

The cause of these variations is well known: the axis of the Earth's rotation is inclined at a considerable angle ($23\frac{1}{2}°$) away from the perpendicular to the plane of the Earth's orbit. If, however, the axis of the Earth's rotation were exactly perpendicular to the orbital plane, then everywhere on Earth exactly half of each day would be daylight, and half night, and there would be no seasons anywhere, although the equator still would be the hottest part of Earth and the poles the coldest.

Among the seasonal shifts in the times of sunrise and sunset, one regularity can be noted on Earth: every day the Sun's center is bisected for an instant by the great circle that passes through the north and south points on the horizon and through the zenith. This significant bisection or crossing occurs not far from noon, when the customary cycle of numbering the hours of morning reaches its end with 12 o'clock.

Our whole day is said to begin at midnight. The Mesopotamian and Egyptian chronometrists, on the other hand, began their days at sundown—the Jewish and the Muslim still do so. Oddly, none of the great calendars begins the day at sunrise.

Airplane flights, train schedules, military operations and scientific events often use the less confusing clock on which 0000 hours represents midnight, 1200 is noon, 1500 is 3 P.M., and so on, all the way to 2359, which is 11:59 P.M.

Adding to the general confusion in timely matters, the English word *day* in a statement like "The week has seven days" means the full period of 24 hours. But, in "Night is followed by day," the same word means a variable period with a mean, or average, of twelve hours.

The ecliptic—that is, the apparent path of the Sun in the sky— deserves mention. It oscillates back and forth in a one-year cycle. An observer 35 degrees north of the equator will see the Sun cross the north-south meridian on December 22 (the winter solstice) at an angle of 58° 30′ south of the zenith, and on June 22 (summer solstice) at 11° 30′ south. The difference is 47 degrees—twice the 23° 30′ angle by which the Earth's axis is tipped, relative to a perpendicular to the plane of its orbit.

An easy way to record changes in the Sun's position in the sky is to observe the shadows cast on the ground by upright posts or markers. This leads to the familiar form of shadow clock called the sundial. Sun clocks have taken various forms, some large and massive, like obelisks or plinths, others small, and even portable, like the hand-staff timepieces of ancient Egypt. A traveler let his staff hang from a strap, pointing straight down. Set to project at right angles near the top of the staff was a small peg, whose shadow was cast along the side of the staff. These staffs had four sides, each with a season's scale for telling the time of day from the position of the lower end of the shadow.

Such portable staffs, having horizontal shadow casters, showed their longest shadows at midday. Shadow clocks with vertical shadow casters, like the familiar sundials, show shortest shadows at midday.

The best sundial, read by an expert, may be accurate within 3 minutes; the average sundial, within 10 minutes. In any case, when it is cloudy, sundials cannot be used.

Other nonmechanical methods of keeping time do not require a sunny day. Some work day or night. Candles can be marked so that they burn down at a rate of one division per hour. Fine sand or dust can be allowed to sift through a small hole, as in hourglasses, which today are more likely to be built for timing 3-minute eggs than 60-minute hours.

At least as old as the hourglass is the water clock, where a measured amount of water leaks slowly from a receptacle. The amount of water lost, or the distance by which the surface of the remaining water has dropped, is supposed to indicate the time elapsed since the flow began.

Suppose we place 10 liter (10,000 cc) of water in a cylindrical glass container standing 1 meter high with an inside diameter of 11.3 centimeter. A valve at the bottom is then adjusted until the 10 liter of water drains out of the cylinder in just 10 hours.

If we mark ten equally spaced divisions along the side of that cylinder from the 10-liter level to the bottom, fill it and watch, we shall find that the water level does not drop precisely one division per hour. At the beginning of the process, the level will drop one division in less than an hour; but when the water level nears the bottom, it will take more than an hour to drop one division. The greater depth of water in the early stages increases the pressure at the bottom and thus speeds the flow through the valve.

There are ways to eliminate or adjust for this source of error, assuming that one has a reliable time measurer to check against. The divisions can be re-marked, unequally. Or the water container can be an inverted cone, 1 meter high on the inside, 20 cm in diameter at the top and only 1 or 2 cm at the bottom, and this would just about permit equally spaced time markings along the side.

But this water clock would not be accurate all year round; as the temperature of water rises from the cold of winter to the heat of summer, the water becomes less viscous and flows out more easily. Water-clock makers, in the days before mechanical clocks, would adjust for this and for the comparable night-day differential by adding a movable inner cone. It could be lowered within the outer cone, thus raising the water level and so speeding up the clock; or it could be lifted, thus slowing down the clock.

Incomparable times—Every timekeeping device is, after all, a device for supplying recurring cycles of motion with which changes, motions or events in the world can be compared. Time measurement is essentially comparison of one motion cycle with another. There are no ways in which one "quantity of time" can be compared directly with another. An amount or period of time cannot be placed in storage, taken out and compared later with another.

These profound truths remain valid whether we are keeping track of time by letting water drip, causing pendulums to swing back and forth, allowing springs to unwind through escapement mechanisms, or dealing with the vibrations of an atomic system.

The characteristics of a good timepiece are easy to state, but difficult to achieve in practice. It should have a very stable, uniform cyclical frequency. Also it should permit precise and unambiguous reading. Some modern watches, designed to allow exact readings at intervals of five or fifteen minutes, violate this latter requirement.

The stability of the cyclical frequency should be a long-term one. Some timepieces run quite uniformly for a day, a week, a month, but eventually wear or other gradual change results in the rate rising above, or falling below, its original level. Rate adjustability thus becomes a desirable, or even an essential, characteristic of a satisfactory timepiece.

Even ordinary clocks and watches seem today to meet these requirements reasonably well. Few of us realize how recent these achievements are. In Europe, around 1300 C.E., the best available timepieces often gained or lost fifteen minutes per day. Within four days of operation without resetting, the error could rise to as much as one hour. Under such conditions the attainment of constant rates remained more urgent than subdividing the major "short" time unit, the hour, into smaller units.

In China, however, from some four centuries earlier—that is, from about 900 C.E.—astronomers and aristocrats had been benefiting from amazingly good clocks with errors of only 2 minutes per day. By 1300 C.E., the most sophisticated Chinese water clocks were accurate to within some 30 seconds per day, or $\frac{1}{30}$ of the error then prevalent in Europe. By ingenious supplementary mechanisms, the Chinese horologists had mastered the most troublesome shortcomings of the water clock.

The West began to catch up only late in the 16th century, when Bungi perfected the so-called cross-beat escapement mechanism. Later, when Christian Huygens applied Galileo's pendulum principles to horology, clock errors were reduced to about 10 seconds per day, or 1.2 parts in 10,000. In our own era, as we shall see, atomic clocks have reduced these errors to about 10^{-7} second per day, which means to about 1 part in 10^{12}, or 1 trillion.

14 THE GREAT SKY CLOCK BY DAY AND NIGHT

Wake! For the Sun, who scatter'd into flight
The Stars before him from the Field of Night,
  Drives Night along with them from Heav'n, and strikes
The Sultan's Turret with a Shaft of Light.
  —EDWARD FITZGERALD,
    Rubáiyát of Omar Khayyám (1859)

Long ages ago thoughtful skywatchers noticed that the great Sun seemed to circle the Earth less swiftly than the stars, set in what Hamlet was later to call the "grave overhanging firmament, the majestic roof fretted with golden fire."

A convincing example can be given. At the time of the March equinox, if we measure the duration between two successive crossings of the upper meridian by the same star, we find that they take place at intervals of 23.934 47 hours, not 24 hours. The difference, in fact, is 3 minutes, 55.91 second.

That difference represents the amount by which a "star day" (sidereal day) is less than a mean solar, or Sun, day. On the average, the Sun moves 360 degrees "around the Earth" in 24 hours, during which a star moves about 361 degrees. Suppose that in spring we see a certain star rising in the east just as the Sun sets in the west. Every day thereafter that star will rise a little earlier, in terms of the Sun-determined time. By fall that same star will be setting with the Sun; and by the following spring it will again be rising at sunset. The star apparently will have gained a day.

In terms of the Mesopotamian-born angular measures, during a solar (tropical) year the Sun appears to revolve through a total of 131,487.19 degrees around the Earth, while a star revolves 131,-852.3 degrees, or 365.11 degrees more than the Sun. This is one full revolution plus 5.11 degrees greater, and is caused by the effect of the Earth's orbit around the Sun, which adds slightly more than one full turn to the 365¼ turns that the Earth seems to make (as seen by an imaginary observer on the Sun). The star-determined day is called the sidereal day (from Latin *sidus*, "star"). It is very nearly 4 minutes shorter than the mean solar day. The sidereal year, on the other hand, is about 20 minutes longer than the mean solar year of 365.242 2 days. The sidereal year equals 366.256 2 *sidereal* days, but only 365.255 9 mean *solar* days, or 8,766.142 mean solar hours. Many a student of astronomical timekeeping, including the author, has fallen for a time into the easy error of assuming that the sidereal year must be 1 (solar) day longer than the mean solar year; but the difference is in fact only the 20 minutes additional for the sidereal, as compared to the solar, year.

The shortening of the daylight period as the year moves closer to winter can be a disturbing and depressing thing, even for moderns. Primitive humankind may have feared, year after year, that daytime would continue shrinking even after December 22, until the Sun vanished forever, leaving darkness and death to quench all life. The relief—even the jubilation—can be imagined when, after the winter solstice, the Sun began its long climb again toward the north. Solstice means "Sun standing still," and it signifies the Sun's halting its seasonal drifting in one direction (northward or southward) and beginning its movement in the opposite direction. Thus, on December 22 the Sun begins its northward drift and daylight lengthens until June 22, when the procedure is reversed.

The most ancient primitive festivals were at, or quite close, to the two solstices. Possibly somewhat later originated the equinoctial festivals, which were linked with agricultural rather than astronomical events and occurred in the autumn and spring, when the Sun crosses the equator in the sky and day and night are of equal duration. Our modern holidays still show these influences—including Christmas near the winter solstice, the Scandinavian Midsummer's Eve, and so many others.

The stars too shift their places of rising and setting with the seasons. In the Northern Hemisphere, as summer approaches, they move northward; then, after the summer solstice, they shift southward.

Ancient astronomers were often hindered by sunlight from calculating exact times of star-rise or star-set. Today, however, these can be determined quite exactly for any observable star. One could speak, indeed, of winter and summer "starstice," separated by 183.127 stellar (sidereal) days, just half the number of days in the stellar year.

Time could, in fact, be kept by stars alone. If our planet had no Moon, and if its inhabitants had some aversion to reckoning time with the help of the sacred Sun, we might then depend solely on stellar days of 23 hours, 56 minutes, 4.09 seconds (in terms of present units). We would then have 3 years in our calendar with 366 sidereal days each, followed by a leap year of 367.

Meaning of a "mean"—We are obliged repeatedly to speak of "mean" or "average" solar days, not because the rotation of the Earth speeds up or slows down noticeably, but rather because the Sun is usually not quite where it should be to tell us about the Earth's actual rotation. This distortion arises from the Earth's slightly elliptical (rather than truly circular) orbit around the Sun.

Early in January the Earth is at its perihelion, or closest point to the Sun; by mid-July it has swung around to its aphelion, or greatest distance from the Sun. The Sun looks about 3.5 percent greater in diameter at perihelion than at aphelion. We in the Northern Hemisphere are cold in January because the Sun, though closer to us then, is not as directly overhead and there are fewer hours of daylight in each full day then than in July, when we actually have better access to the source of our light and heat.

In January the Earth's orbital velocity around the Sun is about 3.5 percent greater than in July. The farther the Earth is from the Sun, the more slowly it moves in its orbit. The imaginary, or "mean," Sun must be based on the fiction that the Earth travels in a truly circular orbit, with the Sun moving always across the sky at the rate of 15 degrees per hour. Instead, the real Sun appears where the mean Sun should be, only four times each year: in mid-April, mid-June, at the beginning of September, and again near the end of December.

On January 1 the real Sun crosses the meridian 3 minutes later than mean solar time would indicate; by mid-February that lag has increased to 13 minutes. By mid-April the real and mean Suns are in step, with neither lag nor advance, but by mid-May the real Sun is 3 minutes ahead of the mean Sun; and so on.

Astronomers prefer sidereal time to solar time, because the star-determined time needs no correction for the Earth's elliptical orbit. The problem has a brief surcease on September 22, for then sidereal time and mean solar time coincide.

The time and the place—Practical timekeeping presents an additional major problem. Suppose we are at 0° longitude, near London, where the Royal Greenwich Observatory is—or *was,* until 1958. If the correct time there is just 10 A.M., then at Portsmouth, 1° (about 45 miles) west, the time should be 9:56 A.M. This follows, because 1 day contains 1,440 minutes, and $\frac{1}{360}$ of that is 4 minutes (of time).

By the same unchallengeable geodesic logic, the time should be 9:52 A.M. at Birmingham, 9:48 at Liverpool and 9:44 at Plymouth.

But such a hodgepodge of local times would be quite impossible for modern living. If every separate site had its own local time, then transportation schedules and TV programs would be thrown into confusion or downright chaos. Travelers would have to make constant readjustments of their watches or clocks, and so on.

At the opposite extreme, if there were but one uniform time imposed on the entire globe, some places would be seeing the Sun set at 8 A.M., while others would be going to bed at "noon."

The convenient compromise between these unacceptable extremes has been the establishment of time zones, running north and south. Usually they have many a zig and zag, unfortunately, to accommodate state and national boundaries, but residents become used to that—in time. Within each zone, all clocks agree (or are supposed to).

At the equator such zones are about 1,000 miles wide, on the average; and less in temperate zones. When one travels westward, one turns back the clock or watch 1 hour on entering each new time zone. In eastward travel, one turns the clock or watch forward one hour every time a new zone is entered.

When it is 10 A.M. throughout the zone of Eastern Standard Time, it is 9 A.M. by Central Standard (the next west); 8 A.M. by Mountain Standard; and 7 A.M. by Pacific Standard Time. California brokers must rise early to catch opening transactions on the New York Stock Exchange.

The most intriguing time-zone boundary lies halfway around the world from Greenwich, England, at 180° longitude E. (and also 180° W.). It is the international date line.

If calendars east of that line show June 8, those to the west of it show June 9. The line is zigged and zagged to avoid cutting through Siberia, Fiji, and islands off New Zealand. Otherwise people living there would be getting their dates mixed unmercifully.

The date line is convenient for those who prefer tomorrow or yesterday to today. The traveler who crosses that line westbound is said to "gain" a day, while the eastbound line-crosser "loses" a day.

Phileas Fogg, the unflappable hero of Jules Verne's *Around the World in Eighty Days,* returned to London dejected, believing that he had taken a day too long to circumnavigate the globe and thus had lost his historic wager. However, having traveled eastward he had, in fact, lost a day on crossing the international date line— and on realizing this, he was able to claim victory, in the very nick of time.

15 PATHS TOWARD PRECISION: DEPARTURE FROM THE DAY AND YEAR IN DEFINING THE BASIC TIME UNIT (THE SECOND)

. . . time is included in the class of continuous quantities
. . . the point may be compared with the instant of time,
and the line may be likened to the length of a certain quan-
tity of time, and just as a line begins and ends in a point, so
too [does] such a space of time.
—LEONARDO DA VINCI,
Notebooks, section on stars and time

Take a little time—count five-and-twenty . . .
—CHARLES DICKENS,
Little Dorrit (1855–57)

Allen V. Astin, former director of the U.S. National Bureau of
Standards, thinks it odd that the founders of the metric system,
when they defined new units of length, volume and weight, did not
"bother" also to define a new unit of time. Instead, they took the
already existing unit, the solar or natural day, and suggested decimal submultiples to replace the established hours, minutes and
seconds.

However, the more we learn about such natural units as the
day, the month or the year, the more complications and contradictions we discover. The historic consequence has been, first, that
the second, not the day, was chosen to serve as the basic time unit;
and then, quite recently, that the second was cut loose from the day

and from all astronomical bases, and became the present arbitrary but wonderfully exact unit for measurement of time.

The second, as a scientific and technical time unit, has ceased to be the fraction $\frac{1}{86,400}$ of the mean solar day, or a defined fraction of any other astronomically based time measure.

During many decades of progress in metrology, mean solar time served to define the adjusted or corrected natural day. Such mean solar time was first determined at the historic Greenwich Observatory in England. At first the mean solar day was computed from one noon to the next. When the beginning of the day was shifted instead from noon to midnight, mean solar time became the same as universal time, abbreviated either as U.T. or as UT.

Strictly speaking, it began as UT0. The added zero became necessary because at least two more elaborate and sophisticated variations were to follow, identified respectively as UT1 and UT2.

As increasingly accurate man-made clocks were used in observatories they revealed that UT0 was slightly but unmistakably in error. Its deviations from actual observations were traced to various irregularities in the rotational motion, or spin, of the Earth itself. The first such irregularity singled out and adjusted for was the so-called "wandering" of the poles.

The Earth's actual rotational axis does not remain precisely fixed. The North Pole and the South Pole both wander in small but highly significant spirals. The geographical North Pole, for example, spirals around within an imaginary circle whose diameter, as measured on the surface of the "top of the world," would be no greater than 20 meters. This polar wandering traces irregular spirals, which tend to return near to their beginning in periods of about fifteen, rather than twelve, months. If it were possible to trace out on the snow or ice of the polar region each successive location of the actual North Pole from one January 1 to the end of the next December 31, the slightly irregular loop might show a total length of as much as 50 or 60 meter. Yet the end point might be only a few meter from the beginning. For example, during the calendar year of 1964 the North Pole showed a net shift of about 6 meter; however, within that period it had passed through positions that were, in some instances, more than 16 meter distant from each other.

These incessant and unpredictable polar wanderings can reach

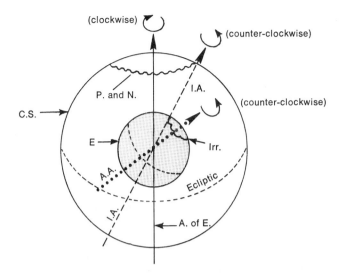

Complexities and irregularities of Earth's rotation motions are pictured here, indicating why our planet provides a poor timekeeper for the present exacting needs of science and technology. The outer sphere, marked C.S. (celestial sphere), is the great sky dome in which the stars seem to be set as they apparently spin around us. The short-dash line marked ecliptic is the track of the Sun as it (seemingly) travels around us, and the vertical line marked A. of E. is the apparent axis of the ecliptic about which the Sun appears to revolve. The direction of that revolution is clockwise, if viewed from a position in space far out to the north.

Within this celestial sphere spins our Earth, E, shown with its equator. Its average or mean axis of rotation, shown dotted and marked A.A., makes an angle of about 23½° with the axis of the ecliptic. However, at any instant the Earth's actual axis of rotation, marked I.A. for instantaneous axis and shown by a dashed line, deviates from the average axis A.A. The irregularities in this deviation cause the I.A. to wander about within a small distance of the A.A., as marked out near the north pole by the curved line marked Irr, for "irregular." For emphasis the irregularities are much enlarged.

Meanwhile, the instantaneous axis wobbles in complicated but generally predictable patterns, indicated by the wavy line marked P. and N., for precession and nutation, at the top of the celestial sphere. Precession is the larger swing around the axis of the ecliptic; nutation is the more rapid smaller vibration which produces the wavy or scalloped effect shown here.

These effects have been exaggerated in the diagram for purposes of clarity. Also for greater simplicity of conception, we have used the old fiction, long held to be a fact, that the heavens (fixed stars) and the Sun all revolve around the Earth, rather than the Earth around the Sun.

Seen from above (north with respect to us Earth-dwellers), the rotation of the Earth is counterclockwise with respect to the celestial sphere. So too is the stately swing or wobble of Earth's axis around the apparent axis of the ecliptic, or track of the Sun in the sky.

In earthbound terms, the Sun appears to move from east to west each day, because the Earth in fact is spinning from its west toward its east.

as much as 0.7 second of arc. That is about 5 parts in 10 million of a full circle (360°), and its effects are by no means negligible to the tasks of astronomers and time metrologists. Though unpredictable, these polar wanderings are measurable, and when the original universal time scale (UT0) was adjusted for them, it became UT1. The correction for the migration of the poles is of the order of 0.05 second (50 millisecond).

Use of ever more accurate quartz-crystal-controlled clocks enabled the watchers of the skies to discover still other irregularities in the Earth's rotation. These include seasonal variations with periods of about one year and also one-half year. They are probably traceable to meteorological causes involving formation and shifts of ice masses, the winds, and so forth. They are partly, but not entirely, predictable on the basis of increasing past records. In still another refinement of time scales these variations were mathematically smoothed away, insofar as this was possible. The resulting scale is designated as UT2. Its correction, relative to UT1, is also of the order of 0.05 second.

In spite of the statistical sophistication of UT2, the fact is that UT1 remains useful to navigators, because it is based on the Earth's angular position and hence on the apparent positions of heavenly bodies as viewed from Earth; but because UT2 has smoothed out the actual anomalies in the Earth's angular position, it is misleading to navigators.

Emphasis on ephemeris—Nearly a century ago the eminent American astronomer Simon Newcombe compiled elaborate tables of the future positions of the Sun, the Moon, and some of the planets. Such astronomical projections or extrapolations are called ephemeris tables. There are several series of such, each with a venerable and esteemed history: France's *Connaissance des Temps,* compiled by the Bureau of Longitudes established in the days of the Revolution; Britain's *Astronomical Ephemeris* tables; and the United States' *American Ephemeris and Nautical Almanac.*

All these ephemeris computers employ the best available observational data and unimpeachably orthodox Newtonian mechanics to compute tables on which navigators and geographers can depend. The separate tables agree remarkably well with one another, but year after year observations of the skies reveal actual positions of the heavenly bodies deviating increasingly from the ephemeris predictions.

The truth was unmistakable. The Earth itself was out of step. Its rate of spin was slowly but persistently diminishing, as we have already seen. This deceleration of Earth's spin is still another irregularity, besides those for which first UT1 and then UT2 had made adjustments.

Still another rotational irregularity—a fourth, in fact—must be mentioned here: nonseasonal shifts in the spin rate which seem strangely linked to activity on the surface of the Sun. Indications are that the charged particles, poured out in the great "solar wind," interact with the magnetic field of the Earth to produce these perceptible changes in its spin rate.

Because of these many rotational irregularities, astronomers and time metrologists concluded that they should change from the UT scales, based essentially on that rotation, to another scale whose unit, also called the second, should be a defined fraction of the Earth's orbital cycle around the Sun—or rather, of one particular such orbital cycle, called a "tropical year." The tropical year chosen was that of 1900, which had been measured with utmost precision in terms of the prevailing second based on the mean solar day.

By international agreement, formulated in 1956 and put into effect in 1960, the second ceased to be defined as $\frac{1}{86,400}$ of the mean solar day and became instead $\frac{1}{31,556,925.974\,7}$ of the tropical year 1900 January 0 and 12 hours ephemeris time. Thus was born ephemeris time (E.T. or ET).

A tropical year is the time interval between two successive vernal (spring) crossings by the center of the Sun of the celestial equator. Such crossings take place at the vernal and autumnal equinoxes. The geometry of such timekeeping is grandiose and fascinating. One plane is visualized as passing through the orbital plane of the Earth—the plane that contains Earth's center as it circles in its ellipse around the Sun. Another plane is visualized as passing through the Earth's equator, perpendicular to the imaginary axis of rotation extending from South Pole to North Pole. These two planes, as projected on the imaginary celestial sphere surrounding us, intersect at two definable points: the vernal and autumnal equinoxes.

The Earth's orbit, as it would be seen from the Sun, is called the ecliptic. To us on Earth the ecliptic appears rather as the path traveled across the sky (the celestial sphere) by the center of the Sun, in its seeming revolutions around the Earth.

The great circle of the celestial equator can be located by an observer who knows his latitude on Earth. For example, an observer in Philadelphia, at 40° north latitude, first points his instrument straight up, at the zenith. Then if he tilts it 40 degrees toward the south it will be aimed at the intersection of the celestial equator with the north-south line called the meridian.

The tilt of the Earth's rotational axis produces an angle of very nearly 23½° between the planes of the celestial equator and the ecliptic. This is another way of saying that the Earth's axis makes an angle of $90 - 23\frac{1}{2} = 66\frac{1}{2}$ degrees with the plane of its orbit around the Sun.

A wilting tilting—The 23½-degree figure for the obliquity of Earth's rotational axis is only approximate. *Smithsonian Physical Tables,* as issued in 1954, showed for the beginning of tropical year 1900 the far more precise tilt of 23° 27' 8.26''. Furthermore, the same source indicated that this tilt was *diminishing* at an annual rate of 0.468 4'', or slightly less than ½ second of arc per year.

According to this, by the beginning of tropical year 1980 one can expect a tilt of 23° 26' 30.788'', or 37.472'' less than when tropical year 1900 began with its vernal equinox. By 1980, observations and calculations may alter the figure for the annual diminution of tilt. The fact is, however, that even so tiny an angular change as 0.468 4'' per year has enormous long-run effects for astronomers and metrologists. If continued through about 180,000 present tropical years it would suffice to eliminate the tilt entirely and bring the plane of the Earth's equator into that of its orbit or ecliptic.

The Earth's actual axial tilt is basic in astronomical time measurements, because an unchanging or constant tilt angle would be required in order that successive tropical years remain equal in duration. We know, by the definition which took effect in 1960, that the duration of the tropical year 1900 was declared precisely equal to 31,556,925.974 7 second. However, no tropical year since then has contained just that number of seconds. Each has been briefer than the one before by about 0.005 3 second (5.3 millisecond).

This, the best current figure for the annual shrinkage of tropical years in terms of an invariable second, is a very tiny proportion. It corresponds to less than 2 parts in 10^{10} (10 billion). Yet if continued during 50 million years it would suffice to reduce the number of seconds in a tropical year by nearly 1 percent, a shortening amounting to more than 3 mean solar days, as we now know them.

Such shrinkage in the duration of tropical years explains why the particular tropical year of 1900, and no other, was designated in that definition of the second which became effective in 1960 and remained so only until 1967, when the present "atomic second" was adopted. That period 1960–1967 was by no means a happy one for metrologists of time. They could not reproduce the quantity of time referred to as the tropical year 1900. Time intervals cannot be stored somewhere and taken out for inspection or comparison. Instead, they were obliged to use their current measurements of solar, lunar and stellar positions and from them calculate backward, so to speak, by means of elaborate and unwieldy equations, to arrive at an equivalent for that long-gone basic tropical year.

The time scale based on this unalterable "tropical" second was called ephemeris time (E.T. or ET). It was not based on the latest orbits of the Earth around the Sun, but only on one past cycle between the vernal equinox of 1900 and that which followed it (the start of the tropical year 1901).

Most of us think that the Earth now completes one full orbit around the Sun in 365.242 2 mean solar days, since that is the present length of a tropical year, from one vernal equinox to the next. However, a true orbital year would be about 1,500 second (25 min) longer than a tropical year. Such a true orbital year is called by astronomers an *anomalistic* year and is measured from one perihelion to the next, the perihelion being the point in Earth's elliptical orbit when it is closest to the Sun.

Why, then, did the tropical year rather than the anomalistic year become basic in timekeeping? Simply because the former can be determined far more easily and precisely by observation than the latter.

The anomalistic year is negligible in importance here, but does serve to remind us that what we call a year, with its pattern of seasons and dates, is the result of the interaction of the Earth's orbit on the one hand and its rotational angle or tilt on the other, and not of the orbital motion alone.

In order that the durations of the tropical and anomalistic years might be equal, it would be necessary that the imaginary rotational axis of the Earth always pointed in the same direction in space. Or more precisely, that the line of the rotational axis at any instant must be parallel to its line at any previous or following instant. This is not the case, not by a very wide margin.

The Earth, which spins like an enormous globular top, also has a pronounced wobble, or *precession,* such as tops and gyroscopes develop when disturbing forces act upon them. This is no new discovery. More than 2,100 years ago, about 130 B.C.E., the Greek astronomer Hipparchus spotted it. He did not describe it as a wobble of the spinning Earth. To him it was the unmistakable movement of the equinoxes, both vernal and autumnal. Relative to the fixed stars they move slowly westward along the path of the ecliptic.

The rate of this significant movement is now 50.26 second of arc per year, or about 1° 24' per century. In a period of 25,800 years, at this rate, the vernal equinox moves completely around the 360° circle of the ecliptic. The effect of this is to make the tropical year shorter by the indicated 1,500 second than the full orbital or anomalistic cycle.

However, even this stately precessional gyration of the Earth's axis does not proceed at an entirely uniform rate. It accelerates very slightly, adding to the 1° 24' angular movement per century an additional 0.022 2″ (second of arc).

This does not complete the complexities of the Earth's motions. Its axis of rotation, while it is slowly changing direction in the gyration of precession, also vibrates in a far smaller cycle called "nutation," with a period of some 18.6 years. Thus the great precessional circle, completed in nearly 26,000 years, shows nearly 1,400 tiny nutational scallops or wavelets. Their scope is less than ¼ of 1 percent of that of the great precession, but they cannot be overlooked by astronomers seeking to use Earth's orbit as a clock.

Nutation is fascinating in another way. It tends to turn the direction of Earth's rotational axis in a direction, or "sense," opposite to that of the great precession and to the senses of the Earth's orbit around the Sun and its spin also.

Suppose we were able to observe the solar system from a region of space far above the Sun. "Above" here means northward. As we looked down we should see near the center of the Sun's disk that portion which from Earth we see at the lefthand edge of the Sun when it rises in the east, as viewed by us on Earth. From such a lofty vantage point in space we would see the Earth orbiting in a counterclockwise sense, and spinning also in that same sense. Even the slow precessional wobble of the Earth's rotational axis would follow that same counterclockwise direction, as we observed it.

However, the tiny nutational vibrations would seem to be trying to turn the Earth's axis in the opposite, or clockwise, direction. (Needless to say, if we observed from the opposite or "southern" side of the plane of the Earth's orbit, the senses or directions would be reversed—orbit, spin and precession would be clockwise; nutation counterclockwise.)

The peculiarities of precession and nutation result from the fact that Earth is not a perfectly round and homogeneous mass of matter. It has a distinct bulge around its midriff. Its circumference around the equator is greater than its circumference measured around the two poles, north and south. The gravitational attractions of Moon and Sun, acting on this additional bulge of matter around Earth's middle, tend to reduce the tilt of Earth's rotational axis, from $23\frac{1}{2}°$ toward $0°$. But since the Earth spins like a great globular top, this deflecting force is transformed into the slow precessional swing of the axis of rotation, with a period of nearly 26,000 years. Accordingly, astronomers call this the *lunisolar* precession.

Earth's sister planets also exert gravitational pulls, though far smaller than those of the Moon and Sun. The net effect of these planetary attractions on the Earth's equatorial bulge is the nutational vibration, which is thus a sort of *planetary* analogue of the lunisolar precession.

These simultaneous variations in Earth's axial tilt, its slow precession, and its smaller, more rapid nutation, combine to unfit our planet to serve as a suitable clock, even in its orbiting movements. Thus, what once seemed unchangingly uniform cycles of rotation and orbit fall short of the exacting needs of time measurement of modern technology and science.

Shrinking years and stretching days—Two great classes of time scales have been based on the movements of our Earth. Those derived from the planet's spin on its axis are all variations of UT (universal time). Those linked to the Earth's period of orbit, as measured from one vernal equinox to the next, are identified as ET (ephemeris time).

ET, by definition, is uniform and unvarying. UT, on the other hand, being tied to the Earth's actual rotation, is admittedly irregular. Hence ET and UT could not keep completely in step. In 1900, ET was about 4 seconds behind UT; but by 1967 ET was about 36 seconds ahead. Thus there was a relative shift of 40 seconds, or an average shift of about 0.6 second per year. Since 1967 this relative shift has been growing by about 0.5 s per year.

Deviations in UT are by no means entirely uniform or predictable. Sometimes within a single year the Earth's rate of spin, and hence the actual length of its mean solar day, will alter by as much as 1 part in 10^7. Shifts of 1 part in 10^8 have been found within a span of a few months.

Peculiar but unmistakable changes, neither very short nor very long in duration, have been observed. Thus, during the forty-year period 1870–1909 the rate of spin increased by an average of about 8 parts in 10^8, or per year about 2 parts in 10^9. However, the following half-century, 1910–60, showed increase in the average rotational rate amounting to between 3 and 4 parts in 10^8.

By 1960, in fact, the average rotational rate had risen to within 4 or 5 parts in 10^8 of its mean value ninety years earlier, in 1870.

Such irregularities, as well as those of far briefer duration, are now relatively easy to detect and allow for, thanks to the extraordinary atomic clocks, whose uniformity and precision vastly exceed those of all previous time-measuring devices.

Thanks to these atomic clocks it is possible also to estimate closely the average shrinkage of the duration of successive tropical years. This shrinkage is very nearly 5.3 millisecond. Each tropical year is thus about 17 parts in 10^{11} shorter than its predecessor.

Since tropical years are slowly shrinking while mean solar days are slowly lengthening in duration, some striking changes may be anticipated in the future—if the deviations themselves do not alter appreciably.

The mean solar day in 1 million years from now, for instance, should be about 20 second, or about 2.3 parts in 10,000, longer than today. By that time, the tropical year should have lost about 1.7 parts in 10,000 of its present duration.

At present, 365.242 2 mean solar days make up the tropical year. By 1 million years from now, this ratio should have fallen to about 365.1; by 10 million years from now, to about 364.1; and in 100 million years to about 351. Cosmologists deal handily with periods of 1 billion years also. By 1 billion years from now, if present tendencies prevail, about 246.5 of the lengthened mean solar days will make up 1 shortened tropical year.

Luna's antics—or seeming lunatics—Earth's many irregularities of movement may unfit it to serve as the ultimate clock, but what about the Earth's great satellite, the Moon? Could not a satisfactory and readily reproducible time unit (the second) be based on the Moon's motions?

Alas, the inconstant Moon too is seriously unsuited for this vital role. Chapter 13 has already noted that the Moon is gradually spiraling in, closer to the Earth. Hence the average duration of a month diminishes slowly because of the more rapid orbital motion of our satellite.

But even on a short-term basis, the carefully measured movements of the Moon present many obstacles to its use as the standard clock. The familiar phase month, or *synodic* month, has now 29.530 59 mean solar days. But this is only one of many possible kinds of month. The phase month arises from the relative angular alignments of the Moon, the Earth and the Sun.

The *tropical* month, however, is measured by the position of the Moon relative to the equinox. A tropical month, with 29.530 40 mean solar days, is just 0.000 19 such days, or about 16.4 second, shorter than a synodic month.

Then there is the *sidereal* month, measured by the Moon's position relative to the so-called "fixed" stars. Its length is 27.321 66 mean solar days. Hence it is more than 2 days shorter than the common synodic month. Also, it is easier to measure accurately than the synodic month.

Still another is the *nodical* month, measured by the Moon's position with reference to the ecliptic (or sun-track) in the celestial sphere. *Nodical* is not to be confused with *nautical*. The nodes are the points where the track of the Moon crosses the track of the Sun as seen from Earth. These nodes move around the ecliptic at the rate of 19.3 degrees per year, completing a full circle in about 18.7 years. The nodical month, with 27.212 22 mean solar days' duration, is more than 2¼ days shorter than the synodic month.

The rotation of the Moon's nodes around the circle of the ecliptic has been known ever since it was described by the Greek astronomer Meton in the fifth century B.C.E. Very likely it had been recognized by the still-earlier skywatchers of Sumer and Babylon in Mesopotamia. Its duration is 235 synodic months.

Finally, there is the *anomalistic* month measured from one perigee of the Moon's orbit to the next. The perigee is the point of closest approach to the Earth during the Moon's elliptical orbit. The anomalistic month, with 27.554 6 mean solar days, is very slightly less than 2 days shorter than the synodic month. At its apogee, or point farthest from Earth, the Moon's distance is about 10 percent greater than at perigee.

Even the Earth's sister planets in their orbiting about the Sun fail to supply an ideal celestial clock. Their orbits are slightly irregular because of mutual gravitational effects. Furthermore, for planets closest to the Sun, such as little Mercury, an effect first recognized by Einstein produces a small but by no means negligible shifting of the orbital ellipse, called "the precession of the perihelion."

Thus, the sky with all its cyclical patterns offers no single satisfactory clock or any possible combination of clocks adequate to the demands of our modern age. The answer, finally, had to be found not in the direction of the great macrocosm of the solar system but in the microcosmic realm of molecules and atoms.

Man-made clocks move ahead—Human ingenuity led from the sundial to the water clock to the pendulum-controlled clock and the spring-driven watch. Then, and rather recently, it was found that quartz crystals can be cut in such a way that they will resonate to precise frequencies of electrical oscillations. Mechanical clocks, corrected and controlled by such crystal-electric interactions, became the next great improvement in timekeeping devices. Indeed, the "quartz clocks" permitted the first reliable determinations of irregularities in the rates of the Earth's spin on its axis.

During World War II, development of sophisticated radar and microwave circuitry multiplied the extent of control over high-frequency electric oscillations. Such man-made oscillations, identical with very short radio waves, could be tuned to correspond to specific energy changes in molecules and atoms. These changes had been carefully measured in terms of the electromagnetic oscillations that they either emitted or absorbed.

During the late 1940s work of this kind leaped ahead. By 1948 the U.S. Bureau of Standards had operated a device in which the microwave frequencies were those that matched energy changes in molecules of ammonia (NH_3). Such molecules had been excited— that is, given an energy content slightly above their normal, or ground, state. Then they were allowed to drop back to that ground state by emitting bursts, or photons, of radiant energy. Each such photon was precisely identical with every other.

Thus, ammonia molecules were used as ultrasensitive regulators of the frequency of the short-wave radio transmitters coupled to the device. The "escapement" of this clock was, in fact, a beam or flow of ammonia molecules, each an indistinguishable twin of every other. The resonant responses of these molecules were vastly more uniform than those of pendulums and escapements, or even of quartz crystals. All such structures are subject to changes, to wear, to distortions, and to minute alterations which affect their time-keeping qualities.

Any oscillation that can be maintained at a uniform frequency can serve as the basis for a clock. This is true whether the oscillation be that of a pendulum, a spring-coupled balance wheel, or a molecule that emits a burst of energy at a frequency determined by its structure.

The operation of molecular-beam clocks was self-regulating. If the radio-frequency oscillator deviated from the peak frequency for the absorption or emission of energy by the ammonia molecules, then a corrective signal was generated that changed the transmitter frequency toward that peak value.

Later and even more effective timekeeping devices employed

beams of atoms rather than of molecules consisting of two or more atoms linked together. The best of the modern devices now make use of the atoms of a common isotope of cesium, known as cesium 133. Others have been operated using hydrogen, iodine, and several other atoms. The atoms of various elements differ drastically in complexity and structure. Hydrogen, lightest and simplest of all elements, has a nucleus composed of a single proton (positively charged) around which circles or vibrates a single electron (negatively charged). Cesium 133, on the other hand, has a massive nucleus composed of 54 protons, 79 neutrons (no charge), and roundabout, arranged in shells, are 54 electrons, the outermost of which is alone in its shell.

In the early 1950s, Harold Lyons and others at the National Bureau of Standards measured to within 1 part in 10 million the resonant frequency of the cesium-133 atom in its transition from ground (or least energy) state to the next-higher energy level. That frequency was found to be very nearly 9.19 gigahertz (10^9 cycles per second). This is a frequency suitably within the range of microwaves, the shortest form of radio waves, somewhat longer than the longest infrared radiations.

In 1957, L. Essen and J. V. L. Parry, of the National Physical Laboratory in Britain, improved the precision of these cesium measurements to a few parts in 10 billion. Their report, issued in *Philosophical Transactions* of the Royal Society of London, was significantly titled "The caesium resonator as a standard of frequency and time." "Resonator" meant, in that context, the cesium-133 atom itself, moving freely in an atomic beam through a vacuum and absorbing or emitting microwaves of the characteristic resonant frequency.

This transition or energy change of the cesium atom results from the fact that when the spin of the lonely outermost electron opposes the spin direction of the nucleus, the cesium atom has its least possible energy content. If, however, it absorbs a precise small amount of energy from surrounding electromagnetic oscillations at the right frequency, then that outside electron flips over and spins in the same sense or direction as the nucleus. The atom has made what is called a "hyperfine" transition to a higher energy state. The energy of this transition, in either an upward or a downward direction, is precisely determined by the invariable structure of the atom of cesium 133. It does not differ an iota from one such atom to another, no matter how many myriads of atoms may be moving in a beam or vibrating within some container.

The amount of energy absorbed or emitted by these cesium atoms in the atomic clocks seems infinitesimal by human standards. It can be measured in the SI energy unit, the joule, which is just the amount of energy delivered by 1 watt of power in 1 second of time. The joule is a small amount of energy—about enough to lift a weight of 1 pound a distance of 9 inches, in the outdated U.S. units. The crucial energy transitions of the cesium atoms in these superclocks represent about 6×10^{-24} joule. This means that 1.6×10^{25} such transitions would be needed to supply energy to light a 100-watt light bulb for just 1 second.

Though almost incredibly tiny, atoms are uniform and invariable in structure. This cannot be true of any devices above the molecular level. It is impossible to make identical wires, rods or gears. However, the form of cesium called cesium 133 always is composed of just 54 protons, 79 neutrons and 54 surrounding electrons. One atom never differs from another by a single particle nor by a fraction of a particle. Were 1 proton too many or few to occupy the nucleus, the structure would not be an atom of cesium 133.

This supreme uniformity of the resonating devices, the atoms themselves, accounts for the otherwise incredible timekeeping achievements of these greatest of all clocks.

Atomic clocks in action—A modern cesium clock, properly adjusted and maintained, can keep time with a stability of about 1 part in 10^{12}. This means that in a year of continuous operation, its deviation from "the right time" should be no greater than 30 microsecond (or 3×10^{-5} s). If it could be kept operative for a full century, the deviation should be no greater than 3 millisecond.

When two such cesium clocks are compared, it is found that they maintain agreement to within about 1 part in 10^{10}. Thus if they are in step at the beginning of a year, by the end of that year there should be no more than about 3 millisecond difference between them. The metrologists of time correctly stress that the *stability* of the frequency of one clock is better than the *reproducibility* of that frequency (or rate) from one clock to another of the same kind. Two atomic clocks, however carefully constructed and controlled, are likely to show tiny differences in rate, of the order of perhaps 1 or 2 milliseconds per year. This is so tiny, by past standards, that only the extraordinary demands of modern technology make it worth mentioning here.

Such superb stability and reproducibility in time measurement had to be recognized and put to work. It was inevitable, then, that in 1967 the 13th General Conference of Weights and Measures substituted the cesium-133 atom for any and all heavenly bodies as the primary definer and basis of the one unit of time.

Here are the words of that historic pronouncement: "The second is the duration of 9,192,631,770 periods (or cycles) of the radiation corresponding to the transition between two hyperfine levels of the cesium-133 atom."

This means, in slightly different words, that the radiation typical of that hyperfine transition has a frequency of 9,192,631,770 hertz, or cycles per second. Its wavelength is about 3.26 centimeter.

What goes on inside one of these atomic-beam clocks, such as the cesium-133 timekeepers at the National Bureau of Standards in Boulder, Colorado, or other great standards institutions of the world? How, in physical terms, does such a superclock work?

A classic recipe for rabbit stew is said to begin, "First catch your rabbit." Cesium was first caught—that is, isolated and identified and described—in 1860. It is a metallic element, naturally occurring almost entirely as cesium 133. There is another stable isotope identified as cesium 135, but it is far less abundant than the 133 isotope. There are also radioactive isotopes ranging all the way from cesium 127 to cesium 145, but their instability makes them quite unsuitable for clocks.

Though the official definition of the second of time specifies that only cesium 133 shall be used to provide the "transition between two hyperfine levels," one of the advantages of cesium happens to be that even if a moderate amount of cesium 135 should be mixed in with the cesium 133, the atomic clock loses relatively little accuracy. Isotopic purity here is not as essential as it is in krypton 86, which supplies the light radiation by which the meter of length is now defined.

Both the cesium-133 clock and the krypton-86 lamp are suppliers of well-defined electromagnetic oscillations, but at vastly differing frequencies. The frequency in the krypton 86 is 50 million times that of the Cs 133. Accordingly, each energy transition in the krypton-86 atoms emits a photon (of light) with about 50 million times the energy content of the photon (of radio oscillations) emitted by the cesium-133 atom in the clock device.

A cesium clock looks like a pipe or an assemblage of pipes, standing either horizontally or vertically. Before operation begins the main pipe contains nothing—or very nearly nothing. It is exhausted of air and other gas to such an extent that the pressure within it is less than one ten-billionth of normal atmospheric pressure.

At one end of this timekeeping pipe is a small electric oven in which cesium 133 is heated to about 100° C, which happens to be the boiling point of water at normal pressure. Cesium itself does not boil, or even melt, at this temperature; but a small percentage of cesium atoms do gain such velocity that they fly off from the rest and move into the pipe. Their velocity is around 100 meter per second as they begin their strange migration.

The following is a deliberately oversimplified summary of their odyssey along the pipe. The first adventure takes place as they move through a magnetic "state selector." Some of the flying cesium atoms are in the lowest possible energy state; others have the slightly higher energy resulting from the cooperative spins of the outermost electron and the central nucleus. The former atoms are drawn toward the south pole of the magnetic device; the others toward the north pole.

Thus they are separated, and the ground-state, or least-energy, atoms are allowed to proceed, while the others are blocked off. The surviving ground-state atoms now enter a chamber or cavity almost free of magnetic field but filled with radio waves (electromagnetic oscillations) maintained as close as possible to the atoms' natural resonance frequency of 9,192,631,770 hertz, or cycles per second. This cavity is sometimes called a transition region, but it can be regarded as a kind of interrogation chamber.

The question asked of each arriving atom is, in effect, "Is this very close to your natural frequency?" If the answer is yes, the outermost electron of that atom absorbs the tiny increment of energy—or photon—and flips its spin direction. Thus, what had been an atom in the lowest-possible, or ground, state becomes an

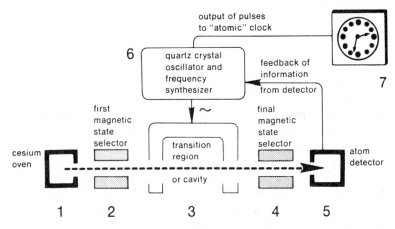

Essentials of the strange events within a cesium-beam atomic clock are indicated in this ultrasimplified schematic diagram. The dashed line from left to right shows the general course of the cesium atoms, which are boiled off in the oven at left (1) and finally arrive at the atom-detecting device at right (5).

En route they pass through at least two magnetic "state selectors" (2 and 3), each of which weeds out atoms in one of the two lowest energy states, and allows those in the other to continue on. (The entire device operates in a high vacuum to allow the cesium atoms to move with least possible interference.)

In the simple type of cesium clock described in the text of this book, the first-state selector (2) lets through only those atoms which are in the very lowest energy state (also called the ground state). These enter the transition region or cavity (3), where they encounter electromagnetic oscillations supplied by the precision quartz-crystal oscillator (6), which operates much like a miniature radio-frequency transmitter. Many of the ground-state atoms absorb a photon of energy in the cavity, thus rising to the next-higher energy level. It is these, and only these, atoms which the second-state selector (4) allows to proceed onward, to the atom detector (5).

If the frequency supplied by the oscillator just matches the natural frequency of the cesium atom, many atoms will make this transition. If the oscillator frequency deviates from that natural atomic frequency, fewer atoms will make the transition. This reduction will be reported, via feedback, from the atom detector, and the quartz-crystal oscillator will at once automatically alter its frequency so as to maintain at maximum the number of atoms reaching the detector.

The oscillations of the quartz-crystal device are converted, by means of a sophisticated electronic frequency synthesizer, into a series of precisely timed pulses which drive the atomic clock (7). Thus the clock itself is not driven by oscillations supplied by the cesium atoms. Rather, the atoms supply the "balance wheel" or regulator which constantly corrects the frequency supplied by the quartz-crystal oscillator.

Basic to all atomic timekeepers is the fact that free atoms, such as those in the cesium beam, are extremely "choosy" about the frequency of electromagnetic waves from which they will accept energy packets or photons. Their natural frequency for absorption is the same as their natural frequency for emission of radiation, just as a tuning fork will resonate to absorb sound waves at just the same frequency as those which it emits when it is struck.

None of the myriads of timekeeping devices used by humans have ever been as accurate and stable as the new cesium-beam clocks. It appears that they will continue until well into the future to represent the very finest in precision timekeeping.

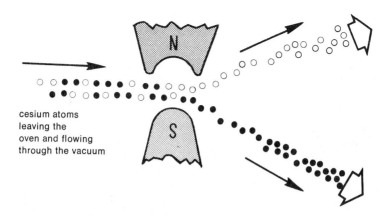

cesium atoms
leaving the
oven and flowing
through the vacuum

This is how cesium atoms in two different energy states are sorted out magnetically, to form beams of atoms in one energy state only. From the left a beam of cesium atoms flows out of the oven. Some (shown black) are in the lowest energy state, or "ground state." Others (shown white) are in an energy state slightly higher, because of a change in the spin of their outermost electron.

The strangely curved magnet poles create a nonuniform magnetic field, to which the ground-state atoms respond by being deviated toward the magnet's south pole, whereas the above-ground-state atoms are deviated toward the magnet's north pole. Thus the sheep are separated from the goats, so to speak, and it becomes possible to decide what to do next with either the low-energy atoms or the slightly higher-energy atoms, or with both of the groups, for that matter.

Modern cesium-beam timekeeping devices use two such magnetic-state selectors, one before the transition cavity, the other after it, as the atoms move along toward the detector at the end of the line.

atom in an energy state very slightly higher—by a "hyperfine" amount, which happens to be about 6.1×10^{-24} joule.

As the atoms move on out of the transition region they arrive at a second magnetic state selector. Once again the atoms in the slightly higher-energy state are drawn toward the north pole of the magnetic device, and those which have not gained the additional energy, or photon, but have remained in the ground state, are drawn toward the south pole. This time it is the latter, the ground-state survivors, whose further progess is blocked, and the slightly more energized atoms whose progress is permitted to the final stage of the strange tube.

That last stage consists of a detecting device which electrically registers changes in the number of such higher-energy atoms that reach it. If the relative number of such arrivals shows a decrease, it means that the radio frequency fed into the transition region (interrogation chamber) has departed somewhat from the natural

Intervals of time and frequencies have been measured by this cesium-beam atomic clock with an accuracy amounting to a deviation of less than about 1 part in 10^{11}, or 1 second in 3,000 years! This strange and sophisticated device, operated at Boulder, Colorado, by the National Bureau of Standards, provides reference control for the time and frequency broadcasts over stations such as WWV, WWVB, WWVH, and WWVL.

The principal events in the cesium-beam clock take place within the long horizontal pipe, one end of which is seen at the right. The boxlike attachment to that end is the small electric furnace in which cesium atoms are boiled off, to begin their journey through the vacuum maintained within the pipe. At the far end, near the attendant at upper left, those cesium atoms that have completed the entire "obstacle course" are received and detected. The actual time and frequency counting devices are placed in an adjoining room. (COURTESY NATIONAL BUREAU OF STANDARDS)

resonance of the cesium-133 atoms. Only when that frequency is at or extremely near the atom's natural resonance will the maximum number of energy transitions take place in that region. Any other frequency inevitably reduces the number of atoms that can make their way through the second magnetic separator and arrive finally at the detector at the end of the line.

A so-called servo mechanism is connected with the detector. If an increase in the frequency of the oscillations fed into the transition region has been followed by a reduction in the number of arriving cesium atoms, the mechanism reduces the frequency slightly. If, on the other hand, a reduction in frequency is followed by such a decrease, the mechanism increases the frequency slightly. Thus the flow of cesium atoms produces the tiny changes of frequency that will maximize that flow; and it is the natural resonant response of the atoms to the impinging electromagnetic oscillations that determines whether the number of atoms running the full obstacle course will show increase or diminution.

This has been a simplified description of one of several possible types of cesium-beam clocks. However, in any other form the basic processes are essentially the same. Details of design are important. The transition cavity should not be too short, else there will not be time enough for sufficient atoms to react to the impinging electromagnetic waves. On the other hand, in a very long transition region the atomic beam tends to become weakened and scattered, because it cannot there be focused and held together by a magnetic field. It appears that a compromise length of about 2.5 to 3 meter may represent the optimum transition region for a laboratory cesium clock.

Portable cesium clocks, now being made in considerable numbers, naturally have shorter transition regions and cannot attain the same degree of timekeeping accuracy as the elaborate laboratory models.

Subdividing frequencies to fit needs—A clock that ticks 9 billion times a second is not suited for the measurement needs of scientists or technicians today. Hence the cesium and other atomic clocks feed their regulated frequencies into sophisticated electronic network devices called frequency dividers. These can be adjusted to emit a pulse after a predetermined number of oscillations have passed through them.

For example, since the basic frequency of a cesium clock is very nearly 9,192,631,770 cycles per second, if the frequency divider is adjusted to emit a pulse for every 9,192,632 cycles, those pulses will provide time intervals of 1 millisecond (0.001 s). If adjusted to pulse once for every 9,193 cycles, those pulses will provide time intervals of 1 microsecond (10^{-6} s). Thanks to the elaborate complexities of such frequency dividers, cesium clocks can convert their fundamental frequency into just about any lesser frequency likely to be useful in the work of the metrologists of time.

Advances in clock accuracy since the middle 1950s have been much greater than those made during the preceding three centuries. Pendulum clocks once were considered good if they accumulated errors no greater than 10 seconds within a day. Now, for any critical work, atomic clocks can reduce error to as little as a few picosecond (billionths of a second) per day. This is a gain of nearly ten billion times, or ten orders of magnitude, as scientists often describe it.

16 THE LIMITS OF PRECISION
IN OUR TIME

I'll put a girdle round about the earth
In forty minutes.
——SHAKESPEARE,
A Midsummer Night's Dream

O Time! consumer of all things.
O envious age!
You destroy all things and devour all things
with the relentless teeth of years,
little by little,
in a slow death.
——LEONARDO DA VINCI,
Notebooks, section on morals

The previous chapter introduced three universal time scales derived from the Earth's rotation (UT0, UT1 and UT2), the scale called ephemeris time (ET) and based on the natural year, and atomic time (AT) based on a second defined in terms of a perfectly uniform and stable physical unit, the cesium atom.

In terms of convenience or derivability, these time scales differ drastically. Universal time can be determined to within a few millisecond per day in a reasonable observation period, but ephemeris time requires nearly a decade of observation and elaborate averagings to be determined to within 50 millisecond per day. In contrast, a well-regulated cesium clock within one minute of operation can determine atomic time to within a few nanosecond (billionths of a second) per day.

The enormous advantages of atomic time become apparent from such practical revelations.

Timekeeping today means frequency setting, as well as supplying answers to questions like "How late is it?" or "How long did this event last?" Precise frequency settings and maintenance are essential for such standard procedures as television, radio broadcasting, and other communications operations.

If the basic second became longer or shorter to match vagaries in the Earth's actual rotation, it could not provide the frequency controls required by TV, radio and the rest. Therefore the AT scale is inescapably necessary, though for navigational needs the UT0 and the UT1 scales cannot be discarded.

A time scale called Coordinated Universal Time, abbreviated as UTC—not CUT—was established in the late 1950s. It had previously been called Greenwich Mean Time (GMT), and is widely broadcast by radio signals as the standard time scale for the United States and other technologically advanced nations.

The UTC year is the mean solar year, but the UTC second is not $\frac{1}{86,400}$ of a mean solar day, as we shall see. Rather, that second is now determined by the atomic (cesium-133) definition already quoted.

The discrepancy between atomic time and UTC is about 32 parts in 10^9. This means that the regular calendar year, widely believed to contain just $365 \times 86,400 = 31,536,000$ second, contains in fact almost $31,536,001$ atomic second.

During the period 1960–71, about 70 percent of this discrepancy was eliminated, or made to vanish, by setting the UTC clock to run slow. The remaining 30 percent of the discrepancy was taken care of by adding "leap intervals," small bits of time, fractions of 1 second, which filled in the remaining gap.

Beginning in 1972, however, the UTC clock was run strictly on atomic-clock seconds, and the entire discrepancy was eliminated by adding leap intervals of whole seconds. This important new procedure deserves description in some detail.

During 1960 and 1961 the UTC clock was deliberately adjusted so that it ran slow, compared with the atomic time scale, by 15 parts in 10^9. The UTC second, during this period, was longer, by 15 nanosecond (or 0.000 000 015 s), than the true (atomic) second.

During 1962 and 1963 the UTC clock was set to run slow by 13 parts in 10^9. Then in 1964 and 1965 the previous retardation of 15 parts in 10^9 was resumed; and finally from 1966 through 1971 that retardation was doubled, to 30 parts in 10^9.

These retardations averaged about 22 parts per billion from 1960 through 1971, accounting for a total shift of 8.4 second in 12 years, or an average of 0.7 second per year. These corrections were, however, insufficient. The discrepancy between the two time scales is not uniform since the Earth's spin is so erratic. The additional adjustments took the form of halting the UTC clock for periods of 0.05 second (in 1960–63) and of 0.1 second (in 1964–71), then allowing it to proceed at the same rate as before these halts.

By these measures the actual discrepancy between the UTC and the atomic time scales was not allowed to exceed 0.05 s from 1960 to 1963, or 0.1 s from 1964 to 1971.

Since the broadcast-time scale was deliberately set slow compared with the basic atomic-time scale, the problem existed of providing some kind of receivable signal that should beat time in true atomic seconds. This led to the periodic broadcast of a special time signal known as Stepped Atomic Time (SAT). Its seconds were the true, unadulterated atomic variety, as provided by the best cesium-beam clocks.

At intervals, these SAT broadcasts would be stopped and restarted, thus keeping SAT within 0.1 second of the UTC signals. In this way, even before 1972 both the adjusted or "doctored" UTC and the correct but deviant SAT were offered to those able to receive the time broadcasts.

Restoration of the right rates—The International Radio Consultative Committee issued an epochal decision in February 1971. Henceforth the UTC time scale was to keep time, second for second, with the atomic scale, beginning January 1, 1972. This results in the mean solar day containing, not the old 86,400 second exactly, but rather 86,400.002 second. The total discrepancy of about 1 second per year would add up to nearly a quarter of an hour in a millennium. However, no such discrepancy is to accumulate, for at predetermined intervals the UTC clock is to be halted, and a full "leap second" will be inserted.

The first such leap second was inserted at the end of June 1972. A bulletin issued by the National Bureau of Standards late in May 1972 announced history's first 61-second minute in these words: "At the recommendation of the Bureau International de l'Heure

(BIH), Paris, France, notice is hereby given that a positive leap second will occur in transmission from NBS radio stations WWV, WWVH and WWVB at the end of June 1972. The positive leap second will begin at 23h 59m 60s UTC on June 30 and end at 0h 0m 0s on July 1, 1972."

Usually there is no "23h 59m 60s," for 23h 59m 59s is followed directly by 0h 0m 0s. Leap seconds will be inserted in future only at the end of June and, if necessary, at the end of December, just before the first second of the incoming year.

A "positive leap second" in the careful language of the NBS is equivalent to stopping the UTC clock for just 1 second, then restarting it at the previous rate. It is unlikely that any future situation will require a *negative* leap second, which would mean jumping the UTC clock ahead by a full second, the opposite of stopping it for a like time.

The day containing a leap second might be called a leap day, just as the years which contain a leap day (February 29) are called leap years. However, this would give us two different kinds of leap day, and the technicalities of timekeeping are complicated enough without such a semantic overlap.

In any case, the analogy between the leap second and the 366th day of a leap year is a sound one, and not a mere figment of speech.

The leap second of June 30, 1972, caused no crisis or confusion. It affected only users of precise timekeeping equipment, and high-precision frequency generators and controllers, such as those in television and radio stations, scientific laboratories, electric power companies, makers of some electronic devices and of sophisticated navigation and radar equipment. All had been amply forewarned about what was to take place.

Timekeepers of today—The official international atomic time scale (IAT) is maintained and issued by the International Time Bureau of France. The relationship between IAT and UTC is carefully monitored. Though the latter runs later than the former, the differences now are restricted to whole numbers of seconds.

Also UTC will be leaped, a full second at a time, in such a way that it will deviate by no more than 0.7 second from UT1, which is the navigator's time scale. One essential advantage of UT1 is that, though it is less uniform than UT2, it does fit the positions of the Sun and the stars as they are actually observed from various parts of the world. And in many places where it is impossible to receive broadcast signals of the atomic time scale or to refer to some nearby cesium-beam clock, observation of the Sun and the stars remains indispensable to practical navigation.

Thus, it is not true that atomic clocks have rendered celestial observations superfluous. The Earth's actual angular positions in space, however irregular, remain of major importance to travelers upon its surface. Thus we live in an era where diverse time scales are not only compiled and disseminated, but will continue to be useful for the foreseeable future.

Two kinds of questions about time—Leonardo da Vinci wrote in his private and secret notebooks five hundred years ago that the measurement of time involves both extension (duration) and position. "How long did it go on?" and "How often does it happen within a minute?" are questions of extension. "When did it start?" or "When did it explode?" are questions of position in time.

If time is visualized as a line extending from the remotest possible past to the present instant, then questions of duration involve measuring the length of a segment of that imaginary time line. Questions of position, on the other hand, require the locating of a point on the line. In English the confusion can be linguistic as well as conceptual. For example, "The miler's time has been announced as four minutes flat" versus "The time is now four o'clock." The former states a duration; the latter a point in time.

The technical or official name for a point in time is an *epoch*. This is an unfortunate usage. "Epoch" to most of us means an extensive period of time, not a point in time. The word *date* would be better to signify a point in time, even though it too has the second meaning of a numbered day (a duration) on the calendar. We prefer here, however, to use "date" to mean a *cut* in time, which separates all that happened before from all that happens afterward.

Synchronization is also an essential time concept. It involves the comparison and possibly the adjustment of two distinct clocks, A and B, in such a way that an observer of the information supplied by A can infer what must be the reading on B, and vice versa.

Dr. J. A. Barnes, head of the time and frequency division of the NBS Institute for Basic Standards, in Boulder, Colorado, has stressed that synchronizers need not be concerned with dates (points in time). So long as their respective clocks show the same point on the time line, it is not necessary that they know precisely what that point is. Practical examples are electronic navigation systems and the sophisticated devices for avoiding mid-air plane collisions. Synchronization of frequencies to a high degree of precision is essential for the color matching of TV programs between the East Coast and the West Coast of the United States.

Dr. Barnes has defined a time scale as "a system of assigning dates to events." Two different time scales will supply different dates (points in time) to the same events. A marker date or arbitrary zero point may be the birth of Jesus, the flight of the prophet Muhammad from Mecca, the beginning of the reign of some tyrant, or the start of a laboratory experiment. From such a marker date to a following event date the duration may be 1,974 years, or 18 microsecond.

Frequency controls for communications—Time signals and frequency patterns are widely broadcast by radio. In turn, radio and TV stations are totally dependent on precise frequency data. One example will suffice to show why. If an FM station has an assigned frequency of 100 MHz (10^8 oscillations per second), and its second deviates by only 1 part in 1,000 from the true atomic second, then its broadcast oscillations will be off pitch by 100,000 Hz.

Broadcasting stations operated by the Bureau of Standards offer varied and unwearying time-signal services—for example, the sequence of pulses at 1-second intervals transmitted from WWVB, Fort Collins, Colorado. The first day of the month, after listeners have been advised what is to take place, adjustments are made, if necessary, up to 0.5 s, thus holding this sequence of pulses to within 100 microsecond of agreement with the time scale called UT2. The data for these broadcasts comes all the way from the U.S. Naval Observatory on the East Coast.

UTC signals have been transmitted from WWVB, Fort Collins, and from WWVH, Kauai, Hawaii, with data from the Bureau

International de l'Heure (BIH), Paris. This is closely coordinated, within 10 microsecond, with data from the U.S. Naval Observatory, which maintains its own UTC scale.

Such time transmissions are not without troubles. In 1970 the National Bureau of Standards announced that its transmissions of 1-second pulses of UTC time were longer by 3 parts in 10^{10} than "one coordinate second." This tiny, but not negligible difference was attributed to an offset, or rate change, in the frequency of the carrier wave bringing the data from France to the United States. The ability to identify and estimate so tiny a deviation demonstrates the extremes of precision attained in today's time technology.

Time signals from Paris to the United States are now transmitted with the help of communications satellites high above the surface of the Earth. The sequence Paris–satellite–U.S. forms a huge inverted V. The pulse thus moves through a much greater distance than if it had followed close to the surface of the Earth all the way.

Atmospheric variations and disturbances, being mostly concentrated near the surface, tend to distort such satellite-forwarded time signals somewhat less than in the old, surface-following radio transmissions. However, fluctuations in density, moisture content, and other meteorological changes do distort the time signals by the time they have been received at the western end of the Atlantic Ocean. The accuracy demanded today of time signals is such that even these seemingly minor disturbances in the atmosphere do impose limits on accuracy.

Basically, time signals travel by means of the electromagnetic oscillations called radio. In contrast, twenty-two centuries ago in the days of the Second Temple at Jerusalem, before the present elaborate Jewish calendar had been developed, the new Moon was not predicted astronomically. Instead, the all-important beginning of another month depended on the observation of a bona fide new Moon, as certified by the Sanhedrin itself. When suitable witnesses had come forward to assure this powerful legislative-judicial body that the new Moon had been seen, the decree was made manifest throughout the little land by flares lighted on high and holy places. Those time signals too traveled at the speed of light from point to point.

Timekeeping attainments—Today the atomic-based second is by far the most reproducible unit in the entire International System of Units (SI).

When independent measurements of luminous intensity are made in units of the candela (cd), they may show as much as 1 percent, or 1 part in 100, discrepancy. When independent measurements of temperature are made in units of the kelvin (K), they may prove discrepant by 1 part in 10,000. Independent measurements of electric current in units of the ampere (A) may fail to agree within 1 part in 100,000. Such measurements of mass, in units of the kilogram (kg), should agree, at present, to about 1 part in 10 million; and independent measurements of length, in units of the meter (m), to within a few parts in a billion (10^9).

However, independent measurements of a given time duration, in units of the second (s), should now agree to within about 1 part in 1 million million (10^{12}), if the best available devices and procedures are used. Thus, if measurement A indicates an event or interaction occupied just 1 second, measurement B should not disagree with that by more than 1 or perhaps 2 picosecond.

Metrologists of time seem to be insatiable, for they aspire to attain even greater reproducibility—to within 1 part in 10^{13}. That corresponds to an error of 1 second in 300,000 years. Such quests for extreme accuracy are not born solely of compulsive perfectionism. They are needed in order to cope with present and planned space expeditions by manned or unmanned satellites or probes.

Distances from the surface of Earth to such man-made vehicles or to some bodies in our solar system can be measured by means of reflected radio waves (radar) or even, under some conditions, by reflected light waves (radarlike laser devices). When a radar pulse is aimed at the surface of the Moon and bounces back to Earth again, greatly weakened but still recognizable, the time of its round trip at the speed of light (c) gives the distance to the Moon. Such a round trip requires about 2.5 second, equivalent to a round trip distance of some 7.5×10^8 m, and indicating that the Moon is very nearly 3.8×10^8 meter distant.

If the precision of timing such a round trip is within 1 part in 10^6, the distance can be determined within 380 meter; but a precision of 1 part in 10^{10} permits the distance to be determined within 3.8 centimeter (less than 2 inches).

During the early 1970s the huge steerable dish antenna at Goldstone, Nevada, beamed signals to the unmanned Mariner VI when its distance from Earth was about 2½ times Earth's average distance from the Sun. Those signals triggered instant response signals from the Earth-directed radio aboard Mariner VI. The total elapsed time between the sending of the signal and the receipt of the responding signal was roughly 43 minutes.

Actually, this elapsed time was measured to within 0.2 microsecond. The uncertainty as to time and distance was thus as tiny as 1 part in 10 billion (10^{10}). The result was that the distance of Mariner VI from Earth could be established to within 30 meter.

The value of such precision was soon demonstrated in a memorable way. After a long series of such timings it was found by a team of scientists from the California Institute of Technology that when the radio signals from Earth to Mariner VI and back had to pass close to the Sun, there was a delay of about 204 microsecond in addition to the time corresponding to the normal speed of light. Such a delay had been predicted by Einstein's theory of general relativity, according to which electromagnetic waves (light, radio, etc.) move more slowly when they pass through regions of intense gravitational field, such as those near the Sun itself.

The velocity of light in empty space is well known: 299,790 km/s or 186,000 mi/s. But here, in the regions near the Sun, the velocity was reduced sufficiently that the round trips of the signals took as much time as if Mariner VI had been about 30,000 meter more distant from Earth than it was known to be. These discrepancies in timing, though tiny, were unmistakable; they were about 1,000 times as great as the least duration-difference that the elaborate Goldstone equipment could detect dependably.

The result was an impressive validation of at least one important part of general relativity theory—an actual experimental confirmation, though the "laboratory" equipment included a great steerable radar antenna in Nevada and a small space probe moving enormously far from the Earth on which it was made and launched.

Another timely test—In 1971 another extraordinary experiment was made, using atomic clocks to test Einstein's relativity theories. Two physicists, J. C. Hafele and R. E. Keating, circumnavigated the globe by air, first eastward, then westward. They bought tickets on commercial jet planes and their essential baggage consisted of four separate portable atomic clocks. This number was taken for the sake of safety and reliability. Even if one should stop and another prove to be grossly in error, the remaining two could provide the correct elapsed time.

Total elapsed times for both the eastward and the westward circuits were compared with the reference atomic clocks at the U.S. Naval Observatory. The results were strange, though not surprising to the physicists themselves. Going eastward, their four clocks had lost an average of 59 nanosecond (59×10^{-9} s). Three of the four clocks, in fact, had shown losses in the range between 51 and 57 nanosecond. The fourth, possibly less well adjusted than the others, showed a 74 nanosecond loss.

Going westward, on the other hand, the same four clocks had *gained* an average of 273 nanosecond. Three of the four showed gains in the range between 266 and 277 nanosecond. The fourth— the same deviator mentioned earlier—showed a 284 nanosecond gain.

Using the two averages of 59 ns loss in the eastward and 273 ns gain in the westward circumnavigation, we find the time discrepancy between the two opposite circuits was $273 - (-59) = 332$ ns. This is a substantial discrepancy, in no way explainable by accidents or chance variations.

Why should there have been such a discrepancy? And how could the physicists, in calculations made before their flights, have arrived at the expectation that a discrepancy of just about this size should result? The answers lie in Einstein's two great and related relativity theories: the special theory, launched in 1905; and the general theory, launched in 1916.

Special relativity includes an inescapable effect commonly called "kinematic time dilation." It means that an observer will find that a clock which is moving will seem to lose time compared with one that is stationary, from his point of view. The great round-the-world experiment actually compared three differently situated timekeeping systems, each of which had a different rate of motion as seen from the framework of fixed stars. First, the system at the U.S. Naval Observatory, which we shall call Clock O. It rotates with the Earth from west to east at about 1,200 kilometers per hour. Then the system being carried eastward in jet planes at about

800 km/hr, which we shall call Clock E. Its total velocity is thus 2,000 km/hr. Finally, the system being carried westward in jets also at about 800 km/hr, which we call Clock W. Its net velocity is about $1,200 - 800 = 400$ km/hr.

Clock E is in most rapid motion, so it loses time compared with Clock O. Clock W is in least rapid motion, so it gains time compared with Clock O. In theory, based on the jet plane speeds and the rate of the Earth's rotation, a net difference of 280 ns was expected between the E and the W timekeepers, because of this kinematic time-dilation effect.

However, there was a further source of deviation, based on the general theory of relativity, which says that physical events take longer—meaning that clocks run more slowly—in an intense gravitational field than in one less intense. Now, it is well known that gravitation is more intense at the surface of the Earth than at altitudes such as 8 or 10 km, where jet transports do most of their flying.

Thus, the atomic clocks were expected to run a little more rapidly when aloft than when on the landing strip of an airport. The physicists who shepherded the clocks used careful logs of the actual flight altitudes of the planes they had taken. From these records and the equations supplied by general relativity, they computed that their eastbound circuit should have caused a gain of about 144 ns, while their westbound circuit caused a gain of about 179 ns. This is the extent of the so-called "red-shift effect" attributable to variations in gravitational intensity during the actual flights.

When these computed red-shift effects were combined with the computed kinematic time-dilation effects, the predicted outcome was a loss of 40 ns in the eastward circuit, a gain of 275 ns in the westward circuit; and a resulting total discrepancy of 315 ns.

The observed discrepancy was in fact 332 ns, differing by less than $5\frac{1}{2}$ percent from that predicted. This deviation was not statistically significant. The results of this first experiment in theoretical physics carried on aboard commercial jets were a striking confirmation of basic assumptions of relativity theory. The project would have been impossible without the great accuracy of cesium-beam clocks—and the convenient portability of the models used.

In summing up the significance of these results, J. C. Hafele has stated, "Our flying clock experiments have proved conclusively that the atoms that regulated the [cesium] clocks had different frequencies at different altitudes." He meant by this that their frequencies became greater at higher altitudes because of the slight decrease there of the Earth's gravitational intensity.

Certainties about uncertainty—How small are the least intervals of time that now can be measured with accuracy and confidence? Let us see what happens as we have to deal with ever smaller intervals. At present the actual duration of a year can be measured to within about 1 microsecond (10^{-6} s). This means that the "relative uncertainty" of such substantial intervals of time is about 3 parts in 10^{14}. An hour can be measured to within 20 nanosecond. Thus the relative uncertainty of comparable time periods is about 2 parts in 10^{13}. But if the interval of time to be measured is as small as 1 ns and the uncertainty of the measurement is 0.1 ns, then the resulting relative uncertainty will be 1 in 10, or 10 percent. The absolute uncertainty sets limits on the least magnitudes that can be measured in a satisfactory way. But recent improvements in time-measuring techniques have extended those limits remarkably.

During the 1970s enormous advances were made in devices and methods for measuring so-called ultrafast and ultrashort phenomena in liquids and solids. The devices are elaborate and intricate adaptations of the laser, capable of generating pulses of light as short as 0.3 picosecond each (10^{-13} s). In principle, such laser systems may eventually permit resolutions of time as brief as 0.01 ps (10^{-14} s). That is a duration just long enough to allow light to move 3 one-millionths of a meter through empty space.

Reports of experiments performed with these superbrief laser pulses are couched in intriguing but esoteric terms. They include findings like the following: The optical phonon life in a calcite crystal is 22 ps when the crystal is at 100 K temperature, but only 8.5 ps when it is at room temperature, around 300 K. Ethyl alcohol has a population lifetime of 20 ps, but a dephasing time of only 0.25 ps. And so on.

We cannot here venture afield to explore optical phonons, population lifetimes, or dephasing times. But we must observe that by the early 1980s techniques may be at hand for measuring, or reliably approximating, durations of events occupying no more than a few femtosecond (each fs being 10^{-15} s).

Yet even if durations between 10^{-14} and 10^{-15} s become measurable, there will remain great realms made up of physical events still briefer by far. Are these ultra-ultrabrief events and interactions forever proof against time measurement or useful estimation?

The answer is definitely *no*. Texts and papers in modern science deal constantly and confidently with durations thousands and millions of times briefer than 10^{-14} or 10^{-15} s. Thus, we can read and believe that a sufficiently excited atomic nucleus will emit an X-ray photon within a period of about 10^{-16} s. Or we learn of newly identified subatomic particles, sometimes called "resonances," whose lifetimes prior to decay into still other particles are 10^{-20} s or even less.

However, such figures result from approaches differing from measurement in the ordinary sense of the word, as the following introduction seeks to show.

Increasingly in these realms of the ultrabrief and ultratiny we are compelled to employ the physical variable of *energy,* the ability to do work of various kinds, measured by the SI unit, the joule (J).

Energy is intimately linked with frequency, which is an inverse form of duration (events or cycles per second, not seconds per event or cycle). All electromagnetic radiations, including radio waves, light, X-rays, gamma rays, and the rest, can be dealt with as floods of small bits or bundles of energy, called quanta, or photons. The energy E of a photon equals its frequency F multiplied by h, which is Planck's constant of action. The relationship is summarized thus:

$$E = Fh$$

The frequency F is also related to the wavelength W of the radiation and to the velocity of light c, in this simple way:

$$F = \frac{c}{W}$$

Combining these two simple equations gives us another:

$$E = \frac{ch}{W}$$

The product of the two constants c and h is 1.986×10^{-25} joule meter. If we substitute for W the known wavelength of some electromagnetic radiation, we obtain the energy content E of the photons composing that radiation. For example, we find that ultraviolet light with wavelength of 0.41 micron (4.1×10^{-7} m) is made up of photons each with an energy content of 4.84×10^{-19} joule.

Spreads, smears, ranges and their uses—Actually, the wavelength of any radiation cannot be measured with complete sharpness. Neither can its frequency, nor the energy content of the photons which compose it. The best we can do in practice is to obtain a spread, or bandwidth, or smear, or uncertainty range, within which our measurements will fall.

What has all this to do with time? A great deal. Generally speaking, the longer the time within which we are able to observe a particular process or interaction, the more precisely and sharply we can define its energy. Conversely, the broader the smear, or range, of its energy, the briefer the length of time within which we have (presumably) been able to observe and measure it.

Thus has arisen a natural and remarkable link between energy deviations on the one hand and lifetimes of events or observation periods on the other. This relationship is one form of a general rule known by such names as "the uncertainty principle" or "the indeterminacy principle."

Let us make it somewhat less uncertain or vague. Suppose the energy of some particular process is found spread between a lower value E_L and a higher value E_H, and that the difference between them is ΔE, to be read as delta-E. The corresponding time interval is symbolized by ΔT, to be read as delta-T.

Now, in the elegant shorthand of mathematics, the resulting relationship takes this form:

$$\Delta E \cdot \Delta T \geqq \frac{h}{2\pi}$$

Between the lefthand side and the right stands a symbol which does not mean "is equal to." Instead it means "is either greater than or equal to," and it implies also "is not less than." The fraction on the right has as numerator the same Planck's constant of action, whose measured value is very nearly $6.626\ 2 \times 10^{-34}$ joule second, and as denominator twice the familiar ratio *pi,* or $2 \times 3.141\ 6 = 6.283\ 2$. This entire fraction is commonly symbolized by $\hbar$, which is read as "h-bar."

Essentially, this is the key with which ultrashort time intervals are unlocked today. It is handiest to use after it has been rearranged slightly, as follows:

$$\Delta T \gtreqqless \frac{\hbar}{\Delta E}$$

In words it means that the typical duration or lifetime is equal to or greater than, but never less than, h-bar divided by the observed energy spread, or smear. Its use is illustrated by the following recent example.

The short, happy life of rho prime—Early in 1973, Dr. George Yost, of the University of California, Berkeley, speaking for a team of Berkeley and Stanford scientists, announced the discovery of the *rho prime,* a particle with a lifetime of about 10^{-24} second. It is thus the shortest-lived subatomic particle thus far identified and described.

No actual rho prime track had been detected on the millions of bubble-chamber photographs studied. With such a short lifetime, the rho prime, even if traveling close to the velocity of light, could move only across the diameter of a proton!

Evidence for its existence came only from tracks revealing the daughter particles into which it had so swiftly decayed. The rho prime, indeed, is said to begin its decay before it is fully formed. High-energy physicists speak of such ultratransients as "resonances," hesitating to call them anything so concrete as "particles."

The rho prime was found to have a mass energy of about 1.5×10^{9} electron volts, equal to 2.4×10^{-10} joule. Yet this was only its average mass energy. Observed energies were spread or smeared over the range from 1.92×10^{-10} to 2.88×10^{-10} J, thus giving a spread or ΔE of 0.96×10^{-10} J. This spread is 40 percent of the average or central mass-energy value.

Using the relationship already stated, Dr. Yost and associates concluded that the lifetime of the rho prime resonance could be no less than h divided by ΔE, or 1.1×10^{-24} second, and there is reason to believe that it is, if more, not much more than that.

Such high-energy particles, which are not after all truly particles, become "curioser and curioser" as the elaborate experiments continue.

The author has felt moved to comment in a verse dedicated to Dr. Yost and associates, for his helpful sharing of insights, but in no way to be blamed on him:

SONG FOR A HIGH-ENERGY PHYSICIST

We seek, we study, and we stare
At particles that weren't quite there.

We cannot even find their trace
Or signature in any place.

They cannot pause to say hello;
They don't arrive before they go.

In fact, there's nothing we can see
Save where they spawned their progeny.

Thus, rho prime's always on its way.
Its mark is that it just can't stay.

Such particles are truly trying:
Before they're fully formed, they're dying.

Yet we describe this missing one
That's gone before it quite gets born.

We calculate its energy
Though it did not quite come to be.

Its "lifetime" we can estimate,
Despite not-quiteness of its state.

Ten to the minus twenty-four!
And scarce a minisecond more.

Envoi:

Ah, humans, would *you* like to be
Senile before your infancy?

And say with first and final breath,
"I've had no life before my death"?

III / The International System (SI) in Action

Matters of fact . . . are very stubborn things.
—MATTHEW TINDAL,
Will (1733)

*Our life is frittered away by details . . . Simplify,
simplify.*
—HENRY DAVID THOREAU,
Walden (1854)

Measurement has meaning only if we can transmit the information without ambiguity to others.
—RUSSELL FOX,
The Science of Science (1963)

. . . whether or not a thing is measurable is not something to be decided a priori *by thought alone, but something to be decided only by experiment.*
—RICHARD P. FEYNMAN,
The Feynman Lectures on Physics (1963)

17 LINKING LENGTHS WITH LIGHT

Arise, shine, for your light has come . . .
—Isaiah 60:1

I shall light a candle of understanding in your heart, which shall not be put out . . .
—2 Esdras 14:25 (Apocrypha)

From late in the eighteenth century until past the middle of the twentieth, the meter was always one particular bar of metal, suitably stored and protected at a selected site in France. The last of these metal, material meter bars is still preserved in honor in the vault at Sèvres, just outside Paris.

Formed from a dense alloy of platinum and iridium, it had an X-like cross section, so that when properly supported it would sag and distort as little as possible. Contrary to general belief, the bar itself was not just 1 meter from end to end. Rather it bore, incised or engraved at right angles to its length, two fine scratches, 1 meter apart. Measurements on this master meter were always made at 0° C, under carefully controlled conditions.

Precisely made and scrupulously protected copies of this one and only meter reposed also in the standards bureaus and laboratories of each nation adhering to the Treaty of the Meter. So, too, did careful duplicates of *the* kilogram mass, likewise enshrined in the Sèvres vault.

The United States received two copies of the meter bar, numbers 21 and 27. They were kept for some time at the old Office of Standard Weights and Measures (Treasury Department), the forerunner of the present large Bureau of Standards (Commerce Department).

After painstaking comparisons, the basic yard in the United States was defined legally as just 0.914 401 83 meter—a far more precise definition than the yard had before. In Britain, on the other hand, the comparison processes resulted in a definition of 0.914 399 2 meter for the British yard, a figure which was later simplified to just 0.914 4 meter. The U.S. yard thus became 1,830 nanometer (nm) longer than the British yard—but this difference is less than 2 millionths of 1 meter, so has little significance.

It was a sign of progress that both the United States and Britain defined their length measures in terms of the meter, in preference to constructing their own yard standards of platinum-iridium. Far more forward-looking, however, was the later definition of the British yard as just 1,420,212 wavelengths of the light identified with the red line in the spectrum of the metallic element cadmium.

Careful comparisons—From time to time the copies of the meter were brought back to Sèvres to determine whether metal creep or other internal alterations had caused deviation. These comparisons always included at least three meter bars. Had only one Sèvres bar been compared with the bar to be inspected, no certainty would have resulted as to which of the two had changed.

Methods were developed which were far more exact than measuring the distance between two scratches on a bar of metal. The eminent experimental physicist Albert A. Michelson (1852–1931) designed a series of extraordinarily precise optical interferometers for making such length comparisons. An interferometer compares distances traveled by two parts of the same beam of monochromatic light that have been split, sent on their separate paths, then reunited, with the resulting bands of light and dark (interference fringes) observed and counted by magnifying eyepieces.

Michelson first used mercury lamps, then cadmium lamps, for his light sources in these interferometers. Always the need was for still more uniform and coherent light sources. The modern answer to this need, the laser, did not then exist and was indeed undreamed of.

In 1945–50 the U.S. Bureau of Standards developed improved length standards using light from mercury-198, an isotope formed by transmuting atoms of gold in nuclear reactors. At the 10th General Conference of Weights and Measures in 1954, representatives from the United States were among those who supported study, and eventual adoption, of a meter defined solely in terms of light waves. By that time, precise measurements had revealed beyond doubt that metal bars, even when formed from the most suitable alloys and kept at uniform low temperatures, tended to shrink slightly during long periods. Old yard bars formed from bronze have been found, for example, to have lost 1 part in 10 million of length.

Light waves, however, never shrink, and the maxima and minima of interfering light, which we see as fringes, can be measured far more accurately than distances between lines scratched into metal bars.

After intensive study of monochromatic light emitted by lamps using different substances, the 11th General Conference of Weights and Measures, meeting in 1960, chose a reddish-orange radiation produced by atoms of krypton-86, and defined the meter as follows:

"The meter is the length equal to 1,650,763.73 wavelengths in vacuum of the radiation corresponding to the transition between the levels $2 p^{10}$ and $5 d^5$ of the krypton-86 atom."

The element krypton was discovered in 1898. It is a gas at ordinary temperatures, and its atomic number is 36. It is all around us, for it comprises about one part per million of the Earth's present atmosphere. Atmospheric krypton has an atomic weight of 83.8, and is actually a mixture of six different isotopes, ranging in weight from 78 to 86. Krypton-86, heaviest among these, forms nearly 60 percent of all the krypton in the air. Thus the meter is now defined in terms of the "majority" kind of krypton. The standard lamps must contain not more than 1 percent of isotopes other than the designated krypton-86.

Krypton's spectrum is composed mostly of brilliant yellow and green lines, each line corresponding to a particular energy transition of one of the 36 electrons that surround the krypton nucleus. The line chosen for the meter standard, however, is reddish orange in color, as noted.

The wavelength of that reddish-orange radiation is 6,057.802 11 angstrom (1 angstrom = 10^{-10} meter), in vacuum. Were the same radiation to be measured at standard air pressure and temperature it would be about 1.675 angstrom shorter, because light moves with slightly less velocity through the atmosphere than through a vacuum. Thus, if the krypton radiation were measured under normal atmospheric conditions, the designated number of wavelengths would equal 0.999 7 meter, not the exact, full meter.

If the natural atmospheric mixture of krypton isotopes were used in the lamp, rather than the 99 percent pure krypton-86, the resulting wavelength would be shortened to such an extent that nearly 460 additional wavelengths would be needed to equal just 1 meter.

Accuracy advances—The eighteenth-century definition of the meter as the ten-millionth part of the quadrant from the pole to the equator was accurate to about one part in 10,000, which is to about one tenth of a millimeter per meter. The French law of 1799 designating a measured rather than a theoretical meter standard improved the definition so that it was accurate to about one part in 100,000. Such a tenfold gain is commonly referred to as 1 order of magnitude.

In 1889, with the new meter bar installed at Sèvres, length comparisons could be made accurate to about 1 part in 1 million (10^6). Today's light-based meter standard permits measurement that is accurate to within 3 parts in 100 million (10^8), and under especially favorable conditions even to 1 part in 10^8.

The krypton-86 lamp does have some drawbacks. Several measurements must be made with it in order to define 1 meter accurately. It permits accurate length measurements within a range beginning at about 10^{-4} meter (0.1 mm) and extending up to about 0.4 meter. Its light is not sufficiently uniform (coherent) to measure a full meter in a single operation. In other words, it cannot provide clearly measurable fringes when two light paths in the interferometer differ by more than 80 centimeter, which suffices to define a 40 centimeter length.

Completely coherent light would remain uniform over an indefinitely long period of time. The light from the chosen krypton-86 radiation, however, changes even within 2.7 nanosecond (10^{-9} s), the time within which light travels 80 centimeter.

Gas lasers today can supply far more coherent light than the best krypton-86 lamp, because the process called stimulated, or induced, emission keeps closely in step the radiations from the atoms in the laser. During the first half of the 1970s, the U.S. Bureau of Standards used a helium-neon-gas laser to measure the meter in a single step, and lengths of tens of meters can now be measured in the same way.

For the time being, however, the krypton lamp rather than some form of the gas laser remains the basis for the meter. In spite of the far greater coherence of light emitted by a well-adjusted gas laser, the krypton lamps are more readily and accurately reproducible. That is, one such lamp differs from another less than (till now) one gas laser differs from another in terms of its performance.

Such problems very likely will be solved before long, and the laser then probably will replace the krypton lamp as the accepted international standard for the meter. To achieve highest possible accuracy, nearly total vacuum will be required over the entire path of the light in such laser length measurers. Problems of maintaining such vacuum conditions, rather than any lack of coherence in laser light, probably will set the limits for the total lengths that can be accurately measured in a single step.

Since the counting of millions of fringes is a difficult and deadly task for humans, electronic fringe-counting devices, able to tally myriads of the light-and-dark stripes swiftly and accurately, will doubtless become standard components of the future laser length-measuring machines.

18 MEANINGFUL LENGTHS MEASURED BY THE METER

Significant lengths, arranged as a descending ladder, form the following table. It begins with a length suggesting the uttermost reaches of the observable universe and diminishes to the least definable distance or distance equivalents of the subatomic realm. Entries in small capitals are SI multiples of the meter and some other length units.

The table shows an enormous range, 10^{43}, of known or definable lengths. The meter itself is close to the middle of this range.

The smallest of the listed lengths, called the Compton wavelength, is not the actual measure of a particle or wave, but indicates the extent of the change in the apparent wavelengths of high-energy radiations, such as X-rays or gamma rays, after they collide with the named structure, in this case the atomic nucleus of the element iron. The proton and the electron, being far less massive than the iron nucleus, have longer Compton wavelengths, respectively, as their entries show, near the bottom of the great ladder of comparative lengths.

Another very short measure is that listed for high-energy radiation, in the range of 1 GeV (10^9 electron volt). This wavelength is not based on direct measurement, as by an interferometer. Rather, it corresponds to an electromagnetic oscillation frequency of about 10^{24} hertz (cycles per second). Such a wavelength of

	APPROXIMATE LENGTH OR DISTANCE (meter)
Cosmic horizon (limit of the observable universe)	1 to 2 $\times$ 10^{26}
Most distant galaxies photographed by 200-in telescope (Mount Palomar)	3 $\times$ 10^{25}
1 MEGAPARSEC (Mpsc)	3.084 $\times$ 10^{22}
To Galaxy M-31 (Andromeda nebula)	1.4 $\times$ 10^{22}
Diameter, our galaxy	5 to 6 $\times$ 10^{20}
1 KILOPARSEC (kpsc)	3.084 $\times$ 10^{19}
1 LIGHT-CENTURY	9.463 $\times$ 10^{17}
To stars in the Great Dipper	5 $\times$ 10^{17}
To Sirius, a relatively near star	6 $\times$ 10^{16}
1 PARSEC (psc)	3.084 $\times$ 10^{16}
To Alpha Centauri, nearest star to our Sun	3 $\times$ 10^{16}
1 LIGHT-YEAR	9.463 $\times$ 10^{15}
From Sun to Pluto	5.9 $\times$ 10^{12}
1 TERAMETER (Tm)	10^{12}
From Sun to Earth	1.49 $\times$ 10^{11}
From Sun to Mercury	5.8 $\times$ 10^{10}
1 LIGHT-MINUTE	1.8 $\times$ 10^{10}
Diameter, Sun	1.391 $\times$ 10^{9}
1 GIGAMETER (Gm)	10^{9}
1 LIGHT-SECOND	2.997 9 $\times$ 10^{8}
Diameter, Jupiter	1.428 $\times$ 10^{8}
Polar diameter, Earth	1.271 4 $\times$ 10^{7}
New York City to Los Angeles	4 $\times$ 10^{6}
1 MEGAMETER (Mm)	10^{6}
1 KILOMETER (km)	10^{3}
Height, human being	1.8
1 METER (m)	1
Length, cat	7 $\times$ 10^{-1}
1 CENTIMETER (cm)	10^{-2}
Length, small insect	7 $\times$ 10^{-3}
1 MILLIMETER (mm)	10^{-3}
Diameter, human hair	8 $\times$ 10^{-5}
Lower limit, unaided human eye	4 $\times$ 10^{-5}
Red blood cell	7 $\times$ 10^{-6}
Small yeast cell	2 $\times$ 10^{-6}
Chromosome, inside a cell	10^{-6}

	APPROXIMATE LENGTH OR DISTANCE (meter)
1 MICRON (μ)	10^{-6}
Wavelength, red cadmium light	$6.438\ 469\ 6 \times 10^{-7}$
Wavelength, red line of krypton 86	$6.057\ 802 \times 10^{-7}$
Small bacterium	2×10^{-7}
Pore, fine filter	2.5×10^{-8}
Virus	10^{-8}
1 NANOMETER (nm)	10^{-9}
1 ANGSTROM	10^{-10}
Diameter, atoms	1 to 2×10^{-10}
Bohr radius, hydrogen atom	$5.291\ 77 \times 10^{-11}$
Compton wavelength, electron	$2.426\ 21 \times 10^{-12}$
1 PICOMETER (pm)	10^{-12}
1 SIEGBAHN UNIT (for measuring wavelengths in X-ray spectroscopy)	10^{-13}
Diameter, atomic nucleus	10^{-14}
"Classical" radius, electron	$2.817\ 77 \times 10^{-15}$
Compton wavelength, proton	$1.321\ 4 \times 10^{-15}$
1 FERMI (unit of nuclear distance) = 1 FEMTOMETER	10^{-15}
Wavelength, very high-energy photon (in the GeV energy range)	10^{-16}
Compton wavelength, atomic nucleus of iron 56	2×10^{-17}

10^{-16} meter seems suitable for "seeing" smallest subatomic particles, such as the electron. However, the same enormous energy that shrinks the wavelength equivalent makes the photon so powerful that it drives away or annihilates any electron that it strikes.

This is another aspect of the "uncertainty principle," referred to in an earlier chapter; the energies necessary to observe the behavior of subatomic particles are so great that they distort or alter the objects observed. Thus, complete data on one particle becomes impossible to obtain.

What are the tiniest lengths that can be measured? The answers here are analogous to those responding to the question: what are the tiniest time intervals that can be measured?

Using indirect strategies, scientists have been able to make estimates closely resembling measurements down to a fraction of 1 fermi (10^{-15} meter). In 1966 the Argonne National Laboratory in Illinois announced that the proton, a component of every atomic nucleus, has a sort of multilayered structure. Data had been gathered by bombarding "stationary" protons in a polyethylene target with swiftly moving protons in a high-energy beam. The 12.5 GeV accelerator ran continuously during more than a month, producing more than 10^{11} proton-to-proton collisions per second. The flying protons bounced away from the "stationary" ones in what physicists call elastic scattering.

Energies of groups of scattered protons were measured by scintillation counters after they had been separated by means of magnetic fields. Graphs drawn of the relative numbers of scattered protons at different energy levels showed two lines: the less-steep slope represented scattering by the innermost proton region or energy shell; the steeper represented the scattering by the outer proton region or energy shell.

In all, three decreasing levels of energy were found, reckoning from the interior of the proton toward its exterior. The innermost, region of most concentrated energy, was found to have a radius of 0.33 fermi. The middle region was found to have an outside radius of 0.5 fermi. And the outermost was found to have a radius of 0.9 fermi. The middle shell thus had a thickness of 0.17 fermi, and the outermost shell a thickness of 0.4 fermi.

This does not mean that all of a proton is packed into a diameter of twice 0.9 fermi, or 1.8 fermi. There may be regions of weak or diluted energy extending beyond such a limit, just as traces of the Earth's atmosphere extend far above the surface of the Earth itself.

A beam of particles with energy of 12.5 GeV corresponds to a wavelength of 0.1 fermi (10^{-16} meter). However, the enormous amount of data accumulated and treated statistically in this experiment enabled the physicists to report with confidence that they could discriminate distance differences as small as about 0.02 fermi or 2×10^{-17} meter, which happens to be just the magnitude assigned to the Compton wavelength of the iron nucleus, forming the very tiniest rung on our extended ladder of significant lengths.

In search of natural length units—Natural units of length can be derived from several of the most important universal constants developed in the course of scientific work. These are units which, presumably, could be derived identically by scientists who inhabit another planet, in some solar system enormously far from our own.

1. Max Planck gave the equation for one such unit, a natural or universal length measure, based on three great constants: h, the Planck constant of action; G, the gravitational constant, sometimes preceded by the name of Newton; and c, the velocity of light or other electromagnetic radiation in space:

$$\left(\frac{Gh}{c^3}\right)^{\frac{1}{2}} = 3.99 \times 10^{-35} \text{ meter}$$

2. If we replace G by m_e, the rest mass of an electron, we can derive another such natural or universal length unit:

$$\frac{h}{m_e c} = 2.43 \times 10^{-12} \text{ meter}$$

3. The combination G, c and e, the negative charge of the electron particle, permits still another natural length unit to be derived:

$$\frac{eG^{\frac{1}{2}}}{c^2} = 1.371 \times 10^{-34} \text{ meter}$$

4. If we substitute for G the Avogadro constant, symbolized by N_A, still another natural length unit can be derived:

$$e\left(\frac{N_A}{c}\right)^{\frac{1}{2}} = 2.113\,69 \times 10^{-5} \text{ meter}$$

5. Finally, using the constants e (the charge on the electron) and m_e (the mass of the electron) together with c (the velocity of light) allows us to derive still another natural length unit:

$$\frac{e^2}{m_e c} = 2.817\,77 \times 10^{-15} \text{ meter}$$

This last is the only one of these derivations once believed to have an actual counterpart in nature. It is still called "the classical radius" of the electron. Formerly, prior to the more revealing insights of quantum physics, it was often referred to as if the electron must have just such a tiny radius.

No real physical aspect of the electron is now associated with this very specific length. It is true that when light is scattered by free electrons, in the effect called *Thompson scattering,* the electrons may behave as though this were indeed their actual radius. However, quantum and relativity theories do not permit such literal interpretations to be taken at their face values.

6. Still another measure of length derived from basic physical constants has a name implying that it is a measure of a very common structure scattered widely in the universe. This is the "Bohr radius" of the hydrogen atom, named for the Danish atomic physicist Niels Bohr (1855–1962). It is symbolized by an a_0, the subscript zero indicating that at this radius the total (kinetic plus potential) energy of a hydrogen atom will be at a minimum. Any change from this radius, either one of expansion or contraction, would require addition of energy to the atom.

The formula for the Bohr radius uses h, m_e and e, as already defined:

$$a_0 = \frac{h^2}{4\pi^2 m_e e^2} = 5.291\ 77 \times 10^{-11} \text{ meter}$$

The Bohr radius is the distance between the proton, which forms the nucleus, and the single electron of the hydrogen atom. If that distance became larger than a_0, then the electron's potential energy (energy of position) with respect to the proton would increase more than its kinetic, or motion, energy would decrease. On the other hand, if the distance became smaller than a_0, then the electron's kinetic energy would increase more than its potential energy would decrease. (As a large-scale analogy, the Earth's kinetic energy results from its motion around the Sun, while its potential energy results from its "height" above the surface of the Sun. In this analogy we disregard the kinetic energy resulting from the Earth's spin on its axis.)

The principle of minimum energy in a stable structure such as the hydrogen atom accounts for the fact that atoms resist forces that tend to crush them or disrupt their structure. This principle helps to answer such homely but baffling questions as this: "If the floor is mostly empty space, why don't we fall through it?"

7. Another natural length grows out of behavior of hydrogen, simplest and most abundant of all the atoms making up the bodies in our universe. When a "resting" electron and proton, drawn together by the negative charge on the former and the positive

charge on the latter, fall into the first Bohr orbit—that is, combine to make a normal hydrogen atom—they emit a characteristic photon of radiation. This radiation has a well-known wavelength, related to the three constants in the Bohr radius and also to the constant c (velocity of light) in this way:

$$\frac{ch^3}{4\pi^2 m_e e^4} = \frac{a_0 hc}{e^2} = 9.117 \times 10^{-8} \text{ meter}$$

The wavelength of this electromagnetic radiation is more than 1,700 times longer than the Bohr radius, but is shorter than the wavelengths of visible light. It is, in fact, ultraviolet light. The common symbol for this particular radiation is λ_R, referred to in words as "lambda sub-R." The R stands for J. R. Rydberg (1854–1919), a Swedish physicist.

If we compare four peculiarly important lengths—the "classical" radius of the electron, the Compton wavelength of the electron, the Bohr radius of the hydrogen atom, and the Rydberg wavelength—we find them all to be derived in various ways from the fundamental physical constants c (velocity of light), e (charge on the electron), m_e (rest mass of the electron) and h (Planck's quantum of action).

All these significant lengths are closely related to the so-called "fine structure constant." This is a dimensionless number, which means that it, like the Mach number, remains the same regardless of what system of units of measurement may be used. The fine structure constant is equal to the ratio of (a) the velocity of the electron orbiting the proton of the hydrogen atom, to (b) the velocity of light in a vacuum. This ratio is 7.23×10^{-3}, which means that the electron moves around the hydrogen nucleus at 0.723 percent of the velocity of light moving through space.

The fine structure constant has a host of sophisticated uses in modern physics, beyond the scope of the present book. However, its close relationship to this group of natural length units, and to the constants from which they are derived mathematically, merits its mention here.

Despite their universal character, these assorted natural length units, as a group or separately, are not threatening to displace the meter as the basis for length measurements. Most of them are inconveniently tiny. For example, a man's height is a little less than 2 meter, but it is about 34,000,000,000 Bohr radii.

Nevertheless, it is important to know that the interactions observed and measured in the physical world can be so combined that they yield units of length that are not limited to Earth alone. The same interactions, similarly related, would yield identical natural length units on a planet circling a sun in some remote galaxy near the observable horizon of our universe.

On Earth, however, the overriding objectives of metrology are to attain maximum precision and reproducibility in the application of units to measurements. This being the case, the meter as now defined in terms of the light of krypton-86 is not likely to be displaced by one of the various natural length derivations that we have displayed here.

19 MATTERS OF MASS, WEIGHT, AND ACCELERATION

Sufficient to have stood, though free to fall . . .
—JOHN MILTON,
Paradise Lost (1667)

Why do ye fall so fast?
—ROBERT HERRICK,
Blossoms (1648)

Brute force without wisdom
Falls by its own weight.
—HORACE,
Odes III (1st cent. B.C.)

Everything rises but to fall.
—SALLUST,
Jugurtha (1st cent. B.C.)

They only fall that strive to move . . .
—OWEN MEREDITH,
LORD LYTTON,
The Wanderer

What has become of the kilogram, once the unit of weight in the metric system? It has been transformed from the metric unit of weight (force) into the unit of mass (amount of matter), an even more basic role in the world of measurements.

Its definition in French:

> *Le kilogramme (unité de masse) est la masse d'un cylindre spé-cial fait d'alliage de platine et d'iridium, qui est considéré comme*

le prototype international du kilogramme, et est conservé par les soins du Bureau International des Poids et Mesures dans une chambre forte à Sèvres, France.

In our unofficial English equivalent:

> The kilogram (unit of mass) is the mass of a particular cylinder, made of platinum-iridium alloy, which is considered to be the international prototype of the kilogram, and is preserved in the care of the International Bureau of Weights and Measures in a vault at Sèvres, France.

Unlike the meter and the second, the kilogram has no present referent in nature; it is an arbitrary amount of mass. Today the meter is no longer the distance between marks on a man-made bar, but a stated number of wavelengths of light emitted by krypton atoms, and the second of time is the duration of a stated number of oscillations of radiation by cesium atoms. It is not impossible that, even before the twentieth century ends, the kilogram also will be redefined, perhaps as a large multiple of the mass of the atom of some designated element. Meanwhile the prototype cylinder, stored under triple bell jars, remains *the* kilogram. Careful copies of it are kept not only at Sèvres, but also in the standards laboratories of the world. Copy number 20 is at the U.S. Bureau of Standards.

While we wait for the choice of a more constant mass standard, traces of microscopic film add infinitesimally to the mass of the kilogram, year by year, despite all the many precautions.

Birth of the first—The earliest kilogram, then designated as a measure of weight rather than mass, was derived, with difficulty, from the first meter of length. In 1795, it was decreed that the weight unit, then the gram, was to be the weight of a volume of pure H_2O "equal to the cube of one one-hundredth part of a meter, at the temperature of melting ice." (That the ice was to melt at normal or standard atmospheric pressure was understood.)

But this had to be modified. First, the cubic centimeter, less than a half inch on a side, is awkward to work with, so the volume was changed to a cubic decimeter (0.1 meter on a side).

Second, Fabbroni and Lefèvre Guinea, in their attempts to carry out the prescribed operations, found they could not cool water to less than 0.3° or 0.4° C (Celsius) without getting a little ice. This mixture would not do, for ice, being less dense than water at the same temperature, floats in the water. The weight of the designated volume of water-ice mixture would have been less than that of pure water.

Water, in fact, is densest not at the freezing point (0° C) but at 3.98° C. Consequently the careful metrologists proposed that the standard weight be determined with water at that maximum density temperature, or rather at just 4° C, which was as close as they were then able to measure it.

This proposal required, and received, approval of the Academy of Sciences. Work was resumed, but not until 1799 was the weight of just 1,000 cc of such water determined, and a metal prototype formed of that same weight. This metal piece became *the* kilogram —the very first—and was stored for safekeeping and periodic comparison in the Archives of the young Republic of France.

Actually, we know now, it weighed not just 1 kilogram but 1.000 028, which was not bad for a first effort. This error was later corrected, but a similar error affecting the liter (the metric unit of liquid capacity) has persisted to this day. The liter has a volume of 1,000.027 cubic centimeter, instead of exactly 1,000 cc, which it was intended to have. It is, in short, too large by 27 parts in 1 million.

From grave to gram—This first prototype of the metric weight standard was not called the kilogramme (the French spelling) when it came into being. Its original name was, in fact, the *kilograve,* the suffix -*grave* meaning "heavy," and linked obviously to the word "gravity." The name was changed after the adoption of the gram as a weight unit.

In 1889, this kilograve-kilogram of 1799 was replaced by a new metal cylinder shaped from a mixture of platinum and iridium, which was regarded as the metal alloy most resistant to wear and tarnish. The height and diameter of this dense cylinder were each about 3.84 centimeter, and its volume was 46.400 52 cubic centimeter. The alloy forming it had a density of 21.551 times that of water at 3.98° C. That density ratio, commonly called its specific gravity, is notably high.

It is this second prototype kilogram that still reigns supreme at Sèvres. Together with subsidiary copies, it is used in the exquisitely precise comparisons of mass, commonly called weighings. These comparisons can now be carried out with a margin of error no more than 1 part in 100 million. That is about the proportion of the weight of the ink in one comma of this book to the weight of the entire book.

Such precise comparisons are made on balance weighing machines, the modern form of the traditional scales or weighing beams. One scale or pan carries the prototype kilogram or one of its surrogates; the other carries the object whose mass is to be ascertained. After such a weighing, the scales are interchanged, and the process repeated: the mass which had been in the right-hand pan is now in the left, and vice versa. Every precaution is taken to guard against air movements, vibrations and temperature variations.

The kilogram is a mass whose magnitude is particularly good in terms of the relative accuracy attainable under such conditions. Comparisons of weight in the kilogram range can be carried out with accuracy of about 1 part in 100 million, as stated. Comparisons of weight in the range of 1 microgram (10^{-6} gram) can be accurate to about 1 part in 10, however. Those in the range of 1 milligram (10^{-3} gram) attain accuracy of up to about 7 parts in 1,000. And comparisons of weight in the range of 10,000 kilogram can be accurate to about 1 part in 100,000.

Weight comparisons of objects over 20,000 kilogram are performed more readily by flotation and displacement techniques than by balances or mechanical weighing machines. When a marine engineer estimates the weight of cargo loaded into a ship by measuring the apparent rise of the water line, he is using a flotation method.

What is weight?—Weight is a force, and force, according to the principle called Newton's second law, equals mass times acceleration. Where weight is concerned, the acceleration is that of gravitation acting on the mass of the body being weighed.

Gravitational effects on the surface of the Earth vary but little. Hence, we tend to think that the weight of an object is a permanent, unchanging characteristic of it, and so to confuse its weight with its mass. However, today almost everyone is aware that a human weighs only one sixth as much on the Moon as he does on the Earth and is often completely weightless during space travel. It is mass, not weight, that remains unaltered. Weight is a force that varies widely, according to the changes in gravitational or other acceleration that act on the mass.

Standing on the surface of Earth, we may not sense its gravitation as an acceleration. However, the skydiver knows it is just that, for his downward velocity keeps increasing. If it were not that the friction (drag) of the air finally sets a maximum to the speed of his fall, he would know it even better, or perhaps he would not be diving at all, since in that case his parachute would not operate to bring him to a safe landing.

The unit of acceleration is the meter per second per second (m/s^2). It can be negative, as when an automobile is braked (decelerated) to a stop. Acceleration is not limited to changes of speed in straight-line motion. Every motion in a curve—that is, motion with changing direction—results in acceleration, even when the speed along the curved path is unchanging. Thus the Earth constantly accelerates toward the Sun, because it orbits in a curve (ellipse). The Moon constantly accelerates toward the Earth. And so on.

An object on the rim of a wheel turning at a uniform rate may seem neither to accelerate nor decelerate. Actually, its path causes it constantly to accelerate toward the center. The extent of such acceleration, whether by an orbiting planet or an object on a turning wheel, can be determined if we know two magnitudes: D the distance from the object or planet to the center around which it turns, and T the time (duration) required for one complete turn:

$$\text{Acceleration} = 39.5 \, \frac{D}{T^2}$$

The constant factor 39.5 equals $4\pi^2$. If D is measured in meters and T in seconds, the result will be a magnitude in the SI acceleration unit of meter per second per second.

Uniform acceleration requires the steady, unchanging application of force to the mass being accelerated. Space vehicles, such as those in the Apollo program, are accelerated by forces that act for relatively few seconds. Then they coast onward, at unchanging (unaccelerated) velocity.

The average or typical acceleration of gravitation on the surface of the Earth supplies an acceleration unit, symbolized by g, useful in making comparisons with nongravitational accelerations. This Earth-bound acceleration unit has become common in our communication: "The racing driver had to stand an acceleration of 5 g." Or we may find expressed in the g unit the shock that a fine watch is supposed to be able to withstand when dropped or bumped.

In the prevalent SI units, the average g is roughly 9.8 meter per second per second, or in the obsolescent old units, 32 feet per second per second (32 ft/s^2). Far more precise is the standard or accepted value adopted by the International Committee on Weights and Measures: 9.806 65 m/s^2 (32.174 ft/s^2). This standard is arbitrary and somewhat unduly complicated, but it is thoroughly accepted and used wherever science and technology dictate procedures.

Actually, the gravitational acceleration measured from place to place on Earth varies to a marked extent from that arbitrary average or standard. These variations amount to as much as 1 or 2 percent. They became apparent first when experimenters compared the lengths of pendulums that would beat seconds—1 second to, 1 second fro—at different places on Earth.

Thus, we learn to distinguish two kinds of gravitational acceleration: the arbitrary, assumed standard, symbolized by either plain g or g_0 or even g_n (where n can be thought of as meaning "normal"); and g_L, the actual local gravitational acceleration, the L representing the all-important fact that it is the particular acceleration measured in *that* locality. The g_0 values of 9.806 65 m/s^2 is sometimes called the standard free fall. The falling, in such case, must be done in a vacuum, to prevent air resistance from distorting the velocity changes.

The g_L measurements made along the Equator tend to be about 100.3 percent of g_0, while those made at the North Pole are about 99.7 percent of g_0. The centrifugal force resulting from the Earth's rotation is most noticeable at the Equator and makes the difference between polar and equatorial acceleration somewhat less than it would be if the Earth turned less rapidly. The greater gravitational intensity at sea level on the Equator than at sea level at or near the pole is caused by the Earth's greater thickness in the plane of the Equator than from North to South Pole.

Altitude is a more important factor in producing variations of local gravitational acceleration. Thus, in a latitude where g_L happens to equal g_0 at sea level, the g_L at 500-meter altitude will be about 1.5 percent less, and at 1,000-meter altitude about 3 percent less, and so on.

Geophysicists and geodesists constantly use a special unit of acceleration called the *gal*, equal to 1 centimeter per second per second, or to 0.01 m/s². It is named after Galileo, and its submultiple, the milligal (0.000 01 m/s²), is probably even more widely used than the gal itself.

Here are two actual measurements of g_L: Karajak Glacier, Greenland: 9.825 34 m/s²; and Batavia, Java: 9.781 78 m/s². The difference between them is 4,356 milligal, 4.356 gal, or about half a percent. If a spring or pendulum scale says at Batavia that a piece of metal weighs just 1 kilogram, it will report a weight of 1.005 kg on the Karajak Glacier.

The same scale would weigh that piece of metal at only 0.17 kg on the Moon, at about 0.4 kg on Mars, and at a heavy 2.7 kg on Jupiter, most massive of the planets. Yet in each case the mass of that piece of metal would have been the same. The differences would be those of the force (weight) with which it was attracted to the great mass of the planet or body below it.

The familiar kilo of force—Everyday use in metric countries continues to link the kilogram with weight rather than mass. Thus, the housewife who buys "a kilo of bread" is thinking of its weight, then and there. This reflects the practice of an older system of weights and measures called the Metric Technical System, and based on the meter of length, the kilogram of force (weight), and the second of time. The metric system proper—that is, the International System of Units (SI)—is a meter-kilogram (mass)-second system. The two kinds of kilogram can be abbreviated as kgf and kgm to distinguish the force (weight) from the mass kind of kilogram. In science and much of modern technology, kg alone is used to mean kgm.

Similarly, the everyday U.S. system based on the foot of length, the pound of force (weight), and the second of time can be contrasted with a foot-pound (mass)-second system. The abbreviations for the avoirdupois pound of force and of mass are, respectively, lbf and lbm.

Systems that are based on force or gravitational units rather than mass units must be supplemented by a separate unit of mass designed to go with the basic unit of weight. In the Metric Technical System, this added mass unit is the *metric slug, or hyl* (9.806-65 kgm). In the foot-pound(force)-second system of the United States, this mass unit is the *slug* (32.174 lbm). In both cases the coefficients in the parentheses are the equivalent values of the standard gravitational acceleration, g_0.

In terms of behavior, the metric slug is the mass which is accelerated 1 m/s² by a force of 1 kgf, and the U.S. slug is the mass that is accelerated 1 ft/s² by a force of 1 lbf.

On the other hand, in the systems based on mass rather than force, there will be a subordinate or derived unit of force. In the SI it is the newton (N), the force that will accelerate a mass of 1 kgm by 1 m/s². A kilogram of force (kgf), sometimes called a kilopond, equals 9.806 65 newton.

In the foot-pound (mass)-second system, called by various names, including the British absolute system, the unit of force is the poundal, equal to 32.174 lbf.

In a variant of the metric system once used in Germany, the kilogram mass was used, but instead of the present newton the unit of force was the kilopond. A more important variant, only recently superseded and still displayed in important textbooks and references, was the CGS (centimeter-gram-second) system, in which the gram of mass was basic, along with the centimeter of length and the second of time. In this CGS, the force that accelerates 1 gram mass by 1 centimeter per second per second was called the *dyne;* and 1 newton = 100,000 dyne.

The energy unit in the CGS was the *erg,* equal to 1 dyne (force) acting through a distance of 1 centimeter. The joule, unit of energy in the SI, is equal to 1 newton (force) acting through a distance of 1 meter. The result: 1 joule = 10 million (10^7) ergs.

Still another system, now an obsolete curiosity of metric history, was known as MTS: meter-tonne-second. That tonne, or ton, was the metric ton of mass, not weight, equal to 1,000 kgm or about 2,205 lbm. The derived unit of force in this system was known as the *sten* or *sthène* (sn), and was equal to 1,000 newton.

Concerning conversions—What about the conversion of present U.S. force (weight) units into SI force units? It may not be realistic to suggest that when the United States finally goes metric, as it must, we shall buy meat, fish, vegetables and other staples by the newton rather than by the kgf (kilogram force) or that we shall state our bodily weight in newtons.

So long as we live and move on Earth, the problem is no great one, since the weight of any one mass does not change much when it is shifted from place to place on this planet. It should be recalled, however, that these everyday kilogram weights (forces) are actually approximations of kilogram masses. If the dietary program of astronauts called for them to consume "a kilo of bread" during some period of lunar exploration, this would mean the same amount (mass), not the same weight, that would be called "a kilo of bread" back home on Earth. Actually, 1 kilo of weight on the Moon would be supplied only by six such kilo loaves from Earth!

A weight-conscious space traveler might be deluded for a time into believing he had shed five-sixths of his avoirdupois when he stepped on the soil of Luna. However, return to the environment of Earth would rid him of this misunderstanding. In a very literal sense, mass is the substance, weight (force) the shadow or variable interpretation of that substance.

Science first pointed the way to escape the traditional confusions between mass and weight. One thoughtful commentator, Eduard Emblick, has urged that all technology take the same course: "The basis of all technical knowledge is the science of physics . . . it seems logical that the kilogram in technology should mean a mass too, and not the attraction caused by another mass, for example a planet."

Our weight (force) measures are inevitably earthbound. The mass measures, underlying advanced measurement systems such as the SI, are a long step nearer to being universal, and so better suited to this amazing and threatening era.

20 MINIMAL MASS UNITS AND THE INSIGHTS OF MASS-ENERGY ("MASSERGY")

Energy is eternal delight.
—WILLIAM BLAKE,
The Marriage of Heaven and Hell (1791)

O amazement of things—even the least particle!
—WALT WHITMAN,
Song at Sunset

Chapter 18 opened with a list of significant lengths in descending order, from cosmic to subatomic. Here is a similar table of significant *masses* in kilograms. Only those objects that are measured on the surface of the Earth will have the same numerical *weight* in kilograms (kgf), and even these equalities are only approximate, for they vary slightly with latitude and altitude of the site at which weight measurement is made on Earth.

The largest mass listed is about 10^{75} times as great as the smallest. That last, least item—the elusive neutrino particle—has a mass which is at best uncertain, and may even be zero. The mass of the electron, on the other hand, is solidly established.

One greatest-possible mass not included on the list is the ultimate: the mass of everything in the observable universe. This must be, for many profound reasons, too uncertain to fix with certainty, but with that twentieth-century device, the "guesstimate," we are tempted to try: 10^{52} to 10^{54} kilogram, including not only galaxies, stars and lesser objects, but even extensive masses of dust and wandering molecules, atoms and particles in the depths of space.

	APPROXIMATE MASS (kilograms)
Large galaxy	10^{42}
Medium-sized galaxy	10^{41}
Moderate-sized galaxy such as M-82	10^{40}
Sun	2×10^{30}
Jupiter	1.88×10^{27}
Earth	5.976×10^{24}
Mars	6.4×10^{23}
Moon	7.37×10^{22}
Earth's atmosphere (air)	5×10^{18}
All human beings on Earth	2×10^{12}
Waterborne commerce, U.S., typical year	10^{12}
Tanker, loaded	10^{9}
Passenger liner	10^{8}
Whale	10^{4}
Elephant	10^{3}
Large man	10^{2}
Dog	10
Rat	1
Insect	10^{-3} to 10^{-4}
Bacterium	10^{-13} to 10^{-14}
Virus	10^{-21}
Molecule, complex biochemical compound	10^{-22}
Molecule, chlorophyll *a*	1.45×10^{-24}
Molecule, water	2.99×10^{-26}
Atom, carbon 12	$1.992\,396 \times 10^{-26}$
N* 3245, nuclear particle (mean life about 10^{-22} s)	5.9×10^{-27}
X$^-$ particle (mean life 10^{-10} s)	2.35×10^{-27}
Neutron (long-lived, neutral particle)	$1.674\,82 \times 10^{-27}$
Proton (stable, positively charged particle)	$1.672\,614 \times 10^{-27}$
International atomic mass unit (amu)	$1.660\,33 \times 10^{-27}$
Mu$^-$ particle (mean life 2×10^{-6} s)	1.9×10^{-28}
K$^+$ meson (mean life 10^{-8} s)	9.8×10^{-29}
Electron (stable, negatively charged particle)	$9.109\,558 \times 10^{-31}$
Neutrino (stable, electrically neutral particle)	5×10^{-34}

Masses of particular particles—The measured mass of the neutron, the chargeless nuclear particle, is $1.674\,82 \times 10^{-27}$ kg; that of the proton, the positively charged particle found in every atomic nucleus, is 0.13 percent less; and the mass of the negatively charged electron, lightest of atomic components, is only about $\frac{1}{1,840}$ that of the proton.

These masses of subatomic particles are so much smaller than the kilogram (kgm) that metrologists devised an atomic mass unit, abbreviated amu. Unfortunately three competing values may be encountered, and one must be certain which of the three is being used:

$$1 \text{ amu (international)} = 1.660\ 33 \times 10^{-27} \text{ kg}$$
$$1 \text{ amu (physical)} = 1.659\ 81 \times 10^{-27} \text{ kg}$$
$$1 \text{ amu (chemical)} = 1.660\ 26 \times 10^{-27} \text{ kg}$$

Each of these units is close to an atomic mass (often called an atomic weight) of 1, and thus near to the mass of a proton or a neutron.

The amu (international), now generally adopted, is defined as $\frac{1}{12}$ the mass of the principal isotope of carbon, designated as carbon 12 or sometimes abbreviated as ^{12}C. Its nucleus contains 6 protons and 6 neutrons, with 6 electrons moving around the nucleus. A fuller abbreviation for this carbon isotope is $^{12}_{6}C$, where the 12 stands for the A, or mass, number, popularly known as the atomic weight; while the 6 represents the Z, or atomic, number, equal to the number of protons in the nucleus. The amu (international) unit is sometimes called the dalton, after the British chemist and theorist John Dalton (1766–1844).

Two carbon isotopes are found in nature, and four others, all radioactive, have been produced in laboratories and reactors. Their half lives range all the way from 20 second to about 6,000 years. Carbon 12, however, makes up nearly 99 percent of the carbon found on Earth.

The amu (physical), instead of being $\frac{1}{12}$ the mass of carbon 12, is $\frac{1}{16}$ the mass of oxygen 16, which constitutes 99.8 percent of the oxygen found on Earth. The full abbreviation for this isotope is now $^{16}_{8}O$, which shows that its nucleus contains 8 protons and 8 neutrons, surrounded by 8 electrons outside the nucleus. Minor quantities of $^{17}_{8}O$ and $^{18}_{8}O$ are found in nature also; and there are artificial, short-lived isotopes designated as $^{14}_{8}O$, $^{15}_{8}O$, and $^{19}_{8}O$. The amu (physical) is 0.03 percent less than the amu (international). They are sometimes referred to simply as physical mass unit and international mass unit, respectively.

The third member of the trio, the amu (chemical), is only 0.004 percent less than the amu (international) and is also based on oxygen, but it is heavier, since it is $\frac{1}{16}$ of the average mass of a mixture of the oxygen isotopes with the superscripts 16, 17 or 18. They are found on Earth in the relative proportions 2,024:4:1, respectively.

Quite different from the atomic mass unit (amu) whose variants we have presented above, is the *atomic unit of mass,* which means the rest mass of the electron, and thus is about $\frac{1}{1,840}$ amu, or 9.11×10^{-31} kg. The atomic unit of mass, or electron rest mass, is symbolized by m_e, which can be read literally as "m sub-e" or interpreted to mean "rest mass of the electron."

It is one of the natural constants from which a complete system, the Hartree atomic unit system, has been derived. The British physicist D. R. Hartree (1897–1958) devised the system, using, besides m_e, also e, the electrical charge (negative) of the electron, and $\hbar$ (bar-h), which—as noted before—is the Planck constant of action divided by 2π ($h/2\pi = \hbar$).

The Hartree unit system has been useful in simplifying computations and concepts in some branches of atomic and particle physics, but is not to be considered a competitor to the metric (SI) system for general technological and industrial uses.

The Hartree mass unit, being equated to the electron rest mass, equals about 9.11×10^{-31} kilogram, an extremely tiny magnitude indeed. The length unit is identical with the Bohr radius, a_o, as previously presented in this book, and can be expressed in this formula:

$$1 \text{ Hartree length} = \frac{\hbar^2}{m_e e^2} = 5.292 \times 10^{-11} \text{ meter}$$

The Hartree time measure takes the form:

$$1 \text{ Hartree time unit} = \frac{\hbar^3}{e^4 m_e} = 2.419 \times 10^{-17} \text{ second}$$

Best-known of the units in this atomic system is the Hartree energy unit, often called simply "the hartree," in honor of its formulator:

$$1 \text{ hartree} = \frac{m_e e^4}{\hbar^2} = 4.360 \times 10^{-18} \text{ joule}$$

The Hartree units are obviously extremely small, relative to units in the SI metric system, as befits a scheme devised to simplify the labors and thinking of atomic physicists. Hartree's proposals followed a direction pioneered by Max Planck, early in the present century. In Chapter 18 we displayed Planck's formula for a natural or universal unit of length, derived from his quantum of action, h (not $\hbar$); from c, the velocity of light in space; and from G, the Newtonian gravitational constant:

$$\text{The Planck length unit} = \left(\frac{Gh}{c^3}\right)^{\frac{1}{2}} = 3.99 \times 10^{-35} \text{ meter}$$

The corresponding formula for a natural unit of mass from the same trio of constants:

$$\text{The Planck mass unit} = \left(\frac{ch}{G}\right)^{\frac{1}{2}} = 5.43 \times 10^{-8} \text{ kilogram}$$

And for the corresponding unit of time:

$$\text{The Planck time unit} = \left(\frac{Gh}{c^5}\right)^{\frac{1}{2}} = 1.33 \times 10^{-43} \text{ second}$$

The Planck pattern, on the other hand, provides a natural unit of energy as enormous as the foregoing units are tiny:

$$\text{The Planck energy unit} = \left(\frac{c^5 h}{G}\right)^{\frac{1}{2}} = 2.6 \times 10^9 \text{ joule}$$

Natural or universal unit systems such as those derived by Planck, Hartree and others have an undeniable fascination. However, for everyday needs of measurement there is little or no reason to prefer such natural units to arbitrarily selected ones, such as those now built into the great SI system.

The day, the month, and the year—astronomical natural units rather than atomic ones—retain importance and will continue to do so. The unit systems derived from various physical constants, though instructive and in many ways revealing, are more likely to remain among the curiosities of metrology for most of us than to attain the status of essentials for practical measurements.

Measuring masses in molecules—A basic measure of mass, especially in chemistry, is the mole, or gram mole (abbreviated mol), formerly called the gram molecule (gmol). A mole of a substance is the mass in gram equal to the molecular mass (or molecular "weight") of the substance, as based on the amu (international). Thus 1 mole of carbon 12 is 12 gram; and 1 mole of water (H_2O) is $1 + 1 + 16 = 18$ gram.

The mole can be defined also as the mass of a substance which contains just as many molecules (or atoms of an element) as there are atoms in 12 grams of carbon 12.

A kilomole (1,000 mole) is the mass in kilogram equal to the molecular mass of a substance, or the mass that contains as many molecules of a substance as there are atoms in 12 kilograms of carbon 12. The abbreviation of kilomole is kilomol or kmol. The abbreviation of its old name of kilogram-molecule was kgmo.

The mole, almost beyond doubt, will be adopted widely as a basic unit of the SI, a position well merited by its usefulness.

The Avogadro constant, symbolized by N_A, is an eponymous memorial to the Italian physicist Amadeo Avogadro (1776–1856), whose essay on "A Way to Determine the Relative Masses of the Elementary Molecules of Bodies . . ." proposed the remarkable theory (Avogadro's law) that under uniform temperature and pressure equal volumes of all gases contain the same number of molecules—or of single atoms if the gas is composed of them rather than of molecules.

N_A is $6.022\ 169 \times 10^{23}$ mol^{-1} (that is, molecules or atoms per mole), or $6.022\ 169 \times 10^{26}$ kmol^{-1} (molecules or atoms per kilomole). Thus, for example, there are $6.022\ 169 \times 10^{23}$ molecules in 32 gram of oxygen gas (O_2).

Possibly Avogadro's law could be used to define the kilogram mass in terms of a volume of gas rather than a metal cylinder. Or progress in solid-state physics and crystallography may someday permit a really precise measure based on a solid substance. Such precision will be attainable only by actually counting atoms or molecules, because these are the most uniform, unvarying objects known.

Mass and energy: Siamese twins—Every consideration of mass must now take account of Einstein's triumphant relativity theories and, thus, of the equivalence and interconvertability of mass (M) and energy (E) expressed in the famous equation $E = Mc^2$.

Here again c is the velocity of light in empty space (2.997 925 × 10^8 m/s). Thus, when M is taken as the SI unit of mass (1 kilogram), E is the large number 8.987 554 3 × 10^{16} joule. The joule, SI unit of energy, is the work done when 1 newton of force produces a displacement or motion of 1 meter in the direction of the force. In still more basic units, 1 joule = 1 kilogram times meter-squared per second-squared (kg m^2 s^{-2}), or, in another order which means the same thing, 1 meter-squared times kilogram per second-squared (m^2 kg s^{-2}).

The gigantic coefficient resulting from squaring the speed of light reveals that what seem to be enormous amounts of energy can be obtained, by means of a nuclear bomb or reactor, from the conversion to energy of a tiny amount of mass.

The same disproportion seems to prevail in the opposite direction. If all the energies moving through the universe in the form of radiations, the heat of the stars, the kinetic energies of the stars and planets, and so on, were converted into mass, this would add not more than about 10 percent, it is believed, to the existing mass we observe now in the form of matter.

Not only can mass be converted to energy, and vice versa, but mass is best conceived as congealed energy, while energy should be regarded as a mobile aspect of mass. The dominant characteristic of matter (mass) seems, to most of us, to be its inertia. Yet every bit of energy is linked also with inertia, just like every bit of mass. Every bit of energy produces gravitational attraction and responds to gravitational attraction, just as does every bit of mass.

One obstacle to an earlier discovery of the mass-energy equation first enunciated by Einstein was the principle of the conservation of mass; another was the principle of the conservation of energy. The first said that in every interaction, the final amount of mass was no more and no less than the initial amount; mass could be neither created nor destroyed. The second said the same things about energy; only the form of the energy could be altered (heat to electricity, electricity to motion, and so on), but the total amount of energy could not be changed.

Both of these influential conservation laws are wrong, when considered separately; but in combination they embody a most basic insight: what is always and everywhere conserved is mass-energy or energy-mass. We will do well to stress this indissoluble Siamese-twin relationship by using a single word, *massergy*.

In every interaction the sum of the active energy plus the con-gealed energy (or mass) remains the same. Massergy *is* always and everywhere conserved.

Betrand Russell wrote, in *Human Knowledge, Its Scopes and Limitations* (New York: Simon and Schuster, 1948): "Mass is only a form of energy, and there is no reason why matter should not be dissolved into other forms of energy. It is energy, not matter, that is fundamental in physics."

In this sense, the kilogram mass is a congealed form of 90 quadrillion joule of energy.

Objectively, it can be added, there is no reason why an atomic-bomb explosion must be regarded as a "vast" release of energy from a "small" amount of matter. The energy is vast in terms of its power to disrupt and destroy living creatures, like ourselves, and the environment we must have in order to survive. However, in numerical terms, which we are entitled to single out in analyzing measurement methods or meanings, the relationship could be shifted rather simply.

Suppose that our basic unit of length was not the meter but a length of 30 kilometer; and that our basic unit of time was not 1 second but just $\frac{1}{10,000}$ second. Light travels at very nearly 300,000 km/s in a vacuum. Hence, with these different units the velocity of light would equal just 1 distance unit per time unit. Einstein's equation would still be $E = Mc^2$, but the numerical value of c^2 would be 1.

We have not, however, supposed any change in the unit of mass. The new form of the equation would amount to just $E = M$, with M measured in kilogram as now, but E (energy) in what could be called ultrajoules, each equal to about 9×10^{16} present joule (or $9 \times 10,000$ terajoule). Thus the equivalent of 1 kilo-gram of matter would be just 1 ultrajoule of energy.

It is our human physiology which persuades us that a lopsidedly large amount of energy corresponds to a disappearingly tiny amount of mass. We can lift 1 kilogram, even toss it about. We may be able to afford kilogram quantities of bread and other foodstuffs. But 1 "ultrajoule," in our hypothetical case, is the energy released by a 20-megaton H-bomb. Such energy not only can lift and toss us, but tear us to bits and vaporize the fragments.

At the minimum end of the scale too, we are physically equipped to sense certain energy changes far more easily than changes in mass. Healthy human eyes can respond to a mere half dozen photons of visible light. Thus we receive a visual sensation from about 2×10^{-18} joule of light (electromagnetic) energy. The $E = Mc^2$ equation tells us that this is equivalent to about 2.2×10^{-35} kilogram of mass ("congealed energy").

However, the smallest change in mass that our muscles can detect is about 0.1 gram, or 10^{-4} kilogram. This is about 4×10^{30} times less sensitive than our visual response to light energy.

The science historian Max Jammers, to whom we owe these bold comparisons and contrasts, has estimated that human sensory equipment is 10^{32} times "more sensitive to energy perceptions than it is to mass perceptions."

The equivalence of energy and mass, or matter, "would have been an obvious fact," if only human senses could have been as responsive to mass perceptions as to energy perceptions. Under such imagined conditions, "the human eye, perceiving light from the Sun, would then also feel the impact of the photons" of which that light was composed. Such perceptions, however, have been denied to us.

Thus, man—meaning both women and men, of course—remains the measure as well as the measurer of all things.

21 MASSERGIES, MOSTLY VERY LARGE

*The baby figure of the giant mass
Of things to come at large.*
—SHAKESPEARE,
Troilus and Cressida

*. . . every time our experimental technique has taken a
leap forward we have found things totally unexpected and
wholly unimagined before.*
—HERMAN BONDI,
Assumption and Myth in Physical Theory
(1967)

Energy has become the currency of modern metrology, the gold standard in terms of which we must evaluate and relate measurements of other physical variables—length, time, and even mass.

Scientific advance makes possible ever more precise measurements, and these in turn contribute to rapid gains in science and technology, as does the development of new units and new methods for the measurement of physical variables and effects.

Aristotle and Archimedes were unable to distinguish between weight (force) and mass. Leonardo da Vinci grappled brilliantly but unsuccessfully with the concept of free fall, but he could not succeed without an understanding of the pattern of motion we call uniform acceleration (that is, velocity changing at a constant rate).

Less than two centuries ago, work and heat were measured in different units and not recognized as two forms of a broadly conceived energy relationship. Heat was long attributed, even by experts and specialists, to a weightless form of matter called caloric, which constantly flowed into and out of bodies, making them hotter or colder.

Only after the epochal insights of Einstein could metrologists begin to apply the concept of massergy, or the complete equivalence of energy and mass. In this chapter, by way of developing that concept in a most meaningful way, we examine the massergy of the Sun and the Sun-Earth relationship.

Energy in the SI system is measured by the joule (J), which equals the work done by a unit force (1 newton) acting through a unit distance (1 meter): 1 N × 1 m = 1 J. (By convention the units named after scientists and mathematicians are capitalized when abbreviated, even though the full names of the units themselves are not.) The unit of power, the watt (W), is equal to 1 joule per second.

Careful measurements of solar radiation received on Earth and by space probes above most of Earth's atmosphere reveal that the Sun must constantly pour out very nearly 4×10^{26} watt of power in the forms of light, heat (infrared), ultraviolet, X-rays, and other electromagnetic radiations.

Of this the Earth intercepts only some 1.73×10^{17} watt and retains about 1.21×10^{17} watt, for about 30 percent of the amount is promptly reflected back into space as earthlight.

The Sun is a normal and stable sort of small star. It is able to continue its vast outpourings of energy only because of thermonuclear reactions, or "burning," deep within its core. There, high pressures and temperatures bring about the incessant fusion of lighter atomic nuclei into heavier ones. *Lighter,* of course, means "less massive" here, and *heavier* means "more massive."

In these nuclear fusions, however, the resulting nuclei have slightly less total mass than did the original nuclei from which they were formed in a chain of rather complex interactions. The process is accompanied by a "mass deficiency," and in keeping with the principle of the conservation of massergy such a deficiency or disappearance of mass must be accompanied by the appearance of a corresponding amount of energy.

Technically speaking, the principal fusion process in the Sun's interior results in the replacement of 4 nuclei of hydrogen-1 by 1 nucleus of helium-4. That new helium nucleus has a mass about 0.7 percent (7 parts in 1,000) less than the combined masses of the 4 hydrogen nuclei. This mass defect is the source of most of the energy now flooding through the body of the Sun and radiating into space from its surface.

Other fusion processes are known to be increasingly important in stars that have passed their primes—that is, stars that have exhausted most of their available hydrogen "fuel." Thus, 3 nuclei of helium-4 are replaced by 1 nucleus of carbon-12. One nucleus of carbon-12 and 1 of helium-4 are replaced by 1 nucleus of oxygen-16. One nucleus of oxygen-16 and 1 of helium-4 are replaced by 1 nucleus of neon-20. And so on. Such fusions into ever more massive atomic nuclei can proceed, under conditions of vastly increased interior stellar temperatures, all the way to the end of the line, which is the formation of nuclei of iron-56. Thermonuclear fusions beyond that stage require the net input of energy from the outside; they do not emit more energy than they absorb.

These transformations, all the way from helium-4 to iron-56, are far less efficient in transforming mass into energy than is the first great jump from hydrogen nuclei to helium nuclei. The entire sequence from helium-4 to iron-56 releases as energy about 0.1 percent (1 part in 1,000) of the initial mass. This is only one seventh of the comparable energy release from hydrogen to helium.

If we consider the entire chain from end to end, with 1 nucleus of iron-56 replacing 56 nuclei of hydrogen-1, we find that the former is 0.8 percent less massive than the total of the latter, but seven-eighths of this loss of mass took place in the first link, or step, from hydrogen to helium.

Mass-loss by the Sun—Since the Sun each second pours out a total energy of nearly 4×10^{26} joule, and since each 9×10^{16} joule corresponds to the vanishing of 1 kilogram of mass, the Sun must be losing mass at about 4.4×10^9 kilogram per second. (The table in Chapter 20 shows that a fully loaded oil tanker represents a mass of around 10^9 kilogram.)

The mass of the Earth itself is about 6×10^{24} kilogram. Hence the Sun is burning away a mass equal to that of the Earth in about 1.36×10^{15} second, or 43 million years. The present mass of the Sun is 330,000 times that of the Earth. This does not mean, however, that in about 14 trillion (10^{12}) years the Sun will have burned away to nearly nothing.

On the contrary, the thermonuclear possibilities, as now understood, allow the Sun to convert into energy no more than 1 percent of its mass by the complex fusion reactions we have noted. And within a cosmically brief period of about 5 billion years, the Sun's internal fires probably will attain far greater intensity, as the initial hydrogen fuel supply approaches its end, and the swifter burning of helium and carbon becomes dominant. The Sun today may be converting its mass into energy about 30 percent faster than it did 5 billion years ago; and 5 billion years from now its conversion rate may be about 30 percent greater than today. When helium-4 and carbon-12 become the principal thermonuclear fuels, the temperatures in the solar core will rise five or ten times beyond the present level, estimated at 20 million degrees C. Solar radiation will then increase greatly, perhaps to 1,000 times its present level. By this stage the Earth will long since have lost all its moisture, and even its solid substances may be melted or vaporized.

The entire posthydrogen phase of solar burning will be crowded into possibly 900 or 1,000 million years; that is, into less than one tenth of the Sun's preceding life as a stable, well-behaved, and moderately shining star.

When nuclear fuel supplies within the Sun are exhausted, gravitational forces will triumph. Within a period of only 10 to 20 million years they will shrink the formerly swollen Sun to a white dwarf star, which cools slowly to the frigid temperature of space itself. Such, at any rate, is the synopsis of the future story of the Sun, as astrophysicists and cosmologists now envisage it.

Myriads of stars, with larger initial masses than the Sun, are believed not to linger long as white dwarfs. Their gravitational forces are so strong that the cooled substance of the white dwarf contracts, first to the state of a superdense "neutron star" and then beyond that to ultimate gravitational collapse, in which a strange "black hole" takes the place of what was once a tiny, shrunken, but massive body.

Birth and youth of the Sun—The same insights that lead to forecasts of the Sun's future have also developed hypotheses of its origin. Great amounts of diffuse hydrogen gas and dust, widely scattered in space, are believed to have condensed under mutual gravitational attraction, within a period that may have been as short as 200 million years, a mere cosmic moment. As this matter clumped ever closer together, the growing ball was internally heated by pressure and the kinetic energy of the arriving particles and atoms. Gravitational energy was finally being transformed into heat at rates hundreds of times that of the Sun's present radiation into space.

This growing ball of incandescent hydrogen gas attained at last very nearly its present mass, 2×10^{30} kg. Temperatures within its center rose so high that hydrogen nuclei were forced together and fused, to form deuterons, a nuclear structure composed of 1 proton and 1 neutron, and slightly less massive than the two hydrogen nuclei (1 proton each) that they replaced. Thus began the thermonuclear "burning" processes, which went on to form the nuclei of helium-4 and cause the Sun to shine as steadily as it has been doing for billions of years.

The gravitational contractions ceased, because the heat generated internally created pressure, which counterbalanced the pull of gravitation. A remarkable equilibrium was achieved: any tendency toward further contraction tended to step up the rate of thermonuclear energy release, which, again, counteracted the gravitational pull. Any expansion of the solar sphere tended to have the opposite effect, reducing the rate of thermonuclear energy release and so allowing gravitational pull to restore the former balance.

Speaking generally, the Sun's size has been about what it is now for some five billion years, and it is likely to remain about the same for another five billion years. The relatively minor loss of mass in the thermonuclear furnaces does not alter this marked stability of size and the comparative uniformity of solar radiations. Only when the existing fuel supply (largely hydrogen) is exhausted will the Sun begin to alter drastically, as described here.

A body with much less mass than the Sun might never attain within its core the necessary "kindling temperatures" for thermonuclear reactions. Hence, it would not shine or become what we call a star. A body with much more mass than the Sun would burn through its available supplies of thermonuclear fuel far more swiftly. Its light would appear blue or blue-violet compared with that of the Sun, and its lifetime would be far briefer.

The total accumulated mass is thus of prime importance in determining whether a star is born and, if so, what kind of star it will be, and for how long.

Other possible paths from mass to energy—Other processes by which mass may be converted into radiated energy are being searched for. Radiations other than electromagnetic waves (light, radio, etc.) may surge through space around us. Much interest is focused on attempts to detect and measure "gravitational waves." Some evidence suggests that from near the center of our galaxy such waves may be emitted with total power equivalent to the conversion into energy of a body as massive as the Sun within half a year.

Since the Sun has 2×10^{30} kilogram mass, and each kilogram is equivalent to 9×10^{16} joule of energy, such a flow of power would be at the rate of 1.8×10^{47} joule every six months, 3.6×10^{47} joule per year, or 1.1×10^{40} joule per second (which means 1.1×10^{40} watt of power). This equals a power 28 million million (2.8×10^{13}) times the power poured out by our Sun in light and other electromagnetic radiations.

Increasingly, theorists look to processes of gravitational collapse of dwarf stars and neutron stars as the events that may lead, directly or otherwise, to such fantastic outpourings of energy in the form of gravitational waves or other radiations.

Astronomical measurements during the past quarter century have revealed extraordinary bodies in space, known as quasi-stellar objects, or *quasars* for short. Their radio emissions prove to be almost incredibly powerful. Whereas a steady, well-behaved star like the Sun radiates about 4×10^{26} watt of power, and an entire galaxy like ours may emit in the range of 10^{37} or 10^{38} watt, quasars have been found to radiate as much as 10^{42} or 10^{43} watt of power. This is from 100,000 to 1 million times the power of an ordinary galaxy, and must represent the results of interactions far different than those known until recently. It is as if a quasar was converting completely into electromagnetic energy within 2 days or less a mass equal to that of our Sun.

Measuring awful energies on Earth—The poet Robinson Jeffers (1888–1962) wrote not long after the announcement that atomic bombs had been dropped on Hiroshima and Nagasaki:

. . . the awful power that feeds the life of the stars
had been tricked down
Into the common stews and shambles.

Though those first bombs operated by nuclear fission rather than fusion, Jeffers was right: the power was awful, both in the original sense of full of awe and in the later, more common sense of terrible and frightening.

The awful as well as the ordinary, or average, must be measured, however. And so the problem arose: how to quantify the "power" (actually, the energy) of A-bombs and of H-bombs also, when the fusion-based nuclear devices had been developed.

The solution lacks elegance, but it is already well established and likely to remain so. The bombs, whether the original A or the later H type, are rated in terms of tons of TNT (trinitrotoluene), which previously enjoyed the reputation of being man's most destructive explosive. Detonation of 1 ton of TNT releases energy to the amount of 4.2×10^9 joule. Hence, 1 kiloton of TNT corresponds to 4.2×10^{12}, and 1 megaton to 4.2×10^{15} joule.

Thus a nonmetric unit of weight became a measure of explosive energy. Perhaps it would be more accurate to call it an approximation rather than a true measure or unit. In any case, we constantly encounter statements to the effect that a new thermonuclear device (bomb) has a yield in the 100-megaton range. The use of "yield" has a flavor of almost bucolic benevolence—much as a husbandman might chortle over the higher yield of some new and better cereal variety.

What analogues can we find for such estimates of destructive capability? Our Earth absorbs each second about 1.21×10^{17} joule of energy from the Sun. This is just about the amount of energy that a 30-megaton H-bomb can be assumed to release within less than 1 second. It is as if a cosmic burning glass focused on a single doomed spot all Earth's life-giving solar radiations, for the purpose of death.

Poets like Robinson Jeffers and other interpreters of our times have often remarked that the hydrogen bomb uses the same process, thermonuclear fusion, that makes the Sun shine on and on. This is true, but subject to certain corrections. The raw material and the product of an H-bomb differ from those in the core of the Sun.

The solar process is essentially a proton-proton reaction. Two protons (each identical with the nucleus of a hydrogen atom) are forced together with such velocity that they form one deuteron (one proton and one neutron), with the emission of a positron (the positively charged antiparticle of the electron) and a neutrino, which flies off at the speed of light and hardly interacts with other particles.

The H-bomb, however, is based on a deuteron-deuteron reaction. Two deuterons, each about twice as massive as a proton, are forced together so that they form a helium-3 nucleus (two protons and one neutron), emitting a neutron.

The Sun's steady burning is the result of an extremely energetic process, but the H-bomb's explosion results from a process completed far more swiftly than that in the Sun. In fact the H-bomb's deuteron-deuteron interaction is consummated about 10^{18} times more rapidly than the Sun's typical proton-proton interaction. If the Sun's shining were based on the more rapid process, it would have a very short life indeed. The Sun would then flare up, disintegrate, and die, rather than shine on through billions of years.

22 MORE ABOUT MASSERGIES, BIG AND LITTLE

Energy is the only life . . .
and Reason is the bound or
outward circumference of Energy.
— WILLIAM BLAKE,
The Marriage of Heaven and Hell (1791)

. . . nobody will object to an ardent experimentalist boast-
ing of his measurements and rather looking down on the
"paper and ink" physics of his theoretical friend, who on his
part is proud of his lofty ideas and despises the dirty fingers
of the other.
— MAX BORN,
Experiment and Theory in Physics (1943)

The conservation of massergy and the equivalence of mass and energy are more apparent in the microcosmic realms of atoms and subatomic particles, and also in the macrocosmic realms of thermonuclear fusion and the stellar universe, than in our everyday experiences on Earth.

A striking evidence of the equivalence of mass and energy in atomic and subatomic measurements is the fact that units of mass on the one hand and energy on the other have attained a certain interchangeability.

For example, the rest mass of the electron ($m_e = 9.109\ 558 \times 10^{-31}$ kilogram) is also used sometimes as an energy unit. Since 1 kg corresponds to 9×10^{16} joule (the SI energy unit), the electron rest mass is equivalent to 8.2×10^{-14} joule.

In the opposite direction, the electron volt (eV), a subatomic energy unit, is used also to measure or compare masses of subatomic particles and to measure high-velocity relationships wherein energy and mass are so inextricably merged that it is almost pedantic to assign priority to either one of them alone.

The electron volt (1 eV), defined as the energy gained by an electron responding to a difference of 1 volt in electrical potential, equals 1.6×10^{-19} joule. If an electron moves from A to B in response to a positive potential (voltage) that is 1 volt greater at B than at A, then that electron gains 1 eV of energy no matter what the distance from A to B.

On the other hand, if the electrical potential at A is positive with respect to that at B, the electron, being negatively charged, will move from B to A, since like charges repel and only unlike charges attract each other. (The volt is further discussed in Chapter 36.)

We can be much more specific about how the electron responds to a potential difference of 1 volt. Assume that an electron was motionless at A, and then a switch was thrown to give a metal electrode at B a positive potential of 1 volt greater than that at A. The electron will move toward B, accelerating as it does so, and arrive at B with a velocity of 646,000 meter per second. If B happens to be just 1 meter from A, the tiny electron will have been accelerated at a rate of 646,000 m/s², which is about 66,000 times g, the normal gravitational acceleration at the surface of the Earth.

Newton's formula for force in relation to mass and acceleration is $F = MA$. Thus a force, even if small, when exerted on an almost infinitesimal mass, like the rest mass of the electron, results in a large acceleration.

The electron volt can also be expressed, like kinetic energy, as $E = \frac{1}{2}MV^2$, where V represents the velocity attained by the electron in response to the potential difference applied to it. If a 1-volt difference accelerates an electron to 646,000 m/s, then a 10-volt difference will accelerate it 10½ (square root of 10, or 3.162) times that much—or, in this case, to 2,042,650 m/s.

The energy gain is, of course, proportional to the square of the velocity gain. Hence, the energy of this arriving electron will be just 10 eV, as we should expect.

Up to this point we make no relativistic adjustments. Those are reserved for velocities which approach significantly near to c, the velocity of light in empty space (roughly 3×10^8 m/s).

When we use the electron volt as an energy unit, we are multiplying an electrical potential (measured in volts) by the fixed electrical charge on every electron, measured in the coulomb, SI unit of charge. The electron charge, often called "the elementary charge," is just $1.602\ 191\ 7 \times 10^{-19}$ coulomb. The important matter here is that a charge (measured in the coulomb) multiplied by an electrical potential (measured in the volt) results in an energy (measured in the joule).

Mass can be looked on as the gravitational or inertial *charge* of matter; and the square of velocity (measured by m^2/s^2 in SI units) is like the electrical potential difference. Thus, just as electrical charge times potential difference gives energy as a product, so does mass times velocity-squared give energy (kinetic energy) as a product. This is one of many analogies that help to link electrical and other units with the more familiar units of mechanical motion, such as length, mass, time, et cetera.

The gravitational or inertial potential (measured in m^2/s^2) is often represented by the Greek letter phi (ϕ). If we divide energy in joule (with dimensions of kg m^2/s^2) by mass (gravitational charge) in kilogram, we arrive at m^2/s^2, the measure of gravitational potential.

The old foot pound, used by Watt and so many others, is a true energy unit, but an earthbound one. It is abbreviated as ft lbf, which indicates that the body raised 1 foot must have a downward force (weight) of just 1 pound (1 lbf). Such a weight exists when a mass of 1 pound (1 lbm) is acted on by gravitational acceleration of g_0 (that is, 32.174 ft/s^2 or its metric equivalent of $9.806\ 65$ m/s^2).

Thus the energy or work content of 1 foot pound is just 32.174 ft^2 lbm/s^2 and, so, equal to $1.355\ 82$ joule. Conversely, $1\ J = 0.737\ 56$ ft lbf.

The metric analogue of the foot-pound measure of energy or work is the meter kilogram (force), abbreviated as either m kgf or kgf m. It too is a thoroughly earthbound measure. It specifies the amount of energy or work that lifts 1 kilogram mass a distance of 1 meter where gravitational acceleration is just g_0, as already specified. Thus 1 m kgf $= 7.233$ ft lbf; and 1 ft lbf $= 0.138\ 26$ m kgf. Also 1 m kgf $= 9.806\ 65$ joule; and $1\ J = 0.102$ m kgf.

An outstanding advantage of the joule as the sole SI unit for all measurements of energy and work is the fact that it does not depend on the Earth and the arbitrary assumption of standard gravitational acceleration (g_0). The joule, as defined in the SI, could be determined even in a state of weightlessness, for instance, in an orbiting satellite, as well as in a laboratory on Earth.

Almost all about free fall—Energy and work units such as the foot pound and the meter kilogram (force) are closely related to the concept of free fall. Such fall takes place when a mass is allowed without interference to respond to gravitational acceleration. To achieve perfect free-fall patterns on Earth would require elimination of the atmosphere or experiments in vacuum, because air resistance does retard velocity gains, especially of small objects.

The effects of air resistance can be largely eliminated. When this is done, we find that freely falling bodies appear to gain velocity at a constant rate, second after second, as they drop. In other words, they accelerate uniformly—provided that they are not dropped over distances so great that we can measure appreciable difference between the (lesser) gravitational acceleration at the altitude from which they were dropped, as against the (greater) acceleration at the level to which they fell.

Gravitational acceleration diminishes in proportion to the square of the distance from the center of the Earth. Thus, we agree that at the surface, just 1 Earth radius from the center, the standard acceleration is g_0 (9.806 65 m/s^2). Then at an altitude of one tenth Earth radius, or 637 km, the acceleration should be g_0 divided by 1.1^2, or 1.21; that is, an acceleration of 8.104 67 m/s^2. Likewise, at an altitude of 1 Earth radius (2 Earth radii from the center), the acceleration becomes g_0 divided by 2^2, or 4—that is, 2.451 66 m/s^2. And so on.

When measuring falls that do not exceed a few hundred meters, such changes in gravitational acceleration are usually tiny enough to be disregarded. We say then that a freely falling body gains velocity in proportion to (a) the elapsed time of the fall, and (b) the square of the distance that it has dropped.

Uniform increase of velocity is synonymous with constant acceleration. The distance dropped, however, rises in proportion to the square of the number of seconds since the fall began. Thus, where g_0 prevails, in its first second of fall a body drops 4.903 325 meter; by the end of the next second it has fallen 4 times as far; by the end of the next, 9 times as far; by the end of the next, 16 times as far; and so on.

Now, the kinetic or motion energy of a moving body increases in proportion to the square of its velocity. Thus, while the velocity is increasing in proportion to the elapsed time of the fall, the body's kinetic energy is increasing in proportion to the square of that elapsed time. The last link in the chain is now here: since the distance moved and the energy accumulated by the falling body both increase proportionately to the square of the elapsed time, they must increase in proportion *to each other*.

Specifically, when a 1 kilogram mass falls freely during 1 second, it attains a velocity of 9.806 65 m/s, a kinetic energy of 48.085 192 joule, and a total distance dropped of 4.903 325 meter—provided that the acceleration is g_0. This amounts to an energy change per meter of fall of 9.806 65 joule per meter. And every meter of fall will add that same amount of kinetic energy, so long as we can say that g_0 prevails.

Thus, we have reached the same relationship underlying the meter kilogram (force) unit. The energy change involved in lowering or lifting a mass of 1 kg when gravitational acceleration is g_0 is always just 9.806 65 joule. Lifting that mass gives it added potential energy, at this rate. Allowing it to drop adds to its kinetic energy at the same rate per meter of movement.

The SI defines the newton of force as that force "which gives to a mass of 1 kilogram an acceleration of 1 meter per second per second." And the joule of energy is defined as "the work done when the point of application of 1 newton (force) is displaced 1 meter in the direction of the force."

Obviously, work is done when a mass of 1 kg drops in free fall. That work amounts to 9.806 65 joule for each 1 meter that the mass drops. When the body is lifted 1 meter, that amount of work goes into increasing its potential energy; when it falls 1 meter, that amount of energy is transformed from potential to kinetic energy, and the falling mass can do work of that magnitude when it hits something at the end of its drop.

All these different approaches lead to the same relationships. Whether we approach energy as the product of force times distance moved, or as the product of the square of velocity times mass times a constant (one half), we come to the same conclusions. The path we choose should be that which best fits the physical situation to which our measurements are to apply.

About energies in orbit—In this era of constant launchings of man-made satellites and space travel, free fall (downward) is not the only motion pattern deserving of attention. We should look also at the orbital motion of a satellite, natural or man-made, around some central body, such as the Earth or Sun. For simplicity, we shall assume fully circular orbits, rather than elliptical ones, which also are possible.

As noted before, an orbiting body actually accelerates constantly —that is, it falls, in a sense—toward the center of the body around which it moves. Meantime, however, its forward motion is just sufficient to keep it in the circular path at constant distance from the center.

Were it not for atmospheric resistance, a satellite could orbit nicely very close to the surface of the Earth, provided no lofty mountains were in its path. The velocity that would just suffice to hold it in such a near-surface orbit would be about 7,900 meter (7.9 km) per second, sufficient to take it around the globe once in about 1 hour 25 minutes.

At that rate, each kilogram of mass in the orbiting satellite would have a kinetic (motion) energy of nearly 31,205,000 joule because of its velocity.

The energy needed to escape an Earth orbit and fly freely into space is always just twice that required to remain in that orbit. This orbiting satellite would need 62,410,000 joule of kinetic energy per kilogram of its mass in order to break free from the Earth's gravitational control. Since motion energy is proportional to the square of velocity, the escape velocity is $2^{1/2} = 1.414$ times the minimum orbital velocity. Thus, the 7,900 m/s would have to be accelerated to 11,171 m/s in order to convert the imaginary surface orbit into a trajectory away from Earth.

More than a century ago, in his classic science fiction work *From the Earth to the Moon* (1865), Jules Verne used a gigantic cannon to fire a spaceship toward the Moon at the necessary escape velocity of nearly 11,200 m/s. Despite his genius, his hypotheses were faulty. Such travelers would not have survived the initial shock, the almost instantaneous large acceleration in the tube of the cannon. Now we attain equal and greater velocities gradually by means of rockets, and keep the accelerations within the range of human survival.

By using energy measures rather than velocities alone, we arrive at simplifications such as the fact that the energy required for circular orbiting is always half the energy required for complete escape from that orbit.

Knowing that the average gravitational acceleration on Earth's surface (g_0) is about 9.8 m/s², we can multiply this magnitude by the Earth's radius (6.37×10^6 meter) and our product will be the gravitational potential (ϕ) of 6.24×10^7 m²/s², typical of our normal surroundings.

How much energy is needed to raise a unit mass (1 kg) to various heights above the Earth? At a distance of 1 Earth radius above the surface (or 2 radii from the center of Earth), the gravitational potential will be just half that on the surface, or 3.12×10^7 m²/s². Hence, 3.12×10^7 joule of energy is required to lift 1 kilogram to a height of 1 Earth radius; 4.16×10^7 joule of energy is required to lift it to a height of 2 Earth radii above the surface; and 4.68×10^7 joule is required to lift it to 3 Earth radii above the surface. However, the ultimate breakaway, the escape from the chains of Earth's gravitational attraction entirely, requires the full amount of energy: 6.24×10^7 joule per kilogram of mass sent on such a voyage of no return.

Massergies of the leading subatomic particles—The use of the electron volt and of the joule itself to compare the massergetic magnitudes of subatomic particles is shown in the following table, which lists the rest masses (masses without motion-energy) of three principal particles, whose names, symbols, and masses in kilograms are given here: electron (m_e), $9.109\ 558 \times 10^{-31}$ kg; proton (m_p), $1.672\ 614 \times 10^{-27}$ kg; and neutron (m_n), $1.674\ 82 \times 10^{-27}$ kg.

	REST MASS (in MeV)	REST MASS (in joule)
Electron	0.511 0	$8.187\ 26 \times 10^{-14}$
Proton	938.26	$1.503\ 27 \times 10^{-10}$
Neutron	939.55	$1.505\ 25 \times 10^{-10}$

The same rest masses apply also to the antiparticles: the positron, opposite of the electron; the negatron or antiproton, opposite number of the proton; and the antineutron, which like the neutron is electrically neutral (uncharged).

When antiparticles are produced in a high-energy accelerator, they speedily encounter particles (positrons encounter electrons, negatrons encounter protons, and so on). In these encounters both are annihilated and a pair of radiation photons appear, shooting away from the site of the encounter in more or less opposite directions. Here again bits of matter (particles) are transformed into electromagnetic energy (photons). Yet photons, from many points of view, are particles too. The old pigeonholes separating matter and energy, or particles and waves, simply fail to fit the facts of the realms of high energies.

The energy E of a photon, or least quantum of a radiation, equals the frequency of that radiation times h, Planck's constant of action (6.626 2 $\times$ 10^{-34} joule second). Thus, orange-yellow light, with a frequency of 5 $\times$ 10^{14} Hz, is composed of photons, each of which has an energy of 5 $\times$ 10^{14} $h = 3.31 \times$ 10^{-19} joule. Since 1 joule equals 6.24 $\times$ 10^{18} eV, each such photon has an energy of about 2 eV. The lowest visible frequencies of red light have photon energies of about 1.6 eV, and the highest visible violet, of about 3.2 eV.

We have seen that a free electron (negative) and a free proton (positive) may "fall together" and form one bound atom of hydrogen, simplest and lightest of the elements. The proton becomes the nucleus, and the electron becomes its satellite or attendant, bound by electrical, not gravitational, forces. Before they formed the atom, the two separate particles (according to the table given earlier) showed masses totaling 938,771,000 electron volts. However, the mass of the resulting hydrogen atom proves to be 13.5 electron volt less than that total. This energy difference was emitted as a photon of ultraviolet light radiation, typical of such hydrogen atom formations.

The same energy of 13.5 eV is the binding energy of the hydrogen atom in its ground state. It is the least amount of energy capable of ionizing a hydrogen atom—that is, stripping away the single electron from the proton nucleus.

The ordinary hydrogen atom (1 proton, 1 electron) has no nuclear binding energy. Its nucleus is just the single proton with no nuclear companions to be bound to. All atoms other than hydrogen have nuclear binding energies, as well as the energies binding the various electrons to the nucleus. Hence, these other atoms show a mass deficiency: the mass of the complete atom is less than the total separate masses of the electrons, protons and neutrons it is composed of.

These relationships clarify the basic processes of the fusion reactions by means of which the Sun and stars shine. When two atoms of hydrogen-1, the usual form of that element, fuse into hydrogen-2 (deuterium), the nuclear binding energy is about 3 MeV per nuclear particle or 6 MeV for the entire nucleus. For hydrogen-3 (tritium), the figures are 2.5 MeV and 7.5 MeV. For helium-4, formed from four atoms of hydrogen, the figures are 6.7 MeV and 26.8 MeV. That means that the formation by fusion of helium releases 26,800,000 electron volts of energy per atom formed.

In more complex and massive atoms, not only does the total nuclear binding force rise, but the average too rises. In the case of oxygen-16, for example, that average is 7.97 MeV for each of the eight protons and eight neutrons composing its nucleus. The mass of such an oxygen atom is about 1.8 percent less than the combined separate masses of the 24 particles composing it (16 in the nucleus, plus 8 electrons around it).

For iron-56, these figures rise to an average of 9 MeV per nucleon (proton or neutron) and 12 percent.

However, when we proceed to isotopes more massive than those of iron or nickel, we find that the average binding energy per nucleon decreases. This is the reason, for example, that the formation of the very heavy transuranium elements in the laboratory requires large amounts of energy input.

Iron in an old star is like the ash clogging an old furnace. It can neither "burn" to produce still heavier elements and set free additional energy, nor can it revert to the lighter elements. Either step would use up energy, rather than set energy free. Thus the stars that reach the iron-ash stage lose the sources of the energies by which they formerly shone. They cool, shrink, collapse into dense dwarf stars, or perhaps even into a tiny ultradense neutron star, and finally, in the ultimate instances, undergo total gravitational collapse, becoming an invisible and unobservable point, void of volume but still possessed of, or by, mass.

Fusion is thus characteristic of elements below the "iron center" of the isotope ladder, while fission, or breakup of massive nuclei, is characteristic of the heavy isotopes well above that center.

We can now form an instructive ladder of typical energy magnitudes measured in electron volts. Heat, being molecular motion, and chemical reactions exhibit energies no greater than about 1 eV. At 2 eV incandescence (emission and absorption of visible light) is achieved, and the most chemically active atoms, such as those of potassium, can be ionized by knocking away an outer electron.

At 13.5 eV there is ultraviolet light, able to tear away the electron from the hydrogen atom. At 1,000 eV (1 keV) we are at the energy level of X-rays, able to disturb the innermost electrons of complex atoms. At 500 keV the ladder arrives at the level of energy of the powerful cosmic rays that reach the Earth from outer space and are able in some cases to annihilate the mass of an electron.

At 1 GeV (1 billion electron volt) the mass of a proton or a neutron may be annihilated, and at 1.5 GeV, the mass of the heaviest known subatomic particles. ("GeV" is used, for the *G* stands for "giga"; it is preferred to "BeV," in which *B* is supposed to mean "billion.")

The great proton synchroton of the National Acceleration Laboratory in Illinois has attained energies of 200 GeV and may soon be increased to 400 or even 500 GeV. None of us can predict what may be discovered with the aid of such enorm.ous massergies.

These powerful, complex and costly particle accelerators use electromagnetic rather than gravitational forces. Despite our everyday impressions, gravitation is usually a very weak force compared with the electromagnetic and nuclear interactions that go on within the atom. By using electromagnetic forces, the N.A.L. adds 200,000 MeV to the proton's original rest mass of 938 MeV. Can the resulting particle still be called "a proton"? Actually, it is necessary to distinguish rather clearly between "1-GeV protons" and "200-GeV protons." The changes in their behavior make this almost mandatory.

The great linear accelerator called SLAC, near Stanford University, accelerates electrons almost to the velocity of light itself.

A table is offered here showing how, as the energy rises to 10,000 MeV, the attained velocity approaches ever more closely to, but does not quite reach c, the speed of light. It shows also how the total massergy of the flying electrons grows. In the old way of speaking, at an energy of some 10^{10} electron volts the apparent mass of the electrons is nearly 20,000 times greater than when electrons are at rest. Effects of the increasing MeV's are more marked on electrons than on protons because the electrons are 1,840 times less massive to begin with.

ENERGY (MeV)	VELOCITY (% of c)	TOTAL MASSERGY (in M_e)
0	0	1
0.5	86.3	1.98
1	94	2.96
5	99.6	10.8
10	99.88	20.6
20	99.96	40.1
100	99.98	196.7
1,000	99.99	1,958
10,000	99.99	19,571

The rightness of relativity—Einstein supplied the equations describing how the massergies of swiftly moving particles should grow as their velocities rose ever closer to the ultimate velocity of light itself. These predictions have been tested millions of times in the operations of the great particle accelerators and have been found to be fully valid.

If we symbolize by M_t the mass of the object in motion, by M_o its mass at rest, by V the velocity at which it is made to move, and by c the velocity of light in space, then the relationship is:

$$M_t = \frac{M_o}{(1 - V^2/c^2)^{1/2}}$$

The denominator of this fraction is a square root, as indicated by the superscript fraction $1/2$.

In everyday life, V^2 is negligible compared with c^2, which has the enormous value of about 9×10^{16} m²/s². Hence, for all practical purposes, $M_t = M_o$, and there are no relativistic effects worthy of mention. However, at high values of V, when attained velocities begin to approach c itself, then M_t grows as shown in the last column of the table on page 235.

The additional mass is not created out of nothing, however. By the law of conservation of massergy, sufficient mass-energy must be supplied to an accelerator to equal the excess mass-energy exhibited by the particles it accelerates. In fact, the mass-energy supplied to such accelerators vastly exceeds this excess, for there are tremendous losses of power in the form of heat and unwanted radiations.

Conservation of massergy sets a limit which even the most efficient and perfected accelerating device cannot pass. If in practice it comes anywhere near that limit, its designers and operators have cause to congratulate themselves. The cost of high-energy experiments comes high; but the unanswered questions are many and basic to our understanding of the physical world.

23 PATTERNS OF POWER, THE TIME RATE OF ENERGY

Soon shall thy arm, unconquered steam! afar
Drag the slow barge, or drive the rapid car;
Or on wide wings expanded bear
The flying chariot through the field of air.
—ERASMUS DARWIN,
The Botanic Garden (1789–92)

The first practical units for measuring energy and power developed largely in response to industrial and commercial needs to price either an amount of energy to be sold (as in the form of coal), or to permit rational sale of power, which is energy developed per unit time. The horsepower became the first important power unit.

It was not, however, the agricultural or recreational or even the transportation uses of horses that brought this about. Rather it was the industrial use of horses. In the eighteenth century horses provided the principal sources of energy in Britain for hauling coal and pumping water in the mines and even for running textile machinery in the mills.

Around 1700, Thomas Savery, a British pioneer in applying the heat energy in steam to pumping uses, suggested that the rate at which a horse does work should be used as a standard or unit for measurement of power. The actual horsepower unit, however, was worked out a good deal later by James Watt (1736–1819), as a standard for the performance of his new steam engines, the first widely useful devices of their kind. Watt's precision in working out his measure of power, in conjunction with his partner Matthew Boulton, was rivaled by his public-relations skill in choosing for the unit a vivid and dramatic name: the *horsepower*.

His prospective customers, the new millowners and mineowners, were less interested in the total amount of energy that Watt's steam engine could produce over a long period than in its energy capacity per day, per hour or per second. Their concentration on maximizing profits was unshakable, as Watt discovered when he worked out changes that reduced the infernal racket made by his first models, only to be assailed by complaints that these improvements reduced the power. He failed to win thanks for his early efforts to abate noise pollution.

Watt shaped his new power unit in terms familiar to such tough customers. Workhorses at that time were usually made to walk around a circular track, dragging one end of a stout lever attached to a central capstan whose shaft, through gears or linkage, worked the pumps or other machinery.

Watt estimated that the average horse pulled with a force of 180 pounds (lbf). Force times distance equals energy or work done. Also force times the velocity at which it acts, equals power. Choosing the latter relationship, Watt calculated that a typical capstan lever was 12 feet long. The circular track around which the horse plodded, accordingly, was $2\pi \times 12 = 75.4$ feet in circumference. An average horse, Watt reckoned, completed 144 circuits an hour, or 2.4 per minute, or 1 in 25 s. This meant a velocity of about 181 feet per minute. Thus the force times the velocity came to 180 lb $\times$ 181 ft/min = 32,580 foot pounds per minute (ft lbf/min). Watt rounded this off up to 33,000 ft lbf/min, which equals just 550 ft lbf/s. He called this new unit a horsepower, as we still do today.

In colloquial speech and informal writing, energy and power are often used as if they meant the same thing. The confusions are by no means limited to English. German scientists use *Kraft* to mean power, but in everyday German *Kraft* is closer to our English "strength." French scientists commonly use *force* to mean power, but *force* in English must be distinguished carefully from *power* on the one hand and *energy* on the other.

From James Watt to the watt (W) of power—Watt defined the horsepower but not the later SI unit of power, the watt (W), which was named to honor him. In the nineteenth century, when dynamos were first used to generate electric power and electric motors to transform electrical into mechanical power, it became desirable to develop a power unit related to electrical units like the volt, the ampere, and the ohm, and also to the joule, the scientific unit of energy or quantity of work.

The watt of power actually was proposed a few years earlier than the joule of energy. In his 1882 presidential address to the British Association, the electrical engineer William Siemens (1823–83) proposed the watt, as now adopted. Siemens, who had been born Carl Wilhelm Siemens in Germany, probably did not anticipate one source of opposition and delay: problems involved in the use of the letter *w* in the French language. Nevertheless, before 1890, international recognition of the watt unit was attained.

At about the same time the joule, which had been proposed in 1888 by the British Association, also was recognized internationally. The joule equals 1 watt second; and the watt equals 1 joule per second. They form a perfectly matched pair.

Today the watt and its multiples, such as the kilowatt, are more widely used in the United States than any other SI units with the exception of the second of time and the electrical units just mentioned (ampere, volt, ohm). Our light bulbs, electrical appliances, radio and TV transmitters are all rated in watt, not in horsepower.

The old "horsepower" persists, however, though on the wane. Its survival undoubtedly reflects the fact that most of those who work with power are engineers and technicians, who have always shown more inertia about units than have scientists and mathematicians.

One of the great advantages of the SI system is its constant use of unity (1) as the coefficient relating two different units. Thus, while the watt is simply 1 joule per second, the horsepower remains 550 foot pounds per second. In the metric (SI) units, no arbitrary coefficient has to be memorized.

From 1908 to 1948, the "international" watt was used. It was based on the standard electrical units: 1 watt $= 1$ ampere^2 $\times$ 1 ohm; also 1 watt $= 1$ ampere $\times$ 1 volt. After 1948 the watt was based upon the joule of energy, as is reflected in the present SI definition: "The watt (unit of power) is the power which gives rise to the production of energy at the rate of 1 joule per second."

This is the "absolute" watt, which is slightly larger than its predecessor, the international watt: 1 absolute watt $= 1.000\ 19$ international watt. The absolute watt is, for all standard uses, the only watt today, and will be so regarded throughout the rest of this book. The adjustment of just 19 parts in 100,000 necessitates a certain caution, however, when dealing with data in books and periodicals published during that 1908–1948 era of the international watt.

For conversion, 1 horsepower = 745.7 watt; 1 W = 0.001 341 hp; and 1 kW = 1.341 hp. For rough estimations, 1 hp can be taken as approximating $\frac{3}{4}$ kW.

Horsepower, horses and human workers—James Watt's estimate of the power output of the British workhorse was generous. During the workday that prevailed in his era—12 hours, at least—few horses maintained 1 hp, even deducting time taken out for oat and water breaks.

A healthy human laborer can work at about 0.1 hp over an extended time. During short spurts, however, he can considerably exceed that figure. The SI equivalent of 0.1 hp is 74.57 watt. This does not indicate that humans are very powerful creatures. Less than 75 joule of work per second is little by the standards of our motor-minded period.

There was a time, however, when this little might give the illusion of being a lot, simply because of the tiny energy unit provided by the CGS (centimeter-gram-second) system, which was prevalent for scientific work before the present MKS (meter-kilogram-second) basis. The unit of energy in the CGS equaled the work done by force of 1 dyne acting through a distance of 1 centimeter. The dyne was just $\frac{1}{100,000}$ of the present newton, SI force unit. And since the centimeter is $\frac{1}{100}$ of the meter, this CGS energy unit, named the *erg,* was just 1 ten-millionth (10^{-7}) of the present joule: 10^7 erg = 1 joule. The corresponding CGS power unit, which had no single name of its own, was the erg per second: 10^7 erg/s = 1 watt.

Thus, if a human laborer averages only 50 watt of power over a lengthy period, he is likewise averaging 500 million erg per second. This is far more imposing, especially to those who have no clear notion how very tiny is the amount of work represented by 1 erg!

The so-called "electrical horsepower" was used to some extent in Britain and the United States. It was set at just 746 watt; hence almost exactly equaled the regular mechanical horsepower. Today, use of the electrical horsepower is infrequent and can even be illegal. Electrical appliances and devices must be rated in the accepted watt, volt, ampere, ohm, and related units, all parts of the great SI unit complex.

Another variant of similar magnitude was the "metric horsepower" (735.5 watt). This is the power which will raise a 75 kilogram mass 1 meter in 1 second at the surface of the Earth. In German this unit was called the *Pferdestärke* (PS), and in France the *cheval-vapeur* (CV or ch).

An independent unit of power that sometimes appears in older works on mechanics and engineering is the foot-pound per second (ft lbf/s), equal to $\frac{1}{550}$ horsepower and to 1.36 watt.

Similar in appearance and sound is the power unit known as the foot-poundal per second. However, it equals only 0.042 1 watt or 42.1 milliwatt, for the poundal is a force equal to 1 pound divided by 32.2, the numerical value of standard gravitational acceleration in feet per second per second (ft/s^2).

Since volt times ampere equals watt, it is possible to calculate the current in ampere used by household devices. If the voltage supplied is the usual electromotive force of 120 volt, then a 60-watt bulb draws $\frac{1}{2}$ ampere, a 600-watt heater 5 ampere, and a 1,500-watt quick-heating kettle 12.5 ampere.

Paying for power in terms of energy—Our electric power bills are actually paid for in terms of electric energy. We do not pay by the watt of power nor even by the joule of electric energy used. Instead, the basic unit is the kilowatt hour (kW hr), which equals 3,600,000 joule. The thousand-times-larger megawatt hour (MW hr) equals 3.6×10^9 joule.

From horsepower to carpower—At least three different kinds of horsepower have been used in the promotion and selling of automobiles and motorcycles. *Brake horsepower* is the most meaningful in terms of performance. It is the amount of power that the engine actually supplies to propel the vehicle and to operate auxiliary devices. It is commonly measured when the engine is rotating at a rather high rate, such as 4,000 rpm, which is equal, in terms more consonant with SI principles, to 66.7 rps (revolutions per second). Brake horsepower measurements should mention the rpm or rps at which they were made.

Other horsepower ratings are based on numbers of cylinders and their dimensions, and consequently do not deserve to be relied on by careful buyers. Thus, there is the *indicated horsepower,* which multiplies four factors—(1) cylinder area in square inches, (2) the piston travel, or stroke, in feet, (3) pressure on the piston in pounds per square inch, and (4) number of cylinders—then divides the grand product by 33,000. Also there is the *nominal horsepower,* which multiplies three factors—(1) the square of the cylinder diameter in inches, (2) the number of cylinders, and (3) the cube root of piston travel, or stroke, in feet—then divides the grand product by 15.6.

Another arbitrary horsepower formula emanated from the Royal Automobile Club, of Britain. During a quarter of a century ending in 1947 it was used to determine the British tax on automobiles. It multiplied the square of the cylinder diameter in inches by the number of cylinders, and divided the product by 2.5.

If the United States goes totally metric, and applies the great SI system throughout its technology, we may someday find power outputs of automobile or motorcycle engines expressed in watts rather than horsepower. Advertising copywriters could then have a carnival. An engine with a brake horsepower of a modest 100 hp could be glorified as surging with more than 74,500 nimble or eager watts.

One incidental and unintentional consequence of such a unit substitution might be the greater willingness of the car-buying public to purchase less arrogant chariots—that is, autos devoid of some of the present superfluous horsepower, which adds more to pollution and petroleum profits than to performance.

Regarding rotational power—The SI definition of the joule of energy implies work performed by motion in a straight line, for it is "the work done when the point of application of a force of 1 newton is displaced a distance of 1 meter in the direction of that force." However, we have seen that Watt derived his horsepower unit from studying the work done by horses that moved constantly in circular paths, and the power of swift-spinning automobile engines and of electric motors is transmitted by rotating shafts, gears and wheels.

Torque, or the magnitude of twisting effect, is important in almost all studies of practical power problems. Watt's imagined average workhorse pulled with 180 lbs force on a capstan lever 12 feet long, thus producing a torque or twisting effect of 2,160 foot-pounds (ft lbf). How swiftly was this applied? A full circle is 2π radian (6.283 2 rad), and since the horse made a full circle within 25 s, it was turning at a rate of 0.251 33 rad/s. This rate times the torque of 2,160 gives 543 foot-pounds per second, identical with the 32,580 foot-pounds per minute which Watt rounded up to 33,000 foot-pounds per minute, basis of the horsepower.

In the SI, the unit of torque is the newton meter; that is, 1 newton of force acting at the end of a lever 1 meter long, and at right angles to the line of that lever. When such a torque moves the lever through an angle of 1 radian, then 1 joule of work has been done rotationally. If that joule has been supplied in just 1 second, the rotary power is 1 watt.

Thus, torque is the rotational analogy of force, and angle turned (measured in radian units) is the rotational analogy of linear displacement, measured by the meter.

What about the kinetic energy of a rotating body, such as a flywheel? We have seen that the kinetic energy E of a body of mass M moving in a straight line at a velocity V, is given by $E = \frac{1}{2}MV^2$. For a rotating body, an analogous relationship exists. In place of mass we use the *moment of inertia* of the body, symbolized by I. And in place of linear velocity, we use its angular velocity, symbolized here by w and measured in SI units of radian per second.

Then rotational energy (E_{rot}) is given by:

$$E_{rot} = \frac{1}{2} I w^2$$

Just as the kinetic energy of linear motion grows in proportion to the square of velocity, so does the kinetic energy of rotational motion in proportion to the square of angular velocity. If a small flywheel rotating at 1 radian per second has just 1 joule of energy, it will have 100 joule of energy when accelerated to a rotational rate of 10 radian per second.

Bodies of different geometrical shapes have different moments of inertia. If a uniform circular plate of mass M has a radius of R, its moment of inertia is $\frac{1}{2} MR^2$ when it is spun on an axis through its center and perpendicular to its surfaces. A uniform solid sphere of mass M and radius R has a moment of inertia of $\frac{2}{5} MR^2$, or slightly less than the disk of the same mass. A thin, uniform rod of mass M and length, end to end, of $2R$ has a moment of inertia of $\frac{1}{3} MR^2$ when spun around an axis through, and at right angles to, its center.

Properly designed flywheels can store surprisingly large amounts of mechanical energy. Such flywheels are mounted on precision roller or ball bearings, spun in a vacuum chamber, and are precisely balanced to rule out vibration. They remain in one place, if need be, are compact, and may someday be far more widely used to store and deliver energy than at present.

Transmissions in transit—In order that cars and trucks may perform properly without engines of excessive size and power, the actual power delivery of their engines must be kept more or less uniform so long as the vehicle is in motion. This means keeping nearly constant the product of the motive force times the speed of the vehicle (in terms of its linear motion), and of the engine turning rate (rpm) times its torque (in terms of the power plant's rotational motion).

By means of gearshift and transmission, force can be made high just as the car begins to move and gather speed, then can be decreased as its velocity rises. On giant trucks there may be a dozen or more gear ratios through which the driver shifts on his way from a standing start to full speed on a level highway. This helps hold down the size and power of the engine needed.

The tachometer measures the engine's rotational rate. When that indicator shows the engine has exceeded a given rpm, the gears are shifted into another ratio, thus reducing the engine rotational rate while increasing the torque the engine exerts on the drive shaft and so on the wheels in contact with the road.

Uniform acceleration means excessive power demands. Suppose, for example, that a force of 5 newton is maintained on a 1 kg mass, producing a uniform acceleration of 5 m/s² (which happens to be about 0.51 g). The following table shows, second by second, the attained velocity, the accumulated kinetic energy of the moving object, and the increase of that energy within the preceding second (measured in watt, since it is actually the power consumed during that second).

TIME (second)	VELOCITY (m/s)	KINETIC ENERGY (joule)	ENERGY GAIN IN 1 SECOND (watt)
1	5	12.5	12.5
2	10	50	37.5
3	15	112.5	62.5
4	20	200	87.5
5	25	312.5	112.5
6	30	450	137.5

Kinetic energy is calculated as $\frac{1}{2} MV^2$. If the engine can supply only 150 watt, there can be little further acceleration.

Again ignoring air resistance, let us assume that gear-shifting results in nonuniform acceleration but constant power, or wattage (at 100 W):

TIME (s)	VELOCITY (m/s)	KINETIC ENERGY (J)	ENERGY GAIN (W)
1	14.1	100	100
2	20	200	100
3	24.5	300	100
4	28.3	400	100
5	31.6	500	100
6	34.6	600	100

The second of the two tables above shows the acceleration gradually flattening out, whereas in the first table the acceleration has been kept constant, at the cost of ever-increasing consumption of power. If the pattern of the second table were continued on to the end of the tenth second, we would find that the velocity of the vehicle is still less than 45 m/s, and that during the tenth second the vehicle accelerated by only 2.3 m/s².

In the second table we see a steady gain in the kinetic (motion) energy of the vehicle. Each second that energy grows by 100 joule. Hence, the power consumed is just 100 watt, all along the line.

Thus we can compare the two extremes: first, constant acceleration with ever-rising power consumption; then constant power consumption and energy gain, with ever-diminishing acceleration.

Meaning of momentum—An important physical variable is *momentum,* the product of a mass times the velocity at which the mass is moving. The SI unit for momentum has no single name; it is 1 kilogram meter per second (1 kg m/s).

If a force of 1 newton acts for 1 second on a 1-kilogram motionless mass, the mass will gain 1 kg m/s of momentum in the direction of the force. If then a force of ½ newton acts for 1 second in the same direction on the same mass, its momentum will rise to 1½ kg m/s in that same direction. On the other hand, if a force of 1½ newton acts for 1 second on that body in the opposite direction, its momentum will be reduced to zero. That is to say, it will be brought to a stop.

Momentum is one of the physical variables, called vector variables, which have both magnitude and direction. The direction must always be taken into account in adding or subtracting any such variable. Vector variables always include a single term in length (L). Thus the vector variables include length itself (L); velocity, or length per time (LT^{-1}); acceleration, or velocity change per time (LT^{-2}); and force, or mass times acceleration $(LMT^{-2}, \text{ or } MLT^{-2})$. Finally, there is momentum, or velocity times mass $(LMT^{-1} \text{ or } MLT^{-1})$.

The nonvector, or *scalar,* variables, on the other hand, have magnitude only, without any definable or essential direction. They are found to have terms in L^2 (length squared). Among them are area (L^2), energy $(L^2MT^{-2} \text{ or } ML^2T^{-2})$, and power $(L^2MT^{-3} \text{ or } ML^2T^{-3})$. These can all be added or subtracted without reference to direction in space.

When we deal with the kinetic energies of bodies moving in straight lines, we find that those energies increase in proportion to the *squares* of the velocities of those bodies. When we deal with the momentums of such bodies, we find those momentums increase in simple proportion to the velocities.

Both scalar and vector variables are indispensable for the measurements constantly required in coping with interactions in the world around us. The overriding importance of energy and its time-rate, power, does not deprive such vector variables as force and momentum of the roles which they alone can properly fill.

24 THERMAL POWER MEASUREMENTS, INCLUDING SOME WITH WEIGHT FOR ORDINARY FOLKS

Measure not the work untill . . . the labor's done.
—ELIZABETH BARRETT BROWNING,
Aurora Leigh (1856)

Are all thy conquests, glories, triumphs, spoils,
Shrunk to this little measure?
—SHAKESPEARE,
Julius Caesar

The preceding four chapters deal principally with velocity, energy, and power measured in bodies moving as wholes and not in a scattered, dispersed manner. Now we turn to measurements of effects arising from the kinetic activities of submicroscopic bits of matter—molecules, atoms, even subatomic particles. These randomized internal movements include vibrations, rotations, and free-flying dashes and collisions.

Such submicroscopic motions constitute what we commonly call heat. We shall first deal with heat in general and with its thermal effects, reserving the thermometric realm of temperature (measurements of the degree of hotness) for Part IV.

By the end of the eighteenth century it was known that different substances behaved differently when exposed to seemingly identical temperature conditions. Thus, if 1 kilogram of water and 1 kilogram of lead are at identical temperatures and are placed over similar gas flames, the lead will have become hotter by about 32 degrees Celsius when the water has gained only about 1 degree Celsius.

We might say that lead responds to heat 32 times as readily as water, the standard substance. Early measurers, on the other hand, said that lead's heat capacity is only 31 thousandths (0.031) that of water. In more formal language: lead's specific heat is 0.031. This can also be taken to mean that raising the temperature of a mass of lead by 1 degree Celsius requires only 0.031 times as much heat as is required to raise the temperature of an equal mass of water by 1 degree Celsius.

Some other specific heats are: platinum, 0.032; mercury, 0.033; copper, 0.093; iron, 0.119; aluminum, 0.21. These specific heats apply to ordinary temperatures. Near absolute zero they become much smaller.

A simple analogy may be used. If a given amount of milk is poured into a narrow tube, it will rise higher than if poured into a wide tube. The fluid, milk, represents the heat absorbed; the narrow tube is like a substance with a low specific heat, like lead; the wide tube to a substance with notably high heat capacity (water). And the height to which the fluid rises is analogous to the resulting level of temperatures.

Analogies like these may have been overly tempting to early scientists, for they came to conceive of heat as an actual substance, actual though weightless and invisible. They gave this hypothetical substance the name *caloric,* which should not be confused with *calorie,* a unit of heat.

With the benefit of hindsight, we can assert today that the caloric hypothesis was unnecessary and incorrect. In both respects it resembled the phlogiston hypothesis about the nature of fire. It was the caloric concept which pictured water as having a large heat "capacity," meaning that it could swallow up large amounts of caloric without showing much increase in hotness.

When cool lead was placed in hot water, both shifted toward the same temperature, which proved to be closer to that of the hot water than to that of the cool lead. Caloric theorists pictured this as a flowing of the caloric substance out of the water and into the lead, until that equilibrium temperature had been reached. Then the caloric supposedly stayed put, until further temperature changes.

But what of the obvious fact that hammering or bending a piece of metal can make it hot? Or that objects can be rubbed until hot? The caloric theorists had to devise all kinds of elaborate and *ad hoc* explanations to account for such "exceptions." This is always a bad sign for the state of health of a hypothesis.

On the other hand, the kinetic theory of heat, referred to at the beginning of this chapter, implies that what we call temperature is the average kinetic energy per particle composing the substance (lead, water, or whatnot). This theory, emerging in a crude form late in the eighteenth century and seriously proposed during the first part of the nineteenth, was considered controversial and even somewhat disreputable by many eminent academic authorities. However, its final victory could not be indefinitely delayed.

We know now that mechanical work causes heat to increase in a body by increasing the energies of the tiny units, molecules or atoms, of which it is assembled. We know, too, that heat always and only is found to flow from a body at higher temperature into one at lower temperature, never in the reverse direction.

If heat and work are both forms of energy, how can they be equated? What is the mechanical equivalent of heat? What is the heat equivalent of mechanical work?

These questions were answered by the patient and stubborn British experimenter James Prescott Joule (1819–89), whose name is already familiar here because the SI unit of energy is named for him. Though beset by enormous difficulties involving the leakage of heat from his apparatus, he carried through a series of historic measurements of how much mechanical energy it took to stir water until its temperature was raised by a measured amount. His method consisted in supplying a precisely controlled amount of work (energy) to the stirring process, then noting the rise in temperature of the previously measured mass of water.

He could not, of course, measure the work done in units of the joule, which had not yet been established. Instead he used the foot-pound, for the work was done by allowing a known number of pounds weight (lbf) to descend a measured distance. Thus Joule determined how much work had to be done to raise by just 1 degree Fahrenheit 1 pound of water that had been at 60° F. (The equivalents in the Kelvin temperature scale, basic to the SI today, would be a 0.56 K rise in temperature of water that had been at 288.71 K. And in the Celsius scale, widely used in everyday applications, the equivalent would be a 0.56-degree rise in temperature of water that had been at 15.56° C.)

Joule's best result, published in 1878, was 772.55 foot-pounds of work at the latitude of Greenwich. That is, 772.55 ft lbf were equivalent to 1 Btu (60° F). Btu, sometimes written BTU, represents "British thermal unit," which is the heat that produces a 1-degree-Fahrenheit increase in 1 pound of water. The 60° F in parentheses specifies that the starting temperature was at that point.

The latitude was mentioned because the use of a unit of weight (lbf) rather than one of mass (such as lbm or kgm) involves the gravitational acceleration, which varies slightly from place to place on the surface of the Earth.

Joule thus demonstrated a true mechanical equivalent of heat: it takes as much energy to raise 1 pound (of water or any other substance) 773 feet as to raise by 1 degree Fahrenheit the temperature of 1 pound of water.

In spite of his nearly fanatical precautions, Joule's figure was almost 1 percent lower than the best present figure, which is 777.77 ft lbf, equal to 1,054.5 joule. Today, mechanical equivalents of heat energy are best stated in terms of accepted SI units. For example: 4.185 5 joule = 1 gram calorie (15 degrees Celsius). That gram calorie, often called just calorie, is the amount of heat needed to raise the temperature of 1 gram of air-free water from 14.5° C to 15.5° C when air pressure is 1 standard atmosphere.

It is necessary to specify the exact range of temperature change (14.5° C to 15.5° C). If the temperature change were from 19.5° C to 20.5° C with everything else remaining the same, the mechanical equivalent in joule would be about 0.1 percent less—that is, about 4.181 3 instead of 4.185 5 joule. This corresponds to the fact that water's specific heat at 20° C is measurably lower than at 15° C.

Water's specific heat—or better, its specific heat *capacity*—reaches a minimum at 34° to 35° C. If this lowest heat capacity of water is taken as 1, then the specific heat of water at 0° C (freezing) and also at 100° C (boiling) is found to be 1.009. Such a difference of 9 parts in 1,000 cannot be disregarded by metrologists and thermodynamicists.

Calories large and small—More in line with the SI system is this form of the mechanical equivalent of heat: 4,185.5 joule =

1 kilocalorie (15° C). This is based on heating a mass of 1 kilogram, rather than 1 gram, of water, and therefore is equal to 1,000 gram calories.

Formerly the two kinds of calorie were distinguished by capitalization: *calorie* meant the gram calorie, while *Calorie* meant the kilogram calorie. Today, however, the prevalent usage is *calorie* for the gram calorie, and *kilocalorie* (kcal) for the kilogram calorie. Also, as will be stressed later here, the common "calorie" of dieticians and dieters is the *kilocalorie,* not the tiny gram calorie.

The constant practical use of the mechanical equivalent of heat should not obscure the fact that the conversion of mechanical work into heat is much easier than the conversion of heat back to mechanical energy, or electrical energy, for that matter. Machines, motors, light bulbs and even living organisms throw off great amounts of "wasted" heat, which is thus made unavailable for conversion to mechanical or electrical or light energy. Today's advanced technologies reduce but cannot eliminate such costly conversion of other energies into scattered and unreclaimable heat. In the long run, all other forms of energy seem to run downhill into heat, so to speak.

However, the precise determination of the mechanical equivalent of heat, or the heat equivalent of mechanical work, is useful for measurements even in these processes of randomization of energies. We know now that each 4,185.5 joule of mechanical energy seemingly lost by friction in some mechanical device must give rise to 1 kilocalorie of heat energy.

Joule himself was well aware that it was far easier to measure the conversion of mechanical work into heat than of heat into mechanical work. As a young man he had devised new and effective methods for measuring the power of electric currents and of losses of electrical energy in the form of heat, generated as current passes through resistances. The phrase "Joule heat" is still used in this connection.

Without the work of Joule and a few other persistent and unswervable measurers and theorists of energy, science would have arrived even later at recognition of the interconvertability and equivalence of energy of all forms, now subsumed under the broader concept of the conservation of massergy. One of the classic true stories in the fascinating annals of metrology concerns Joule's nuptial journey. He was nearly forty when he married. The honeymoon of the new Joule couple was routed to the province of Haute Savoie in France. There at Chamonix, under the white majesty of Mont Blanc, Joule unpacked his sensitive thermometers, and the newlyweds went to work measuring temperatures of the waterfalls, both above and below the fall itself. Joule hoped to detect small temperature gains that would provide additional evidence for the mechanical equivalent of heat. In this he was disappointed. Apparently there was so much spray that the water as it fell was cooled by evaporation. Detectable temperature differences between top and bottom of the fall simply could not be established to meet Joule's exacting experimental standards.

Varieties of the Btu—Besides the Btu (60° F), which equals 1,054.5 joule, there are variants which differ significantly. The Btu (mean), equal to 1,055.8 J, is the 180th part of the energy that will raise 1 pound of air-free water all the way from 32° F to 212° F; that is, from freezing to boiling point under standard atmospheric pressure. The International Table British thermal unit, symbolized by Btu (IT), equals 1,055.1 J. The thermochemical Btu equals 1,054.4 J.

Besides these confusing contenders for the Btu title, there are others still, whose mechanical equivalents are also in the range of 1,054 to 1,055 J.

Variants exist also of the calorie (15° C) and the kilocalorie (15° C), equivalent respectively to 4.185 5 and 4,185.5 joule. These variants include the following, all listed for brevity in terms only of 1 calorie values: (1) the International Table calorie, with 4.186 8 joule; (2) the thermochemical calorie, with 4.184 joule; (3) a very low-temperature calorie, based on the energy that raises from 3.5° to 4.5° C a single gram of air-free water, and is equal to 4.204 5 joule; (4) the mean or average calorie, based on the hundredth part of the energy that raises from 0° to 100° C a single gram of air-free water and equals 4.189 7 joule.

For most computations, the thermochemical calorie (4.184 J) seems to be preferred. In making rough estimates and approximations, a working calorie equal to 4.2 J is sometimes useful. Between the smallest and the largest of the five calorie variants listed above, the difference in joule equivalents is less than ½ of 1 percent. This would apply also to the difference in the joule equivalents of the corresponding kilocalories.

Since the calorie and the Btu are both units of energy, the calorie per second and the Btu per second are units of power. Some important equivalences are:

$$1 \text{ calorie } (15° \text{ C) per second} = 4.185 \text{ 5 watt}$$
$$1 \text{ kilocalorie } (15° \text{ C) per second} = 4,185.5 \text{ watt}$$
$$1 \text{ Btu } (60° \text{ F) per second} = 1,054.5 \text{ watt}$$

Caloric cautions—This book will use the term *kilocalorie* (kcal) rather than *Calorie* for the unit used to measure the energy and heat contents of foods, diets, and other weighty subjects. In this way we reserve the word *calorie* for the gram calorie, which is technically correct; also we emphasize that the kilocalorie is no trivial matter for weight watchers.

A glass of milk contains about 170 kilocalories (or old "Calories"), which means 170,000 gram calories (or old "calories").

Let us consider the case of a man who is getting fat. Once, perhaps, his expenditure in work and heat averaged 3,000 kilocalories per day and his intake in food and drink likewise averaged 3,000 kilocalories per day. In that period, no doubt, his weight remained virtually steady. Now, however, a more sedentary way of life has reduced to 2,500 kilocalories his daily output of heat and work, but he has begun to pile in alcoholic beverages, gravies, pastries, and whatnot to an extent that has increased his average daily intake to 4,300 kilocalories.

His weight has begun to increase noticeably and distressingly. Can he compensate for extra food and drink by means of exercise alone? He has vowed one full hour of rigorous workouts per day. He is able to quote authorities on nutrition who declare it is *not* true that exercise is ineffectual in weight control. The exerciser, they maintain, does not inevitably experience so great an increase in appetite that he or she consumes additional food equal to that which has been "burned off."

Insights from the modern world of measurements show us that this exercise-minded overeater deceives himself when he expects additional muscular activity to compensate for an average daily excess food-energy intake of 1,800 kilocalories, or 7.5 million joule of energy.

His actual bodily needs, we have seen, are about 2,500 kilocalories, or 10.5 million joule, per day. This is an average power use of 122 watt around the clock. The fact is, to be sure, that during a sleep-rest period of some eight hours, he consumes about 1.5 million joule, an average power of roughly 52 watt. During sixteen active waking hours he uses the remaining 9 million joule, at an average rate of 156 watt, or about three times the resting average.

The ambitious exercise plan assumes that in one hour of intensive workout he can compensate for those 7.5 million excess joule. To do so would require that his additional power expenditure during that time rise to more than 2,000 watt *above* his hourly waking average of 156 watt. In short, that he increase that power rate more than 14 times.

In fact, even though he is strenuous, persistent and clad in sweatpants and sweatshirt all during that hour, he is unlikely to be able to expend more than 4 or 5 times his normal average of 156 watt. Even at 5 times that normal rate, or 780 watt, he can consume only 2.8 million joule of energy per hour, or about 2.4 million more than he would have used in that time without the special exercising.

The sad fact is that, even making every possible allowance and a few extra, about two thirds of his excess caloric, or kilocaloric, intake remains uncompensated. Even Superman in a sweatshirt would hardly have the stamina, even if he had the time, to approach that supposed solution to steady overeating. Exercise alone, though regular, intensive and exhausting physically, is not the long-run answer to increases in body weight and other ills resulting from continued overeating.

Not without heat—All such energy-power measurements of human or animal organisms at work require consideration of the fact that for every joule of energy appearing in the form of mechanical work, the body converts about 3 to 4 joule additional into internal heat energy. Humans and other higher animals can be measured as heat engines, fueled by food. Mechanical work output represents, at most, a minor fraction, one fourth, one fifth, or even less, of the total energy transformations in their metabolic processes.

Even the wonderfully adapted birds are no more than 20 to 25 percent efficient in producing mechanical work from the kilocalories of chemical energy contained in the food that they consume. The rest appears directly or indirectly as body heat, or as chemical energy remaining in the wastes which they excrete, or as stores of bodily fat.

Accumulation of excessive deposits of fat in a human or animal body is a signal that the body is not burning as many kilocalories of energy as it is assimilating in its digestive processes. The result is the storing away, largely in the form of fat, of the difference or surplus.

The only possible way to take care of such surpluses is to eliminate them at the source; that is, to reduce caloric intake to fit the body's actual needs. Health reasons aplenty exist for regular and ample physical exercise. But elimination of body weight gained because of overeating is not really among them. Even that admirable and desirable daily hour or two of walking will have little effect on excess weight, if other routines remain unchanged.

We can continue the chain from kilocalories to joules to watts, and arrive at conclusions combining metrology and practical moralizing: only two truly effective forms of exercise lead to reliable reduction in body weight. One is repeatedly shaking the head from side to side when urged to take a second or third helping. The other is pushing yourself away from the table before dessert time, if it is highly kilocaloric!

The chemical, or food, energy packed away into 1 kilogram of animal fat may be as much as about 9,000 kilocalorie, or nearly 38 million (potential) joule. Carbohydrates, including starches, and proteins, which contain essential body-building substances, are less concentrated in energy content. Typically, they contain per kilogram about 4,000 kilocalories, or nearly 17 million joule of combustible energy.

Terms such as "combustion" and "burning" are justified here. The release of the chemical energies of foods assimilated in the body takes place when they are combined with oxygen. The bodily combustion takes place at moderate temperature, however, thanks to a splendid regulatory system and the mediation of catalysts (enzymes).

The immediate source of body heat and the work done by the bodily muscles is the food recently eaten and digested. Only when recent intake has lagged behind total energy needs does the body begin to draw on—that is, consume—its stores of fats and carbohydrates. And only when such reserves have been largely exhausted, as after prolonged fasts or starvation, does it begin to burn up its own protein structures: muscles and other tissues.

These complex and marvelous metabolic processes, it must be stressed, result primarily in bodily heat rather than mechanical work. This is most easily apparent when the body is in a state of complete rest, with muscular activity reduced to heartbeat, respiration and other essentials. In such a basal or minimum metabolic state an adult transforms energy at a rate of about 1,000 kilocalorie per day, equal to nearly 4.2 million joule per day or a power of between 48 and 49 watt. All but a minor fraction of this power appears as internal bodily heat.

On the other hand, an athlete during a short burst of physical exertion may increase his rate of energy conversion to 750 joule/s (that is, 750 watt). Even then substantially more of that energy appears as body heat than as mechanical work done by the body.

Consider an athlete of extraordinary strength and stamina, such as a Channel swimmer, churning through cold waters toward a distant coast. Such a human motor may metabolize food-fuel at as much as 300 kilocalorie per hour, corresponding to nearly 350 watt of power. Even then, it is unlikely that more than about 70 watt (or 20 percent) can be assigned to actual work done by the swimmer's muscles. The rest (80 percent) increases, or maintains, the body's internal heat.

To call that 80 percent "wasted" heat would be a gross oversimplification and misunderstanding of physiology. Without the processes producing that heat flow, life and motion itself could not be maintained.

The chained oarsmen who propelled Roman galleys, miners lashed to work until exhausted, the Nazi regime's slave laborers who were worked literally to death—even in such extreme instances of misuse of humans as sources of mechanical power, we find that not more than 10 to 20 percent of the food-fuel supply was accounted for by work done. The rest appeared as body heat, so long as life itself survived.

Cold-blooded creatures, such as reptiles, generate less "wasted" heat than the warm-blooded, including humans. However, when temperatures outside are low, reptiles are sluggish and unable to muster much energy for movement. When humans become very active, in a mechanical or work sense, they generate much excess internal heat. Then their bodily cooling systems are called into play: sweat glands pour out moisture whose evaporation cools the skin and the blood that flows beneath it.

Not for nothing did the Book of Genesis speak of man laboring in the sweat of his brow.

Man-made engines do not necessarily generate more wasted heat than humans, horses or camels. Only about 10 percent of the gasoline-derived energy of a modern automobile is absorbed in pushing the car along the ground and through the air. Another 1 or 2 percent is consumed by the generator that charges the battery, by the fan that draws air in to cool the engine, by the water pump (if any), and by such adjuncts as power steering (if any). The remaining 88 or 89 percent is dissipated as heat. Some of this heat is emitted in the exhaust gases that emerge, with their burden of pollutants, from the muffler. Some is radiated away by the engine's water- or air-cooling system.

The advantage of the internal-combustion engine, whether gasoline or diesel type, is not in its avoidance of wasted heat, but rather in its ability to operate hour after hour without rest, and in its truly superhuman strength.

Nutrition and numbers—In warning against superfluous kilocalories and overweight we do not forget the hundreds of millions who are hungry and malnourished, both in the affluent United States and, to a far greater extent, in the underdeveloped world.

The following table shows, approximately but clearly, that children, though they need less food than adults, do need more food per unit of body weight, if they are to escape the lifelong effects of childhood malnutrition. The numbers shown are daily nutritional requirements (average) in kilocalories per kilogram of body weight:

AGE (years)	CHILD	WOMAN	MAN
0–½	120		
½–1	100		
1–3	108		
4–6	78		
7–9	73		
10–12	69		
13–15		53	63
16–19		44	57
25		40	46
45		38	43
65		31	36

A man or woman in the mid-twenties needs about 30 percent more food energy per unit of body weight than forty years later, at retirement age. At all these ages, of course, from birth to three score and more, kilocalories alone are not sufficient for health. Essential, too, are specific nutrients and body-building materials such as proteins, vitamins, and minerals.

Because of its high food-energy requirements per kilogram, a baby's power consumption in watts represents a respectable amount. A 4-kilogram (8.8-lb) infant consumes daily about 480 kilocalories, or 23 watt in power terms. This means a total of almost 2 million joule per day, some of which is used for growth, but most for other current needs.

A modern jet plane carries vast quantities of kerosenelike fuel, which is burned in oxygen sucked in from the atmosphere. The burning of 1 kilogram of such fuel releases about 47 million joule of energy.

A rocket to be fired in space must provide its own oxidant as well as fuel. A common choice for the former is liquid oxygen, and for the latter liquid hydrogen. The burning of 1 kg of hydrogen releases 143 million joule of energy. The burning of 1 kg of carbon, principal ingredient of coal, releases only 33 million joule.

Among man's basic food components, we have seen that 1 kg of fat when oxidized releases about 38 million J, or 9,000 kilocalories. Carbohydrates, on the other hand, when oxidized release 17 or 18 million J per kilogram, or 4,000 to 4,300 kilocalories. Proteins, when oxidized in the body, supply per kilogram about the same energies as carbohydrates, though in some instances proteins may supply 10 or 15 percent more. Typically, the body burns protein only incompletely, reserving some of these complex substances for use as building-blocks of tissue.

Adequate and desirable human diets are varied. A typical mixture is 50 percent carbohydrate, 33 percent fat or fatlike substances, and 17 percent protein. A healthy body builds up and maintains impressive reserves. Thus, during fasting or starvation a healthy human of normal weight can draw on a physiological "savings account" of between 300 and 350 million joule of stored energy-rich substances. The first to be used are the carbohydrate stores, then the fats, and finally the proteins, even those making up the muscles.

With physical activity reduced to a minimum, the body's energy consumption can be cut to about one-third of normal; that is, to about 4 million joule daily. This sets a limit on starvation survival of less than 3 months.

If we exclude, rather arbitrarily, the food intake that maintains the body's uniform internal temperature of 37° C (98.6° F), we find that a strikingly small amount of food suffices to account for an imposing amount of physical effort. Thus a slice of bread, with 65 kilocalories, supplies sufficient kinetic energy, by means of the muscles, to lift 2,000 kilogram a distance of 1 meter. But for the fact that gravitational acceleration diminishes with height, we would express it as lifting 1 kilogram a distance of more than 2 kilometer. This last statement is, in fact, correct: at 2 kilometer height above the surface, what had been 1 kg of weight (1 kgf) would be somewhat lighter, hence easier to lift still farther.

We humans, not being the most active of animals, are not very large eaters compared with some of our fellow creatures. We tend to consume daily as food about 1 to 2 percent of our body weight. A hummingbird or a small mouse must eat each day more than its own body weight in food.

At an opposite extreme are such huge vegetarian animals as the elephant and the hippopotamus. They must consume such vast amounts of low-kilocalorie fodder that they spend most of their waking hours foraging and eating.

Such curiosities, as well as the crucial facts of human nutrition, become measurable and meaningful thanks to Joule's establishment of the mechanical equivalent of heat, and the development of units that may be applied to the processes of living organisms and to the characteristics of inanimate chemical compounds or elements, relating energies and powers of many different kinds. The underlying unity of life with its nonliving environment is emphasized and elucidated by the world of measurements.

25 LIGHT ON LIGHT: MEASUREMENTS AND COMPARISONS

> *But, chief of all,*
> *O loss of sight, of thee I most complain!* . . .
> *Light, the prime work of God, to me is extinct* . . .
> *O dark, dark, dark, amid the blaze of noon,*
> *Irrecoverably dark, total eclipse*
> *Without all hope of day!*
> *O first-created Beam, and thou great Word,*
> *"Let there be light, and light was over all,"*
> *Why am I thus bereaved* . . . *?*
> *Since light so necessary is to life,*
> *And almost life itself, if it be true*
> *That light is in the soul,*
> *She all in every part, why was the sight*
> *To such a tender ball as the eye confined,*
> *So obvious and so easy to be quenched,*
> *And not, as feeling, through all parts diffused,*
> *That she might look at will through every pore?*
> > —JOHN MILTON,
> > *Samson Agonistes* (1671)

> . . . *All this long eve* . . .
> *Have I been gazing at the western sky,*
> *And its peculiar tint of yellow green:*
> *And still I gaze—and with how blank an eye!*
> > —SAMUEL TAYLOR COLERIDGE,
> > *Dejection* (1802)

The photometric units in the International System deal with the sensations that we designate by such names as light intensity, illumination, brightness and brilliance.

Measurement of electromagnetic radiation without regard to its seeability is radiometry. Measurement of the seeability of radiation is photometry, which has been called with some justice an aspect of physiology rather than one of physics. Photometry, in the strict sense, deals solely with radiations whose frequencies lie within the visible spectrum. The word light, indeed, means those electromagnetic oscillations to which the human eye responds. There is a certain semantic contradiction in such familiar phrases as "black light" to indicate ultraviolet radiations, or "invisible light" with reference to either ultraviolet or infrared radiations.

In Chapters 45, 46 and 47, sound, another physiological realm, will be examined. Light is related to electromagnetic radiation much as the sensation of sound is related to the vibratory waves of compression and rarefaction in matter.

There is an old philosophical teaser or paradox: If a tree falls in a forest where no one is at hand to hear, does it make a sound? The answer can be given: No, that falling tree causes compressional waves to travel through the air, but no ears and brain are present to interpret those waves as sound. Sound, in this sense, exists only within an organism. Similarly, if the forest burns, it emits radiations, but these do not become light unless someone perceives them—that is, sees them.

This, though a valid precaution, has little practical consequence. We still speak freely of "the speed of light in space" with reference to the speed of any and all electromagnetic radiations, from the longest radio waves to the most energetic (highest-frequency) gamma rays.

In terms of frequency, this is an enormous range or bandwidth. The lowest-frequency (longest) radio waves begin at about 10,000 hertz, where 1 hertz means 1 cycle per second. Gamma rays have been measured at energies corresponding to 3×10^{19} hertz or even greater. Thus electromagnetic radiations range from about 10^4 to 10^{20} hertz, a ratio of 10^{16} times.

Within that vast range we perceive as light only radiations between the red light of about 4.3×10^{14} Hz (430 TH or terahertz) and the violet light of about 7.5×10^{14} Hz (750 TH). Thus the ratio between the highest and the lowest visible frequencies is about 1.75—less than 2—against a ratio of 10^{16} (10,000 million million) for the full range of electromagnetic radiations which modern science detects and measures.

Every frequency has a corresponding wavelength, for frequency times wavelength must equal the velocity of light (c, or 2.997 925 $\times 10^8$ m/s). The wavelength limits of the visible spectrum are about 4×10^{-7} m (400 nm, symbol for nanometer) at the shortest violet; and 7×10^{-7} m (700 nm) at the longest red.

Our eyes, however, do not respond equally to all colors of light within this narrow visual bandwidth or spectrum. We perceive only feebly the longest red and the shortest violet. The peak of the curve of human visual perception lies almost exactly halfway between, near the 550 nm wavelength. The spectrum displays the familiar modulated rainbow sequence of colors: red, orange, yellow, green, blue and violet. Our peak sensitivity point lies in the green sector, but closer to the yellow sector below than to the blue above.

Different sources still supply slightly different wavelengths for this peak sensitivity point, all the way from 550 to 556 nm. Such uncertainties are typical of many measurements in the subjective field of photometry. They do not, however, seriously hamper the principal tasks. Here we have chosen 554 nm as the most probable peak, and note that it corresponds to about 5.41×10^{14} Hz in frequency.

The "greenest" green in the spectrum lies at about 520 nm in wavelength, and the "yellowest" yellow at about 575 nm. Hence the visual sensitivity peak of 554 nm lies nearly two thirds of the way between the middle green and the last color that most observers see as more green than yellow.

Shorter than the last visible violet of the spectrum are the ultraviolet radiations, which we cannot see. And longer than the last visible red are the infrared radiations, which we cannot see either.

It is true that we may sense them in other ways. When our skin reacts to sufficient infrared radiation, it sends to the brain signals meaning "There is heat here." And when exposed to enough ultraviolet, the skin reddens or tans, and may send to the brain distress signals meaning, "This hurts!"

Also, radiations far more energetic than those in the visible spectrum may injure our bodies (X-rays and gamma rays), while radiations far less energetic, such as radio waves, may affect our bodies little or not at all.

Information extracted from nonvisual radiations has enormously increased human knowledge during the past century. For example, astronomy today must be classified as optical astronomy, radio astronomy, X-ray astronomy, and even as infrared, ultraviolet and gamma-ray astronomy. The telescope is today by no means the only tool of the searchers of the skies.

In the submicroscopic, molecular, and even subatomic realms, great and continuing advances are made with microwaves, infrared and ultraviolet rays, X-rays and gamma rays. Probably the bulk of the basic scientific advances of the past several score years have become possible only with the aid of such nonseeable radiations and the precise methods that have been developed to deal with them quantitatively.

In this chapter, however, we are concerned with human, and essentially subjective, responses to the effects of light proper, rather than with objective measurements of electromagnetic oscillations and radiations in general.

Concerning the candela—The basic photometric unit in the International System (SI) is the candela (cd). It is still often, but erroneously, referred to by the older name of candle. The candela is the unit of luminous intensity. The French words *intensité lumineuse* have been used for it; also, earlier, the French word *brillance*, akin to the English word *luminance*, definable as "the property by virtue of which a surface emits more or less light in the direction of view." Critical care must be used with all these terms and concepts.

(Photometry is beset by many problems of nomenclature and precise definition. The common French word *brillance,* for brilliance, or brightness, is not quite right for the candela, and similar problems appear repeatedly in English.)

The candela, called at first "the new candle," was adopted in 1948. It was the latest of a long line of photometric units, varying from nation to nation and period to period.

The first photometric standards, more or less inevitably, were actual light sources, or standard candles. Thus the British standard candle was formed from $\frac{1}{6}$ troy pound of spermaceti wax, so shaped that it would burn (supposedly) at the rate of 120 grains per hour during an eight-hour period, since $120 \times 8 = 960$ grain $= 2$ troy oz $= \frac{1}{6}$ troy pound. The definition did not specify the size of the wick nor the purity of the air in which it was burned, although both of these factors influence the resulting luminous intensity or candlepower.

A later international unit was one tenth the luminous intensity of the Carcel lamp, in which colza oil was the fuel. Such lamps actually emitted from about 9.4 to 10 times the luminous intensity of the British standard candle. There were later also various standard lamps burning pentane (a hydrocarbon) and named for their designers, among them Harcourt, Simmance, and Dibdin.

German photometrists used the *Hefnerkerze* (HK), or Hefner candle, which yielded about 10 percent less luminous intensity than the international candle. The French had their own *bougie décimale* ("tenth of a candle"), with a luminous intensity between 95 and 100 percent of the international candle. The standard candle of the United States gave out about 1.5 percent more luminous intensity than its British opposite number. In short, the familiar unitary hodgepodge prevailed, and in an area where measurements and concepts were difficult at best.

The SI candela does not depend on the burning of any lamp or candle, but is defined by the luminous intensity of a theoretical perfect radiator, or blackbody, which glows incandescently at a particular high temperature.

The temperature is that at which pure molten platinum, under standard air pressure, begins to freeze, or solidify. It is 2,045 K (kelvin, in the SI temperature scale) or 1,772 degrees Celsius.

Precise measurements showed that when platinum was at this carefully defined temperature, each square centimeter of its surface emitted light at luminous intensity of 58.9 international candles. This was rounded off, and the standard area was designated as $1/60$ square centimeter, or $1/600,000$ square meter.

This rounding-off resulted in the present candela being about 1.8 percent smaller in luminous intensity than the old British international candle. Such a difference was relatively unimportant under the circumstances. The old unit had varied more than that percentage, from measurement to measurement. The candela is far more uniform; however, as mentioned earlier in this book, the photometric unit is still the least accurate of the basic SI units.

The process by which the candela is "realized" or produced under laboratory conditions reveals much about the nature of light radiations. If a flat surface of freezing platinum were measured, it would be found to emit with an intensity of only about $1/3$ candela for each $1/60$ square centimeter of its area. This results from the fact that platinum is only about $1/3$ as emissive as a perfect blackbody. In order to ensure that the freezing platinum will behave more nearly like such a theoretical perfect radiator, it is placed at the bottom of a narrow enclosure or well, whose sides are formed of thoria (thorium dioxide), a very white and very heat-resistant substance. The radiations emitted by the incandescent platinum are reflected back and forth repeatedly; many of them reflect back to the surface of the metal itself, which thus absorbs some of its own emissions, then reradiates them, and so forth.

Thus the small platinum surface at the bottom of the thoria well behaves very nearly like a complete radiator, or blackbody. It is kept at the critical temperature by intense microwaves, generated on the outside, which cooks the metal much as a potato is baked in a modern microwave oven. The platinum is first melted, then allowed slowly to cool until it attains its freezing temperature.

At that instant, the emitted light, streaming up the thorium tube whose cross section is accurately known, is directed by a reflector-and-lens system into one side of a photometer head, a device for comparing very precisely two illuminations. The illumination of the other side of the photometer head is supplied by a precision electric lamp, whose filament has a variable voltage supply. Adjustments in voltage are made until the two illuminations appear to be equal. Then the lamp is standardized as a 1-candela lamp, with the help of an equation that takes into account voltage, distance to the photometer head and optical characteristics of the reflector and lenses. Such a standardized 1-candela lamp can then be used to measure other luminous intensities, without the need for again melting the platinum at the bottom of the thoria well.

Formerly photometer heads required decision-making by trained observers to determine the point at which the two illuminations could be considered equal. Today such comparisons are commonly made by means of photoelectric cells, which perform more rapidly and dependably. By means of filters these cells have first been made to respond to luminous intensities in the same way as the human eye—or as close to the same as photometrists are able to attain.

Luminous intensity, measured in the candela unit, is commonly symbolized by the letter I. It is supplemented by a related physical variable called luminous flux, which is the rate of transfer of light radiation over a given surface, and for which a special unit, the lumen (lm), is provided. Luminous flux is commonly symbolized by Greek capital phi (Φ).

Looking at the lumen—The lumen is defined by the International System as "the luminous flux emitted in a solid angle of 1 steradian by a uniform point source having an intensity of 1 candela." Thus the lumen can be measured only after the candela is first established.

The steradian (sr) is the SI supplementary unit for measuring solid angle. It is thus the analogue of the radian (rad), which measures plane angle. The angle of 1 radian in a circle is defined by an arc of r on the circumference, where r is the radius of the circle. The length of the entire circumference of every circle is $2\pi r$. Hence the radian corresponds to an arc that is $\frac{1}{2}\pi = 0.159$ times the whole circumference.

Now, in a sphere with the same radius r, we imagine the solid angle of 1 steradian as formed between two planes that intersect along a line through the center of the sphere, just as the plane angle of 1 radian was formed between two lines intersecting at the center of the circle. The portion of the surface of the sphere lying between the two planes is r^2 (just as the length of the arc between the two lines to the circumference of the circle was r). But the total surface of a sphere is $4\pi r^2$. Thus the steradian corresponds to the fraction of the sphere's entire surface equal to $\frac{1}{4}\pi = 0.079\,6$ of that surface. The result is that the surface of a sphere contains $4\pi = 12.566$ steradian.

The SI definition of the lumen requires that 1 lumen of luminous flux appear within each solid angle of 1 steradian around the point source whose luminous intensity is 1 candela. There are 12.566 steradian of solid angle in a sphere completely surrounding such a source. Hence the total luminous flux from a source with luminous intensity of 1 candela is 12.566 lumen.

One prominent example is that of our Sun. True, it is not a "point" source, for it has a very apparent disk. However, its light flux is 2.56×10^{28} lumen; and its luminous intensity can thus be considered to correspond to about 2.04×10^{27} candela.

Light sources used in our homes and offices are not pure "point sources" nor do they emit light flux uniformly in all directions. However, it is possible to measure the luminous flux in various directions and compute an over-all average. Thus a typical 100-watt tungsten-filament bulb emits about 1,700 lumen, indicating a luminous intensity of about 135 candela.

Photometrists sometimes make use of a variable called quantity of light, which is the product of luminous flux, measured in lumen, and the duration of its action, measured in the second of time. If a bulb emitted an average of 1,000 lumen of luminous flux over a period of 100 hours, it would have supplied a quantity of light of $100 \times 3,600 \times 1,000 = 360$ million lumen-second.

Considerations of quantity of light are close to some consumer questions faced by most householders. For example, a typical 75-watt tungsten bulb emits about 1,200 lumen when its total light flux is averaged. Another 75-watt bulb, of the type called "long-life," may be usable from 5 to 10 times as long as the first. However, because its longer life is the result of the lower temperature at which its filament glows, it delivers only 750 lumen of light flux, or 37 percent less than the other. Obviously, the long-life bulb delivers more lumen-second units before it burns out. On the other hand, since both bulbs have like electrical power rating, the longer-life bulb uses more watts per lumen, though in the long run it may use a smaller ratio of electrical energy (watt-second or watt-hour) per lumen-second, or quantity of light supplied, before it burns out.

Measuring a light in the round—Measurement of the luminous intensity of a light bulb is done by a series of separate photometric measurements at different so-called zone angles related to the line of the bulb's central axis. Thus a zone angle of 0° may correspond to the direction of the base of the bulb (which cuts off much light); 90° corresponds to sideways light from the bulb (likely to be the direction of maximum luminous intensity); and 180° is the round end of the bulb (considerably less luminous intensity than at the maximum).

Here is a rather typical group of 10 different readings in candela of luminous intensity at intervals of 20° in zone angles:

ZONE ANGLE (degrees)	LUMINOUS INTENSITY (candela)
0	50
20	65
40	127
60	147
80	148
100	146
120	138
140	128
160	125
180	120

The arithmetic mean of these 10 measurements is 119.4 candela, but by graphical and geometric methods it is found that the so-called "spherical" mean, or average all around the bulb, is 133 cd. This corresponds to an average luminous flux of about 1,670 lumen.

The question of the luminous efficiency of light sources is an inevitable and important one. How much mechanical or electrical power, measured by the watt, is required to produce 1 lumen? Or, the other way around, how many lumen of light flux can be obtained from 1 watt of power, electrical or mechanical? These units are properly comparable, for just as the watt is a rate of flow of energy, so the lumen is a measure of the flow or flux of luminosity or luminousness.

The maximum visual effect, we have seen, is given by monochromatic light in the yellowish-green, at a wavelength of 5.54×10^{-7} meter. The response of average or typical human eyes is such that if 1 watt of power were converted completely into light at this wavelength, such light would show 683 lumen of luminous flux. Thus we can say that maximum luminous efficiency is 683 lumen per watt, or lm/W.

Though this would be the most efficient and economical light source, it would be a most unsatisfactory one, for everything we would see by it would be green. Some objects would be darker green, some lighter, but other colors would be lacking.

What, then, would be the efficiency of a light source that supplied us a mixture of light at wavelengths (colors) like those composing daylight? It would supply about 35 percent as many lumens per watt as that maximum efficiency all-green light, or about 239 lumen/watt. This supposes, rather unrealistically, that the power goes entirely into visible light in the indicated mixture, and not into invisible heat radiations.

When we turn to the realities of hot (incandescent) objects as light sources, we find the efficiencies fall still lower. Thus, if we had a perfect blackbody radiator heated to 6,000 K (just about the temperature of the surface of the Sun), its visible radiations would be mixed much like daylight. However, its efficiency would be only 14 percent that of the imaginary all-green maximum-efficiency source. Thus it would attain about 96 lumen/watt.

Suppose it were possible to create a light source that would emit neither heat (infrared) nor ultraviolet radiation, but poured out all its power in the same mix of visible wavelengths that we find in ordinary daylight. It would have a luminous efficacy of about 239 lumen per watt, or 35 percent of that theoretical maximum-efficiency standard (683 lm/W).

Another purely hypothetical light source would put equal power into all visible wavelengths from the longest red to the shortest violet. Its luminous efficacy would be about 200 lm/W, or 29.3 percent of that theoretical maximum.

The actual light sources around us are substantially lower in luminous efficacies. A typical gas-filled tungsten filament bulb of 100-watt power provides about 17.5 lm/W, or 2.6 percent of the theoretical maximum. Larger tungsten lamps, in the 500-to-1,000-

watt range, reach about 33 lm/W, or nearly 5 percent of that theoretical maximum.

Fluorescent lamps, especially the so-called "cool white" types, between 20 and 200 watts in power, attain luminous efficacies from about 63 to 84 lm/W, or 9.2 to 12.3 percent of the theoretical maximum. Colored (nonwhite) fluorescent lights vary widely in their luminous efficacies. A 40-watt green tube may reach 112.5 lm/W, or 16.5 percent of maximum, but a red tube of the same power attains less than $\frac{1}{20}$ of that luminous efficacy.

Among the halide type of high-intensity discharge lamps that are coming into increasing use, luminous efficacies as high as 115 lm/W, or 16.8 percent of the theoretical maximum, have been attained.

Luminous efficacies of candles or oil lamps, or even of gas-mantle lamps, are lower than those of the tungsten incandescent bulbs.

To guide consumers, modern lamps should be marked clearly, not only in terms of their power requirements in the watt unit, but also in terms of their output of luminous flux in the lumen unit, and even in terms of their luminous efficacies, measured by lumen per watt.

Any marked improvements in the present luminous efficacies of artificial light sources probably will have to be attained in the direction of some kind of "cold light," which, like the bioluminescent emissions of fireflies and many other living creatures, emits little infrared radiation, but channels its power into the visual spectrum.

Measuring luminance—The SI definition of the lumen assumes a point source of light. However, the distant stars, as we see them from Earth, are about the only light sources that approach a pointlike state, with little or no visible disk. Our common light sources have apparent and even measurable emitting areas. Hence, the SI provides for the physical variable called luminance. It is measured as luminous intensity per unit area, or candela per square meter, in unit terms. The area is always the apparent area of the emitting body or surface as seen by the observer making the measurement.

Metric luminance units are the *nit* (nt), with 1 nit = 1 candela per square meter (cd/m^2), and the *stilb* (sb), with 1 stilb = 1 cd/cm^2. The word "nit" derives from the Latin *nitere,* "to shine." It is not nit-picking to note that the nit has not made much headway as a unit in the United States or Britain. Even though, unlike the stilb, it fits into the MKS system, it is inconveniently small for most practical photometry. The far larger stilb derives its name from Greek *stilbein,* also meaning "to shine."

Luminance as a photometric variable is often symbolized by *L,* and sometimes by *Lv*. It is the measure for comparing the relative amounts of light an observer receives from surfaces.

Another physical variable in photometry is illuminance or illumination, commonly symbolized by *E*. It must carefully be distinguished from luminance. An early unit of illuminance was the foot candle, sometimes called the candle-foot. This was the illumination resulting when 1 lumen of luminous flux was uniformly distributed over an area of 1 square foot. Obviously, this is not a metric unit combination. In the SI the equivalent is the *lux* (lx), being the illumination when 1 unit of luminous flux is uniformly distributed over an area of 1 square meter. Since 1 m^2 = 10.76 ft^2, 1 foot candle = 10.76 lux.

Its abbreviation is best avoided, for lx looks much like "one x," and does not even save appreciable space. Like the nit, the lux is rather small for the photometric tasks it is needed for. Use is often made instead of the *phot,* a unit equal to 10,000 lux, or even of the milliphot (10 lux).

Some significant luminances—Clear blue sky on a smogless day shows luminance in the range from 0.2 to 0.6 stilb. A typical candle flame is luminant in the same range: about 0.5 stilb. The natural brightness of the moonless night sky at the zenith is about 2×10^{-8} stilb (provided there are no nearby urban centers whose lights cause luminous "pollution" of the sky). Thus, the night sky between the visible stars shows a luminance of about 1 ten-millionth that of the blue sky by day.

Swiftly increasing amounts of light reaching the sky from the illumination of cities and urban areas are causing huge problems for optical astronomers in many an observatory. Thus, in parts of Southern California, light streaming upward provides interfering illumination up to 20 percent as strong as the natural brightness of the night sky. A host of important and delicate observations are hindered as the lights of heaven are swamped and submerged in the unwanted light from Earth.

It has been reliably estimated that outdoor public lighting, measured in lumens, has been increasing at fantastic rates—the "lumen growth" reaching more than 20 percent per year in some important urban areas. It results not only from increased use of electric power for such lighting, but from the installation of the new high-pressure sodium (Lucalox) lamps with their high luminous efficacies of 110 to 120 lumen per watt.

Luminance of a typical fluorescent tube light is about 1 stilb. A sheet of white paper under the noonday sun may show 3 stilb or more. Sodium vapor lights are measured in the range of 5 to 10 stilb. A frosted (diffuse) incandescent light bulb shows as little as 30 sb, compared with 200 to 1,000 sb for a similar bulb of clear glass. The crater of a carbon arc lamp, or searchlight, shows 15,-000 to 16,000 sb; ultrahigh-pressure mercury vapor lamps, up to 30,000 sb; and the Sun itself, 150,000 to 200,000 sb.

Such luminance figures, measured in sb units, show the brightness of the light source per unit area of that source. Illuminance figures, on the other hand, being measured in terms of unit area of the target or illuminated object, show how well or perhaps excessively illuminated it is, as the observer sees it.

A corridor or hallway may be dimly lighted with 20 to 30 lux. A room not intended for reading or writing may be adequately lighted with 200 lux, but up to 500 lux should be supplied for close reading or writing. A workshop in which tools and instruments are to be used requires about 1,000 lux, if the work is not too fine or exacting. However, for precise drafting or engraving, that should be doubled to 2,000 lux—about the same as the illumination under a clear blue sky on a smogless day.

The illuminance of a sheet of white paper may be as great as 70,000 to 80,000 lux when it is held flat under the unclouded noonday sun. This is 35 to 40 times that which the eyes require or feel comfortable with. The situation is much like that with clean white snow under intense sunlight. It may not seem very bright, but can cause eye discomfort and worse if stared at, even when dark glasses are being worn for protection.

Extraordinary differences in levels of illumination, as measured in the SI lux unit, can be coped with by some of today's most sophisticated electro-optical devices. For example, the "low light level" television systems developed for the United States intervention in Viet Nam and more recently offered in modified versions for domestic surveillance, security and snooping applications. They include cameras mountable on poles high above the ground, turnable as desired by remote control, and able to register images in almost any kind of light. Thus they can operate at about 0.001 lux, on a moonless night, or even at about 100,000 lux, in bright sunlight. This represents a range of light levels between 1 and 100 million.

Photographers and some illumination engineers in the United States have often used another luminance unit, the *lambert* (L), named for the photometrist J. H. Lambert (1728–77). It is the brightness of a uniformly diffusing surface that radiates or reflects 1 lumen per square centimeter. The *millilambert* (mL) is of more convenient size. The *apostilb* (1 lumen per square meter) equals a tenth of 1 mL. For scientific work, however, units such as the nit or the stilb, expressing candela per unit area, are preferable.

The nonmetric foot-lambert unit is the luminance of a uniformly diffusing surface that radiates or reflects just 1 lumen per square foot of area. One foot-lambert is equivalent to 3.426 25 candela per square meter, or nit.

Photographers could reduce confusions by adopting SI units for their light meters (exposure meters). One type of such meter is called a "reflected light" meter. It is pointed at the scene to be photographed. Hence such meters could be calibrated in units of the nit or the stilb. The second type, called incident light meters, are pointed at the light sources illuminating the scene to be photographed. Thus, they measure illuminance, and could be calibrated in units of the lux, or lumen per square meter.

However, even use of the best available equipment does not solve all problems in the tricky field of photometry. The photographer, after getting an averaged or integrated reading, may have to hold his light meter close to highlights and hot spots to get local readings on extremes. These and other rule-of-thumb procedures will modify his interpretations of the meter's over-all readings.

The case of "color temperature"—Photographers using color film measure what they call "color temperature" in units of the kelvin, the SI unit of absolute temperature. Actually, they are not measuring degree of hotness or coldness; rather, they are indicating the mix of "hot" (blue) and "cold" (red) colors in a flashbulb or other light source.

When a piece of iron or other substance with high melting point is heated ever higher, it begins in time to glow a dull red, then becomes a bright red, orange, yellow, and finally "white hot."

All bodies, including the myriads of stars, can range from red-hot to white-hot, or even blue-white-hot incandescence. The relatively colder colors are those at the red (lower-frequency) end of the light spectrum; the hotter ones are at the higher-frequency, or blue-violet, end.

As temperature is raised, the total radiated power (including all electromagnetic emissions, visible or otherwise) increases steeply —as the fourth power of the absolute temperature. At first, before incandescence begins, power is radiated only in the infrared region, hence invisibly. Then, as the temperature mounts, it moves increasingly through the visible spectrum and even into the ultraviolet region beyond.

Thus the "mix" of any kind of light is likely to reveal temperature conditions under which it was emitted. Sunlight is "whiter" (farther from orange-yellow) than the light from a tungsten bulb. Hence sunlight's color temperature is higher. On the other hand, light from the unclouded blue sky is even higher than sunlight in its color temperature.

We see the sky as blue because the particles in the atmosphere scatter back to our eyes far more of the high-frequency (short) blue waves than of the longer red waves. The relative amount of such scattering, in fact, increases as the fourth power of the frequency of the light itself.

The atmosphere, however, tends to screen out the very short, high-frequency ultraviolet and far-violet wavelengths. In sum, the very high "color temperature" of the blue sky does not reflect actual temperatures up in the air. It reflects, rather, the selective way in which the atmosphere scatters back to our eyes the wavelengths typical of higher, rather than lower, temperature origins.

In terms familiar to millions of eager photographers the world over, we can say that a color temperature of 2,500 to 3,500 K characterizes light supplied by tungsten spotlights or floodlights; a color temperature of about 6,000 K characterizes the familiar mix we call daylight; while the light from the blue sky alone corresponds to a color temperature distinctly higher still.

Flashlight bulbs may be rated at 3,300 K, indicating they are suited for color film whose emulsion is designed for tungsten-light color temperatures. Or they may be rated at 6,000 K, indicating they are suited for use with the daylight type of color film. The maximum luminous flux supplied by a flashlight bulb is expressed today in lumen units. Some may have a peak luminous emission of 900,000 lm; others of 2.5 million or 3.5 million lm.

Measuring normal sight patterns—Definitions of SI photometric units do not mention the human eye and its peculiarities. Nevertheless scientific photometry is based on extensive testing of the visual responses of a large number of subjects with normal vision. Much of this fundamental measurement took place early in this century when K. S. Gibson and W. P. T. Tyndall, of the U.S. Bureau of Standards, established a scale of the luminous efficiencies of various wavelengths in the visible spectrum.

Their results, generally confirmed by later investigators, show how rapidly the eye's response falls off as the wavelength becomes either shorter (more blue) or longer (more red) than the maximum response wavelength of 5.54×10^{-7} meter, synonymous with a frequency of 5.41×10^{14} Hz.

For example, at a frequency of 6×10^{14} Hz, only about 11 percent greater, the resulting green light has lost 27 percent of the effectiveness of that maximum point. At 7×10^{14} Hz, in the blue violet, and about 29 percent higher in frequency, about 95.5 percent of the peak effectiveness has been lost.

In the opposite (red) direction the fall-off is similarly steep. At 5×10^{14} Hz, or about 7½ percent less than the peak frequency, the yellow light has fallen off more than 67 percent in visual effectiveness. At 4×10^{14} Hz, close to the infrared, 99.95 percent of peak effectiveness is lost.

It is by no means an accident that our peak sensitivity of vision is found just about at the center of the visible spectrum, nor that the peak of the Sun's emissive power is quite close to the peak of our visual sensitivity, whether we measure the light in terms of wavelengths or frequencies. Human eyes, like the eyes of other creatures, evolved in response to the available light.

Some insects, such as bees, may respond to frequencies far in the violet to which humans are almost blind. In every species, however, the visual equipment and response is beautifully adapted to the light "mix" most likely to be useful to those creatures in their struggles to survive and reproduce their kind.

The Sun as a light source—Even the luminous efficiency of the Sun itself can be estimated. Its surface temperature is nearly 6,000 K, or about three times the freezing temperature of the platinum used in establishing the basic candela unit of luminous intensity.

We have noted that the Sun's total radiated power at all electromagnetic wavelengths, from radio waves to the shortest detectable X-rays, is about 4×10^{26} watt. Since its surface area is 6×10^{18} square meter, this indicates a radiation of 6.7×10^7 watt per m².

About half of the Sun's radiation lies within the wavelengths of visible light. However, its luminous efficiency is not 50 percent, for some colors are more visible than others. That is to say, if the Sun's total visible emission consisted of green light at just 5.41×10^{14} Hz, its luminous efficiency would be about 50 percent. In fact, however, the Sun's actual efficiency figure is probably no greater than 9 to 10 percent. This is a little better, but not much better, than the luminous efficiency of a fluorescent-tube lamp.

Specifically, if all the Sun's total power of 4×10^{26} watt were converted into luminous flux at the maximum rate of 683 lumen per watt, the Sun would emit a total luminous flux of 27.32×10^{28} lumen. We have noted, however, on an earlier page that its actual luminous flux totals about 2.56×10^{28} lumen. Hence, its luminous efficiency is about 9.4 percent.

The Sun radiates a total power of about 66 million watt per m² of its surface. The freezing platinum, referred to before, radiates a total power of nearly 1 million watt per m² of surface. However, the luminous efficiency of that platinum is only about 1.1 percent, lower even than that of a tungsten filament light bulb. This is about what we should expect because that tungsten filament is operated at a temperature higher than that of the freezing platinum.

It may seem arrogant to rate the Sun in terms of efficiency in supplying the visual sensations that human eyes perceive as light. However, by definition, photometry deals with effects and sensations experienced by normal eyes.

Some modern interpreters of metrological practices maintain rather eloquently that luminous intensity is not truly a measurable physical variable, in the sense that a meter of length or a kilogram of mass is measurable. True, it must be compared with a subjective norm. The photometers employed in determining the basic candela of luminous intensity must be manned by human observers, or monitored by photo-optical cells whose responses, by means of filters, have been made to approximate closely the responses of average human eyesight. Thus, the candela and every other unit derived from it is bound to be a subjective standard. However, humans do not live by objective frequencies and wavelengths alone. Their lighting arrangements, their photographic procedures, their sensations of security or lack of it, all depend on photometry for important data.

Photometric measurements probably will continue to lag behind measurements of other physical variables in precision and reproducibility. This is all the more reason to select and use consistently the units that are as well defined and rational and consistent as the limitations of the human sight and human brain will permit.

IV / This Thermal Universe

> . . . *the most dynamic, distinctive, and influential crea-*
> *tion of the western mind is a progressive science of nature.*
> —C. C. GILLISPIE,
> *The Edge of Objectivity* (1960)

26 ON THE TRACK OF TEMPERATURE

Some say the world will end in fire . . .
—ROBERT FROST,
Fire and Ice

"The kelvin, the unit of thermodynamic temperature, is the fraction $\frac{1}{273.16}$ of the thermodynamic temperature of the triple point of water."

Thus is the International System's temperature unit defined, since the adoption of Resolution Four at the Thirteenth General Conference on Weights and Measures in the autumn of 1967.

This definition establishes a *unit,* rather than a scale and its degree, to measure temperature. The previous definition, dating from the Tenth Conference, had not established a true unit, for it stated:

"The thermodynamic Kelvin degree is the unit of temperature determined by the Carnot cycle with the triple point of water defined as exactly 273.16° K."

Today, for scientific purposes, temperature is measured by the kelvin unit, abbreviated simply as "K," without any degree sign required or even permitted. Temperatures are constantly expressed in terms of degree Celsius (°C) and even degree Fahrenheit (°F), but these have become secondary, and the K is king.

Moreover, the K unit and its current definition can serve as a key to help unlock one of the most teasing and fascinating tangles of metrology. What *is* temperature? Above all, what is this particular type called in the definition "thermodynamic temperature"?

Humans are constantly aware of temperatures around them, in the vaguely qualitative way that our perceptions permit. Centuries ago began the slow struggle to work out systems and scales for quantifying temperatures. But even while progress was made in these directions, the nature of temperature remained obscure. In this chapter we shall seek to reveal enough of that essence of temperature to make clear why the kelvin unit exists today, and why it takes precedence over the other scales that still merit mention, because they are still in use around the world.

Temperature's cryptic qualities—Temperature is a cryptic kind of physical variable. In the first place, it cannot be measured directly. Indirect and relative approaches must be employed. Our temperature-measuring devices—the thermometric mechanisms in actual use—depend on such consequences of temperature changes as the expansion of substances resulting from increasing temperature; the increased electrical resistance of substances resulting from increasing temperature; the changes of electromotive force (voltage) when junctions of unlike metals are subjected to changes of temperature; the velocity of sound waves in substances under various temperature conditions; and so on.

Literally dozens of different physical effects have been used or could conceivably be used as indirect indices of temperature changes. Yet the actual measurements are not measurements of temperature as such: they are measurements of the length of a column of mercury; of the resistance in ohms of a strip of metal; or the velocity of sound waves in some medium; and so forth. The numerical changes in such nontemperature physical variables are then interpreted as an index to the temperature level with which the sensing device or mechanism is in contact.

When two masses or systems of matter are placed in thermal contact, a form of energy called heat flows from the one at higher temperature into that at lower temperature. This energy transfer continues until the temperatures are equalized. The mass that was "hotter" has lost temperature; the mass that was "cooler" has gained temperature. The former temperature difference has been eliminated. Once that equilibrium is attained, no further net transfer of heat energy takes place.

Hence, temperature is a name for some condition in matter which determines the direction and extent of transfer of the kind of energy we call "heat." Heat is one form of energy, but by no means the only form. The temperature of a body can be increased by allowing heat to flow into it from a warmer body next to it, but

the temperature can be increased also by doing work on the body
—as by pounding it, stirring it, rubbing it. Sending an electric
current through the body may raise its temperature. Radiating its
surface with infrared or light waves also may increase its temperature.

Each such change increases the body's temperature and so increases the internal energy that it is able to share, in the form of
heat, with other bodies—provided that the essential temperature
difference exists. The measurement of the "heat content" of bodies
goes on constantly. But the internal energy being measured did not
necessarily enter those bodies in the form of heat, nor is it certain
that it will all later leave those bodies in the form of heat.

Heat, in fact, should be used only with reference to the energy
inflows and outflows that take place in consequence of *temperature
differences* between the body in question and other bodies or systems of matter. Heat and temperature are thus virtually inseparable, but by no means identical.

Thermodynamics is the area of physical science concerned with
relationships between the energy-in-motion called heat and the
mechanical motions of sizable masses of matter. Thermometry
might be called the measurement of the effect of these ever-shifting
relationships on the tiniest particles or structures within matter.

The tiniest particles or structures within ordinary matter all
around us are atoms and molecules, each molecule being an assemblage of two or more atoms. What do these atoms and molecules *do* in order to cause the mass of matter to show a change in
temperature?

They move, on the average, more swiftly with each increase in
temperature, and more slowly with each decrease in temperature.
But that is an upside-down way to state it. In the first place, the
thermodynamic temperature is proportional not to the average
velocity of the atoms or molecules, but to the average motion
energy per atom or per molecule, as the case may be.

Motion energy, usually called kinetic energy, increases in proportion to the square of the *velocity* of a moving object, big or
little. Thermodynamic temperature increases in proportion to the
average kinetic energy per atom or molecule composing the matter
whose temperature is being measured.

Most substances around us are mixtures of atoms and molecules of widely different masses. For example, each atom of the mercury in a thermometer is far more massive than the molecules of oxygen, nitrogen, and other gases composing the air whose temperature that mercury is used to measure. Yet when the mercury is "at" the temperature of the surrounding air, the average kinetic energy of the mercury atoms must be neither greater nor less than the average kinetic energy of the oxygen molecules, the nitrogen molecules, and the rest of the less abundant molecules in the air.

Clearly, at any such equality of temperature the average or typical velocity of the more massive mercury atoms must be far less than the average velocity of the lighter oxygen molecules, and thus we must stress the *energy* per atom or molecule in preference to the average *velocity* per atom or molecule.

Temperature reflects the average disposable or transferable motion energy of the least or tiniest bits of which a substance is composed. It is the average energy-of-motion-per-smallest-component. Temperature is not the *total* of the kinetic energies in any mass of matter. If it were, the larger the aggregate mass, the higher its temperature would register. A 1-gram bit of iron at the boiling temperature of water (100° C) and a 100-kilogram chunk of iron also at 100° C are in equilibrium, thermally speaking. The big piece contains 100,000 times as many iron atoms as the small one, but in each the *average* kinetic energy per iron atom must be the same.

The word *average,* and its synonym *mean,* constantly appear in discussions of thermodynamic temperatures. At any instant some atoms will be moving faster than the average speed and others more slowly. Hence some have less kinetic energy than others, but they are so tiny, so numerous, and so constantly interacting that their energies are continually being redistributed. When a body of gas, of liquid, or of solid is found to be "at" a particular thermodynamic temperature, this means that everywhere within the space it occupies the average kinetic energy per molecule or atom is equal to that everywhere else.

Thermometric measuring devices always take *averages* of the kinetic energies of myriads of atoms or molecules. Temperature is indeed a statistical variable. It summarizes average, or representative, conditions among enormous numbers of individual and interacting components. A single atom or molecule cannot have a "temperature." Indeed, millions or billions are involved in every meaningful temperature measurement.

Temperature, being a statistical average, is analogous to some social indices, such as "per capita income" in a large population. Some individuals have income many times that average, others have little or even no income. Yet it is possible to state that nation A has a higher per capita income than B; or a lower one; or even an identical one. An individual citizen of nation A or B has an annual income, just as a single atom at each instant has a kinetic energy. But no one citizen of A or B has a "per capita income," and no single atom, or even few score atoms, can have a "temperature."

Unmeasurable "temperatures"—Knowledge of the fact that thermodynamic temperature is directly proportional to the average motion energies of atoms or molecules in a large assemblage of matter is enormously enlightening and useful. It helps to understand why, in some highly unstable and swiftly changing states of matter, temperature cannot properly be measured at all. A certain stability and continuity is required in order to make possible the interactions and comparisons that yield numerical temperatures.

Atoms and molecules move in various ways within the containers or the shapes that confine them. The state best suited for illustrating the kinetic or motion theory of temperature is that of a gas. Its atoms and molecules fly about, every which way, constantly bumping into one another and into the walls of the confining receptacle. Each encounter results in a bouncing-away, followed a tiny instant later by a new collision.

Helium gas, composed of single atoms, comes about as close as any substance to the hypothetical "perfect gas." Its atoms have a diameter of about 1.9×10^{-10} meter. This means that more than 5 billion of them would have to be lined up to attain a total length of just 1 meter. The colliding units are thus extremely tiny.

Also they are numerous almost beyond comprehension. The atomic mass of helium is 4. This means that 1 mole (gram molecular mass) of helium has a mass of 4 gram. The universal constant called Avogadro's number tells us that 1 mole of any substance contains $6.022\,2 \times 10^{23}$ separate bits (atoms or molecules, as the case may be). This means that 1 gram of helium contains $1.505\,5 \times 10^{23}$ atoms, and 1 kilogram contains $1.505\,5 \times 10^{26}$ atoms.

With so many atoms in motion the numbers of collisions are enormous, and the frequency of the collisions is, at first, difficult to conceive. For example, in a volume of helium gas at standard atmospheric pressure (equivalent to about 14.7 lb/in^2 or *psi*) and at the relatively cool temperature of freezing water (0° C or 273 K), the average number of collisions per atom is about 4,800 million (4.8 billion) per second.

The average velocity per atom is, under these conditions, about 1,202 meter per second (1.2 km/s). Thus, on the average, each atom travels only about 2.51×10^{-7} meter between successive collisions. In other words, in each meter of total travel, a helium atom averages about 4 million collisions and bouncing-away movements. A single helium atom travels, on the average, about 1,300 times its own diameter in distance between successive collisions.

A perfect gas would be one whose atoms were so small that they behaved like mere points of mass. Also it would be one whose atoms did not attract each other, no matter how closely squeezed together they were. No such perfect gas exists. But the hypothesis of the behavior of an imaginary perfect gas has helped to make the kinetic theory of temperature the useful tool that it is.

The atoms of a nearly perfect gas such as helium move in straight lines between collisions. These are called *translational* motions. The velocity of the atom is constant from one collision to the next. At any instant between collisions it has a definite kinetic energy, equal to one half the product of its mass times the square of its velocity.

The *average* of all these per-atom kinetic energies is the variable that is measured by the thermodynamic temperature figure. How does this work out in the case of helium gas at 1 atmosphere of pressure and at the freezing temperature of water (0° C)? The average velocity per atom, as stated above, is about 1,202 meter per second. Since about $1.505\,5 \times 10^{26}$ such atoms have a combined mass of 1 kilogram, the mass per atom is about 6.64×10^{-27} kilogram, and half of this mass times the velocity squared gives us an energy average of 4.8×10^{-21} joule per atom.

In other words, under these conditions the combined kinetic

energies of about 210 million million million average helium atoms would together amount to 1 joule of energy.

The average energy of 4.8×10^{-21} joule per helium atom is a kinetic kind of way of expressing the meaning of a temperature of 0° C, which equals about 273 K. If we were to seal off a small volume of the helium gas at this pressure and temperature, and then warm it until its temperature had doubled, to 546 K, we would find that the average kinetic energy per atom likewise had doubled, rising from 4.8 to 9.6×10^{-21} joule.

And, even more interesting, if we measured the pressure at the doubled temperature, we should find that it too had doubled, to 2 standard atmospheres (or about 29.4 *psi*). The pressure exerted by a mass of perfect gas confined in a constant volume is directly proportional to its thermodynamic temperature. It is not proportional to its temperature as measured on the Celsius scale, in °C, or as measured on the Fahrenheit scale, in °F. It is, however, proportional to the temperature as measured in the SI unit of the kelvin, K. This is but one of the many meaningful relationships which come to light when, and only when, a thermodynamic temperature is used.

Amontons's early insights—The earliest glimpse of thermodynamic temperature was caught by a remarkable French pioneer in thermometric measurements, Guillaume Amontons (1663–1705). Just after the dawn of the eighteenth century he summarized measurements showing that if a volume of ordinary air was sealed off and heated from "room" temperature to the temperature at which water boils, its pressure always rose by about one third.

Room temperature, we can assume, was about 280 K, in modern terms, and the boiling point of water about 373 K. Hence the increase of about one third followed, applying to the exerted internal pressure on the container, and also (though Amontons could not think of it in those terms) to the average kinetic energy per molecule of the confined air.

We are obliged to refer to *molecules* of air, not atoms. Most of our atmosphere is nitrogen, whose molecules are each formed of two nitrogen atoms (N_2). The next most abundant gas in the air is oxygen, likewise in the form of a 2-atom molecule (O_2). The masses of these molecules are substantially greater than the mass of the single helium atom. This is reflected in the fact that at 273 K and 1 atmosphere pressure, where the average velocity of helium atoms is 1,202 meter per second, that of nitrogen molecules is 453 m/s and of oxygen molecules 425 m/s. The still-heavier molecules of carbon dioxide (CO_2), found in lesser proportions in the atmosphere, have an average velocity under these conditions of 360 m/s. Yet each of these different groups of molecules attains the identical kinetic energy average of about 4.8×10^{-21} joule per molecule.

Each kind of molecule collides with each other type of molecule, so their average kinetic energies are rapidly and continually being equalized. It was not necessary that Amontons make his measurements with pure nitrogen, pure oxygen, or any other pure gas.

With insight well ahead of his era, Amontons suggested that a new temperature scale could be established. Its basic principle is often called Amontons's Law. It can be stated in somewhat simplified form like this. If T_c represents a change in temperature as measured on a truly thermodynamic scale, and P_c is the resulting pressure change, then $P_c/T_c =$ a constant. That constant depends on the particular thermodynamic temperature scale that is chosen (for there are endless possible choices). In the case of the kelvin, the actually adopted SI unit, the constant is $\frac{1}{273.16}$ per K, or 0.003 66 per K.

Amontons was sure that somewhere along the line of pressures that dwindled as temperatures were reduced, a point or region would be reached where pressure itself became zero, and so temperature too would become zero. Amontons, in other words, wanted to measure temperature by changing gas pressures, rather than by changing expansions or volumes of mercury or other liquids.

It is interesting to see how close to modern concepts Amontons's crude measurements were able to come. He set up a rough and arbitrary scale of pressure units, assigning 73 to the temperature at which water boiled (about 373 K) and 51 for that at which water froze (about 273 K). His "thermodynamic" unit was thus somewhere between 5.1 and 5.3 times as large as the present kelvin or K, depending on which "fixed point" one uses.

His measurements indicated that ice melts at about 70 percent of the thermodynamic distance between absolute zero and the temperature at which water boils under standard atmospheric pressure. Today's data indicates that it is actually about 73 percent of that distance. A mere 3 percent discrepancy under the conditions prevalent in the early 1700s is truly a remarkable achievement in early thermometry.

Like many another early thermometrist, Amontons came to his concern with temperature principally by way of studies of weather and climate. He was but one of a series of eminent meteorologists who expanded into general metrology of temperature.

By means of his rather makeshift temperature system he assigned 58 of his pressure units to the "utmost summer heat" that he could measure. This would seem to correspond to about 311 K or 38° C or 100° F. He seems to have had greater success with his laboratory tests (freezing and boiling points) than with his weather analyses.

Since the kelvin (K) is the universally accepted scientific unit of thermodynamic temperature, we must stress that temperature differences measured by the K unit are always and exactly equal, in a numerical way, to their measurements in the scale of the degree Celsius (°C). For example, from the "triple point of water" (273.16 K) to the boiling point of water under standard pressure (373.15 K) is just 99.99 K. It is also the difference between 0.01° C and 100° C, namely, 99.99° C.

In this sense it is true that the temperature interval of 1 K equals the temperature interval of 1 degree Celsius. (The "triple point" of water will be explained later.)

What, then, is the "perfect gas" ratio of change in average kinetic energy per molecule or atom to the resulting change in thermodynamic temperature, as measured by the kelvin temperature unit? We have shown that at 273 K, the average kinetic energy per helium atom (a "nearly perfect" gas) is about 4.8×10^{-21} joule. That is the same as 480×10^{-23} joule.

Now, if 2.1×10^{-23} joule additional kinetic energy are added to that average amount per atom, then the helium will show a temperature of 274 K, not 273 K, a gain of just 1 K.

A small seeming discrepancy can be found here. If Amontons's Law were to hold completely, then a move from 273 K to 274 K would seem to require the addition of just $\frac{1}{273}$ of the average energy per atom corresponding to a temperature of 273 K. In this case, that would be an addition of about 1.8×10^{-23} joule, not of 2.1×10^{-23} joule.

The difference is a result of the fact that not even helium is a "perfect" gas, and of the further fact that, at very low temperatures, between about 30 K or 40 K and absolute zero, all substances show a rapid falling off of their heat capacity. This means that, in those lowest or cryogenic temperature levels, the energy increment that raises by 1 K the average kinetic energy per atom is *less* than the energy increment required for a 1 K increase in the temperature levels to which we are accustomed from our environments.

The heat capacity, per atom or molecule, at room temperatures is typically 2.1×10^{-23} joule per K for nearly perfect gases; but near absolute zero this heat capacity too drops rapidly toward zero.

The heat-capacity patterns of typical 2-atom molecular gases, such as oxygen, nitrogen and hydrogen, differ distinctly and significantly from those of the "nearly perfects," like helium and argon, composed of single atoms only. The differences result from the shapes, structures and consequent behaviors of the multiatom molecules.

A 2-atom molecule, such as that of hydrogen gas (H_2), is shaped like the conventional dumbbell, with one H mass at one end, the other H mass at the other end. A dumbbell can be thrown in such a way that it flies without spinning or turning. Its kinetic energy then equals half its mass times the square of its velocity. However, a dumbbell can be thrown also so that it spins end over end as it flies. In that case it will have additional energy because of its rotary motion or spin. At the same time, half the velocity of its center of mass times its total mass is still a translational energy, not a rotational energy.

What we commonly call thermodynamic (or absolute) temperature is traceable to the translational energies of the molecules making up gases, not from their rotational energies. Thus, when a unit of energy (such as 1 joule of energy) is added to a mass of gas whose molecules revolve at the same time that they fly about, less temperature gain will be registered per molecule than if the same amount of energy had been spread among a like number of atoms that do not rotate as they fly.

A dumbbell, if its two ends are joined by a spring, can move in still another way: its two ends (the two H atoms, in the case of hydrogen gas) can vibrate, toward and away from each other. This too drains off some of the energy, diverting it from the center-of-mass motions that are reflected in temperature increases.

The result of these additional kinds of motions, which become ever more demanding as temperatures rise, is this: the "heat capacities" of diatomic gases like hydrogen and the rest increase considerably beyond the heat capacity of a nearly perfect gas such as that made up of helium atoms.

For example, at a pressure of 1 atmosphere and temperature of 288 K, to increase by 1 K the temperature of 1 gram molecular weight of hydrogen requires nearly 1.4 times as much energy as the same increase in 1 mole of helium. The ratio of chlorine (Cl_2), another 2-atom molecule, to helium is more than 1.6; and in the case of 3-atom molecular gases such as carbon dioxide (CO_2) and nitrous oxide (N_2O) the ratio is nearly 1.9.

This shows that substantial amounts of energy input are "soaked up" to produce rotational and vibrational movements *within* molecules, and so are not reflected in corresponding increases in the average per-molecule kinetic translational energy that is reflected in and measured as *temperature.*

The energy that spins and vibrates the molecules is not lost. It is delivered up again when such molecules are cooled. But so far as producing changes of *temperature,* our concern here, that portion of the energy input has seemingly vanished.

Solids generally soak up even more energy than complex gases as their atoms and molecules are raised to higher temperatures. Copper, for example, needs about 2.4 times as much energy per atom as helium gas for each kinetic energy gain corresponding to a 1 K rise in temperature; mercury needs about 2.7 times as much, and zinc chloride ($ZnCl_2$) needs about 3.8 times as much. Even higher thermal capacities, or energy demands per 1 K of temperature rise, are typical of such multiatom compounds as lead sulfate ($PbSO_4$) and calcium sulfate ($CaSO_4$).

The list of the variations and (seeming) inconsistencies in the thermal capacities of various molecules is enormous. In each case the relation between the temperature response and the energy increment (or reduction) is explainable in terms of the peculiar internal structure of the molecules involved, and (at still higher temperature ranges) of the structures of the constituent atoms that are linked together in the molecules.

The important thing to recall always is this: If two dozen totally different compounds, all "at" 300 K, show two dozen different thermal capacities, the fact nevertheless remains that the average kinetic or translational energy per molecule is the same in each of the two dozen. Each has the same disposable or transferable kinetic energy per molecule, on the average; and thus all these so-different molecular structures and assemblages can remain in thermal equilibrium. If two are placed in contact or close by each other, each will give back to the other just the same amount of energy, on balance, that it receives from the other.

This is true despite the fact that, relative to absolute zero so far down the scale, each will have soaked up a quite different amount of total energy per molecule in being raised to that 300 K level of temperature.

How liquids behave—We can readily picture, in most cases, that gaseous atoms and molecules fly about, constantly colliding. When the average kinetic energy per molecule is sufficiently reduced, the attractive forces between the molecules pull them together into a different kind of relationship. The molecules slide easily past one another, but only seldom accumulate energy sufficient to escape and fly free of the whole mass. This is the liquid state, and those molecules that do gain energy enough to escape are the ones that evaporate from the liquid surface.

Even as they slide and glide, the molecules of a liquid have kinetic movements. The average of such kinetic energy per molecule of water at 288 K (15° C), for instance, is exactly equal to the average per atom of helium gas at the same temperature of 288 K. It is possible, taking into account the differences in the masses of the helium atom and the H_2O molecule, to estimate closely what must be the mean or average velocity of motion of the latter.

At still lower energies-per-molecule—at colder temperatures, that is—most liquids freeze and become solids. In the solid state the individual atoms and molecules tend to vibrate about more or less fixed positions. It is much like a spring that vibrates back and forth or in many directions, while one end remains fixed in position. But again, the average molecular kinetic energy in a solid must equal the average kinetic energy per atom in the nearly perfect helium gas, when both solid and gas are "at" the same temperature.

Thus, in a mass of iron at 273 K (0° C) the average velocity per atom vibrating in its latticelike structure corresponds to about 3.2 meter per second.

The proportions of added energy that result in temperature increases vary rather widely in different parts of the temperature scale. This can be illustrated most vividly in the case of the 2-atom molecular hydrogen gas (H_2) already mentioned. When its heat capacity per molecule is carefully measured in the low temperature range of about 25 to 60 K, hydrogen is found to behave much like helium, with no signs of rotational or vibrational molecular movements. Thus it shows a 1 K rise for each 2.1×10^{-23} joule increase in average energy per atom in this level.

Beyond about 60 K, however, more and more of added increments of energy are absorbed by rotational movements of the H_2 molecule. By about 250 K, added energy is being channeled about 40 percent into rotations, and only 60 percent into higher velocities or translational movements.

In terms of common measurements, this means that at "room" temperatures hydrogen gas has about 1.5 times the molecular heat capacity it had at 50 K.

But the diversion of energy from the temperature sector becomes even more marked at still higher temperatures. Over about 500 K, more and more energy goes into the vibrational, in-and-out movements of the two hydrogen atoms. By about 5,000 K, the heat capacity of the hydrogen molecule has climbed to about 2.3 times its magnitude at the 50 K level.

Each joule of energy added to a mass of hydrogen at about 5,000 K is used to the extent of about 43 percent in increased translational movements (that is, in raising the temperature), about 28.5 percent in increasing the spin movements of molecules, and another 28.5 percent in increasing the vibrational movements of molecules.

Determination of these patterns of change in the application of energy increments to various types of possible movements has required enormously precise and painstaking measurements. Such measurements have depended on satisfactory ways to quantify temperature on the one hand and heat energy on the other. Today the units of the kelvin (K) and the joule (J) are best used for these tasks, aided when necessary by such measures of mass as the mole (gram atomic mass).

27 TEMPERATURE SCALES, FANCY AND PLAIN (I. ROEMER AND FAHRENHEIT)

> . . . [through] this distemperature we see
> The seasons alter: hoary-headed frosts
> Fall in the fresh lap of the crimson rose.
> —SHAKESPEARE,
>   A Midsummer Night's Dream

> Were there one whose fires
> True genius kindles, and fair fame inspires . . .
> —ALEXANDER POPE,
>   Epistle to Dr. Arbuthnot

Amontons's early insights into what we now call thermodynamic temperatures were not followed up and indeed were largely neglected and forgotten until more than a hundred years after he made his proposals. Eighteenth-century science was not yet ready to receive and apply the concept of an absolute temperature scale, which, without being tied to some particular substance, such as water, at both ends, would run from a universal absolute zero at its bottom to some one chosen fixed temperature point, such as today's 273.16 K, the "triple point" of water. Instead, the efforts to measure temperature, often clumsy and halting, took a different direction.

Amontons's contributions had been offered nearly a century after the first crude thermoscopes were demonstrated. Galileo and Santorio were among several inventors and developers of these primitive devices. They must be called "scopes" rather than "meters," for at best they could be depended upon only to indicate whether one temperature (A) was higher or lower than another (B), but not how much, in any quantitative or reliably numerical manner.

An early thermoscope, for example, might indicate that condition B was at a higher temperature than A; and that condition C was at a lower temperature than B. But whether A was at higher or lower temperature than C might nevertheless remain uncertain.

Besides improvements in sensitivity and reliability, the thermoscopes needed also some kind of agreed-on and consistent scale, with numbers assigned to its subdivisions or degrees. By a combination of both kinds of improvements, the technical and the quantitative, something resembling genuine temperature *measurement* might be attained.

One of the giants of science history contributed ideas here, as in so many other areas. In a paper that did not even carry his name, Isaac Newton made a series of pregnant observations. He suggested several fixed temperature points, starting with the freezing temperature of water (his 0°) and rising to the temperature of soft coal burning "in a little kitchen fire," fanned by a bellows (his 192° point). Between these extreme points he also indicated levels for the melting of wax and the boiling of water under ordinary air pressure.

Thus, like Amontons, Newton saw validity in choosing the temperatures of the two great changes of state of that most useful and indispensable substance on the face of our Earth (H_2O). It may seem obvious or inevitable now that water's freezing and boiling temperatures should serve to determine scales. Vast numbers of measurements have established that when the ambient pressure is held at a certain level, pure water will always freeze at one consistent temperature (273.15 K) and boil at another (373.15 K). However, in the seventeenth and early eighteenth centuries some influential natural philosophers doubted or even denied that water always boiled at the same temperature level. Their confusion arose from the considerable changes in such temperature that result from changes, noticed or not, in the surrounding air pressure.

Early thermoscopy originated mainly in Italy. The practical and quantitative developments that transformed it into thermometry took place principally in lands farther north. Without pretending to detailed completeness, we shall focus attention first on Copenhagen during the early years of the eighteenth century.

It was there that the eminent Danish astronomer Ole Roemer

(1644–1710) undertook to prepare his own thermometers for a series of temperature measurements.

Roemer was a man of many distinctions: royal mathematician, professor of astronomy at the university, and holder of such high public offices as mayor of his city, chief of police, and privy councilor to boot. His personal prestige and influence did not distract him from his own creative contributions to science. In this he supplies an admirable contrast to more than one modern scientist who has been lured from his own specialty by the bitch goddess Success.

Roemer's high rank in science is reflected by his most famous discovery, that light travels with a finite velocity, not instantaneously. He demonstrated this by means of a beautiful series of measurements of the times at which moons of Jupiter were eclipsed behind the bulk of the giant planet, as the distance between it and the Earth altered in consequence of the orbital motions of both.

In the design and construction of observational and measuring devices Roemer showed notable ingenuity. He was a master of useful and goal-oriented mechanisms. This was the savant who in his middle fifties made serious efforts to equip his thermometers with suitable scales. They were devices that depended on the expansion of "spirits of wine" (alcohol) in glass tubes.

In terms of sensitivity and reliability they left much to be desired. Yet Roemer prepared to use them as the best at hand. The awkward and obscure scale pattern that was formulated by even this able and experienced experimenter illustrates admirably how far thermometry had to travel in order to attain the clarity and simplicity of the scales in which scientific measurements are expressed today.

An Italian, Carlo Rinaldi, in 1694 had proposed that temperature could be measured on a scale with only two fixed points, that of the freezing of water, and that of its boiling. Roemer was interested, or fascinated, also by two other fixed points (though today we would hardly regard either as truly *fixed*). One was the normal heat of the human body (blood heat); the other the most intense ascertainable cold, or lowest temperature then attainable. This latter temperature he produced by mixing salt with water and ice in such proportions that, so far as he could tell, there was no other mix that would create greater cold, or lower temperature.

Each of these four points he then marked off on his alcohol-filled thermometers. He seems to have been impressed by certain approximate distance relationships. The smallest length was that from the lowest (water-salt-ice mix) to the freezing temperature of pure water. The next length, from the latter to the temperature of the human body, appeared to be twice the first length. And from there (blood heat) to the boiling point of water seemed to be a length just about five times the first length. Thus, his four key temperature levels seemed to establish the ratios 1 to 2 to 5. The total, top to bottom, temperature distance was thus eight times that smallest, bottom interval.

When it came to choosing his actual scale subdivisions, Roemer made a choice that reflected his experience and way of thought—the number 60. He made 60 equal degrees from the lowest, or coldest, mark to the highest, or hottest.

The sexagesimal system seemed almost an essential element in his astronomy. Did he not measure stellar angles in degrees divided 6 times 60, or 360, times into a full circle? And did not each such degree include 60 minutes of angle; and each minute 60 seconds of angle? Furthermore, his timing of the transits of stars across the zenith meridian was expressed in hours of 60 minutes of time each, with each minute consisting of 60 seconds. From the Mesopotamian plains formed by the Tigris and Euphrates to Denmark was a long distance in kilometers or in years and centuries of time. Yet the roots of Roemer's scale of 60 degrees seem clear; they reach from one to the other, beyond doubt.

On this new scale the freezing point of water was located at just about 7.5 degrees, while the temperature typical of the human body was given as 22.5 degrees. Each Roemer degree was just about 1.9 times as large as the present kelvin (K) or the Celsius degree (1° C), which is just the same size as 1 K. Also, the new Roemer degree was about 3.4 times as large as the later Fahrenheit degree (1° F).

A Daniel comes to Copenhagen—In 1708, while Roemer was still at work on these thermometric calibrations, he received a youthful visitor, Daniel Gabriel Fahrenheit (1686–1736), then only twenty-two. Born in Danzig, far to the east, Fahrenheit had studied "natural philosophy" in Holland, and there he made his home and the scientific instruments that he sold for a living.

When young Fahrenheit returned to the Netherlands he carried with him important recollections of the thermometric efforts of Roemer. Though the temperature scale of the Copenhagen savant fell into disuse, it survived, strangely altered or distorted, in the continuing Fahrenheit scale.

How are Roemer's scale-making efforts to be interpreted in modern temperature terms? It is safest to assume that his 7.5-degree point corresponded to the present 0° C (or 273.15 K) and his 60-degree point to the present 100° C (or 373.15 K).

If this was true, then his zero point corresponded to about 14.25° C and to 287.4 K today. However, Roemer's assignment of 22.5 degrees to body heat corresponds to about 28.5° C, or 301.65 K today. This is too low by a mere 2.2° C or K, for the average blood heat of healthy humans is about 30.7° C (303.85 K). Indeed, one can say, in terms of the modern kelvin thermodynamic scale, that Roemer came within about three quarters of 1 percent of average blood temperature—another instance of the metrological merits of this distinguished contributor to our understanding of the universe around us.

Young Fahrenheit returned to his Amsterdam workshop and studies more than ever concerned with improving the devices and patterns by which he and his customers could measure and compare temperatures. His first work, like that of Roemer, was done with alcohol-in-glass thermometers. These he improved by the simple but essential expedient of sealing off their tops, fully enclosing the tubes. Thus the contents were protected against evaporation. But more important still, they were protected against the inevitable variations in the ambient atmospheric pressure. The early thermoscopes, being open to the air, often had served simultaneously and undesirably as combination thermometric and barometric devices.

As an expansion fluid in a thermometer tube, alcohol has some advantages, but also important disadvantages. It has a relatively large coefficient of expansion, about eight times that of water, which is an advantage, for it means that any given temperature increase will cause in a given volume of alcohol a volume increase eight times that caused in a like initial volume of water.

Further, the heat capacity of alcohol is less than that of water.

In more common measurement terms, alcohol has a relatively low specific heat (that of water always being taken to equal 1.0). This means that, to effect a given rise in temperature in a mass of alcohol, less heat energy, measured in joule or calorie, is required than in an identical mass of water. Consequently, alcohol responds more rapidly, as well as more markedly, than does water to temperature changes around it.

Alcohol, however, boils—that is, vaporizes—at temperatures a good deal lower than those at which water boils under the same ambient pressure. An alcohol thermometer could not be used, for example, to measure temperatures between 90° and 100° C without the help of some difficult high-pressure preparation of the confined alcohol.

Finally, alcohol and water share one troublesome characteristic. Their molecular structures cause them to "wet" or cling to the walls of the tube in which they are confined. Capillary effects compound these difficulties, causing the columns of liquid to crawl higher than the temperature itself justifies. Also, when temperature diminishes, liquid columns of alcohol or water drop down more slowly and reluctantly than they should.

Practical obstacles such as these are typical of problems always facing thermometrists. They reveal again the inescapable fact that "temperature" is not a variable that can be measured directly, in the sense that a yardstick or meter bar can measure length. Every thermometric device is obliged to measure changes in some other physical variable (such as the length of a liquid column here), and then interpret or impute a temperature from this, by means of a predetermined and sometimes utterly arbitrary scale.

Even the most sophisticated modern thermometric devices partake to some extent of the limitations underlying the story of the practical nurse who was asked how, without using a suitable thermometer, she expected to know whether the baby's bath was at the proper temperature.

"Oh, that's simple," she answered. "If the little darling gets blue, it's too cold; but if he gets red, it's too hot!"

Fahrenheit, one of northwestern Europe's most proficient instrument makers, continued to search for a baby more sensitive and suitable in temperature response than either alcohol or water.

Among the widely and weirdly varied thermoscopes that had been constructed in the preceding fifty or sixty years was one made about 1659 by an astronomer named Ismael Boulliau. For his expansion fluid he had used the strange liquid metal called "quicksilver" (mercury) because of its brilliant appearance. The device itself, however, had been crude and unsatisfactory, its mercury dirty and impure. Also, the expansion coefficient of mercury is less than that of either water or alcohol.

Nevertheless, Fahrenheit turned now to mercury for his new thermometer. He realized that he could compensate for the lower expansion coefficient if he enclosed a fairly large amount of mercury in a lower bulb or pocket, and forced the expansion from here to travel up a glass tube with small interior bore. Even a small temperature rise might then produce sufficient upward climb of the silvery opaque mercury.

Next, Fahrenheit had to solve the problem of providing uniform inner bores or diameters in the glass tubes. If, for example, the cross section of part of such a tube were one third smaller than the rest, then in that narrower section the mercury would climb 1.5 times as much as it should, and here a rise of 2 degrees would appear to be one of 3 degrees. Fahrenheit managed to produce glass tubes of reasonably uniform small bores.

However, impurities in the mercury would block narrow tubes. To get rid of such pollutants, Fahrenheit developed a method of filtering the mercury through leather. The quicksilver, thus purified, had several manifest advantages. Being opaque, it could be seen more easily through the glass tube than either water or alcohol. Since mercury does not wet the tubes that confine it, its top makes a small domelike shape, rather than a concave one as in water or alcohol. Also, mercury hangs together, and when it contracts it slides down the tube in a coherent mass, leaving clear glass above it.

Its temperature range is gratifying, too. It does not turn solid until about minus 27° F, nor does it boil until heated to over 640° F.

By about 1714 Fahrenheit had produced a real mercury-in-glass thermometer vastly superior to any previous device for temperature measurement. Though it was not free of weaknesses and limitations, it became the model for hundreds of millions of thermometers that have served and are still serving daily all over the world. Had it not been that his thermometers were so sound and practical, the Fahrenheit scale that he devised for them would never have become widely known. As it is, and rather ironically too, Fahrenheit is famous today chiefly because of that same peculiar and highly illogical scale, whereas many of his more substantial contributions to scientific measurement are seldom mentioned.

How did Fahrenheit arrive at his strange and almost bizarre scale? He began by building on what Roemer had been doing in Copenhagen half a dozen years before. However, Fahrenheit found the Roemer degree too large or coarse for the measurements he knew his thermometers should be called on to make in practical use.

Accordingly, Fahrenheit decided to quarter the Roemer degree —that is, he divided each one into four parts. Thus, at the outset the 60 Roemer degrees became 240 Fahrenheit degrees. Roemer's zero point, set by the water-salt-ice mixture, apparently was somewhat lowered by Fahrenheit, who managed to make a still colder mix with the addition of some sal-ammoniac (NH_4Cl).

Now, corresponding to the four Roemer points, Fahrenheit had his 0° F; the 32° F (slightly more than four times the old 7.5 Roemer degrees) for the freezing point of water; the 90° F for human body temperature; and the 240° F for the boiling point of water.

Such was the first Fahrenheit scale. However, with his more accurate thermometers he appears to have made measurements which convinced him that this pattern did not conform sufficiently to the actual relations of the four different fixed points. In particular he found that the 90° F for the body temperature was too low, and changed it to 96° F instead. Also he found that the 240° F for the boiling point of water was too high, by about 28 of the new Fahrenheit degrees. He changed it, accordingly, to 212° F. By this curious and even crabbed process was evolved the strange scale on which water freezes at 32° and boils at just 180 degrees higher.

Roemer's original scale, crude and tentative, located the freezing point of water about 12.5 percent of the way from its zero to its boiling-point temperature of 60 degrees. In Fahrenheit's final scale that relationship is 15 percent. In the Roemer scale, human body heat was located about 37.5 percent of the way up; in the Fahrenheit scale the original 96° F level for blood heat was 45 percent of the way.

Taking the range from the freezing temperature of water up to its boiling point, we find that Roemer placed body heat about 43 percent of the way up, while Fahrenheit placed it 53 percent of that way. These differences are large, but not overwhelming. They reflect the growing pains of thermometry.

Even Fahrenheit's assignment of 96° F to body heat is outdated. The blood temperature in a healthy person lies in the 98.4° to 98.6° F range as a rule, although many a perfectly healthy individual shows temperatures typically in the range between 98.0° and 98.4° F.

Discrepancies between the Fahrenheit and Roemer scales are no longer important, save to show the shifts, adjustments, devices and compromises made in the efforts to arrive at some of the pioneer scales.

The continuing widespread use of the Fahrenheit scale, primarily in the United States and in those regions where American bases and enclaves are established, maintains the importance of this strange and illogical thermometric structure.

Little or nothing can now be said on its behalf in any logical or scientific sense. The one possible exception appears to be a pleasant tongue-in-cheek commentary by Morton Mott-Smith, whose science writings remain unfadingly enjoyable. In his *Heat and Its Workings,* first published in 1933, he had this to say, after noting that scientists tend to show contempt for the Fahrenheit scale and to regard its usefulness as past:

> People are always exaggerating temperatures. If the day is hot they add on a few degrees; if it is cold they deduct a few. No one ever gives the air temperature to a fraction of a degree, but only to a whole degree. Now on the Fahrenheit scale, on account of the small size of its degree, these whoppers are only about half as big as they are on the other scales.

Those other scales are identified with the names of Réaumur (°R) and Celsius (°C).

Fahrenheit's well-founded fame—Fahrenheit's total contributions to thermometry, meteorology and general metrology are vastly greater than encompassed by his strange scale and the practical thermometers to which it was attached. His attainments were considerable, and they well justified the honor accorded him by the prestigious Royal Society of England in electing him a Fellow, in 1724, just a decade after he had presented a paper describing his new mercury-in-glass thermometer.

Fahrenheit invented also a related device called a hypsometer with which he performed historic measurements. Hypsometry has a peculiar double meaning. On one hand it refers to the use of devices to measure distances or elevations above sea level—a kind of altimetry. On the other hand, it refers to measurements of the reliability of thermometers in terms of the boiling points of various liquids. Now, boiling points diminish as external pressures are reduced; and external pressures are reduced as one ascends to greater distances above sea level. Hence the two seemingly disparate measuring intentions are actually connected quite closely.

With his hypsometers Fahrenheit made his greatest discovery, one that would have deserved a Nobel prize, had that award been made in his lifetime. He proved by detailed measurements that the boiling points of liquids vary with the ambient air pressure. The lower this pressure, the lower the temperature (average molecular kinetic energy) needed to produce the escape of molecules from the liquid into the free gaseous state.

For example, under a pressure of 1 standard atmosphere (equal to 760 mm of mercury or to 101,325 newton per square meter) water boils at 212° F. But if the pressure is reduced to 0.774 atmosphere (588.4 mm of mercury or 78,426 newton per square meter) water boils at only 199.4° F.

On the other hand, if the pressure is increased to 1.05 atmosphere (798 mm of mercury or 106,391 newton per square meter), then water does not boil until heated to 224.6° F.

The rise in boiling-point temperatures with pressure continues far beyond the levels that Fahrenheit himself could attain. Under a pressure of 225.7 atmospheres, water boils at 705.5° F, and at that critical point the density of the water vapor (steam) has become as great as that of the liquid water. Such combinations of high pressures and temperatures deprive the common distinctions between the liquid and the vapor states of water of their customary meanings.

Increase or decrease of as little as 1 part in 5,000 or 10,000 in the air pressure produces measurable shifts in the temperature at which boiling of water takes place. Indeed, from a state of standard pressure of 1 atmosphere, a shift of only 1 mm of mercury (or $\frac{1}{760}$ part of the pressure) in either direction produces a change in the boiling point of about 0.07° F.

The process goes to true extremes. If the external pressure is reduced to about 4.5 mm of mercury, then water "boils" at 32° F, which is the ice point for ordinary standard pressure! And a further reduction of pressure to about 1.3 mm of mercury reduces the temperature of boiling to 27° F, or 5 degrees F below that usual ice point.

Fahrenheit led the way to the important concept of "vapor pressure"—that every temperature has its corresponding pressure of vapor from water or from other important liquids. The lower the temperature, the lower that vapor pressure, and vice versa. If external pressure is less than vapor pressure, the liquid "boils."

Most modern automobiles make use of a so-called pressure radiator cap. Some of these caps are calibrated to remain sealed until the vapor pressure against them is 8 psi (pounds per square inch); others until it is 12 psi. The effect in either case is to allow the water in the engine's cooling system to rise to higher temperatures than it otherwise would, without boiling away—and so tend to attain greater engine efficiency, by wasting a little less of the engine heat.

The 8-psi cap allows the pressure below to increase to about 1.5 standard atmosphere; the 12 psi cap, to about 1.8 standard atmosphere. The boiling point of the water within is correspondingly forced upward in each instance.

If pressure is boosted to 10 atmospheres, water will not boil until heated to a temperature above 593° F, and at about 218 atmospheres pressure it must be heated to about 705° F before it becomes steam (superheated steam).

Thanks first to Fahrenheit, engineers and scientists learned that pressure definition is essential for any definition of the "fixed points" based on changes of state of familiar forms of matter—as of the vaporizing or the freezing of water.

These basic dependencies exist also in the lower temperature effects involving the freezing of water and other liquids. Water under normal atmospheric pressure, as it cools, grows slightly denser (contracts) until it reaches its maximum density point, at about 39° F. From there until 32° F it expands slightly, and when it becomes ice at 32° F it expands a bit more. Hence the ice is less dense than the water around it and floats.

However, if the surrounding pressure is raised sufficiently, water can be chilled well below its usual 32° F freezing point without leaving the liquid state. At a pressure of about 159 atmospheres it can be cooled some 2 degrees F below its usual ice point, for example, without becoming ice. At about 2,142 atmospheres pressure it can be supercooled to about − 8° F, which is 40 degrees below its normal freezing point.

At that pressure, if water is chilled even lower, it becomes ice— but not the sort of ice we are familiar with. It becomes ice of a different structure, denser than water, for in freezing it has been forced to contract, not allowed to expand in the usual way.

In the 1740s Fahrenheit transmitted details of his hypsometric devices, or "boiling-point thermometers," to the Royal Society in England. But that did not end his inventions and improvements. One of his final innovations was a so-called thermobarograph. Its intent was to make life easier for meteorologists by enabling them to keep track of prevalent temperatures and pressures without confusing the two.

Fahrenheit was an admirable combination of technician and scientific experimenter; of practician and seeker after principle; of mechanic and man of creative imagination. One may wonder whether, had he not visited the great Ole Roemer during his early, impressionable twenties, he might not have evolved a scale less cumbersome and crabbed for his wonderfully useful new thermometers!

28 TEMPERATURE SCALES
(II. RÉAUMUR AND CELSIUS)

Heat not a furnace for your foe so hot
That it do singe yourself.
—SHAKESPEARE,
Henry VIII

H_2O, Hg, C_2H_5OH . . .

Each substance symbolized here—water, mercury and alcohol —has played a leading role in the development of methods and units for measuring temperature on the one hand, and heat, or thermal energy, on the other.

The third of these substances, as shown in the chemists' shorthand, is not alcohol in general. It is but one of the alcohol family, the kind called ethyl alcohol, known in the past by such descriptive titles as "spirits of wine," "Cologne spirits," "fermentation alcohol" and "the curse of the working classes."

Regardless of the influences for good and ill of potable alcohol upon humankind, its part in the progress of metrology has been undeniable. We shall see its influence again in the subsequent pages on density measurements, much of which proves to be an overlapping of fiscal policy and alcoholimetry. Gasoline and alcohol, it has been observed with unchallengeable rightness, do not mix. However, metrology and alcohol have combined in amusing and sometimes surprising ways.

About a dozen years after Fahrenheit in the Netherlands had given his contemporaries the first truly admirable mercury-in-glass thermometers, a notable series of measurements was undertaken on mixtures of alcohol and water. The measurer was a distinguished, versatile and indefatigable French natural philosopher, René Antoine Ferchault de Réaumur (1683–1757).

Réaumur's primary aim was to measure the extent to which

temperature increases cause expansion of the volumes of mixtures of alcohol and water in various proportions. His method was both systematic and simple. He would prepare the alcohol-water mixture in some proportion of interest to him, then place it in a bulblike container terminating in an expansion tube. The position of the liquid at the foot of this tube when the bulb was cooled to the temperature of freezing water (32° F) provided his zero point.

Réaumur considered it logical to graduate his expansion tube in such a way that each expansion of the original volume by 1 part in 1,000 (or by 0.1 percent) should increase his reading by 1 point or degree. Thus, assuming the initial volume at water's freezing temperature had been just 1,000 cubic centimeter, his graduated tube would reach the level marked as 1 when the volume had increased to 1,001 cc, and the level marked 2 when it was up to 1,002 cc, and so on.

Réaumur noted that when he heated the mixture in the bulb halfway between the freezing and boiling points of water—that is, to about 122° F—the liquid in the expansion tube stood about forty marks or units higher than at the start. He decided, accordingly, that he could justifiably assign the number 40 to this temperature, and that on the same scale the boiling point of pure water would be found at 80 degrees.

It was thus from the more or less accidental coefficient of expansion of a particular alcohol-water mixture that Réaumur arrived at the scale destined to preserve his name in history: a scale with the ice-point of water at 0° R and the steam point of water at 80° R, always under normal air pressure of 1 atmosphere.

Thus Réaumur supplied eighty equal intervals for the temperature range for which Fahrenheit had finally supplied 180 equal intervals. Accordingly, 1° R equals 2.25 times 1° F, and 1° F equals $\frac{4}{9}$° R, or 0.444° R.

Some historical accounts have been written as if Réaumur actually heated the alcohol-water mixture in his experiment all the way to the boiling point of water, his 80° R and Fahrenheit's 212° F. This interpretation flies in the face of certain undeniable physical facts. Alcohol has a markedly lower boiling point than does pure water. Pure methyl alcohol boils at less than 52° R, under standard atmospheric pressure; and pure ethyl alcohol at less than 63° R. Mixtures of water and alcohol, depending on their relative proportions, boil at various temperatures between these and 80° R.

It is not quite so simple, however—as any operator of a still can testify. The fact is that, as the mixture is heated hotter and hotter, it is the alcohol that tends to vaporize to a greater extent than the water. The result is that water becomes an increasing proportion of the liquid, while alcohol becomes the principal source of the vapor over the liquid.

Suppose we mix water and alcohol in the proportions expected to produce a liquid with boiling point of about 65° R. By the time it is heated to that temperature, we find that the vapor is approximately 80 percent alcohol, whereas the remaining liquid contains roughly equal parts of alcohol and water.

A mixture of about 91 percent alcohol and 9 percent water by weight, when heated to 63° R, produces vapor with roughly 92 percent alcohol, 8 percent water. As the temperature is raised, the proportion of alcohol remaining in the liquid decreases rapidly, while that in the vapor increases.

At a temperature of about 72° R, though only 10 percent of alcohol and 90 percent of water remain in the liquid, the vapor driven off from it contains about 60 percent alcohol and 40 percent water. By the time 80° R is attained, the last of the alcohol has been boiled out of the former mixture, and what is left in liquid form is only water, ready to be fully converted into steam.

In short, it was not possible for Réaumur to have heated any particular liquid mixture of alcohol and water all through the temperature range from his 0° R to his 80° R without first much reducing and ultimately eliminating the portion of alcohol remaining in the liquid.

The present author concludes that Réaumur noted the expansion rate of his mixture up to a temperature only halfway or even less than halfway between the ice point and the final boiling point of pure water, and that he then projected or "extrapolated," as physicists and statisticians love to put it, all the way to the boiling temperature.

Regardless of such technical details, the important thing about Réaumur's achievement was that he created the first consistent and widely influential thermometric scale in which both of the only key points or levels of temperature were derived from well-defined changes of state of that most common and indispensable of compounds on Earth's surface: pure water. Though led to his choice of just eighty divisions by the behavior of a water-alcohol mixture, he spread those eighty equal intervals between temperatures that reflected changes undergone by water alone, minus any alcoholic or other admixture.

Réaumur's scale, unlike those of Roemer and Fahrenheit, was in no way involved at its inception with such nonwatery and imprecise temperature concepts as average human body temperature or a supposed "coldest possible" mixture of ice, water and salt, or salts. With Réaumur, water assumed preeminence as the defined substance for producing and reproducing the fixed temperature levels between which the scale was subdivided.

Though its inception had been, so to speak, baptized by a water-alcohol blend, the Réaumur scale was not used solely or even mainly for thermometers in which the expansive fluid was alcohol. Fahrenheit had shown the world quite effectively that for most daily needs, the mercury-in-glass instruments, when properly made, had undeniable advantages. The Réaumur scale was used far more often with such mercury thermometers than with alcohol thermometers, though it could serve either, except for that range of temperature above which alcohol no longer remains liquid in the tube.

One indubitable and inescapable fact must be noted with respect not only to the Réaumur scale but to every scale. The nature of the expansion fluid in a thermometer makes a great difference. It would be possible, for example, to calibrate two thermometers, one with mercury, one with alcohol inside, in such a way that at the ice point they would both show 0° R; and at some other temperature, say at 40° R, they would each show precisely 40° R. Yet if the mercury thermometer is now placed in a liquid whose temperature it indicates at 20° R, this does not mean that the alcohol thermometer will also indicate just 20° R when placed in that same liquid at the same time.

The fact is that the expansion rates of mercury and alcohol over the same temperature ranges are not strictly proportional to each other. Nor is the expansion rate of either strictly proportional to the expansion rate of a helium-gas thermometer operated at constant pressure. The helium rate, because it is a "nearly perfect" gas, is the most consistent and uniform of all, but mercury, alcohol, water, and other expansion liquids deviate from it and from each other.

For ordinary household uses these discrepancies seldom matter, but scientific precision requires better agreement than can be obtained with expansion liquids, except by extraordinary precautions and corrections.

Réaumur's breadth and range as a scientist was astounding. He contributed substantially to biology, physiology, zoology, icthyology and entomology, as well as to the physical sciences. His contributions to metallurgy were massive, and he earned the right to be regarded as the father of the steel-making and tin-plate industries of France. Heat and temperature had concerned him in many investigations before he evolved his characteristic and somewhat quaint thermometric scale.

As an enthusiastic meteorologist he had gathered and interpreted weather data and was one of the first to study in a scientific way the direct evaporation, or "sublimation," of snow crystals into the air without any intermediate pause in the liquid state between the solid and vapor forms. In his weather studies he was said to have used "his own thermometer."

The portals through which he entered the immortality conferred by his thermometric scale were, however, his careful studies of how the thermal behaviors of various liquid mixes changed as the proportions of their components were altered—as in the alcohol-water case. Réaumur established that such changes in proportions shifted the freezing as well as the boiling points of the resulting mixtures.

He measured also changes in the specific heat (the relative energy addition required to bring about a given temperature increase), in the density and in the expansivity of the mixed liquids.

Réaumur's temperature scale was widely used for generations not only in France but also in Germany. Many a classic of science, including some of the most basic measurements of thermal effects, uses the °R throughout.

Celsius of Sweden—The first of the remembered founding fathers of thermometry whose life was lived entirely in the eighteenth century was Anders Celsius (1701–44). He was born in Uppsala, Sweden, of a family that had included several prominent astronomers among his immediate forebears.

Anders Celsius followed the family tradition and became a measurer of the skies and the earth. He was in his middle thirties when he joined an expedition of French astronomers and geodesists going into the far north of Sweden to learn more about the size and shape of the Earth. During a difficult year, with months of bitter cold, they completed a series of triangulations. Their many other measurements included that of local gravitational accelerations, by means of "seconds" pendulums, as well as various magnetic, astronomical and weather observations.

During that intensive and enlightening journey Celsius became acutely aware of the need for better thermometers to improve the quality of meteorological data. He began experiments along this line after his return to Uppsala, where he now headed the astronomical observatory.

By 1741 he had in hand his first thermometer equipped with the 100-degree scale now identified with his name all over the world. It was, however, totally different from the Celsius thermometers now in general use. He had chosen to set his zero on top, at the temperature of boiling water, while his 100° point was below, at the ice point!

He reported on the new instrument and its (seemingly) upside-down scale in 1742, in the *Transactions* of the Swedish Academy of Sciences. His paper carried a significant title, translatable as "Observations on Two Constant Degrees on One Thermometer." Those constant degrees referred to what is now commonly called the fixed temperature points—namely the 100° ice point and the 0° steam point, both based on the behavior of pure water under normal or standard atmospheric pressure.

The next year, 1743, the French physicist Jean Pierre Christin (1683–1755), of Lyons, produced some thermometers in which a 100-degree scale was placed "right side up"—that is, with the ice point at 0° and the steam point at 100°. Christin may indeed have been the first in the world to make and use such a thermometer, even though it is now named for Celsius.

Meantime, in Sweden the Celsius instrument continued to be used in its upside-down form until after Anders Celsius' untimely death in 1744. His successor, Martin Strömer, some years later returned to the problem of the scale and in 1750 reversed it so that the zero point was below and the 100° point above. In this revision of the Celsius pattern Strömer was aided by Daniel Ekström, a competent instrument maker associated with the Academy of Sciences headquartered in Stockholm.

This similarly named pair of thermometrists, Strömer and Ekström, had the acuity to note the practicality of the idea that underlay the original Celsius scale. Their inversion retrieved it and led to its final triumph in the world of measurements. Today more temperatures are measured and recorded in °C than in any other temperature scale, and perhaps more than in all other scales combined.

Furthermore, the accepted thermodynamic scale, based on the unit of the kelvin (K), is derived, so far as concerns its unit size (formerly known as its degree size), from the degree of the Celsius scale, which in turn began as the hundredth part of the mercury-expansion distance between the ice point and the steam point at normal atmospheric pressure.

The name of Celsius was missing for some time, so far as this newly inverted scale was concerned. In Uppsala and elsewhere in Sweden it was called either the Ekström or the Ekström-Strömer thermometer. It became the preeminent device for scientific and meteorological temperature measurement in Sweden and apparently also in all Scandinavia. Elsewhere in Europe scientists commonly referred to it simply as "the Swedish thermometer" during the latter half of the eighteenth century.

Then took place one of the errors or accidents of attribution that are no less common in the writing of science history than in other kinds of history writing. A distinguished Swedish chemist, Jöns Jakob Berzelius (1779–1848), wrote a successful textbook. Under its German title, *Lehrbuch der Chemie,* it became widely influential, going through five editions between 1803 and 1848. In its third part it contained the statement that Celsius himself had devised the scale setting the ice point of water at 0° and the steam point at 100°.

Accepted textbooks become autonomous powers in the realms

of learning that revere them; and because of Berzelius rather than the actual events, Celsius fell heir to credit that belongs, in large part at least, to Christin, Strömer and Ekström.

Metrology is an area of science and technology beset by a good many quarrels and quibbles about priorities and credits. Narrowly nationalistic rivalries have tainted more than one problem of unit selection and nomenclature. The truth of the matter is that the idea or combination of ideas incorporated into the present so-called Celsius scale was very much in the air of that era. Had the sequence of events Celsius-Christin-Strömer-Ekström not taken place, it is more than likely that someone else within a few years would have come up with the very same 100-degree thermometer based on the two obvious "water points."

For a long time the Celsius name was not officially attached to the scale, though its symbol was the °C. That letter stood for the word *centigrade,* meaning "hundred-degree." Only in 1948, after the end of the Second World War, did the Ninth General Conference on Weights and Measures make the change to *Celsius* official. *Centigrade* indeed has quite another meaning. The French angular unit the *grade* was the hundredth part of a right angle, and hence the fraction $\frac{1}{400}$ of a complete revolution or turn. The *centigrade* must be the hundredth part of the *grade;* hence the $\frac{1}{40,000}$ fraction of a full turn.

It was convenient, to say the least, that the same capital *C* could symbolize both the discarded *centigrade* name and the restored *Celsius* name—or even, for those who insist on the full glory of France, that it can symbolize the *Christin* scale, surreptitiously, so to speak.

Celsius as man and scientist well deserved some sort of lasting recognition, whether by means of a thermometric scale or otherwise. He has been called the real "founder" of Swedish astronomy, for he worked to rid Uppsala University of domination by proponents of the invalid Ptolemaic theories and hypotheses. He helped to break the yoke of theologians who stifled scientific progress, and he served as a progressive force in Scandinavian science generally.

Independence of mind and a skeptical spirit are attested to by the "last words" anecdote that has survived him. Tuberculosis cut short his most useful life. As he lay dying and aware of it, he was visited by a clergyman who had been sent by his uncle Olof Celsius, a dignitary of the church. The clergyman held forth about immortality and the eternal life that awaited Celsius after his approaching demise.

"Interesting!" Celsius responded. "Well, soon I shall attain the state in which I can determine whether or not it is true . . ."

Characteristics in common—The three great traditional, nonthermodynamic temperature scales now share several common characteristics. Regardless of the considerations that prevailed when they were first devised, all are now anchored firmly to the two temperatures at which water, and no other substance, changes its state under standard atmospheric pressure: its ice, or freezing, point below; its steam, or boiling, point above.

The former provides the 0° point for the Réaumur and Celsius scales, as well as the 32° point for the Fahrenheit scale. The latter provides the 80° point for the Réaumur, the 100° point for the Celsius, and the 212° point for the Fahrenheit scale.

The distance, then, between the boiling and freezing points of water is given as 180 Fahrenheit degrees, 100 Celsius degrees and 80 Réaumur degrees. Thus, the Fahrenheit degree is the equivalent of 0.555 degree Celsius and of 0.444 degree Réaumur. The Celsius degree is the equivalent of 1.8 degree Fahrenheit and of 0.8 degree Réaumur. And the Réaumur degree is the equivalent of 2.25 degree Fahrenheit and of 1.25 degree Celsius.

The three scales share one basic defect. None is a thermodynamic scale, whose zero point is the absolute zero, or absence, of temperature. Consequently the measurement of a temperature, whether expressed on the Fahrenheit, Réaumur or Celsius scale, does not supply a quantitative indication of the average kinetic or motion energy of the atoms and molecules composing the substances at that temperature. Such an indication, useful or even essential for solving major problems of science and technology, requires a thermodynamic scale, like that adopted by the SI and associated with the kelvin (K) unit.

Every temperature scale in current use is defined at one basic level by the behavior of water in that peculiarly critical and sensitive condition called the triple point, placed at 0.01 K and at 273.16° C. It is well worth while, accordingly, to ask just what water is and whether one water sample always behaves just like any other water sample. Water, we know, is chemically symbolized by H_2O, meaning that each molecule consists of two atoms of hydrogen bonded electrically to one atom of oxygen.

Tables used in modern science show the mass of one average water molecule to be 18.015 3, on the atomic scale which assigns the even number 12.0 to the mass of the common carbon atom called carbon-12, symbolized more formally by $^{12}_{6}C$. This means that it is the kind of carbon with 6 protons and 6 neutrons in its nucleus, and 6 electrons moving outside the nucleus.

One mole, or gram-molecular-weight, of normal water thus has a mass of 18.015 3 gram, within which must be found more than 6×10^{23} complete molecules—a staggering number. These molecules each include one oxygen atom and two hydrogen atoms, else they would not be molecules of water.

Recent recognition and measurements of isotopes enable us to point to possible differences and distinctions between different kinds of water molecules. Every gram or kilogram of normal water includes two different kinds or isotopes of hydrogen, and three different kinds or isotopes of oxygen. Statistically this means that there can be as many as *nine* different kinds of water molecule. Each will have a mass differing from the mass of each of the other eight kinds. Yet each is equally entitled to be called water, chemically speaking.

Why *nine* possible combinations? An anology helps to provide an answer. Suppose we can choose any one of three different precious metals (silver, gold, or platinum) and any pair of precious stones from a supply of rubies and diamonds. How many different kinds of rings could we choose from? Here they are: (1) silver with two rubies; (2) gold with two rubies; (3) platinum with two rubies; (4) silver with one ruby, one diamond; (5) gold with the same; (6) platinum with the same; (7) silver with two diamonds; (8) gold with the same; and finally (9) platinum with the same.

In this analogy, ruby and diamond represent two possible isotopes of hydrogen, hydrogen-1 (also written ^1_1H) and hydrogen-2 (also written ^2_1H), while silver, gold, and platinum represent three stable isotopes of oxygen, oxygen-16 ($^{16}_{8}O$), oxygen-17 ($^{17}_{8}O$), and oxygen-18 ($^{18}_{8}O$).

Consider first the two possible hydrogen isotopes. In normal or average water, among each million hydrogen atoms, about 999,850 will be atoms of the lighter hydrogen-1, sometimes called protium. The other 150 atoms will be hydrogen-2, also known as deuterium. Hydrogen-1, overwhelmingly more abundant in nature, has an atomic mass of 1.007 825, while hydrogen-2 has a mass of 2.014 0, virtually twice as great.

Among all possible kinds of water molecules in nature, by far the most numerous are those in which two atoms of hydrogen-1 are linked or bonded to one atom of oxygen-16. This lightest or least massive of the nine possible kinds of water molecule shows a molecular mass close to 18.010 6. However, molecules in which an atom of hydrogen-2 replaces one of hydrogen-1 have molecular mass about 5½ percent greater. And molecules in which two hydrogen-2 atoms are bonded to one oxygen-16 atom have molecular mass about 11 percent greater than that lightest, most common kind.

At any given temperature, the more massive molecules are less likely than the commonest and lightest ones to attain velocities that enable them to escape—evaporate or vaporize—from the mass of water in which they are moving. The physical result is of prime importance in precision thermometry. The greater the proportion of deuterium (hydrogen-2) in a water sample, the higher the temperature at which it will attain its freezing and triple points.

In fact, if we could prepare a sample of extremely heavy water in which half of all the hydrogen atoms were deuterium, while the rest were the common hydrogen-1, its ice point would be raised to about 2° C (275.15 K), instead of the standard 0.01° C (273.16 K).

Such drastic shifts from behavior of normal water samples are not found in the masses of H_2O obtained from the oceans, seas and lakes around us. But it is possible to bring them about by extensive and expensive processing of such water in laboratories, to increase the proportion of deuterium in the molecular mass.

When we compare the possible oxygen components of water, we find that oxygen-16, with atomic mass of 15.994 91, is by far the commonest kind in nature, comprising about 997,590 out of every million oxygen atoms in ordinary water. However, about 2,040 oxygen atoms per million are in the oxygen-18 isotopic form, with atomic mass about 12½ percent greater than oxygen-16. And the remaining 370 oxygens per million are in the oxygen-17 isotopic form, with atomic mass about 6¼ percent greater than that of oxygen-16.

These mass differences are due to the presence of one extra neutron in the nucleus of oxygen-17, which has 9 neutrons and 8 protons; and to two extra neutrons in the nucleus of oxygen-18, which has 10 neutrons and 8 protons. All three of these isotopes of oxygen are stable and long-lived; none is "radioactive."

A sample of water containing more than normal proportions of oxygen-17 or oxygen-18, or both, will show triple-point behavior at a higher temperature than does a sample of normal or average water. Also, a sample of water containing fewer than the normal complement of these heavier oxygen atoms will show triple-point behavior at a lower temperature than the normal 0.01° C (273.16 K). The greater the shifts from the normal proportions of these more massive molecules, the greater the resulting temperature displacements in the triple point, the freezing point, and also the steam or boiling point of the water sample.

The most massive conceivable water molecule would be the kind in which two atoms of hydrogen-2 are bonded to one oxygen-18. This is at most a highly improbable combination, occurring in nature no more than about 46 times in 10^{15} (1,000 million million) molecules of water. Yet in a 1-mole mass of normal water (slightly more than 18 gram), as many as 28×10^9 (28 billion) such ultraheavy molecules might exist among the grand total of about 6×10^{23} separate molecules.

If we could measure the mass of such an ultraheavy water molecule it would come to about 20 units, or about 25 percent greater than the mass of the most common, the lightest of the nine possible isotopic combinations that we call water.

The fact that the more massive water molecules freeze and evaporate or boil away less readily than the less massive ones results in small but measurable differences in the behavior of various kinds of water found around us on Earth. Thus, water samples taken from close to the surfaces of lakes in hot, dry parts of the world are likely to show more than the normal proportion of such heavy molecules, and hence to record overly high triple-point temperatures. The active evaporative processes have tended to leave behind greater than normal concentrations of these "heavies."

On the other hand, waters from streams flowing from the bases of permanent glaciers or from streams on the leeward sides of cold mountains tend to have lower-than-average proportions of the heavy molecules, and thus they record unduly low triple-point temperatures.

The boiling, or steam, points are similarly displaced in such abnormal water specimens; however, the actual displacement, measured in K or Celsius degrees, tends to be about one third as much as for the corresponding displacement of the triple point in the same water sample.

"Heavy water" shows other peculiarities too. This becomes apparent in density measurements made on laboratory-produced masses of especially concentrated heavy water. Ice formed from ordinary water floats on that water. However, ice formed from heavy water sinks when placed in ordinary water. And ice formed from ordinary water floats higher than usual when placed in heavy water. In short, heavy water is markedly denser than ordinary or normal water.

The result of all this is that the official specification or definition of the critical triple point of water requires that the water used in the triple-point cell shall have "the isotopic composition of sea water," meaning the normal mixture of water found thoroughly mixed in the depths of the Earth's oceans. Needless to say, this means chemically pure water, completely free of salt and other chemicals dissolved in actual ocean water.

Though metrologists of temperature are careful to use only such normal or average water mixtures, there are important modern scientific and technical uses for which the heavier isotopic forms are purposely concentrated, by means of elaborate laboratory procedures. At an installation connected with the Weizmann Institute at Rehovot, Israel, for example, oxygen-17 is extracted and sold, at a price equivalent to about $400,000 per kilogram. Approximately the same price prevails for oxygen-18. The usual per-gram price is about $400.

So-called "heavy water," chemically known as deuterium oxide, is extracted or concentrated and sold at rates currently ranging from $70 to $1,000 per kilogram, depending on the purity of the product and the total quantity purchased. The most modern evaporative extraction processes can prepare heavy water in large quantities at costs probably well under $70 per kilogram. The procedure is that of multiple distillation. Beginning with ordinary water, the normal light molecules are increasingly boiled off, leaving behind ever higher concentrations of the water molecules formed with deuterium rather than with hydrogen-1 atoms.

Because of its high deuterium content, heavy water plays special roles as "moderator" or neutron-slower in nuclear power and explosive processes.

Testing by means of the triple-point temperature—The temperature effects of relatively small changes in the proportions of heavy molecules in a water sample can be estimated quite precisely largely because of the extreme sensitivity of the unique equilibrium state called "the triple point."

A modern triple-point cell requires about three days before its content of normal water arrives at a stable equilibrium, showing the necessary presence of the water-ice boundary and also a volume of vapor; and likewise showing that none of these three states (liquid, solid and vapor) is increasing at the expense of the other two.

Such final stability is attained, as noted, at just $0.01°$ C (273.16 K), and once it is attained it is retained by keeping the triple-point cell in an ice-water bath. Thus the cell can be held at an almost incredibly constant temperature steadiness for months at a time, varying not more than $0.000\ 1°$ C. Since the triple point of water is at 273.16 K, this means a variability of temperature of less than 1 part in 2 million.

Even more important, triple-point cells, when properly in equilibrium, closely reproduce one another's actual temperature level. Between many pairs of such cells, no temperature difference greater than $0.000\ 2°$ C has been found. This is an agreement in temperature to within 1 part in 1 million.

Such physical facts make clear why it is that the triple point of ordinary water is the unquestioned choice for the fixed temperature that supplies the practical lower point for the Celsius scale and the upper point for the Kelvin thermodynamic scale today.

In summarizing the merits and weaknesses of the classical non-thermodynamic temperature scales, it is well to recall that no scale, however logical in concept or numerically neat in design, can operate better than the performance of the substances whose changes in response to temperature variations it measures.

Typically, these scales have been used with mercury-in-glass thermometers, to a lesser extent with alcohol-in-glass instruments, and rather rarely with other expansion liquids. Each of these liquids has its own peculiar irregularities as it responds to temperature changes.

Water serves as a particularly patent instance of such irregularities of expansion. Consider the behavior of water in the 10 degrees Celsius above its normal freezing or zero point. From 10° C down to 4° C (or 40° F) every drop in temperature is accompanied by a small shrinkage in the volume of a mass of water. However, from 4° C to the freezing temperature at 0° C, each further decrease of temperature is accompanied by a slight *expansion* in the volume of that same mass of water.

In other words, the behavior of a column of water seems to show a temperature *rise* during that last interval, when in fact the temperature is falling.

Alcohol and even mercury exhibit less pronounced but by no means unimportant irregularities in their expansion patterns. Helium, being a monatomic gas, is far better suited to indicate temperature by its expansion when it is used under conditions of constant pressure. But even fine gas thermometers, filled with helium or perhaps hydrogen, are not fully free of small irregularities.

Expansion is but one of the several physical effects that provide indirect indications of temperature. None of the other effects that have been pressed into thermometric service is perfect. Each has drawbacks as well as advantages.

29 FROM PRINCIPLES TO PRACTICALITIES IN TODAY'S THERMOMETRY

Fire burn, and cauldron bubble.
—SHAKESPEARE,
Macbeth

The first widely used temperature scales—Fahrenheit, Réaumur and Celsius—provide degrees of various sizes for comparing one temperature with another. But none of them permits measurements in magnitudes proportional to the average atomic and molecular kinetic energies of the substances in question. Only a thermodynamic scale can do this. And only a thermodynamic scale provides a true unit of temperature, such as the kelvin (K).

The possibilities, the principles and finally the procedures for establishing a thermodynamic scale suited to scientific work, we owe to a number of imaginative and remarkable men, the great "energeticists." They include the young French genius Sadi Carnot (1796–1832), the German physician Julius Robert Mayer (1814–78), the German thermodynamicist Rudolf Clausius (1822–88), and in Britain especially James Prescott Joule, already mentioned, as well as the influential scientist-engineer William Thomson (1824–1907), better known by his later title of Lord Kelvin.

The foundations of the difficult scientific specialty called thermodynamics were laid mainly in the decade and a half from 1850 to 1865. It was in this era of great advances that, thanks above all to Kelvin's ability to summarize and draw conclusions from earlier work, the essentials of the present thermodynamic temperature scale were clearly formulated and recognized as deserving to be adopted.

Only one "fixed" temperature point is required to set a thermo-dynamic scale, for nature itself supplies the other. This other is the universal absolute zero of temperature, meaning the absence of all average transferable kinetic energy per molecule or atom.

It was inevitable, by the beginning of the last half of the nineteenth century, that the upper fixed point would be one of the two great state changes of pure water: its freezing temperature under normal air pressure. But where, with respect to that fixed temperature, did the absolute zero lie?

By the time Kelvin made his specific proposals, the approximate location of the absolute zero was already rather well known, in terms especially of the Celsius scale. As Amontons had noted so long before, the pressure exerted by a fixed volume of gas diminishes very nearly in direct proportion to the reduction of its temperature. As measured in the principal early scales, the following is the general pattern of pressure diminution per degree of temperature drop: Starting at a room-temperature level (about 70° F, 16° R or 20° C), a reduction of 1 degree Fahrenheit is accompanied by a drop of about $\frac{1}{530}$ of the previous pressure; a reduction of 1 degree Réaumur by a drop of about $\frac{1}{234}$ of the previous pressure; and a reduction of 1 degree Celsius by a drop of about $\frac{1}{293}$ of the previous pressure.

As the temperatures from which the pressures are calculated continue on downward, the proportionate changes per degree grow greater. Thus, from a temperature set by the ice point of water, each further reduction of 1 degree F results in loss of about $\frac{1}{492}$ of the previous pressure; each further reduction of 1 degree R brings a reduction of about $\frac{1}{218}$ of the previous pressure; and each reduction of 1 degree C brings a reduction of about $\frac{1}{273}$ of the previous pressure.

If these tendencies are charted and extrapolated to the limit, they all convey the same conclusion, even though the numbers involved differ because of the differing sizes of the degrees in the F-, R- and C-scale systems. This conclusion is that all measurable pressure, and hence all detectable temperature, should vanish at a point which can be designated either as $-459.67°$ F (roughly $-460°$ F), or $-491.67°$ R (roughly $-492°$ R), or $-273.15°$ C (roughly $-273°$ C).

In all three scales the measurements point to the same place or temperature distance below the ice point of water, and below the nearby "triple point" of water. All three of these classic scales agree as to the location of the great universal absolute zero of temperature; they simply express that agreement in units of different sizes.

Each of the three scales can become the progenitor of its own thermodynamic (or absolute) scale by setting that absolute zero as its zero point, and then applying its own particular degree for the interval between there and some suitable fixed point, which is now chosen everywhere in the world as the triple point of pure water, very slightly above the ice point of water under normal air pressure.

In this way the Rankine thermodynamic scale was formed out of the Fahrenheit scale. The 0° Rankine lies at just −459.67° F, and at 489.67° Rankine lies the ice point of pure water under standard atmospheric pressure, commonly designated as 32° F. The sizes of the degree Rankine and the degree Fahrenheit are identical. No corresponding thermodynamic scale was formed from the Réaumur nonthermodynamic scale, which is now obsolete in any case.

From Celsius to Kelvin scales—The present Kelvin thermodynamic scale was formed from the Celsius scale. For generations its unit, the kelvin, was called a degree and was symbolized by the conventional degree sign (°K). Thus students learned that the triple point of water was at the temperature of just 0.01° C or 273.16° K, the two being the identical temperature. They learned too that the ultimately unattainable absolute zero could be symbolized either by 0° K or by −273.15° C.

Today the kelvin, symbolized solely by "K," has been elevated to the status of a true unit, a unit in the same sense as the meter is a unit of length, the kilogram a unit of mass, the second a unit of time, the joule a unit of energy, and so on.

From its foundation, the kelvin was indeed a unit in that sense. However, the tradition of the degree sign and the degree concept was strong. Also another factor retarded the recognition that the kelvin was indeed a unit. So many strange and unexpected physical events were found to happen within the last few kelvins above absolute zero! It is as if the range from zero itself to just 1 K contains as many events and relational changes as the range from 1 K to 10 K, or in turn as many as the range from 10 K to 100 K. Indeed, these exponential relationships seem to continue, like a nest of Chinese boxes, for each of these three lesser temperature ranges appears to include about as many significant temperature-related physical situations as does the range from 100 K to 1,000 K, and so on up to the maximum temperatures apparent in our varied and vast universe.

Nothing quite like this is found with the other basic units. The length range between zero and 1 meter does not appear to be packed with significance on a par with that between 1 and 10 meters, and so on. There is no mysticism attached to temperature, but our lives are lived in environments many times higher in temperature than 10 K, and the cryogenic surprises within that least-energy domain still astound us, being in many instances quite at variance with what is often called "common sense."

Once the Kelvin-proposed thermodynamic scale was adopted for the purposes of science, the practical problem had to be solved: How can it best be "realized"? That is, with what instruments and procedures would specific temperatures be measured in relation to the fixed temperature of the ice point of water, or the nearby triple point of water?

The thermometric device employed for this realization effort was the constant-volume gas thermometer, using hydrogen, helium or some other more or less suitable gas. Such thermometric devices tell temperatures in terms of pressure changes and are operable from about 10 K to between 1,000 K and 1,200 K.

However, the more distant the temperature of interest is from the one real fixed point—the water triple point, for example—the less accurately even the best gas thermometers can measure the Kelvin thermodynamic temperature. Temperatures within a few degrees of 273.15 K can be measured to within perhaps 1 part in 500,000. But temperatures around 100 K, or about 173 K below that fixed point, can be measured only with a margin of error as great as about 1 part in 1,000. This means that a temperature indicated to be at 100 K might be in error by as much as 0.1 K.

The same falling-off in accuracy is apparent at temperature

levels well in excess of the fixed point mentioned. For example, at around 1,000 K, measurements may deviate as much as 0.1 K from the correct level.

Not far above 1,000 K, gas thermometers cease to be practically useful, and their places must be taken by optical pyrometers in the effort to realize the Kelvin thermodynamic scale in the higher ranges. These pyrometers measure or estimate temperatures by comparing the light emitted by the matter at the temperature of interest, with the light emitted by a standard filament or hot body.

The attainable accuracy with these light-comparing thermometric devices drops off so rapidly that at about 2,000 K measurements cannot be made, with any certainty, closer than about 2 K to the true temperature. By the level of about 7,000 K, the practically attainable accuracy has diminished to the extent that the margin of error is now about 1 percent. This means that a reading may be as much as 70 K off from the true temperature. At still higher temperature levels, in the 9,000- to-10,000-K range, the errors may be greater than 100 K.

Above 10,000 K, the pyrometric temperature indicators no longer can be used effectively. Various complicated spectroscopic techniques must then be employed in the effort to arrive at some justifiable temperature estimates. The accuracy now becomes even less than at lower levels. A temperature computed spectroscopically to be 11,000 K may be off by as much as several hundred K.

In the other direction, toward the cryogenic depths, at less than about 11 K, the actual realization of the kelvin thermodynamic scale was likewise quite unsatisfactory. From about 11 K down to 0.2 K the basic thermometric device was the ultrasonic thermometer. This measured the velocity with which very short, high-frequency vibrations moved through the substances in question, whether solid or liquid. The less the velocity, the lower the temperature.

Typically, an uncertainty of about 1 part in 1,000 mars such measurements. Thus, a temperature computed to be 10 K may be as much as 0.01 K above or below that. This may seem tiny enough; but at a level of about 1 K such an uncertainty amounts to 1 part in 100, and at a level of about 0.1 K it amounts to 1 part in 10.

In short, though the Kelvin thermodynamic scale was and remains the logical and scientifically sound approach to temperature measurement in principle, it was insufficient, alone and uncorrected, to serve the constantly more rigorous demands of practical precision temperature measurements. Practical methods had to be established for extending measurement accuracy into the temperature regions far above and substantially below the narrow zone of satisfactory accuracy around the triple point of water (273.16 K).

Introducing the IPTS—What happened may seem like a retreat or an abandonment of principle. It was neither. It was merely a recognition of the realities of operation and technical possibilities. The effective compromise took the form of the "International Practical Temperature Scale," or "IPTS," which utilized a number of fixed points based on readily reproducible changes of state in various substances other than water alone. To each of these defined changes of state was assigned a temperature, just as the temperature of 273.15 K had been assigned to the ice point of water under standard atmospheric pressure, and the temperature of 273.16 K to the triple point of water.

Each of these additional assigned temperatures has been placed as close to its true thermodynamic Kelvin temperature as can be determined. Specific measuring devices are prescribed and interpolative techniques or formulas are provided for measuring the relatively small intervals between these fixed points and the temperatures being examined.

There are in fact two "Practical" scales today, with a fixed numerical relationship between them. One is the International Practical Celsius Temperature Scale (IPCTS), which places the triple point of water at just 0.01° C. The other is the International Practical Kelvin Temperature Scale (IPKTS), which places the triple point of water at just 273.16 K, and the absolute zero at 0 K. Each IPKTS temperature measurement is precisely 273.15 higher, numerically, than the corresponding IPCTS temperature measurement. The kelvin, unit of the IPKTS, is precisely equal in magnitude to the degree Celsius of the IPCTS.

The present fixed points specified for both scales are arranged in a following table with those at highest temperatures on top, those at lowest below. The nature of the equilibrium state is indicated in parentheses. Thus, "(solid/liquid)" indicates the temperature at which solid and liquid portions of the substance coexist, and neither gains at the expense of the other. This is scientifically a more precise temperature condition than the closely similar "freezing point" or "melting point."

Also, "(liquid/vapor)" equilibrium means a balance between the processes of vaporization and condensation, and it is a condition more precisely definable in temperature terms than the "boiling point" or the "condensation point."

The four highest temperature events in the list of the defining fixed points of the IPTS are all solid/liquid equilibriums of metals, observed under standard air pressure. Of the remaining nine temperature events, six are liquid/vapor equilibriums of substances that we think of as gases, because in our ambient temperature ranges they are in the gaseous state. Five of these six events are also observed at standard atmospheric pressure to obtain the indicated temperatures. The sixth event is described in a following paragraph.

This leaves three other events, each marked by an asterisk. These are the triple-point events, involving equilibrium of all three states (solid/liquid/vapor) for the compound water (6) and for the elements oxygen (8) and hydrogen (12). Triple-point equilibriums are necessarily observed in the absence of pressure. More precisely, the triple-point cell contains no air. It contains only the substance whose triple point is to be measured thermometrically. Whatever gas pressure exists within the cell is supplied by the vapor of that substance itself.

Thus eight of the dozen defining events for the IPTS are specified as at standard atmospheric pressure. Three other defining events must take place at a "built-in" pressure, supplied by the equilibrium of the three states of the single substance contained in the prepared triple-point cell.

The one special case, so far as prevailing pressure is concerned, is that with number 11, next to the last in the list. This is the liquid/vapor equilibrium of hydrogen under a pressure less than one third that of the standard atmosphere. The necessary pressure is specified quite precisely: it must be 25/76 of the standard atmosphere, which is 33,330.6 pascal, or 0.329 standard atmosphere. Only at this pressure will the specified equilibrium be attained between the liquid and vapor forms of hydrogen when the temperature is 20.28 K or −252.87° C. Were this pressure to be increased, the liquid form would gain at the expense of the vapor form; were it to be decreased, the vapor form would gain at the expense of the liquid form. In either case, equilibrium could not continue to exist.

| | ASSIGNED TEMPERATURES | |
SUBSTANCE AND EQUILIBRIUM STATE	IPKTS (in kelvin)	IPCTS (in °C)
(1) Gold (solid/liquid)	1,337.58 K	1,064.43 °C
(2) Silver (solid/liquid)	1,235.08	961.93
(3) Zinc (solid/liquid)	692.73	419.58
(4) Tin (solid/liquid)	505.118 1	231.968 1
(5) Water (liquid/vapor)	373.15	100.0
(6) Water (solid/liquid/vapor)*	273.16	0.01
(7) Oxygen (liquid/vapor)	90.188	−182.962
(8) Oxygen (solid/liquid/vapor)*	54.361	−218.789
(9) Neon (liquid/vapor)	27.102	−246.048
(10) Hydrogen (liquid/vapor)	20.28	−252.87
(11) Hydrogen (liquid/vapor, at pressure of 0.329 standard atmosphere)	17.042	−256.108
(12) Hydrogen (solid/liquid/vapor)*	13.81	−259.34

Comparison of events numbered 10 and 11 illustrates beautifully the dependence of the assigned temperatures on the pressure actually prevailing. The liquid/vapor-equilibrium temperature of hydrogen at 1 standard atmosphere pressure is 20.28 K, but when that pressure is reduced to about 33 percent of the standard amount, the equilibrium temperature drops 3.238 K to 17.042 K.

Reduction of prevailing pressure to 32.9 percent of its previous value reduces the equilibrium temperature to about 84 percent of its previous value. This is not the pattern of change one might expect from the mythical "perfect gas," but hydrogen in the cryogenic region below 30 K departs significantly from that hypothetical perfect gas pattern.

This tabulation of a dozen assigned temperature levels to define the International Practical Temperature Scale (IPTS) shows that, from the unattainable absolute zero (0 K) to nearly 1,340 K, there are now twelve precisely described and rather readily reproducible events. The average temperature interval between each pair of these events is only 120 K or 120° C.

Some of the actual intervals are fairly large, such as the 542 K between events 2 and 3. However, this and other possible problems are taken care of by the addition of some 23 secondary or subsidiary temperature reference points. These supplement the basic dozen defining points, extend their uppermost temperature level to more than 3,653 K (3,380° C), and greatly broaden the base on which practical thermometry now operates.

Eight of these additional reference events have assigned temperatures higher than the gold event (number 1) placed on top of the previous table. The hottest of these events is the melting point of tungsten, a most refractory metal, to which the assigned temperature of 3,380° C is given. At intervals below it are the following seven solid/liquid equilibrium points: iridium, 2,443° C; rhodium, 1,960° C; platinum, 1,769° C; palladium, 1,552° C; cobalt, 1,492° C; nickel, 1,453° C; and copper, 1,083° C. In each case standard atmospheric pressure is specified.

Other secondary temperature reference points include the solid/liquid equilibrium of aluminum, at 660.1° C; and the liquid/vapor equilibrium of mercury, at 356.58° C.

About a dozen different elements and half a dozen different compounds are specified for use in these secondary reference points of temperature.

Worldwide temperature standards—International agreement assures that the same basic defining points and secondary points will be used in every standards and research laboratory in the world. From time to time, as more precise measurements are made and validated, slight modifications are issued for the assigned temperatures of these basic physical events. These modifications bring their agreed IPTS temperatures more closely into line with the theoretical thermodynamic Kelvin temperature scale.

In all, with the basic and the supplementary points, there are about thirty-five physical events, each with its agreed or specified temperature, and the average temperature interval between adjacent temperature points is little more than 100 K or °C. Any possible temperature situation from 4,000 K down to the unattainable absolute zero lies within a couple of hundred kelvin of one or another of these points. It is thus relatively easy to interpolate from one of these defined temperature points to some as yet unknown temperature that is being measured.

The instruments and the procedures to be used in such interpolations are specified quite precisely by the IPTS. Little or nothing is left to chance or to the arbitrary decisions of any individual laboratory.

Three specified types of thermometric devices are to be used, each within a clearly delineated temperature range. The platinum-resistance thermometer is to be used for all temperature measurements between 13.81 K (the triple point of hydrogen) and 903.89 K (the solid/liquid equilibrium point for antimony). Such a thermometer consists of a small strip of platinum, whose resistance to the passage of electric current rises as its temperature increases. Thus, its resistance, measurable in terms of the ohm, serves as an index of the temperature to which it is exposed.

The quantitative relationship between the observed resistance and the temperature into which it is to be converted is clearly indicated by equations, or "recipes," designed to allow the thermometrists to move with certainty from a specified temperature (one of the approved and defined events) to the temperature being measured. Different equations are supplied for different major segments of the large temperature range over which the platinum-resistance thermometer is the specified instrument.

From 903.89 K to 1,337.58 K (the gold point) a different thermometric instrument is required. This is a *thermocouple,* a device that generates small electric currents as a result of temperature differences at two junctures of its double wires. One of these wires is pure platinum, the other is an alloy of 90 percent platinum and 10 percent rhodium. One junction is kept at the ice point (0° C, or 273.15 K), and the other is brought to the temperature to be measured. Here again specific equations are provided for interpolating from the reading at some nearby IPTS temperature point to the reading at the temperature being measured.

A still higher temperature range, above that gold point, requires the use of measurements of the radiations emitted by the incandescent body whose temperature is being measured, and the interpretation of such measurements by means of equations originated by Max Planck, the physicist who first arrived at quantum concepts. The matter, actively radiating in these temperature levels, is compared with an imaginary full or blackbody radiator, and its temperature is deduced from its observed radiation pattern.

The extraordinary detailed specifications that today make up the IPTS reflect the fact that this is the system in which actual temperature measurements are made, regardless of whether their results are reported in terms of the kelvin (K) or in terms of the Celsius scale. The kelvin unit supplies the theoretical pattern and the essential thermodynamic or absolute scale, which is anchored to absolute zero at its bottom. But in the world of practical temperature measurements today, the IPTS is supreme.

This assures that whether temperature measurements are made in Teddington, England, or Boulder, Colorado, or Moscow, U.S.S.R., or Paris, France, or Berkeley, California, a high degree of consistency and comprehensibility will result, for the painstaking prescriptions and patterns of the IPTS supply a worldwide guide.

The degree of precision attainable today in temperature measurements may be judged from the following. At the temperature of water's "steam point" (373.15 K or 100° C), the uncertainty of temperature measurement is considered to be about 0.005 K, which is very nearly 1 part in 100,000. At the freezing point of zinc (692.73 K or 419.58° C) the uncertainty is estimated to be less than 1 part in 10,000.

Such uncertainties should be reduced still further in the future. As the precision of temperature measurements becomes greater, many investigations will benefit, including especially those in the fascinating and astounding cryogenic, or ultracold, temperature domain.

The price of this impressive progress based on the IPTS has been a curious but undeniably ambiguous relationship between it and the thermodynamic Kelvin temperature scale, which remains the basic scientific pattern of the International System (IS). The fact is that the two scales or scale systems, IPTS and thermodynamic Kelvin, completely coincide at only one place on the entire temperature ladder—that is, at the triple point of water, 273.16 K or 0.01° C. Elsewhere, the degree of coincidence is variable and problematical.

The situation is well summarized by R. D. Huntoon, former director of the Institute for Basic Standards of the U.S. National Bureau of Standards. Surveying the status of standards for physical measurement in the mid-1960s, he noted that the actual relationships between the universally used IPTS scales and the corresponding thermodynamic scale are "not precisely known."

The size of the kelvin (K), the unit of thermodynamic temperature, is presumably uniform from the bottom of that thermodynamic scale to the top. But, as Huntoon noted, in the IPKTS and the IPCTS, the size of the degree, whether the kelvin in the former or the degree Celsius in the latter, is "not in principle constant throughout the scale."

Indeed—though it may sound more serious than it is in practice—"the relation between a degree on the thermodynamic scale and a degree on the practical scale is slightly different in different parts of the scale."

These differences, however, are constantly being reduced by the repeated refinements and adjustments undertaken in the key points of the practical scales. Complete and assured conformity of degree size in the practical and the thermodynamic scales will remain a goal like that of arriving at the absolute zero of temperature—ever more closely approached, but never fully attained.

The current situation has been well evaluated by H. S. Hvistendahl, an able British engineer and metrologist—"In most calculations based on the tabulated values of the properties of substances, any error of the International Practical Scale temperature is of no consequence." The fact is, as he points out, that no matter what may be the "true" corresponding temperature on the thermodynamic scale, any stated temperature on the practical scale means the same thing both to the users of the tables and to the compilers. Thus there is no misunderstanding.

It is true that at higher temperatures the use of the values of the IPTS may result in some error, yet Hvistendahl notes that this "is likely to be insignificant compared with errors due to other causes."

An explanatory analogy—The positive and undeniable need for the IPTS to supplement the theoretical thermodynamic scale of temperature may be made clearer by means of an analogy. Let us perform a "thought experiment" involving comparable situations with regard to the physical variable of length, rather than temperature.

Suppose it were decided that the zero point for measurements of latitude on Earth would be set at the North Pole, instead of at the Equator. In a sense, it is at the poles that the effects of the motion (spinning) of the Earth are least noticeable. Also, it is not easy to reach either of the poles. So the analogy is fairly good.

Next, to make a scale we shall need another fixed point. We might take a flagpole at Point Barrow, Alaska, and decide that the distance from the North Pole to that marker would be divided into just 273.16 units. Thus we would have, or think we have, our unit safely in hand.

But how many such units will we measure between the Pole and Kodiak, Alaska, which lies a good way south of Point Barrow? And how many from the Pole to the City Hall of Hilo, Hawaii? (Hilo is about 70 degrees south of the Pole in our present angular units.)

We must suppose that we cannot use astronomical observations or radio or sonar or artificial satellites to help us with these practical measurements. Under such conditions our ability to "realize" or measure accurately the distances from the Pole, or even from Point Barrow, to the various places of interest would be much restricted. We would have very real difficulties in getting from one place to another, measuring as we go. There would be rugged terrain between Point Barrow and Kodiak, and large unmarked distances across the Pacific between the Pole and the Hawaiian Islands.

We would find that our only hope lay in locating various "bench marks," or places not too far from the points of interest—places to which we could assign some definite measurements in terms of our adopted length unit. Then, from those places to the points of interest to us, we could more readily measure the differences.

The analogy is rough, but it does suggest why practical temperature measurement, ranging so widely up and down the scale, needs more than the one fixed point called the triple point of water.

In one respect, temperature measurement is unique in the world of modern metrology. No other important physical variable is measured, or "measured toward," by means of so many different physical effects and interactions. We noted early that temperature cannot be measured directly. Some other physical variable, related to temperature, must be measured, and the temperature then deduced from those quantitative findings, in terms of the temperature scale that has been selected.

We have already mentioned the thermometric use of expansion of substances, of changes in electrical resistance, and changes in thermoelectric effects, as with thermocouples. From about 0.002 K (very close to absolute zero) to 2 K, changes in the magnetic properties of various salts are used to measure temperature changes. From about 2 to 14 K, changes in the vapor pressure exerted by helium are used to measure changes in temperatures.

Gas thermometers working at constant pressure indicate, by their resulting changes of volume, temperature changes all the way from about 4 K to more than 1,800 K. Or the gas thermometer may be designed to operate at constant volume and indicate temperature changes by its alterations in exerted pressure. In such case, the indicated temperature range is about 20 K to 1,500 K, somewhat more restricted than with the constant-pressure method.

Changes in the electrical resistance of a strip of metal can be used to measure temperature changes from less than 1 K to more than 1,800 K. From about 70 K to 500 K, the varying voltage drop in a diode may be used to measure temperature changes. The thermocouple effect, already mentioned, provides voltage variations as a guide to temperature changes from about 25 to nearly 1,700 K. The velocity of sound or echoes in a mass of matter can be used to measure its temperature over a surprisingly large range. The method called thermography permits measurement of radiation emission to be used as a temperature indication from about 270 K to 500 K. Changes in the electrical resonance of quartz crystals can be used to measure temperature changes from about 190 K to 500 K.

Spectropyrometers, based on radiation absorption, can be used to indicate temperatures from about 500 K to levels in the tens of thousands of K. Radiation analysis by means of spectroscopic techniques can be used to measure, however roughly, the temperatures of glowing masses of matter from about 10,000 K on up.

Even so-called molecular "noise" produced by the random motions of molecules in matter can be used to indicate temperature changes from a few K to more than 2,000 K.

"And so it goes," in a phrase immortalized by a contemporary author, Kurt Vonnegut, Jr.

From the original primitive thermoscopes of Galileo and some of his contemporaries, the measurement of temperature has pursued paths of increasing ingenuity, sophistication and complexity. Yet temperature remains in its innermost essence the average molecular or atomic energy of the least bits making up matter, in their endless dance. Matter without motion is unthinkable. Temperature is the most meaningful physical variable for dealing with the effects of those infinitesimal incessant internal motions of matter.

30 INTRODUCING ENTROPY, THE MEASURE OF MIXED-UPNESS

> . . . *throughout the universe*
> . . . *I find alone Necessity Supreme.*
> —JAMES THOMSON,
>   *The City of Dreadful Night* (1874)

All matter is incessantly in motion within itself, even though outwardly it may seem static and motionless. Motion and matter cannot long be considered apart. Even the unbudgeable rock— "this rock shall fly as soon as I!"—is formed of myriads of molecules vibrating at rates that seem incredibly rapid to us humans.

What we consider to be deep-freeze and ultracold conditions still are marked by abundant molecular and atomic motion. Matter behaves in characteristic and diverse ways because of the varied choreographies of the incessant dancing and jiggling-joggling of these fantastically tiny and numerous entities that compose it.

The higher temperatures rise, the wilder grow the dances of these ultimate particles, the atoms and the molecules formed from atoms. The dances gyrate finally far enough to break the bonds that have hitherto held the parts in certain configurations or relationships.

Thus, when the molecules of solids vibrate far and fast enough they tear free and begin to slide past each other, and we coarse-eyed observers say that the solid substance has melted and become liquid. Later, with still higher temperatures, the liquid molecules swing so widely that they cease to slide past one another and begin to fly free through all available space, changing direction only when they collide with one another or with the retaining walls of the vessel that confines them. Then we say that the liquid has vaporized and become a gas.

Amidst these endless tarantellas of the tiniest bits of matter, it might seem impossible to measure or estimate the comparative effects of so much ultramicroscopic whirling and swirling. Yet we have seen already that thermodynamic temperature provides a precise and meaningful measure of the *average* kinetic energy per molecule or atom, and that heat energy transfers from one body to another only when the one transferring shows a higher temperature than the one transferred to.

Now, after some necessary preliminaries, we are about to meet still another measure, which is in effect an index of how much deviation and disorder exist among the individual molecular energies, compared with the mean (average) temperature-related energy per molecule.

There are statistical measures of inequality and deviation from some central mean, or average. In the physical world of thermal phenomena the measure of such disuniformity or irregularity is called *entropy.*

It is fairly apparent that, as the dances of the molecules grow madder and more energetic, the possible differences in the energies of individual molecules, from instant to instant, grow greater, both with respect to one another and to the over-all average per molecule that we call thermodynamic temperature. But how much greater? And how can some meaningful number be found to measure this greatness, a number that will help us ferret out the patterns by which heat energy and known masses of particular substances interact to produce temperature effects?

To such difficult questions, the strange variable of entropy provides answers, not always clear or simple, but the best answers available.

Narrowly considered, entropy is simply a special relationship between the amount of heat energy absorbed or emitted by a body, and the absolute temperature at which such income or outgo of energy takes place.

In the International System (SI), energies of all sorts are measured in terms of joule (J), and thermodynamic temperature is measured in terms of kelvin (K). Hence it is hardly surprising to find that entropy entails the unit combination of the joule per kelvin (J/K), subject to certain important precautions.

We have already met and used a physical variable called *heat capacity*. Its measurements are made in terms of the joule of heat energy absorbed per single kelvin or degree Celsius of temperature increase in the absorbing substance.

Heat capacity, however, is not entropy, though the two are related, and entropy estimates or calculations would be virtually impossible without accurate heat-capacity measurements on which to base them.

Heat capacity—and this may be repeated more than once, for safety—is an amount of heat energy associated with a change of just one kelvin or degree Celsius at some specified temperature level. Thus, if a body X at 300 K reveals a heat capacity of 400 J/K, it means that 400 joule increases its temperature from 300 K to 301 K, or perhaps even from 295.5 K to 300.5 K. In either case, the change is just one kelvin, or one degree Celsius, which means the same thing.

One can express heat capacities handily without reference to the thermodynamic or kelvin scale. In the case mentioned, body X can be said to have, at 27° C, a heat capacity of 400 J/°C. Always understood is the fact that it is a difference of just one degree Celsius that is brought about by the addition of the specified heat energy of 400 joule.

To further distinguish heat capacity from entropy, let us mention some of the ways in which we meet heat capacity. It appears in abundant and precise tables under such names as specific heat, mass heat capacity, molar heat capacity, molecular heat and atomic heat.

The heat-energy unit used in most of the literature is not the joule, but rather the *calorie,* in one of the variations that have been previously mentioned. Each of the calorie-family members is the amount of heat that produces a one-degree-Celsius temperature change in a single gram of pure water under certain specified conditions, such as 1 standard atmosphere of pressure and an initial temperature, such as 14.5° C.

The result is that when any other substance's heat capacity is expressed in calories, a comparison is being made between that substance's heat capacity and that of water under specified conditions. Thus, it is true that specific heat is the heat capacity of the named substance relative to, or compared with, that of water, just as specific gravity is the mass of a body relative to or compared with the mass of a like volume of water. In each case, the temperature level at which the measurement has been made must be specified in order to avoid confusion.

Large changes in heat capacity are found at different temperature levels of the same substance. For example, compared with the heat capacity of water at 15° C, that of ethyl alcohol at 0° C is found to be 0.535 (its "specific heat"), while the same ethyl alcohol at 100° C shows a specific heat of 0.824, more than 1.5 times as great.

Obviously the dance of the alcohol molecules at the higher temperature is such that they require a larger addition of heat energy in order to produce the same relative increase in their average molecular kinetic energy, commonly called thermodynamic temperature.

Strictly speaking, it is not proportionally the same increase in that temperature. A gain of 1 K—that is, from 273.15 K to 274.15 K—means an increase of about 36 parts in 10,000 at the lower (0° C) level; but a gain of 1 K—from 373.15 K to 374.15 K—at the higher level represents an increase of only about 27 parts per 10,000. The ratio of temperature change is greater at the lower level, where the specific heat is smaller.

Though we speak of heat energy being added in the case of such specific heat or heat-capacity measurements, it is also possible that the energy could be added in the form of an equivalent amount of work—such as stirring, shaking, or otherwise belaboring the alcohol or other substance whose heat capacity is being measured.

If we use the preferred SI unit to measure heat capacity, we find that alcohol at the 0° C point requires about 2.24 joule of energy per gram of mass for a 1 degree increase in temperature, while at the 100° C level about 3.45 joule are required.

We have been comparing alcohol with water on the basis of masses of 1 gram of each. However, a molecule of water has a mass of about 18 against that of about 46.1 for a molecule of alcohol. In short, the least unit of the latter is about 2.6 times as massive as the least unit of water.

Most of the really meaningful calculations of heat capacities are based on the heat capacity per gram molecular weight (per mole), sometimes called for short just the *molecular heat* of the substance. For alcohol at 0° C, for example, the molecular heat is 103 J/K while that of alcohol at 100° C has a molecular heat of 159 J/K. In contrast, the molecular heat of water at 100° C is only some 75.3 J/K. Thus the molecular heat of 0° C alcohol is about 1.4 times that of the water standard, while that of 100° C alcohol is about 2.1 times that standard.

Such measurements of molecular heat, or heat capacity per mole, do not mean the heat capacity per single average molecule. Each mole of a substance contains 6.022 2 × 10²³ individual molecules or atoms (as the case may be), as revealed by the important Avogadro number or constant.

"Perfect" gas as a standard—Is there perhaps a more significant standard of comparison for molecular heat capacities than that of water at 15° C? We might make some use of the smallest molecular or atomic heat that is theoretically observable. This would be the atomic heat of the hypothetical perfect monatomic gas, measured under a constant pressure of 1 atmosphere.

Such a perfect gas, never fully realized in nature, has been described as one whose atoms have no attraction for each other and almost no volume or bulk per atom. They merely fly about, bounce back from each collision with perfect elasticity, and always remain in the gaseous state, because no matter how slowly they move about, there is no mutual attraction to cause them to clump together and form either liquids or solids.

Such a perfect gas would have only translational kinetic energy —energy resulting from the place-to-place velocities of its myriads of atoms. It would have no energy in the form of the rotations or spins of its individual atoms, or in the form of vibrations of the structures of those atoms. Since the perfect gas of necessity is composed of single atoms, not molecules, it could have neither molecular spin nor molecular vibration energy.

Suppose, for metrological purposes, that such a perfect gas actually exists and that we place a measured mass of it in a closed, rigid container to assure an unchanging volume. Then slowly, one degree or unit (1 K) of temperature at a time, we raise the temperature of this confined gas, keeping careful record of how much heat (in joule) is required for each such 1 K rise.

What would our measurements show? This can be answered with full confidence: they would show just 12.47 joule of heat absorbed per mole of mass for each successive unit of temperature increase. In other words, the atomic heat of a perfect gas under constant volume is just 12.47 J/K per mole.

But another kind of measurement is possible also. The gas might be placed in a nonrigid container, subject only to the normal pressure of the atmosphere. With each 1 K rise in temperature, its volume would then increase slightly. And our careful measurements would show that under these conditions of constant pressure (not constant volume), the atomic heat is about 20.8 J/K per mole of gas mass—an increase of some 67 percent over 12.47 J/K per mole.

This difference is significant. It reflects the important fact that in the constant-pressure case a substantial part of the added heat energy is diverted into doing work, the work of expanding the volume of the gas against the uniform surrounding pressure. Indeed, 40 percent of the added heat energy is thus diverted, leaving 60 percent, or 12.47 J/K per mole, for increasing the average kinetic or motion energy of the atoms of this hypothetical perfect gas. (They have more room in which to fly about now, in the constant-pressure case; whereas in the constant-volume case that room remained unchanged.)

Thus we have two possible comparative standards of atomic heat to use—the constant-volume standard of 12.47 J/K per mole, and the constant-pressure standard of 20.8 J/K per mole. Each of these is admirably constant regardless of the temperature level at which the imaginary measurements are made, for this constancy is one of the prime perfections of the utterly perfect-gas concept.

The preference here is given to the constant-pressure standard of 20.8 J/K per mole, and for a very practical reason. Most measurements of the atomic and molecular heats of the quite imperfect gases, liquids and solids around us are made under conditions of constant pressure, rather than of constant volume. Indeed, constant-volume measurements of the molecular heats of liquids and solids are difficult, if not downright impossible, in practice.

It seems more reasonable to compare actual constant-pressure measurements with a hypothetical constant-pressure standard rather than with a hypothetical constant-volume standard.

Accordingly, using this atomic heat capacity of 20.8 J/K per mole we can state that the molecular heat of water at 15° C is about 3.6 times as great; that of alcohol at 0° C is about 5 times as great; and that of alcohol at 100° C about 7.6 times as great. Thus the molecular heats of familiar substances show marked departures from the standard of the hypothetical perfect gas.

What of the heat capacities exhibited by the various "imperfect" gases that surround us and on which our lives and beings depend? They are greater than the perfect-gas standard, but not so much greater as those of the water and alcohol already mentioned. At ordinary or "room" temperature, hydrogen gas (H_2) shows a molecular heat of some 28.9 J/K, or nearly 1.4 times that of the perfect gas.

Generally, the more complex the molecules of a gas, the greater its molecular heat. Thus, the gas symbolized by C_2H_6, each of whose molecules include 8 atoms, displays a room-temperature molecular heat nearly 2.6 times that of the theoretical perfect gas under constant-pressure conditions.

Meaningful variations in molecular heat—Molecular-heat capacities of myriads of compounds and the atomic-heat capacities of numerous elementary substances are among their most important and revealing characteristics. The magnitudes of such atomic and molecular heats differ substantially not only from substance to substance but within the same substance at various temperature levels.

The most drastic and even startling changes take place, in fact, in the cryogenic and ultracold temperatures, especially below about 50 K (about −223° C). No substance actually remains a gas all the way down to the uttermost attainable colds. All the common gases condense to liquids, and then freeze to form rigid crystalline solids.

Only helium, the closest approach to the imaginary perfect gas, remains stubbornly gaseous, but finally even helium liquefies under 1 standard atmosphere of pressure at 4.2 K. However, at this pressure it never does freeze to form a solid. Instead, at 2.19 K it transforms into a liquid so fantastically strange in its behavior that it is called helium II and acts like nothing else observed on earth. It becomes superfluid, loses all viscosity, and otherwise seems to flout the normal conventions obeyed by all other forms of matter.

Every solid crystalline substance reveals sharp shrinkages in its heat capacity as it is chilled ever closer toward absolute zero. Heat capacity finally falls off to nothing at or quite near that ultimate and unattainable absolute zero of temperature (0 K).

This means that here, at the bottom of the great temperature ladder, an extremely tiny addition of energy makes more difference in terms of numerical temperature rise than anywhere else up the ladder. The reasons for this important vanishing of molecular heat at the ultimate zero of temperature relate to quantum theory and lie outside our scope here.

Thus, heat capacity, though so basically important in the worlds of thermal measurement, is different from and to be distinguished from its first cousin, entropy.

The simplest kind of entropy-indicating picture is given by drawing a graph based on paired observations for the particular substance being investigated. The horizontal scale of the graph shows the absolute temperature, from zero at left to the maximum at which measurements were made, at right. The vertical scale shows the corresponding figures for—at that temperature—the substance's heat capacity *divided by* that same temperature in kelvin units.

For example, in the case of the form of carbon called graphite, the curve on this graph would pass through one point marked by 30 K and by a height of 0.008 4 J/K (the K in that denominator stands, of course, for 30 K). Another point would be at the intersection of 70 K and at a height of 0.013 J/K. A third point lies at the intersection of 100 K and 0.017 J/K. And so on, up to the intersection of 298 K and 0.029 J/K.

Such a graph, when all its located points are marked and smoothly connected by a curve, shows by the area between that curve and the horizontal base line, the total or absolute entropy from 0 K at left to the highest temperature reached by the curve at right.

In the case of the graphite form of carbon, this kind of graphic analysis reveals a total entropy of about 5.7 J/K per mole of that substance at a temperature of 298 K. A corresponding curve for diamond, another configuration of the same carbon atoms, shows a corresponding total entropy of only 2.5 J/K per mole at the same temperature.

The use of graphic methods to convert heat-capacity measurements into entropy equivalents is a reflection of an all-important fact. Entropy is a measure built up like a mosaic from an enormous number of tiny bits and pieces. Each is a tiny increment of heat energy divided by the absolute temperature at which it was gained (or lost). In case of a gain, the entropy bit is added; in case of a loss, it is subtracted from the previous entropy total. Entropies are summations, or accumulations.

Entropy measurements are usually made on the basis of differences from some arbitrarily chosen level, rather than being taken all the way from absolute zero. Some convenient temperature is treated as the zero, or starting temperature, and from it entropy changes are accumulated until some other higher temperature is reached. The entropy accumulated en route is the entropy change for that particular temperature alteration in the measured substance.

In much the same way we use sea level as a zero point or standard from which to measure the elevations of various features on land, or the variations in depth of various points on the sea bottom (below sea level). An absolute standard would require that we give all elevations on Earth in terms of the distance to the unattainable center of the Earth.

If we begin with ordinary air at 0° C and raise 1 mole of it to 1,000° C, that mass will gain 47.6 J/K of entropy by 1,000° C; 68 J/K by 2,000° C; and about 81.6 J/K by 3,000° C. Each of these is an increase-of-entropy measurement relative to 0° C. Air at 0° C, of course, has a substantial entropy relative to absolute zero, 273 K below. However, so long as clarity is maintained, these comparative entropy measurements serve admirably.

Entropy: a mystery?—Much mystery and even some mysticism has become attached to the term and concept of entropy. It is widely regarded as inhumanly subtle and horrendously hard to handle. The present author does not deny its difficulties, but is well aware also of its indispensable importance in the world of measurements.

A preliminary analogy may help to clarify some of the relationships involved with entropy. This analogy is based on the realization that temperature may be compared to a mechanical force. A force does work—transfers energy—when it moves a distance (length) in its own direction.

We are surrounded by endless numbers of downward forces that we call "weights." Lifting any one of these in the direction away from the center of the Earth means a measurable amount of work. Thus if we lift 1 pound a distance of 1 foot (1 ft-lbf) we do an amount of work of just about 0.74 joule. Likewise, if we lift by a distance of 1 meter a mass of 1 kg that is subject to the standard gravitational acceleration of 9.806 65 m/s^2, we have done 9.806 65 joule of work.

Now, by means of some suitable rocket device, we wish to raise such a 1 kg mass to a considerable distance above the Earth. In fact, we put into that lifting a total work or mechanical energy of 18.3 million joule (18.3 megajoule).

If gravitational intensity remained unchanged as it rises higher and higher above the surface, that amount of work should suffice to lift it 1,870 km. However, the downward force or "weight" of that 1-kg mass diminishes in proportion to the square of its distance from the center of the Earth. Hence the stated amount of work suffices to lift it to 2,640 km, or 1.4 times as far.

When it has reached that 2,640-km distance, its "weight"—the force tending to draw it earthward again—is down to just half of the value it had at the surface. There in space it "weighs" only 0.5 kg.

Because of the manner in which the gravitational force on this mass has changed, we can see that each kilometer of rise from Earth at the start of its journey required at least twice the energy absorbed in the last kilometer of that journey—from 2,639 to 2,640 km overhead.

Thus the total energy or work expenditure on the operation is the summation of a series of tiny increments of forces times distances, the forces diminishing from 1 kilogram of "weight" at the surface to ½ kg at the end of the rise; and the distances adding up to the grand total of 2,640 km. The energy total here, too, is a sort of mosaic of innumerable tiny bits summed up to make the whole.

Now for a rough thermal counterpart to our great weight-lifting example. We take a mass of 812 kilogram of hydrogen gas under low pressure at the temperature of 0° C. Bit by bit we add to it a heat energy of 18.3 million joule. (This energy is chosen equal to that in the weight-lifting case, just to remind again that heat and mechanical energies are interconvertible in terms of the one unit that measures both—the joule.)

As we add the heat energy, the temperature of the gas is raised until finally it reaches 1,000° C. (By means of effective heat insulation we must be sure to prevent almost all heat loss during this process!)

As the temperature goes up, so does the entropy of the mass of gas. We call its entropy zero at 0° C, and find that 1,846 million J/K units of entropy have been added by the time 1,000° C is reached. Since 1 mole of hydrogen gas has a mass of 2 grams, our 812-kg mass represents 406,000 moles, and the entropy increase per mole of gas is thus 4,600 J/K.

The total energy added to the gas is the summation of all the products of tiny entropy additions each multiplied by the temperature at which that entropy was added. Just as a distance times a force has a product called a mechanical energy, so an entropy times a temperature has a product called a thermal energy. In other words, as temperature corresponds to force, so does entropy correspond to the length through which a force is moved.

But in one important respect the thermal process has differed from the mechanical one in our linked examples. As the thermal process went on, the temperature rose ever higher, reaching its peak at the end. As the mechanical process went on, the gravitational force constantly diminished, reaching its lowest level at the end.

Consequently, the actual total increase of entropy was much less than it would have been, could the entire 18.3 million joule of energy have been poured into the gas while it was at or close to the 0° C level. Could that have been done, the total entropy increase would have been more than 3 times that actually noted. (It cannot be done, however!)

On the other hand, we have seen that the diminishing gravitational force accounted for the distance traveled being greater than if gravitational force had remained uniform all the way.

At the start of the thermal process, the entropy increased more per 1-degree Celsius rise in gas temperature than it did near the end of that process. An opposite relationship existed in the great weight-lifting event.

Both these examples are instances of energy transfer in which the total transfer is the product of two physical variables, each of which is a function of the other. In the mechanical case, the total work done is the sum of a long series of tiny distances moved times the gravitational force against which each bit of motion was made. In the thermal case, the total heat energy transferred can be looked on as the sum of all the products of tiny entropy increments times the temperature that prevailed when each such increment was added to the gas.

If we measure the effects accompanying unit additions of work, 1 joule at a time, we find that in the weight-lifting case, the accompanying distance moved is inversely proportional to the gravitational force opposing that motion. If we measure the effects accompanying each unit of thermal energy added to the gas, we find that the entropy change is inversely proportional to the temperature at which that change was made. The gravitational force is diminishing en route; but the temperature, its analogue, is constantly growing, from 0° C to 1,000° C.

Considerations such as those toward which this analogy points suggest why entropy measurements and calculations are important to chemists, engineers and thermodynamicists. Entropy forms indeed an essential link between the changes in the internal heat energies of substances and the temperatures, or average molecular kinetic energies, at which those changes are observed.

Origin of entropy—Entropy is a relatively late addition to the arsenal of physical variables in the great area of thermodynamics.

We owe it to Rudolf Clausius (1822–88), one of the half dozen greatest thermodynamicists. It was he who first suspected, then detected, demonstrated and named this most subtle and oftentimes esoteric physical variable.

At the outset Clausius bestowed on his new concept the formidable German name of *Verwandlungsinhalt*. This means "transformation content" or even "metamorphosis content." Perhaps fortunately, he decided later to drop that polysyllabic label and replace it with the trisyllable *entropy,* which he had put together from two Greek roots which mean much the same as "turning into" in English. Entropy even might be considered as a sort of thermodynamic equivalent of the concept of "becoming." It emphasizes the idea of change: what was *that* becomes *this* and later on will turn into *that other*.

The sound of the word seems to suggest the first syllables of both *en*ergy and *tropi*sm, which together would mean a movement or wandering of energy. That would not be distant from the concept Clausius uncovered, but it does not reflect the actual etymology of the now-famous designation *entropy*.

During most of its life of more than a century, the entropy concept has been applied only to gains or losses in entropy from some previous undefined or unmeasured level. Its basic pattern thus can be presented with the help of the small triangular symbol for *delta* (Δ), which stands for "a small change of."

The symbol "S" has been traditionally used to represent entropy. "Q" commonly represents a quantity of heat energy. And Greek *theta* (θ) represents the physical variable of absolute temperature. ("T" is already preempted for the physical variable of time, just as "E" is for the variable of energy.)

Then ΔQ represents a small change in heat energy in a particular mass of matter; and ΔS represents the resulting change in the entropy of that mass. The resulting recipe or equation is

$$\Delta Q/\theta = \Delta S$$

That is, the heat-energy increment divided by the absolute temperature at which it takes place equals the resulting change in entropy. If the change in Q is a decrease, then the associated entropy change will likewise be a decrease, not an increase.

One simple rearrangement of these three variables converts the equation into a new definition of absolute or thermodynamic temperature itself:

$$\theta = \Delta Q / \Delta S$$

That is, the thermodynamic temperature is the ratio between a change in heat content of a body and the resulting change in its entropy. Or, in other words, thermodynamic temperature is the heat-content change per unit change in entropy.

In the preferred SI units, as we have noted, thermodynamic temperature is measured by the kelvin unit, heat quantities by the joule unit, and so entropies may be measured by the SI unit combination of joule-per-kelvin (J/K). But always, it must be noted, the K value is the actual temperature level above absolute zero. (If the heat transaction has taken place at a temperature of 173 K, that is the number used in the denominator; and if the temperature was 1,717 K, then *that* must be the number in the denominator.)

Though any precise entropy change (ΔS) is actually a sum of many infinitesimal entropy changes, for practical purposes such summations can usually be neglected if only a few joules of energy have been moved into or out of the mass being measured.

Entropy changes in an extraordinary manner, Clausius showed. When any heat exchange or transaction takes place, if all the resulting entropy changes are measured and added together—all the gains as well as all the losses—the net result is almost always a net increase of entropy; and it is *never* a net decrease. In a few very rare and special instances, the net entropy remains virtually unchanged, but these cases are so unusual that they do not dent the ironclad and unmistakable tendency: Total entropy has a built-in tendency to grow! The few exceptional situations in which entropy seems to remain virtually unchanged are called "reversible" processes. In reality just about every process is irreversible and so is attended by increasing total entropy.

Other physical variables can be increased in magnitude and diminished again. Entropy alone, considered as a whole, typically changes in one direction only: toward increase.

Clausius coined a double aphorism, a trifle oversimplified, but summing up enormous areas of thermodynamics. The total energy of the universe, he said, remains constant. But the total entropy of the universe tends toward a maximum. It is ever growing.

The constancy or conservation of energy is exemplified by our use of the same unit, the joule, to measure energy and work of all kinds—mechanical, electrical, chemical and heat. The forms of energy repeatedly change. Kinetic energy becomes potential, or position, energy. That in turn is transformed to mechanical energy plus heat (as through friction and turbulence). Always the energy totals, if properly added together, remain the same for any over-all situation or system of matter that has not received added energy from without or transferred some of its energy elsewhere.

In the incessant energy interchanges, more and more other forms of energy are scattered, dissipated or "degraded" into the form of heat energy within matter. As this happens, entropy increases. Indeed the extent of the entropy increase, as related to the mass of matter involved, is itself a measure of the extent to which this degradation of energy has taken place.

Other energy forms can be converted largely or totally into heat energy. Heat energy, however, cannot be reconverted totally into other energy forms. Entropy also serves as a measure of the extent to which heat energy has been made unavailable to do work for us.

Ever-increasing entropy—The tendency of entropy to increase is inherent, not incidental, to our physical world. A simple example helps to prove this. Suppose we place in thermal contact two identical masses of matter, A and B. The only measurable difference between them is that A is at a temperature of 300 K and B is at a temperature of 310 K. They are well insulated from the rest of the world, so that the only substantial flow of heat will be from B to A, in response to that 10 K difference in temperature.

Designate as "Q" the amount of heat that is consequently transferred from B to A. Thus the entropy change for B will be a loss of Q/310 K, and the entropy gain for A will be Q/300 K. Now, Q/310, having the larger denominator and the same numerator, must be the smaller of the two. Thus A gains more entropy than B loses. And the net change, for A and B together, must be an increase in their combined entropies.

Even if the heat insulation of the A-plus-B combination is faulty, the same principle works. There will also be entropy changes for other bodies, C, D, E, and so on. The more widely the supply of heat is scattered, the greater the total entropy increase of the bodies participating in the heat exchanges.

The only process that could prevent such increases in the combined entropies of the interacting bodies would be a flow of heat energy from a body at *lower* temperature to one at *higher* temperature. And such flows cannot take place. That is one of the basic laws of thermodynamics.

Heat energy flows only downhill, speaking in terms of temperature differences: from the higher to the lower. A temperature difference, when accurately measured on the thermodynamic scale, provides important information, closely related to the entropy concept. Work (mechanical energy) can be extracted by heat engines that are supplied with a fluid, such as steam or hot air, at a temperature above that at which they exhaust or dispose of that fluid. But *how much* work can be extracted?

It is possible to compute the total thermal or heat energy in 1 kilogram, or in 1 mole, of steam at, say, 500 K (226.85° C). But how much mechanical work can be extracted from that steam by the most efficient possible heat engine? The answer was given by Sadi Carnot, the youthful genius whose insights were developed by Clausius and other later thermodynamicists. If the steam or other thermal fluid is exhausted or "rejected" by this engine at a temperature of 300 K, then the very *least* proportion of the heat energy that entered the engine (at 500 K) which the engine cannot utilize is 300 K/500 K, or 60 percent.

Stated in other words, the utmost theoretical efficiency with which such an engine can extract mechanical energy from the heat energy that flows into it is 200 K/500 K, or 40 percent. The 200 K is the temperature drop, from 500 K to 300 K, that the thermal fluid undergoes in its passage through the engine.

Every such engine, no matter how modern and sophisticated its design, wastes far more heat than such a theoretical limit. The fact that an engine and its accessories become hot shows that it is radiating away heat energy, which is thus lost and cannot reappear as mechanical power from the engine.

The perfect heat engine is a practical impossibility. However, that concept is a useful guide in thermodynamic analysis. Such a theoretically perfect engine would produce no increase in entropy. For example, if 100 joule of heat energy are supplied to the engine at 500 K temperature, that represents an entropy of 100 J/500 K or 0.2 J/K. If that engine were perfectly (impossibly!) efficient, this same mass of steam would leave it through the exhaust, still containing 60 joule of heat energy at 300 K, for 60 J/300 K equals 0.2 J/K, the same entropy with which the process began.

In actual practice, even the finest heat engines attain no better than about 30 to 40 percent of their theoretical maximum efficiencies. Thus, with a temperature drop from 500 K to 300 K, instead of deriving 20 joule of mechanical energy out of each 100 joule of thermal energy that flows into it, the engine would produce only 40 percent as much mechanical energy—or 8 joule of energy.

This means that 92 joule, or 92 percent, of the entering heat energy has escaped in one way or another, without being converted into mechanical work. Obviously most of the fuel burned to generate the steam has been wasted. The theoretical efficiency rises when the entering steam is made as hot as possible and the steam that is exhausted has been cooled as close as possible to the temperature prevailing outside the engine. The greater the drop in absolute temperature of the thermal fluid, in relation to its absolute temperature when it enters the engine, the greater the possibility of heat engine efficiency. This is the reason that high-performance steam turbines are fed superheated steam, and why the steam is given opportunity to expand fully before it is finally exhausted to the condenser.

Thermodynamic analysis in terms of absolute temperatures in kelvin units, and by means of entropy concepts, is not confined to heat engines alone. It is essential also to an understanding of mechanical refrigeration and air-conditioning devices, so numerous and increasing in our comfort-oriented culture.

Questions of cooling—Much misunderstanding exists about the energy relations in such cooling processes. In part this is caused by the miserably mixed-up and outdated units used by refrigeration engineers and technicians in the United States. In part, also, it is attributable to the fact that every mechanical room cooler or refrigerating device operates as the opposite or inverse of a heat engine.

The heat-engine analysis must be, so to speak, stood on its head to make refrigeration understandable in terms of energies and powers.

A heat engine extracts mechanical power from some of the heat in a hot gas or vapor that is supplied to it. An air conditioner uses mechanical (or electrical) power to extract some of the heat energy from the matter (gaseous, liquid or solid) in one place and transfer that extracted energy to matter in another place.

The result of successful refrigeration is the reduction of temperature in one locality or enclosure and the inevitable increase of temperature in some other locality or space. Thus the refrigerating machine shows a lower temperature at its intake than at its exhaust. The heat engine, on the contrary, depends on having an intake substantially higher in temperature than its exhaust.

A refrigerator or air conditioner can be said to pump heat energy from where it is not wanted to some place where its presence will matter not at all or only a little to the persons for whose benefit the operation is being carried on.

In fact, refrigerating devices, when properly made, can function also as heating devices. When they function so, they are commonly called "heat pumps." Thus a reverse refrigerator can extract heat energy even from a cold (or relatively cold) outdoors, and shift that heat energy into a house, which is thereby warmed. This becomes possible entirely because of the mechanical power such a heat pump derives either from electric current or from some comparable source. Mechanical energy is converted into heat, quite the opposite of the conversion of heat into mechanical energy in a heat engine, such as a steam turbine or an automobile motor.

How is the cooling capacity of a refrigerating device commonly measured? Its electric power consumption is always measured in units of the watt or a multiple such as the kilowatt. However, its thermal performance is commonly measured even now in one of two antiquated unit combinations: (1) the mean British thermal unit per hour (Btu/hr), or (2) the "ton."

A performance rating of 1,000 Btu/hr means that when operating uniformly during 1 hour the device removes or withdraws a total of 1,000 Btu of heat energy. Since 1 mean Btu equals 1,055 joule and since 1 hour contains 3,600 seconds, this means that the machine in our example extracts heat at a rate of

1,055,000 joule in 3,600 second, or 293 watt (each watt being by definition equal to 1 joule per second). The unit combination 1 Btu/hr equals 0.293 watt of heat removal. Both are units of power, not of energy.

Occasionally one encounters another misleading unit combination, the Btu/min. It is equal to just about 17.576 watt. To convert in the other direction: 1 watt equals 3.414 Btu/hr and 0.057 Btu/min.

Larger, more powerful refrigerating devices than the common room air conditioners are often rated for performance in terms of the "ton." Actually, it should be called the ton per day, for it means the removal of as much heat in 24 hours as would be absorbed by the melting of 1 ton (2,000 lb) of ice in that same time. When 2,000 pounds are melted in 86,400 second (1 day), that is a rate of 0.023 15 lb/s, or 10.500 7 gram/s. The melting of 1 gram of ice at 0° C (273.15 K) absorbs 333.586 4 joule of heat energy. Hence to melt 10.500 7 gram, 3,502.891 joule are required. In short, the strange "ton" unit is equivalent to a refrigeration power of approximately 3,503 watt, or 3.5 kilowatt.

It is noteworthy that "1 ton of cooling" is about 11,955 times 1 Btu/hr of cooling. And refrigeration engineers generally are accustomed to considering their "ton" to be 12,000 times the size of their "Btu" unit.

Debabelizing refrigeration—Encumbered by such "Btu" and "ton" combinations to measure cooling and refrigeration rates, even serious students often fail to realize that such heat withdrawal is just as much a matter of power as is the electric-current consumption of the motor that drives the device. The cooling or heat-removal capacities of refrigerators and air conditioners should be stated in terms of the watt or the kilowatt. Then, and only then, can the performance of the device be compared easily and informatively with the power the device consumes.

When refrigerating devices are being chosen for purchase and installation, a most relevant question is: How many watt of cooling power will we receive for each watt of electric power used? Questions such as these are increasingly crucial in this era of power "brownouts," and continual threats of more serious future shortages of electric power. In answering these questions, some rather surprising facts emerge about the nature of refrigerating and air-conditioning devices, small or large.

A power-driven refrigerator or air conditioner is not a simple reversal of an electric heater. Such a heater, when its radiating coils are using 1 kilowatt of current, can emit only 1 kilowatt of heat power. Any output of heat power greater than the consumption of electric power would violate the universal principle of the conservation of energy.

However, refrigerating machines *can* produce much more cooling power than the power they consume in electric current. This becomes clear from the basic equation for maximum refrigeration performance, which is an interesting reversal of the basic equation for maximum theoretical heat-engine efficiency, already stated.

Four basic measurable quantities are involved. P stands for the power, electrical or mechanical, driving the refrigerating machine. R stands for the rate of heat-energy removal, which means the cooling power of the device. Two basic temperatures are expressed in an absolute scale, such as the kelvin. The lower temperature, T_L, is that to which the machine reduces its cold coil; the higher temperature, T_H, is that to which it raises its hot coil or radiator.

Now, if there could be a totally flawless and perfect refrigerating device, the very best it could do is shown by:

$$R = \left(\frac{T_L}{T_H - T_L}\right) P$$

The numerator of the fraction is the coldest temperature in the refrigerating device, and the denominator is the difference between its highest temperature and that lowest one.

Consider an example. A refrigerator operates with its cold coil at 272 K (30° F) and its hot coil at 322 K (120° F), a difference of 50 K (90 degrees F). The fraction in this case is 272 K/50 K, equal to 5.4. Hence R equals 5.4 P, meaning that the theoretical maximum of efficacy is 5.4 watt of cooling per watt of power used. In this and other examples, the cold coil is absorbing heat energy from the room or the space being cooled, while the hot coil is emitting heat energy into the space outside (the outdoors or the kitchen, as the case may be).

Example 2 is the same refrigerator working less hard. Its cold coil is at 277.6 K (40°·F), and its hot coil at 310.9 K (100° F). The temperature difference is 33.3 K (60 degrees F), and the resulting fraction of 277.6 K/33.3 K equals 8.3. Thus the theoretical best performance is 8.3 watt of cooling per watt of power used. Example 3 shows the same refrigerator working even less hard. Its cold coil is at 283.15 K (50° F) and its hot coil at 310.9 K (100° F), a difference of 27.8 K (50 degrees F). The telltale fraction is now 283.15 K/27.8 K, indicating a theoretical maximum performance of 10.2 watt of cooling per watt of power used.

How is it possible that, even theoretically, 1 watt of electric power can remove many watts of heat energy from within a room or a refrigerator? The heat energy withdrawn from the cooled room or the freezer interior is not destroyed or negated. It is merely transported elsewhere, as we can perceive when we feel the hot blasts of air that have been blown by a fan past the high-temperature coils.

A refrigerating machine is actually a *heat pump*. That name, however, is usually reserved for refrigerating-type devices that have been made to operate backwards, so to speak, in order to warm an interior with heat energy withdrawn · from a colder exterior. They serve, in short, as space heaters, and may produce from each watt of electric power several times as much heat (in watt units) as does an electric heater. The heating effect of such a heat pump includes the heat equivalent of its electric power consumption *plus* a larger amount of heat that has been moved indoors from the colder outdoors.

Practical performance results—No refrigerator, air conditioner, or even heat pump device actually operates at its maximum theoretical efficacy. At best, the attained efficacies fall well below half the theoretical maximum efficacies.

Refrigeration engineers, habituated to working in the old units, often use a rough rule of thumb. They figure that a large, well-working and well-designed refrigerating installation should deliver about "1 ton" of cooling in return for 1.5 horsepower driving it. Since 1 ton of cooling signifies 3,503 watt of heat withdrawal, and 1.5 horsepower equals 1,118.6 watt, this means that each watt of power should be reflected by 3.13 watt of cooling effect. The *theoretical* maximum efficacy for such a device is probably about three times as great, or around 9.5 watt of cooling per watt of power.

Another kind of practical rating formula for actual efficacy appears in some consumer-oriented publications. They counsel prospective purchasers to make choices only after considering the EER (energy efficiency rating) of various available models in the size desired. Thus, a consumer advisory column published in July 1973 in the *Los Angeles Times* explains that the EER "can be figured by dividing the electrical power consumption (watts) into the cooling capacity or British thermal unit (BTU) rating" for each device.

Two more or less typical examples are given, each rated as having 18,000 Btu/h (sometimes written Btuh) cooling capacity. The first consumes 3,000 watt of electric power, the second 2,000 watt. The first shows an EER of six (18,000/3,000), and the second of nine (18,000/2,000). Obviously the second has an operating efficacy 50 percent better than the first.

EER ratings, however, are like comparisons of apples with oranges, since the Btu/hr is not identical with the watt, though both deal with truly comparable kinds of power. More meaningful results come from converting the Btu/hr measures into the equivalent watts of cooling, and comparing that with the watts of electric power consumption. The resulting ratio is a number sometimes called the actual C.O.P. (coefficient of performance).

Here is a table for converting typical EER numbers into their approximate C.O.P. equivalents:

EER NUMBER	ACTUAL C.O.P.
5	1.5
6	1.8
7	2.1
8	2.3
9	2.6
10	2.9
11	3.2
12	3.5
13	3.8
14	4.1

Thus the air conditioner with EER of six provides 1.8 watt of cooling per watt of electric power used, while that with EER of nine provides 2.6 watt of cooling for the same power use. A large refrigerating installation that actually delivers 3.13 watt of cooling power per watt of power used has an EER of 10.7. Even this is sometimes surpassed by the most effective modern devices, which provide 3.6 watt of cooling power per watt of electric power, an EER of 12.2. Above that it is difficult to go, at least in the present state of the art. Such levels are attained only when equipment is in tip-top condition and when the air in the space to be cooled has an average humidity of about 50 percent of saturation.

More goes out than comes in—Every mechanically driven refrigerating device expels, or "rejects," more heat energy than it takes out of the space that it cools. The difference is accounted for by the additional heat energy resulting from the work done by the pumps that compress the refrigerant fluid and by the fans or blowers that cool the hot coil and circulate air chilled by the cold coil.

In the case of the 18,000-Btu/hr air conditioner with 3,000-watt power consumption and a coefficient of performance of 1.8, the total heat it ejects is nearly 56 percent greater than that which it removes from the space it cools.

At the other extreme is a large, well-functioning cooling device with coefficient of performance of 3.6. It ejects an amount of heat about 28 percent greater than that which it has withdrawn from the space cooled. Clearly it is only half as wasteful of electric power as the preceding device.

Our concern here is not with the specific physical changes that take place within the coils and valves of refrigerating devices. It is worth noting, however, that today almost none of them employs air as a refrigerant. Instead they employ one of a limited number of condensing gases, also called "cold gases." These include Freon-12, identified also by the formidable polysyllabic name of dichlorodifluoromethane ($C\ Cl_2\ F_2$); methyl chloride ($C\ H_3\ Cl$); carbon dioxide ($C\ O_2$); and ammonia ($N\ H_3$). Such refrigerants are liquid at low temperatures, as when they enter the cold coil;

but when heat is transferred into them they evaporate. At high temperatures, when heat is transferred out of them, they condense again.

To this extent they behave differently from the thermal fluid (water vapor) of the common steam heat engines, for that vapor remains gaseous from intake to exhaust. If it is condensed again to water, this takes place outside the engine itself. In such an engine, energy is taken in at high temperature and exhausted (rejected) at a lower temperature. In a refrigerating device, on the contrary, energy is taken in at a lower temperature and exhausted (rejected) at a higher one.

The efficacy of any refrigerating device is reduced as it is called upon to establish and maintain a larger temperature differential between the high of its hot coil and the low of its cold coil. Any valid measurement of cooling or refrigerating efficacy must include all of the power consumed by blowers and fans as well as that consumed to operate the primary pumps which compress (do work on) the refrigerant fluid. Some popular kitchen refrigerators that include the "frost-free" feature consume nearly 60 percent more electric power than others which do not have that feature but are otherwise comparable in performance. This is in part attributable to power used to circulate air within the cooled spaces.

The case for consistent unit usage—Actual efficacies of air-conditioning and refrigerating devices must remain obscure or confusing so long as their capacities are expressed in such units as "the ton" or the Btu/hr, while their power usage is measured in watt and kilowatt units. It is one of a number of situations in which "what you don't know" about units not only can hurt you and your financial interests, but help realize incessant threats of energy shortages every hot humid day, when tens of thousands of air conditioners are switched on and driven to their utmost.

The author is privileged here to quote a comment by Bruce B. Barrow of the Technical Activities Board, Institute of Electrical and Electronics Engineers, on this very timely topic:

> Today the only internationally acceptable unit of energy is the joule. The practice of using different units to measure mechanical

energy, electrical energy and thermal energy is obsolete, and both the British thermal unit (about 1,055 joule) and the calorie (about 4.187 joule) should be avoided in technical writing. Likewise, there is only one acceptable unit for power or heat transfer, the watt (or joule per second). When energy and power are expressed in joules and watts respectively, many hidden relationships immediately become obvious.

Barrow's comments were offered early in 1973 in connection with an article on energy needs, published in *Science Weekly* late in 1972. That study had analyzed the efficacies of a wide variety of room air conditioners, declaring that this "ranges from 4.7 to 12.2 Btu of cooling per watt-hour of electricity." This, Barrow pointed out cogently, is better expressed as a cooling efficacy (ratio) of 1.4 to 3.6. When interpreted into this consistent form, it "implies that 100 watt of electrical power buys you from 140 to 360 watt of cooling, which would be a very good buy indeed, if there were no shortage of primary energy."

31 FURTHER ADVENTURES IN ENTROPY

The greatest adventure of man—the adventure of penetrating into matter, the material universe.
    —ROBERT FROST,
        Writers at Work (1963)

The name of Clausius was eponymously perpetuated in a unit called the *clausius* (Cl), proposed as a single-named measure for entropy. It equaled 1 kilocalorie per kelvin (1 kcal/K) and thus was 1,000 times as large as the calorie per kelvin, and 4,184 times the size of the SI unit combination that now has preference for this purpose, the joule per kelvin (J/K). The clausius never caught on to any substantial extent, however.

Far more widely used was that smaller *calorie per kelvin*. In its molar form, as the calorie per kelvin per gram molecular mass, it has been widely called "the entropy unit" and has been abbreviated as "eu." It is advisable to recall that 1 eu equals 4.184 J/K mol; and the latter, the SI entropy unit, in turn equals 0.239 eu.

Detailed measurements of the molecular heat or heat capacity of many important elements and compounds have been made, and from these patterns of heat capacity, the entropies have been worked out by the graphic method or calculus computation previously referred to.

One among thousands of possible examples is the common compound sodium sulfate (Na_2SO_4), with a molecular mass of 142.06. It is a solid with crystalline structure at temperatures below about 1,157 K.

We tabulate here its molar heat capacity at eight different temperatures, under constant pressure of 1 atmosphere. The second column indicates its approximate gain in entropy per rise of 1 K in absolute temperature at that point in the temperature scale. The third column indicates, at three key points only, the total or absolute entropy—the summation of all preceding entropy increments—at the corresponding temperature.

One word for the sake of clarity. The assignment of zero heat capacity and entropy to the absolute zero of temperature is based on one of the recognized laws of thermodynamics: every crystalline solid as its temperature is reduced to zero acquires zero entropy. Also, the molecular heat capacity of such a substance dwindles rapidly to nothing in the final descent to absolute zero.

The Thermal Behavior of Sodium Sulfate
Under Constant Pressure (1 atmosphere) from 0 to 298.15 K

ABSOLUTE TEMPERA- TURE	MOLECULAR HEAT CAPACITY (JOULE PER KELVIN PER MOLE) J/1K MOL	AVERAGE ENTROPY GAIN PER RISE OF 1 K IN TEMPERATURE	ABSOLUTE ENTROPY (JOULE PER KELVIN PER MOLE)
0 K	0		0 J/K mol
14 K	0.84 J/1K mol	0.06 J/K mol per K	0.25 J/K mol
20 K	2.5 "	0.125 "	
40 K	17.5 "	0.43 "	
60 K	36.4 "	0.61 "	
100 K	66.5 "	0.67 "	
140 K	85.8 "	0.61 "	
220 K	110.9 "	0.5 "	
298.15 K	127.2 "	0.43 "	149.5 J/K mol

The table shows that from absolute zero to 14 K, the average entropy increase per K is 0.018 J/K mol. From 14 to 298.15 K, however, that average increase is much larger—about 0.53 J/K mol. In contrast—and again to help stress the differences between entropy and heat capacity—at 220 K, where the rate of entropy gain is about 0.5 J/K mol per K, the molecular heat capacity of the substance is 110.9 J/1 K mol. (That number 1 is inserted to remind that this is the heat associated with an increase of *just 1 K* in the previous temperature; the heat, in this case, that would raise the temperature from 220 K to 221 K in just 1 mole of the compound.)

The third column shows that the rate of entropy growth reaches a sort of peak near 100 K, then dwindles from there on up.

This same pattern of diminishing rate of entropy growth would be continued far beyond the highest listed temperature (298.15 K).

Most measurements of absolute or total entropy for chemical and engineering uses are made under constant pressure at 298.15 K. Most, though far from all, industrial operations are concerned with entropy variations in the temperature levels above 200 K.

The universal tendency of entropies as a whole to increase in any thermal transactions has been mentioned. With any one substance, absolute entropy will be found to increase to some extent with each rise of temperature; however, the proportional increase in thermodynamic temperature may much exceed the proportional increase in the associated entropy.

The need always to specify the amount of mass that is being measured is obvious. The greater the mass of a given substance at any temperature, the greater its total entropy. The table shows that 1 mole of sodium sulfate at 298.15 K has an absolute entropy of 149.5 J/K mol. Hence, 2 moles would have just twice as much entropy, 3 moles three times as much, and so on. Each mole of sodium sulfate represents 142.06 gram of that compound.

Sodium sulfate has a heat capacity of 127.2 J/1 K mol at 298.15 K. The hypothetical perfect gas, on the other hand, would have a heat capacity of 20.8 J/1 K mol throughout a large temperature range, including this particular temperature, as already mentioned. Thus the heat capacity of sodium sulfate is about 6.1 times as great as the "perfect gas" standard used before.

The average entropy gain for sodium sulfate at this temperature of 298.15 K is shown on the table to be 0.43 J/K mol per K. The corresponding entropy gain for a hypothetical perfect gas at the same temperature would be just about 20.8/298.15, or very nearly 0.07 J/K mol per K. Sodium sulfate, per gram molecular mass, obviously gains entropy about 6 times as fast as does the theoretical perfect gas at this temperature level.

Tracing absolute entropy—Absolute entropies of some substances have been measured all the way to fairly high levels of temperature. An example is mercury (Hg), the metal often called quicksilver. Mercury's atomic mass is 200.6. Hence 1 mole of mercury has a mass of 200.6 gram.

The following table shows in five stages how the absolute entropy of mercury increases from the lowest possible point at absolute zero of temperature. In the first stage (S), mercury is a frozen solid; in the second (M), it melts at a constant temperature; in the third (L), it is in liquid form; and in the fourth (V), it vaporizes, again at constant temperature.

The Thermal Behavior of Mercury
Under Constant Pressure (1 atmosphere) from 0 to 630 K

STATE OR SITUATION	INCREASE IN ABSOLUTE ENTROPY, IN UNITS OF J/K PER MOL
Frozen solid, at or very near absolute zero	0
(S) Solid, from 0 to 243.2 K	59.9 J/K mol
(M) Melting at 243.2 K	9.9 "
(2,330 joule of heat energy required)	
(L) Liquid, from 243.2 to 630 K	26.2 "
(V) Vaporizing at 630 K	94.0 "
(59,300 joule of heat energy required)	
Total (absolute entropy of mercury gas at 630 K)	190.0 "

In the solid stage (S), the entropy increased an average of about 0.25 J/K mol for each kelvin rise in temperature. In the liquid stage that increase averaged only about 0.07 J/K mol, less than a third as much. The decline in the average entropy gain per temperature increase of 1 K would continue above 630 K as well.

Most impressive are the two great entropy jumps that take place with the changes of state—with the melting (M) of the frozen mercury at 243.2 K, and especially with the vaporizing (V) of the liquid mercury at 630 K. Nearly 55 percent of the total absolute entropy acquired all the way from zero to 630 K comes from these two great transitions, the second of which (V) adds about 9.5 times as much entropy as the first (M). The actual heat energy absorbed in vaporizing 1 mole of liquid mercury is more than 25 times that absorbed in melting 1 mole of solid mercury.

Entropy gains obviously do not always require temperature changes. A change of state, as when water at the boiling point is converted into steam at that same temperature, is the most obvious and dramatic form of an isothermal entropy jump.

Another dramatic entropy jump would take place if we allowed

1 mole of steam (water vapor) at, say, 500 K to expand into twice the volume it occupied before and then heated it back to the previous temperature, 500 K. In each case we would have the same number of molecules with the same average kinetic energy per molecule. But the entropy of the ensemble would be twice as great in the second case.

This begins to indicate the physical nature of entropy. It is a measure of the extent of freedom of motion, of absence of restraint and constraint, on the atoms and molecules in a measured mass of matter. Steam molecules at 373 K dash about more widely and wildly than do water molecules at the same temperature. The doubling of the volume within which molecules of a gas can bounce about and collide also doubles the extent to which they can become "mixed up."

Entropy has been called a measure of molecular confusion, of disorganization, of chaos, of unpredictability and of disorder. It is all of these, and the only danger is that such descriptions may seem to have a pejorative connotation, as if gain in entropy were somehow a loss or damage to the substance or to the humans who seek to make use of that substance.

This much is fundamental: the greater the absolute entropy, the less we are able to specify *where* the molecules of a substance will be found within the physical limits of the substance itself or its containing vessel. Also, the greater the absolute entropy, the less complete is our information or certainty as to the range of kinetic energies within which molecules will be moving. We can estimate or calculate from the thermodynamic temperature what is the *average* kinetic energy per molecule. But as entropy increases, so does the spread of the distribution of the actual energies—the deviations from that central mean.

No wonder that Josiah Willard Gibbs (1839–1903), the great American thermodynamicist, once referred to entropy as a measure of "mixed-upness." Many a time "mixed-up" has been applied also to the mental states of students who sought to understand entropy, and somehow gain a physical "feel" for it.

An eminent physical chemist, J. H. Hildebrand, of the University of California, has recalled how a group of students studying engineering thermodynamics approached their professor with an appeal that he explain entropy for their benefit.

"Oh, I don't understand entropy," he replied. "Nobody understands entropy. You just use it, that's all!" Hildebrand himself assured his own students that entropy was basically another name for molecular disorder.

Entropy also has an interpretation that links it with probabilities. This insight we owe to Ludwig Boltzmann (1844–1906), an Austrian physicist who pioneered in work on the kinetic theory of gases. He showed that the inevitable increase of entropy that Clausius had first delineated was at the same time an increasing degree of disorder in the atoms or molecules of the substances measured. Greater disorder means greater and greater numbers of ways in which the total energy content of a mass can be distributed among the myriads of atoms or molecules of which that mass is composed. The most probable patterns of energy distribution will always be those which maximize the spread or deviation of individual molecular energies from the mean energy per molecule.

In short, the equilibrium state for any given mass with a certain total internal energy content will be one of the myriads of states in which this spread of molecular energies is maximized. Each of these states is indistinguishable, seen from the outside, and each of these indistinguishable states is a state of maximum entropy for the mass under the given conditions. Following any gain or loss in its internal energy, a mass of substance swiftly attains its new thermal equilibrium—a state of maximum entropy under the new circumstances.

The numbers of separate molecules are so enormous and their rates of interaction so rapid that the entropy-maximizing equilibrium states are attained rapidly and retained indefinitely, so long as the energy content of the mass of matter is not altered.

Even though Boltzmann supplied the mathematical expression that linked entropy with probability (of the states of a mass of molecules), he realized sadly that the entropy concept remained rather alien to most who studied it and used it. He lamented that the human mind remained clumsy when it could not lean on analogies to things such as those that "we see and touch directly."

What makes entropy "strange"?—Temperature does not seem strange. We can touch things that are cold, tepid or hot. Weight and mass do not seem strange. We can "heft" a box, and judge whether it is empty (light) or filled (heavy). We can try pushing a covered wagon and estimate from our muscular signals whether it is more or less massive. Our physiological processes give us some sense of the passage of time. Our bony structure and muscular equipment enable us to compare lengths. And so on.

But for entropy, we have no sense organs and no common stock of experiences that we can compare and share. This lack may be made up in part by imaginative comparisons or analogies, such as have been attempted in the previous chapter. We begin with the familiar variable of temperature and with the variable of energy, most easily understood by virtue of its identity with work. No one would question that temperature and energy are indispensable thermal variables, nor that heat energy flows between bodies in response to temperature differences between them.

We can readily realize that a temperature difference functions as a sort of force and that the amount of heat energy transferred is related in some significant way to the magnitude of this tempera- ture force. When we have to deal with a mechanical force (for which the SI unit is the newton) and a resulting energy or work output (for which the unit is the joule), we know what the in- between link is: for *force* times the *length* through which it moves determines the *work* performed. In units, the newtons of force times the meters of movement give us the resulting joules of energy.

Thermally speaking, what must we multiply by the temperature difference (in kelvin) to obtain the resulting heat-energy transfer (in joule units again)? The answer is simply: entropy, measured by the SI unit combination of joule per kelvin (J/K). To reiterate, entropy is to temperature as length or distance is to mechanical force. Each pair, multiplied together, have as their product a result- ing energy.

Without the physical variable of entropy we would lack the valid and indispensable link between temperature and thermal energy. Entropy is indeed more than a "missing link," but because it is not part of our direct sensory equipment, it is commonly defined in terms of the two more familiar variables: temperature and heat energy.

The most vivid instances of how entropy grows with the addition of heat energy to a body are the great transitions from the solid to

liquid state (melting) and the liquid to gaseous state (vaporizing).
They are especially simple and clear-cut because during these
transitions the absolute temperature of the substance remains unchanging until the entire mass of atoms or molecules has gained the
kinetic-energy-per-atom or per-molecule required for its motion in
the liquid or the gaseous state.

The example of water melting at 273.15 K and vaporizing at
373.15 K is somewhat overfamiliar. So, let us look at the tempera-
ture-and-entropy pattern of another compound, methyl chloride,
each of whose molecules contains 3 atoms of hydrogen, and 1
each of carbon and chlorine. Its atomic mass is 50.49, and the following measurements apply to 1 gram molecular mass, or mole, of
50.49 gram.

At ordinary temperatures methyl chloride is a gas. We trace its
entropy gains during the four previously named stages: solid (S),
melting (M), liquid (L), vaporizing (V), and also through part of
another stage, gas (G), up to the standard entropy-comparing
temperature of 298.15 K.

The Thermal Behavior of Methyl Chloride
Under Constant Pressure (1 atmosphere) from 0 to 298.15 K

STATE OR SITUATION	INCREASE IN ABSOLUTE ENTROPY IN UNITS OF J/K PER MOL
Frozen solid, at or very near absolute zero	0
(S) Solid, from 0 to 175.44 K	77.3
(M) Melting at 175.44 K	36.7
(L) Liquid, from 175.44 to 248.94 K	26.1
(V) Vaporizing at 248.94 K	86.5
(G) Gas, from 248.94 to 298.15 K	7.0
Total (absolute entropy of gas at 298.15 K)	233.6

Of the total entropy of 233.6 J/K mol at the final temperature,
nearly 37 percent is gained during the vaporizing phase (V) at the
unchanging temperature of 248.94 K. And some 16 percent is
gained during the melting phase (M) at the unchanging temperature of 175.44 K. The other 47 percent of the absolute entropy is
acquired during the three phase forms: solid, liquid and gas, as
the temperature is steadily increased by additions of heat energy.

But how much heat energy is required to melt one mole of the substance at 175.44 K? Since the entropy gain at that temperature is 36.7 J/K mol, the energy absorbed must be 175.44 times 36.7, or 6,439 joule. In just the same way, if we are told that a force of 175.44 newton has moved 36.7 meter in the direction of that force, we at once know that the work done is 175.44 times 36.7 or 6,439 joule. In the methyl chloride case, we deal with the joule of thermal energy. In the mechanical case we deal with the joule of mechanical energy, or work. They are indeed equivalent, even though, while the mechanical energy can be totally converted into the heat-energy form, the reverse possibility does not exist.

What of the vaporizing transition (V), in which at the temperature level of 248.94 an entropy increase of 86.5 J/K mol is indicated? The relationship is once again the same: 248.94 times 86.5 equals 21,533 joule of heat energy added to effect this change of state in 1 mole of the substance. It is the analogue to a mechanical force of 248.94 that has moved 86.5 meter in the proper direction, and so has done work of the magnitude of 21,533 joule.

The analogue between the two trios (force times length equals work, and temperature times entropy equals energy) has one more important counterpart. It should be mentioned here, even though the presentation of the basic units for electrical and magnetic variables is yet to come. Just as force is a mechanical actuator or animator, and temperature is the thermal actuator or animator, so we can look on electrical potential or voltage as the electrical actuator or animator.

What is the link between a potential, measured by the volt unit, and a resulting energy transfer, measured by the joule of electrical energy? It is *charge,* measured by the unit of the coulomb.

In response to a potential difference of 1 volt, a charge of 1 coulomb acquires or gains an energy of 1 joule. We have seen the same relationship on a vastly smaller scale with the unit called the *electron volt.* Each electron carries a negative charge, or is a negative charge, of just $1.602\ 2 \times 10^{-19}$ coulomb. When an electron responds to 1 volt of potential difference, it gains an energy of 1 electron volt (1 eV), which equals just $1.602\ 2 \times 10^{-19}$ joule.

Thermal energy too could be measured by means of the very

tiny electron-volt energy unit. However, consistency with the principles of the International System gives preference to the joule (J) as the unit to measure energy, and to the watt (W), or joule-per-second, as the preferred unit to measure power of any kind.

In our rough-and-ready analogy, the entropy (measured in units of J/K) is to the charge (measured by the coulomb) as the temperature (measured by the K) is to the electrical potential (measured by the volt). Just as voltage times charge equals electrical energy, so does temperature times entropy equal thermal energy.

This new electrical analogy leads also to a further extension. A flow of 1 coulomb per second is a current of 1 ampere. And a current of 1 ampere moving as a result of a potential of 1 volt represents electrical power of 1 watt. This is the familiar rule: volts times amperes equals watts.

Since 1 entropy unit is analogous to 1 charge unit in electricity, what is the thermal analogue to an electric current of 1 ampere? Simply a flow or gain of entropy of 1 entropy unit per second. In unit symbols it becomes 1 J/K s.

If a temperature difference of 1 K results in an entropy flow of 1 J/K s, then a thermal power of 1 watt is being transmitted. Let us make this even more specific. Suppose measurement shows that 1 mole of methyl chloride is converted in 10 seconds from liquid at 248.94 K to gas at the same temperature. This means that 21,533 joule of heat energy have been supplied within 10 seconds, or at the rate of 2,153.3 thermal watt (2.153 3 kilowatt).

Analogies are illustrations, not proofs. Nevertheless, a physical variable which has the same relationship to absolute temperature that length has to mechanical force, or that charge has to electrical potential (voltage) is something more than an esoteric brain-twister or an abstraction that we would do better not to bother with.

Entropies and configurations—Many substances can exist in more than one structural configuration. The most brilliant instance again is that of carbon, which we find in the relatively lowly form of graphite, or in the hard and glittering form of diamond. The atoms are identical. Only their arrangements differ.

Graphite at the standard comparison temperature of 298 K has a measured absolute entropy of 5.7 J/K mol. Diamond, however, has an entropy of only 2.5—one of the lowest of all entropies at this temperature level. Diamond is so highly regular, so lacking in atomic confusion or disorder, that its entropy is notably low, less than 44 percent that of graphite. These entropy figures reveal to us the fascinating but impractical information that if we want to make a really hot fire we would do better to burn a pound of diamonds than a pound of soft coal, also almost pure carbon!

The entropy of the product—carbon dioxide gas—would be the same in each case. (At 298 K, CO_2 has a total entropy of 213.8 J/K per mol.) The extent of the entropy increase in the case of the diamond fuel would be substantially greater than with the bituminous coal.

A rough rule is that minerals and other substances that are hard or rigid have notably lower entropies than those that are soft or loose. The greater resistance to deformation of the highly regular crystals shows itself both in regard to the hardness scale and the entropy scale. It would not be farfetched to say that, entropically speaking, graphite is more than twice as mixed-up as diamonds are, or that diamonds are more than twice as organized and ordered as graphite.

The element sulfur has two structural forms or patterns of assembly. One is the rhombic form, the other the monoclinic. The former, with an entropy of 31.9 J/K mol, is about 2 percent less "mixed up" or disordered than the latter, with entropy of 32.6 at the same temperature.

It is well known that at some temperature, such as 298.15 K, a liquid and its vapor may exist together in the same enclosure. The absolute entropy of liquid water at that temperature is 70 J/K mol; that of its vapor form is 2.7 times as great. Thus the molecules of vapor dashing about freely are that much more "disordered" or that much less predictable in terms of individual molecular energies, than the molecules of the liquid form—and at the same temperature or *average* energy per molecule of each.

Instances of this kind can be multiplied at length. Oxygen gas exists commonly in the form of molecules of 2 atoms each (O_2), yet it can exist also under certain conditions as a monatomic gas (O), and even as a triatomic gas (O_3), better known as ozone. The absolute entropies of the 1-, 2- and 3-atom forms are respectively 161, 205 and 238 J/K mol, at the same comparison temperature. Obviously the more complex the molecule or least-unit of the substance, the greater the entropy is likely to be.

Absolute entropy is a physical variable appropriate not only to a particular element or compound, but to a particular configuration of such element or compound. Indeed, the thermodynamic temperature level can be and probably should be considered a part of the configurational pattern: thus 1 mole of iron at 500 K is really not configurationally or structurally identical with 1 mole of iron at 700 K.

Only with the help of the physical variable called entropy can we operate in such a way that the more familiar variables of temperature and thermal energy are adjusted to fit these ever-changing patterns of atoms and molecules in motion.

One final instance illustrates what might be called a kind of "entropical schizophrenia." Tin, a familiar metal, exists in two different forms, known as gray tin and white tin. Gray tin is stable from extreme cold up to 292 K. Above that rather critical transition temperature only white tin is the stable form. In order for 1 mole of gray tin to rearrange its atoms into the structure of 1 mole of white tin, 2,238 joule of heat energy must be absorbed. This means an increase in entropy of about 7.7 J/K mol. The white tin thus has the higher entropy at this transition temperature.

Now, it is possible, by taking suitable precautions, to "persuade" white tin to exist at temperatures below 292 K. Thus its behavior can be compared at these temperatures with that of the more stable gray form. All the way down that temperature ladder the two forms of tin show different heat capacities and different absolute or total entropies.

At those lower temperatures, under 292 K, the white form may spontaneously transform itself into the gray form. The reverse, however, cannot take place. The gray does not spontaneously become white below that same 292 K transition temperature.

The mysteries and the "moods" of substances are strange and surprising. Matter takes many forms, and its interactions often refuse to fit into the neat pigeonholes created by habit or "common sense." Only with the aid of the subtle and sometimes baffling tool of entropy is the chemist, the engineer, the thermodynamicist and the theoretical scientist equipped to deal with many of the most important of the infinite numbers of transactions between substance and energy.

"Nothing," said that prince of experimenters and admirable human being Michael Faraday, "is too wonderful to be true." Some of the wonder and much of the truth in the world of atoms, molecules and their motions depends on this weirdly wonderful concept that Clausius, after much consideration, chose to call *entropy.*

32 MORE TOOLS FOR THERMAL MEASUREMENTS—A MIXED BAG

I have multiplied visions, and used similitudes.
—HOSEA 12:10

Many thermal measurements are made with unit combinations different from, but related to, those already mentioned. The principal components in this mixed but meaningful bag of further physical variables are (1) thermodynamic temperatures or temperature differences, symbolized by θ, measured in terms of the kelvin (K); (2) quantity of heat, often symbolized by Q, and measured in the SI terms of joule (J); (3) entropy (Q/θ), measured in terms of the SI joule per kelvin (J/K), where K represents always the *total* absolute temperature, never just a temperature difference of 1 K!

Thermal measurements are often related directly to specified masses of substance. Thus, we find the quantity of heat per unit mass, measured in joule per kilogram (J/kg). This unit combination takes care of variables known by names such as (1) specific quantity of heat, (2) specific heat of reaction (chemical or otherwise), and (3) thermodynamic potential, among others. All are similar if not analogous, for the joule per kilogram is the specific heat of a process or interaction in which 1 kg of matter gains, or gives up, heat equal to the mechanical energy of 1 joule.

The same physical variables have also been measured in such older and obsolescent unit combinations as the calorie per kilogram or pound, the Btu per pound, and so on.

The molar equivalent measure is, of course, the joule per mole (J/mol). It applies to processes or interactions in which 1 mole of substance gains, or gives up, 1 joule of heat. For many important purposes, measurements expressed in J/mol are more useful and meaningful than those in J/kg.

Entropy itself may be related either to the kilogram-unit mass or to the molar mass. The general physical concept in either case is quantity of heat (Q) per absolute temperature (θ) per designated unit of mass.

The entropy unit combination incorporating the kilogram is joule per kelvin per kilogram (J/K·kg). It is best thought of as the variation in entropy of a substance with 1 kg mass during a process at a mean temperature of nK in which a quantity of heat equal to nJ is added to or taken from the substance. Here n stands for some positive number.

Following are three operations, each of which would be interpreted as causing an entropy change of 1 J/K per kilogram: (1) at a mean temperature of 50 K, 50 J of heat are added to or taken from 1 kg of mass; (2) at a mean temperature of 250 K, 250 J of heat are added to or taken from 1 kg; and (3) at 575 K average temperature, 575 J of heat are added to or taken from 1 kg mass. Preceding pages have sufficiently introduced the measure of entropy change per gram molecular mass (J/K·mol).

Heat capacities—the amount of heat (Q) associated with a unit change in temperature—are similarly related (1) to some physical system as a whole, in which case no mass is specified or implied; (2) to a unit mass, as the kg; or (3) to a molar mass, as to 1 mole.

For the first the unit combination is simply J/1 K, with the number 1 inserted as a reminder that just 1 kelvin of temperature change is in the denominator; or J/1° C, if the Celsius degree rather than the kelvin is employed.

For the second the unit combination is J/1 K·kg, meaning joule of energy per 1 kelvin change per 1 kg mass. And for the third the unit combination is J/1 K·mol, meaning joule of energy per 1 kelvin change per gram molecular or atomic mass.

Special names are used for some of these physical variables. In the case of gases, the entropy per kilogram may be called the "specific gas constant" and is the measure of an ideal gas with a mass of 1 kg performing 1 J of work when its temperature is increased by 1 K (its pressure meantime being kept constant).

Temperature, distance, and time—Temperature differences are constantly being compared with the physical distances between which they are measured. For example, consider a metal bar just

0.25 meter long, one end of which is chilled to 0° C and the other heated to 100° C, while the bar between is jacketed with insulation to reduce heat loss to the outside. Within that bar a temperature gradient of 100° C per 0.25 meter exists, equal to 400° C/m. Temperature gradients are especially important in geology, geophysics and heat-engineering practice. The symbol of the physical variable itself is heat difference per unit length (θ/L).

Time, measured by the universal unit of the second (s), appears also in many thermal variables. Often an energy, heat or mechanical, must be related to the length of time in which it is supplied or transferred. It can be expressed in the unit combination of the joule per second (J/s), though the SI provides also the well-known power unit of the watt (W), which means precisely the same thing: 1 J/s.

A flow of heat is best measured in terms of the watt. Older and obsolescent unit combinations for this purpose bear many different names—calorie per second, minute or hour; Btu per second, minute, hour; and so forth.

It is almost always a convenience and sometimes a real boon to be able to deal with heat flows, electrical-energy flows, and mechanical-energy flows all in the same interconvertible unit—the watt, or its common multiples, the kilowatt (kW) and the megawatt (MW), as well as such submultiples as the milliwatt (mW) and the microwatt (μW).

The alternative, increasingly unwieldy and unacceptable in a world of measurement striving to become just *one* world, is to measure electric power by the watt, thermal power by the calorie, or Btu per some time unit, and mechanical power by horsepower or whatnot; and so on, ad nauseam.

Heat flow is often related to the surface area of the substance from which the heat passes, or to the cross section (at right angles to the line of heat transfer) of the substance through which the heat flows. That flow—be it noted yet again—always goes from the higher to the lower temperature level. The SI unit combination for heat-flux surface density or cross-sectional density is simply the watt per square meter (W/m^2).

A further refinement of this same physical variable converts it into an index of the relative extents to which different substances emit or transmit heat.

Under names such as heat-exchange coefficient or heat-transfer coefficient, the power of heat flux is related both to the area and to the temperature difference in response to which the heat flow takes place. It is, in fact, the physical variable of quantity of heat per unit time per unit area of transmission per unit temperature difference between origin and destination of such a heat flux. Its appropriate unit combination in SI terms is the watt/meter^2 K, where K stands for the temperature *difference,* not for the absolute temperature above ultimate zero. A measurement of 1 W/m^{2}K means that for each 1 K of temperature difference there is a heat flux intensity of 1 W/m^2.

A further amplification of this physical variable leads to thermal conductivity. This compares the flow of thermal power not only with the temperature difference that causes it, but also with the two relevant dimensions of the substance through which that flow passes: (1) the cross-sectional area of the substance; and (2) the distance or length in the direction of the heat flow.

The actual unit combination, in SI terms, is simply W/K·m for thermal conductivity, but the concept that leads to it is somewhat more complex than that fraction reveals at first sight. The basic picture is that of a cube of the heat-transmitting substance, 1 meter on edge. One face of this cube is assumed to be at a temperature just 1 K higher than the opposite face. Hence the distance through which the resulting heat flow passes is also 1 meter, and the cross section through which it moves is just 1 m^2. The greater the cross section, the *greater* the heat flow, other things remaining equal; also the greater the distance that the heat has to flow, the *less* the heat flow, other things remaining equal. Hence the full fraction is W·m^2/K·m. When the m^2 (area) in the numerator cancels with the m (length) in the denominator, there is left only the m (standing for m^2/m) in the numerator.

Later we shall examine an important electrical variable called *electrical conductivity,* as well as another variable called *electrical conduction,* which is not quite the same thing. The inverse, or reciprocal, of conduction is called *resistance.* The inverse, or reciprocal, of conductivity is called *resistivity.* Conductivity measures, or should measure, the willingness of a particular substance to transmit the form of power, whether electrical or thermal, with which our measurements deal. Resistivity, its inverse, measures the *un*willingness, or reluctance, of such a substance to allow such transmission. We can form a unit combination for measuring *thermal* resistivity, merely by inverting the fraction for *thermal* conductivity, and thus obtaining K·m/W as the unit combination: kelvin-meter per watt of heat flow.

Thermal conductivity can also be defined in a way that reveals it to be a combination of two physical variables previously presented—temperature gradient (K/m), and heat-flux surface density (W/m²). When, under stable conditions, a temperature gradient of 1 K/m is established by a heat-flux surface density of 1 W/m², then unit thermal conductivity, or 1 W/K·m, exists. The opposite applies to thermal resistivity, a physical variable less likely to be encountered or needed than thermal conductivity.

Variations in thermal conductivities—Measurements of the actual thermal conductivities of common substances reveal enormous variations. Some metals, especially silver and copper, are outstandingly good conductors of heat, or outstandingly bad resistors of heat flow. Many ceramic substances, and substances containing vast numbers of tiny air pockets, are outstandingly poor conductors or outstandingly good resistors of the passage of heat.

The table on page 380 is based on measurements made at 293 K, because the thermal conductivities of many substances change when their internal temperatures change. The central column is headed by *k,* the common symbol of thermal conductivity. The column at right is headed by 1/*k,* indicating that it is the inverse of thermal conductivity. SI units are used throughout.

Air and common gases are poor heat conductors, hence good thermal insulators, if only they can be kept from flowing in convection currents. Thus the finest heat insulators are fibrous and cellular substances that confine air in immense numbers of non-communicating pockets and traps. In order to use air to convey substantial amounts of heat, its convective motions are artificially increased—as by the fan that draws air through the radiator of the automobile, or the fan that spews masses of air past the hot electric coils on heaters.

To insulate houses and factories against expensive heat losses, a great many outstandingly poor conductors of heat have been fabricated. Among those with lowest k and hence highest $1/k$ are such purposeful artifacts as kapok held between paper layers or between burlap surfaces; felted cattle hair; chemically treated fibers of wood; rock wool (fibrous material made from rock); and various low-density forms of "corkboard."

The table clearly shows a definite loss of thermal conductivity as between pure iron (73 W/K·m) and steel (47 W/K·m). This indicates the extent to which the more markedly crystalline and rigid structure of the steel retards the transmission of the molecular vibrations that we call heat in solids.

SUBSTANCE	THERMAL CONDUCTIVITY (k)		THERMAL RESISTIVITY ($1/k$)	
silver	418	W/K·m	0.0024	K·m/W
copper	380	"	0.0026	"
iron	73	"	0.0137	"
tin	62	"	0.0161	"
steel (low carbon)	47	"	0.0213	"
mercury (liquid)	8	"	0.125	"
graphite	2.4	"	0.417	"
carbon	2	"	0.5	"
window glass	1	"	1.0	"
water (liquid)	0.6	"	1.67	"
building brick	0.6	"	1.67	"
methyl alcohol (liquid)	0.2	"	5.0	"
oil, lubricating (liquid)	0.15	"	6.67	"
plywood	0.1	"	10.0	"
paper	0.06	"	16.7	"
felt	0.04	"	25.0	"
eiderdown	0.021	"	47.6	"
silk	0.018	"	55.6	"
felted cotton wool	0.015	"	66.7	"
air (gas)	0.03	"	33.3	"
oxygen (gas)	0.02	"	50.0	"
water vapor (H_2O gas)	0.02	"	50.0	"

If we are not concerned with the characteristics or dimensions of the particular substances through which a thermal or electric power flow is to take place, then we may be content to measure merely the total *resistance* of the system or arrangement. This is the common situation in electrical measurement, in which resistance, measured in terms of the ohm, is specified for a circuit or for some component in that circuit.

A typical example follows: Your automobile battery probably has a potential of 12 volt across its end terminals. Hence if you place in circuit a light bulb with 24-ohm resistance, the resulting current through it will be just $12/24$, or 0.5, ampere. Furthermore, 0.5 ampere flowing because of 12-volt electrical pressure means a power consumption (in the light bulb) of 0.5 A times 12 V, or 6 watt.

Similarly, if we have a temperature difference of 12 K, then it should push through a thermal resistance of 24 units, a power of just 6 watt, or 6 joule of heat per second.

But how shall we define thermal resistance? Electrical resistance can be defined as voltage squared divided by power delivered (V^2/W). In our example voltage squared is 12^2, or 144, and power is 6 watt; hence the resistance is $144/6$ or 24 ohm.

Thermal resistance can likewise be defined as the square of the temperature difference (K^2) divided by the resulting thermal power flow (W). If we arrange a temperature difference of 12 K and wish a resulting thermal power flow of 6 watt, then we should provide a thermal resistance of 24 K^2/W. The unit combination of thermal resistance is, accordingly, K^2/W, where K represents the temperature *difference* or potential.

The inverse of resistance, thermal or electrical, is *conductance*. Hence the unit combination for measuring thermal conductance is W/K^2. Again K represents the temperature difference.

Relatively little use is made of measurements or estimates of thermal conductance or resistance. Thermal conductivity and resistivity, being related to the heat-carrying characteristics of specific materials, are far more widely used. However, it is possible to make some interesting and dizzying comparisons by applying the concept of thermal conductance, as measured by the thermal *mho* or inverse ohm (W/K^2), to the conditions believed to characterize the Earth and the Sun itself.

A temperature of 6,000 K seems a reasonable estimate for the center of the Earth, the heart of its great iron core. Just under the surface of Earth, however, temperatures on a global basis average about 5,700 K less. And the total heat flow that reaches Earth's surface from within is about 3.2×10^{13} watt. Hence the total thermal conductance of the Earth (under present conditions) appears to be about 10^6 (1 million) thermal *mho,* or watt per kelvin squared.

At the thermonuclear heart of the Sun, the temperature is estimated to be about 20 million K, as against about 6,000 K at the Sun's surface. The total power that flows through that surface and into space, via radiation, is about 4×10^{26} watt. Hence the Sun's apparent thermal conductance is roughly 10^{14} thermal mhos, or watt per kelvin squared.

Why should the Sun apparently conduct thermal energy 10^8 (100 million) times more "willingly" or effectively than does the Earth? The following chapter should help make clear one principal reason: internal *radiation* is the means by which almost all power flows through the Sun, whereas at the much lower temperatures inside the Earth, radiative transfer of energy is relatively insignificant, especially in the cooler outer layers of Earth.

Constantly, like a great wind from its core, electromagnetic radiation flows outward through the Sun. It consists of bundles of energy (photons) moved from particle to particle, from atom to atom. The Earth's flow of thermal power, on the other hand, is moved almost entirely by ordinary heat conduction plus convection, caused by the movements toward the surface of masses of matter that have expanded and become less dense because of increased temperature.

The unit combination suited to measure thermal resistance is K^2/W, and since the watt (W) is identical with the joule per second (J/s) we can also use the unit combination $J/K^2 \cdot s$ to measure thermal resistance. This happens to be the product of two fractions, $J/K \cdot s$ and $1/K$. The former is simply the familiar measure of entropy flow (entropy change per second), and the latter corresponds to "per kelvin" of temperature change.

The electrical analogue of entropy is charge, measured in terms of the coulomb; and the analogue to entropy per second is electric current, measured in terms of the ampere. Also, as noted before, the electrical analogue of temperature difference (measured in terms of the K unit) is electric potential difference (measured in terms of the volt).

One of the basic relationships in electrical measurement is that power (in watts) is the product of the electrical potential (in volts) and the current it causes to flow (in amperes). In short, again, amps times volts equals watts. If our thermal analogues are valid, then the thermal ampere (J/K·s) times the thermal volt (K) should equal the thermal power which flows. The product of those two thermal measures is indeed J/s (joule per second), which is synonymous with W (watt of power). Hence we can say that when a temperature difference of 1 K produces an entropy change (current) of 1 J/K per second, a thermal power of 1 watt is flowing.

Confusions of nomenclature—The rather neat and intriguing correspondences between the electrical and thermal units are often obscured because of mixups in nomenclature. The names given to physical variables are unfortunately inconsistent and confusing, especially in the great realms of thermal and radiative relationships. It would be comfortable, but dangerous, to omit mention of some of these inconsistencies.

For instance, the names "thermal conductance" and "thermal resistance" are sometimes applied to variables in no way analogous to electrical conductance and electrical resistance, respectively. Thus "thermal conductance" sometimes, on closer examination, proves to be something better called "heat-exchange coefficient" or "heat-transfer coefficient." Its proper measure is the watt per kelvin of temperature difference per square meter (W/K·m²). And "thermal resistance" sometimes turns out to be just the inverse of the foregoing fraction, namely K·m²/W.

The coefficient of heat transfer is a very useful measure, when properly understood. It relates the power of heat flow to the area through which it passes and also to the temperature difference that drives or causes that flow. However, it is not the true analogue to the electrical variable of conductance, which is the inverse or reciprocal of electrical resistance (measured in terms of the ohm).

So long as there is a clear statement of the unit combinations in which measurements are expressed, it should be possible to avoid confusions that would otherwise arise from the illogical or overlapping names given to the physical variables themselves. When in doubt, we must check the contents of the unit combination in which a measurement is stated. It is best to remember always that joule, calorie and Btu are units measuring energy amounts; watt, calorie per second and Btu per second, etc., are units measuring power amounts; and temperature, whether expressed in kelvin (K) or in Celsius degree (°C), may represent a difference of temperature levels or, in important instances, the total absolute temperature. An example of this last use is the statement that water freezes at 273.15 K and boils at 373.15 K under standard atmospheric pressure.

Still another electrical analogy has led to the suggestion of a unit combination to measure thermal "inductance." This unit corresponds to the electrical henry (H) unit of inductance, which is sometimes defined as 1 volt-second per ampere (1 V·s/A).

If a thermal analogue of inductance is to be valid, it should include in its numerator the kelvin of temperature (for the volt), and the second of time; and in its denominator the unit of entropy flow per second (for the ampere). This leads to the fraction

$$\frac{K \cdot s}{J/K \cdot s}$$

which reduces to $K^2 s^2/J$ or $K^2 s/W$. Thus, thermal "inductance" is properly measured by a combination of units which, when divided by the unit of time (1 s), reduces to K^2/W, the unit combination for measuring thermal resistance. In electrical measurements also, 1 ohm times 1 second equals 1 henry.

Likewise in the electrical realm 1 henry equals 1 volt-second per ampere (1 H equals 1 V·s/A). The corresponding relationship exists also with these thermal analogues, for:

$$1 \ K^2 \ s^2/J = \frac{1 \ K \cdot s}{1 \ J/K \cdot s}$$

Still another electrical analogue is the so-called "thermal capacitance." It corresponds to the physical variable measured by the electrical unit called the *farad* (F), which has been defined also as the ampere-second per volt (A·s/V).

The thermal farad, or capacitance analogue, proves to be measurable by the unit combination of the joule of energy per square of the kelvin of temperature difference, J/K^2. It can also be expressed as the watt-second per kelvin squared, $W·s/K^2$. This happens to be identical with the unit of thermal conductance multiplied by the second of time—W/K² times s equals W·s/K². The same relationship is found among the electrical variables—the mho of conductance, when multiplied by the second of time, becomes the farad of capacitance.

Tracing thermal-to-electrical analogies may seem merely a subtle sort of game, and one that runs risks if carried too far. Analogies must never be pressed to the limit, lest absurdities follow. The value of these analogies is that they help to illustrate relationships, not to prove their validity. For most readers the electrical relationships are likely to seem more familiar, especially those relating amperes, volts, watts and even ohms.

We know that current will not flow unless voltage difference is present to drive it. It is helpful also to remember that the rate of entropy change, corresponding to electric current measured in amperes, depends on the temperature difference, corresponding to the voltage.

Mechanical as well as electrical analogues can be used to clarify thermal relationships. It has been stressed already that a force corresponds to a temperature difference, and that the distance through which the force moves corresponds to an entropy change. Force times distance equals work done (measured by the joule); and temperature difference times entropy change equals thermal energy (also measured by the joule).

The greater sense of familiarity that analogies can give provides the best reason for their appearance here.

Not all the physical variables used for thermal measurement have obvious or easy analogues in the electrical and mechanical realms. One example is the measure called *thermal diffusivity* and formed from three principal ingredients, two of which are thermal measurements and the third of which is the familiar physical relationship of mass per unit volume, called density.

The thermal diffusivity of a substance is measured by its thermal conductivity (in W/K·m) divided by its heat capacity per unit volume (J/K·m³), where K represents a temperature change of just 1 kelvin. However, heat capacity per unit volume of substance is a measurement not commonly available in tables. Consequently it is derived as the product of two other physical variables: heat capacity per unit mass (J/K·kg), and density (kg/m³). The result, rather obviously, is the measure of J/K·m³.

The complete fraction representing thermal diffusivity looks at first like this:

$$\frac{W/K·m}{(J/K·kg) \times (kg/m^3)}$$

But when all the possible and permissible cancellations have been completed, it reduces to this brief remnant: m²/s. Only that, and nothing more! In short, the unit combination for measuring thermal diffusivity is simply an *area* per *second*. All the thermal ingredients, including the K and the J and the W, have been eliminated.

Thermal diffusivity is used in calculations of the results to be expected from so-called unsteady or irregular heat flow through specific substances. When heat flow is steady—uniform—it is fairly easy to compute the pattern of temperatures that will be found at various points along the route of its flow. But when heat flow fluctuates, as from a variable temperature situation, then there are complications.

How will the temperature vary in time and in magnitude at different locations along the thermal path? Such practical problems can be solved only with the help of measures of thermal diffusivity.

Hence it is not strange that thermal diffusivity is measured by a unit combination equivalent to a rate of spreading—an area per unit of time. The greater the thermal conductivity of a substance (other things being equal), the greater will be its thermal diffusivity. However, anything that increases the substance's heat capacity per unit of volume will increase also the rate at which it "soaks up" heat that comes its way. Hence the denominator of the fraction which has the thermal conductivity in the numerator.

Some standard thermal symbols—Engineers and thermodynamicists are prone to employ particular symbols to represent their most commonly used thermal variables. Thus, the common symbol for thermal conductivity is the lower-case italic k. Similarly C is a common symbol for heat capacity per unit mass, and the Greek small *rho* (ρ) is often employed to symbolize density. Using such symbols, one may compactly indicate the concept of thermal diffusivity as $k/C \cdot \rho$, while noting that in the International System its appropriate unit combination is simply square meter per second (m^2/s).

The advantage of the SI units, when consistently applied, is illustrated if we look at some of the earlier and obsolescent thermodynamic references. A famous German text, Ernst Schmidt's *Thermodynamics,* in its English translation states that the "dimension" of thermal diffusivity is ft^2/hr (square feet per hour). This is not a misstatement, but it is reminiscent of jokes about unit combinations such as "furlongs per fortnight." One could indeed express thermal diffusivities in unit combinations such as hectares per weekend or acres per year, but why belabor an obvious point? Incidentally, the SI combination of 1 m^2/s equals very nearly 38,750 ft^2/hr, which needs to be increased only by some $12\frac{1}{2}$ percent to equal 1 acre per hour!

Still other important thermal measures are as strikingly simple in appearance as at first thermal diffusivity seems complicated. For example, there is the measure symbolized by Q, the magnitude of heat, for which the SI unit is simply the joule (J). The heat associated with specific physical events is of utmost importance. Sometimes it is related to the mass of a substance, as with J/kg; sometimes it is related to the volume of a substance, as with J/m^3. But there are times when the needs of the measurers are satisfied simply by the total amount of heat involved (J), without relating it to either mass or volume.

Among the important measures of heat are those known somewhat misleadingly as "latent heats." These are the amounts of energy required for the great isothermal changes, such as freezing/ melting or vaporizing/condensing. Such measurements are often related to the gram molecular or atomic mass of the substance tested, and thus the resulting measurements are expressed in units of joule per mole (J/mol).

The correct name for these heats of transition when related to unit mass (J/kg) is specific latent heat. When related to molar mass (J/mol) the correct name is molar latent heat. A more accurate name would be heat of transformation, since that is the great outward evidence of the event—the change from the gaseous to liquid or liquid to solid state, or vice versa.

Calorific value is a variable used especially to relate fuels either to their masses or to the volumes they occupy before being burned or reacted. Such measures may use either J/kg or J/m³ unit combinations. "Calorific" here is an adjective that refers to the potential heat content—the "stored" or available heat in the fuel, awaiting release when it is combined with oxygen or some other suitable reactant.

"Heat release" is a name often given to the quantity of heat (Q) related to the length of time within which it is released, *and also* to the volume of space within which that release takes place. Since quantity of heat or energy per time unit is power, heat release can be measured by the units appropriate to power density—watt per cubic meter (W/m³). Measurements thus expressed are applied commonly in the analysis of the design problems and performances of furnaces, burners and jets.

In the older and now obsolescent units, they were often expressed in such combinations as calories per cubit foot or Btu per cubic inch, and the like.

Coefficients related to temperatures are used in the measurements of various important physical changes. Each such temperature coefficient is a "per K" or "per °C" or "per °F" relationship. Thus, there are three different kinds of expansion coefficients: (1) the cubic, or volume, coefficient; (2) the surface, or area, coefficient; and (3) the linear, or length, coefficient.

The volume-expansion coefficient of a substance is the fractional increase in its volume divided by the temperature rise which caused that increase. The temperature level at which the change took place is usually stated, and commonly it is noted that the change occurred under constant-pressure conditions.

Thus, at a temperature of 293 K, paraffin has been found to have a volume-expansion coefficient of about 0.000 6 (6 parts per 10,000) for each 1 K of temperature change. On the other hand, at 313 K, diamond shows only 0.000 003 5 (35 parts per 10 million) volume-expansion coefficient. Thus, paraffin is about 170 times as "expansive" as diamond; and diamond, in turn, is about twice as expansive as emerald.

Area- or surface-expansion coefficients are ordinarily not important. They tend to have values about two thirds those of the volume-expansion coefficients for the same substances.

The linear-expansion coefficient is a substance's fractional increase in length per degree of the temperature change that caused the expansion. Linear coefficients tend to be about one third the size of the volume coefficients for the same substances, and about half the size of the area coefficients.

The metal lead (at 283 K) has a relatively high linear-expansion coefficient—about 0.000 028 (28 parts per million). Platinum, in contrast, expands only about one third as much. This helps to account for the high esteem enjoyed by platinum and its related metal rhodium in the making of such old unit standards as the "meter bar" and the kilogram cylinder.

Pressure coefficients are also measured in relation to the temperature changes that cause them. Most gases, at or near "room temperature," and at starting pressures of 1 standard atmosphere, show coefficients of expansion between 365 and 385 parts per 10,000 (0.003 65 to 0.003 85) per rise of 1 K. This is about what the gas laws would call for, since a change of just 1 K at a temperature level of 273 K is a fractional change of about 0.003 7 in the thermodynamic temperature.

Such gas measurements are performed usually under conditions of constant volume. The reason is revealing. If the gas were allowed to increase its volume against unchanging outside pressure, then it would be doing work, and some of its internal energy would be absorbed in doing that work; besides which, the purpose is to determine the ratio by which the pressure itself increases within the gas, not the ratio by which its volume increases while its internal pressure remains equal to that of the air outside.

33 THE ROLE OF RADIATION

That God is Coloring, Newton doth show,
And the Devil is a black outline,
all of us know.
      —WILLIAM BLAKE

. . . the wonderfulness of the sundown, or
    of stars shining so quiet and bright,
Or the exquisite delicate thin curve
    of the new moon in spring,
These with the rest, one and all,
    are to me miracles,
The whole referring, yet each distinct
    and in its place.
      —WALT WHITMAN,
      Miracles

Temperature first reveals itself because its differences are accompanied by the flow of the kind of energy called heat from body to body or from site to site within the same body. Yet the really enormous transmissions of energy in the universe, and even to and from our own planet Earth, take place by means of quite another kind of interaction called *radiation* in which the flow of energy crosses empty space between bodies and particles.

It is radiation, not conduction or convection, that makes possible the enormous outpourings of power underlying the essential processes of our small world and great universe. Radiation has already been introduced in the chapter on light. But light is only one very small segment or part of the enormous range of electromagnetic interactions called, collectively, radiation.

In that chapter on light it was—or should have been—made clear that all our major sources of visible light are only partial and inefficient providers of the energies that our eyes and brains are able to "see." Whether we study candle flames, tungsten-filament bulbs, fluorescent tubes, or even the lights of the heavens, such as the Sun and the myriads of other stars, we find that they "waste" vast parts of their power outputs in invisible radiations, whose effects are sometimes summarized as "heat, not light."

We humans tend unconsciously toward a kind of visual chauvinism. Radiation that we cannot "see" seems either nonexistent or an unfortunate waste. That is an unjustified assumption, to which this chapter, at least, will not fall victim, if the author's intent is realized. Radiation here refers to each and every describable kind, all the way from the low-frequency (long-wavelength) radio waves, and on up the scale through infrared, through the slender sliver of visible light, through the great range of ultraviolet, into the range of X rays, and beyond that through the extremely energetic gamma rays, finally to the violent realms of ultra-energetic radiations that are more like bullets than waves, more like powerful particles than undulating oscillations.

Each and every rung in this enormous radiation ladder represents a form of electromagnetic launching of energy from a material base into space. This is, in barest summary, the sequence in radiation emission. And in its converse, radiation absorption, that sequence is reversed: energy that has traveled through space (with the cosmic constant speed of light, c) impinges on matter and imparts to that matter an additional energy.

Each act or process of emission, it must be remembered, is the launching into space of a specific amount, or quantum, of energy. Electromagnetic waves are continuous, yet the radiation they embody is discontinuous, discrete, packaged in photons, whose total energy is proportional to the frequency at which they are emitted. If we compare radiations A and B, each being monochromatic, or emitted at just one frequency, then the ratio of the energies of the individual photons in A to those of the photons in B must be the same as the ratio of the frequency (in cycles per second, or *hertz*) of A to that of B.

Here, perhaps fortunately, we are concerned less with measurements of photon energies than with measurements of the total energies and powers of radiations that include many different frequencies (or wavelengths) and so have broad bandwidths rather than monochromatic, or single-frequency, character.

Temperature—which means thermodynamic or absolute temperature—has even greater influence on the energies and powers of radiation than on the energies and powers of conductions of heat through matter. Very roughly speaking, the amount of heat carried by conduction from body A to body B placed in intimate thermal contact will be proportional to the difference in temperature between A (higher) and B (lower). However, the total radiation power that can be emitted from the surface of a body increases as the *fourth power* of the temperature of the atoms and molecules in that surface. If we represent the absolute temperature by θ and the maximum possible emitted radiation power by P, then we know that $P \propto \theta^4$. The symbol $\propto$ means "is directly proportional to."

Magnitudes increase with steep rapidity in proportion to the fourth power. Thus if temperature θ doubles, radiated power P increases 16 times; if θ trebles, P increases 81 times; and so on.

Maximum possible radiated power, sometimes called "full radiation," is the radiation at all frequencies and wavelengths of a so-called "blackbody." Good radiators are good absorbers. A body that absorbed each and every radiation, of whatever frequency, that fell upon it would be a *blackbody,* or perfect absorber. This perfect *absorber* must also be a perfect, or maximal, *emitter* of radiation. Such is a blackbody, the standard of maximum emission power by which all actual substances and surfaces must be judged.

Here we limit our generalizations to the hypothetical blackbody radiators and absorbers, just as in examining the kinetic theory of heat we dealt with hypothetical perfect gases rather than the more or less imperfect ones actually measurable around us. Our Sun is not completely a blackbody radiator nor are our artificial light sources, such as tungsten lights, mercury or sodium-vapor lamps, and so on. However, the blackbody concept gives us the tools with which the great generalizations relative to radiation and absorption can best be approached. Without blackbody analysis, radiation measurements would be lame indeed!

The fourth-power relationship between surface temperature of a blackbody and the energy flux of the radiations from that surface is best measured by the watt per square meter (W/m^2). If we represent by K the absolute or kelvin temperature of that blackbody surface, and by P the power density of the resulting radiation, then we can make a very useful statement of the relative magnitudes:

$$P = K^4 \times (5.735 \times 10^{-8}\,W/m^2)$$

This relationship is called the Stefan-Boltzmann law. A simple example will show how it may be applied. Suppose we have an inner metal cylinder with a metal surface of 1 m² area at a temperature of 300 K. Wrapped around it at a small distance away is another metal cylinder, with area of 1.2 m² at a temperature of 302 K. Between the two cylinders is only air, or even no air at all—a vacuum, in fact. The surface of each cylinder, we shall assume, behaves like a perfect blackbody. Since the outer cylinder is starting at the higher temperature, we know that its radiation intensity in watt per square meter must be greater than that of the inner cylinder. The law already given allows us to compute how much greater.

According to that law, the surface of the outer cylinder must radiate at $302^4 \times 5.735 \times 10^{-8}$, or 477.05 W/m². Since the area is 1.2 m², that cylinder emits into the space between the two surfaces a total radiation power of 572.5 W.

On the other hand, the surface of the inner cylinder begins radiating with the intensity typical of 300 K temperature; namely, at $300^4 \times 5.735 \times 10^{-8}$ W/m², or 464.5 W/m². Thus its 1 m² of surface emits 464.5 W of radiation into the between-cylinder space, and this amount plus the 572.5 W from the outer surface makes a total of 1,037 W of radiation power pouring into that space.

When we assumed that each cylinder's surface acts like a perfect blackbody, this meant that it must absorb all radiation that falls on it and emit as much radiation as its surface temperature makes possible (per the Stefan-Boltzmann law). Any surface whose radiation emissions exceed its absorptions must show a diminishing surface temperature, for it is losing thermal energy faster than it is acquiring it. If, on the other hand, the absorptions exceed the emissions, the surface must show a rising temperature, for it is gaining thermal energy faster than losing it.

Equilibrium, or stability of surface temperature, is attained only when a surface absorbs just as much radiation energy as it emits, and emits just as much as it absorbs. In our example, such stability could be attained when each square meter (of the total 2.2 m² of metal surfaces exposed to the between-cylinder space) is emitting and absorbing at a rate of 471.4 W, for this is 1,037 W/2.2 m².

Let us assume that such an equilibrium is attained. Then the between-cylinder space receives 565.6 W from the 1.2 m² surface of the outer cylinder plus 471.4 W from the 1 m² of the inner cylinder. The total radiation in the space still shows a power of 1,037 W; however, the power radiated by the outer cylinder has diminished about 1.1 percent while that from the inner surface has increased about 1.4 percent.

In temperature terms, the result would be that each of the two facing surfaces has reached a temperature of about 301.14 K, which represents a loss of 0.86 K for the surface of the outer cylinder and a gain of 1.14 K for the surface of the inner cylinder. If we now allow a small sample beam of radiation to emerge from that between-cylinder space, it will show the characteristic blackbody radiation pattern for 301.14 K, and for no other temperature. Any deep, well-like cavity or receptacle in a substance that has reached and maintains a uniform temperature becomes filled with the blackbody radiation typical of just that particular temperature. This is, in fact, how blackbody radiation measurements are made and how the data were obtained from which the Stefan-Boltzmann law was deduced.

If we began with blackbody radiation at 300 K, showing its typical surface power density of 464.5 W/m², and then doubled the temperature, to 600 K, the surface power density would increase 2^4 (equal to 16) times, and so would rise to 7,432 W/m². This means that the relative increase in power radiated is 8 times the increase in the thermodynamic (or "absolute") temperature, since the power has been multiplied by 16 while the temperature has been multiplied only by 2.

Radiation is constantly emitted and absorbed by surfaces all around us, even those that seem cold or quite frigid by human standards. The power density, measured in W/m² in the standard SI units, diminishes steeply as the temperature approaches the absolute zero (0 K). However, even at cryogenic temperature levels, such as those at which oxygen and nitrogen cease to be gases and become liquids, blackbody radiation rates of several watt per square meter are typical. Such rates are small, but by no means negligible.

The following table presents ten typical temperature levels with the corresponding blackbody radiation power densities for each.

ABSOLUTE TEMPERATURE (in kelvin)	SUBSTANCE AND STATE DETERMINING THE TEMPERATURE (under standard pressure)	BLACKBODY RADIANT EMITTANCE RATE (in watt/meter²)
6,173 K	Tungsten, boiling point	83.3 million W/m²
3,643 K	Tungsten, freezing point	10.1 million "
2,047 K	Platinum, freezing point	1.01 million "
373.15 K	Water, boiling point	1,112. "
310.15 K	Human "blood heat" (typical)	530.7 "
293.15 K	"Room temperature" (70° F)	423.6 "
273.15 K	Water, freezing point	319.3 "
234.26 K	Mercury, freezing point	172.7 "
90.16 K	Oxygen, boiling point	3.79 "
77.36 K	Nitrogen, boiling point	2.05 "

The table shows that the radiation rate at the boiling point of tungsten (a temperature near that at the surface of the Sun) is about 40 million times as great as the rate at the boiling point of nitrogen, which (at 77.36 K) lies nearly 196° C below the freezing point of water. The former temperature is about 80 times the latter, in thermodynamic terms. Hence the Stefan-Boltzmann law allows us to expect that the ratio of radiation rates will be about 80^4, or roughly 40 million times; and so it is.

The universality and importance of energy transfer by means of radiation calls for the following additional recapitulation of its essentials.

Radiations all around us—Every surface around us is constantly emitting and absorbing radiations. This happens even inside a freezer, at temperatures far below the ice point of water. Whenever a body's surfaces absorb more radiation power than they emit, their temperature rises; when they emit more radiation power than they absorb, their surface temperature falls.

A typical radiation-measuring instrument is the *bolometer*. In one useful form it consists of a slender strip of blackened platinum, exposed to the radiation to be measured. Most of that radiation is converted into heat on striking the blackened metal, and the resulting rise in temperature is measured as an increased resistance of the metal to tiny electric currents sent through it. The cycle is this: absorbed radiation power becomes increased temperature, which becomes increased electrical resistance, which is measured in order to find the equivalent of the radiation power . . .

It is true, of course, that such bolometric devices cannot convert into heat *all* the radiations that reach them. For example, some long-wave radio waves will have little or no effect. Likewise, at the surface of the Earth few ultraviolet radiations arrive from outer space, because the atmosphere absorbs or scatters most of them before they can reach the bolometer.

For this reason, bolometric and similar measurements made on orbiting satellites above the Earth's atmosphere are of indispensable importance in measuring the actual radiation that reaches the Earth from the Sun. Such measurements show that the Sun's radiations have a power of about 333.4 calories per second for each square meter of cross section through which they pass on their way into the Earth's atmosphere below.

This is equivalent to about 1,395 W/m^2. The cross section of the Earth, as seen from the Sun, is about 1.275×10^{14} m^2 (or 127.5 million million square meters). Hence it is certain that about 1.73×10^{17} watt of power constantly reach the Earth from the Sun. This can also be stated as 173,000 million MW (megawatt).

This is the Earth's power supply, so far as its thermal budget is concerned. It is divided among the 5.101×10^{14} m^2 of the actual surface of the Earth. Since the Earth is a sphere, not a flat round disk, its surface area is four times that of the area of one side of a disk with its same cross section.

With 1.73×10^{17} watt of power from the Sun, and a surface area of 5.101×10^{14} m^2 on Earth, the average power of solar radiation per square meter of Earth surface is about 340 watt. Yet by no means all of the radiant power from the Sun actually reaches the surface of the Earth here at the bottom of our "ocean of air."

Of the Earth's total intercepted solar radiation power (1.73×10^{17} W) about 30 percent is reflected directly back into space at just about the same frequencies it had as it approached the Earth. Such large-scale reflection of sunlight is what enables us to see the other planets and what makes our dead Moon appear to shine, as if with a light of its own.

The remaining 70 percent of total solar radiation is absorbed and "processed" by the Earth, including the atmosphere that belongs to it. This remainder amounts to about 1.21×10^{17} W, of which a very substantial portion consists of energies within the visible frequencies, or within the upper infrared frequencies relatively near the visual band.

The Earth, including its atmosphere and watery surfaces, acts much like a huge bolometer. That is, the arriving radiation from the Sun raises the surface temperatures, and those surfaces—land, water, and even the surfaces of clouds and water vapor in the atmosphere—radiate back again into space, but primarily at the longer wavelengths (lower frequencies) characteristic of those Earth-bound temperatures.

This back-radiation from the Earth into space accounts for about 8.1×10^{16} W, or just about two thirds of all the solar radiation power that was not reflected away from the Earth in the first place.

Why is the Earth-to-space radiation so different in its composition from the Sun-to-Earth radiation? The answer lies in the enormous difference of the average temperatures of the surfaces of Earth and the surface layers of the Sun. The latter radiate just about like a blackbody at 6,000 K. The former, roughly speaking, radiate about like a blackbody at not more than 5 percent of that temperature!

Let us use the best available estimates to make this contrast more concrete. The average power radiated into space per square meter of the surface of the Sun itself is about 6.4×10^7 W (64 megawatt per square meter). What should be the average power radiated into space per square meter of the surface of the Earth—if it behaves like a perfect blackbody radiator? The mean, or overall average, surface temperature on Earth is not far from 273 K (0° C). This takes into account the cold of Arctic and Antarctic regions, the fact that half the Earth's sphere is always in a state of night, and so forth.

Thus the Sun's surface is at a thermodynamic temperature about $6,000/273$ or 22 times as high as the Earth's surface on the average. The Stefan-Boltzmann law shows that the surface density of radiation from the Sun should thus be some 22^4 times that from the Earth—or roughly 234,000 times as great per square meter of surface.

At that rate, the Earth should radiate power into space with a density of about 273.5 W per square meter of its surface, simply because of the temperatures prevailing on that surface. Since the Earth's surface area is close to 5.1×10^{14} m², its total back-radiation into space should amount to a power of about 1.4×10^{17} W. We have shown, however, that the best estimates of geophysicists place this back-radiation at a power level of about 8.1×10^{16} W for the Earth as a whole. This is about 59 percent of the blackbody radiation for a sphere of the Earth's size and with its typical average surface temperature.

The missing 41 percent is readily accounted for by the fact that the Earth's exposed surfaces are far from being full radiators or blackbodies. For instance, some 70 percent of Earth is covered by the great watery surfaces: the oceans, seas and lakes. The land areas also are by no means all blackbody in their radiation characteristics, though human industries and ecological outrages have devastated vast regions and covered thousands of square miles with blacktop and concrete surfaces!

Truly, the great measures of radiation in relation to temperatures enable us to analyze situations both vast and tiny, near and distant. These tools permit also revealing looks in the other direction—at the heat flow from the Earth's interior toward its surface. We have noted before that this internal heat flow is estimated at a level of about 3.2×10^{13} W for the entire Earth. Power of this magnitude, in the form of heat, is constantly arriving just below the surfaces of the Earth as a whole.

How does this compare in magnitude with the Earth's back-radiation into space, the long-wave radiation emitted primarily because of the temperatures built up on the surfaces by the absorption of solar radiation from the sky? The latter radiations, we have noted, probably have a total power of about 8.1×10^{16} W; the former, the internal heat flow, has a power of "only" 3.2×10^{13} W. Hence, the internal heat flow to the surface is about $\frac{1}{2,500}$ of the Earth's surface radiation power into space.

The Earth as a radiator is thus almost totally dependent on the energy it receives from the Sun. Its own inner heat resources are trivial in comparison, though they undoubtedly have enormous effects in producing convective motions in the outer layers of Earth, such as those which lead to earthquakes and to great continental and ocean-floor shiftings through the ages.

Placing the radiation peaks—We are now ready to find an even closer link between surface temperatures and the resulting radiation frequencies or wavelengths in the great processes of emission. For each thermodynamic temperature of a blackbody surface there is a different distribution of power at the various radiated frequencies or wavelengths. Each thermodynamic temperature has a characteristic peak point, at which power is greater than anywhere else in the spectrum.

The location of that peak point is given, for blackbody radiation, by a relationship called Wien's displacement law. It is usually stated in terms of the displacement of the wavelength, L_{max}, at which the maximum power is found:

$$L_{max} = A/\theta$$

The symbol θ stands for the thermodynamic temperature, while A represents a constant: 2.9×10^{-3} meter per kelvin.

Let us apply this simple recipe to the case of the Sun, with a thermodynamic surface temperature of some 6,000 K. The displacement law shows that its peak-power wavelength should be at approximately 4.8×10^{-7} meter. This is just about the wavelength in the green part of the solar spectrum at which maximum power is actually found.

It is possible also to transform this displacement law so that it will show the peak-power position in terms of frequency rather than wavelength. If F_{max} stands for this peak frequency and θ again represents the thermodynamic temperature of the blackbody surface, then the relationship becomes:

$$F_{max} = B\theta$$

And the constant B is 1.035×10^{11} hertz per kelvin. The hertz here as always is the SI unit representing 1 cycle per second. In the case of the Sun, this recipe indicates that the peak frequency should be just about where it is found on the spectrum—at 6.25×10^{14} hertz, in the green portion.

If we apply this form of the displacement law to the approximate average surface temperature of 273 K for the Earth as a whole, we find that it indicates a power maximum at a frequency of about 2.83×10^{13} hertz. This is well into the infrared, far below the visible-light frequencies.

Thus the great energy conversion or "degradation" accomplished by the Earth's surfaces means a down shift of the most powerful frequencies from the 6.25×10^{14} hertz of sunlight to 2.83×10^{13} hertz for the back-radiation of the Earth into space.

Satellites orbiting the Earth pick up unmistakable evidences of this lower-frequency, longer-wave radiation. The great transformations of energy in which the Earth and all heavenly bodies are involved are thus intimately tied to temperatures, to radiation frequencies, and to wavelengths. Comparisons of radiated powers alone will not tell the entire tale; there must be comparisons also of the so-called "spectral" distributions of energies. This has nothing to do with "specters," everything to do with spectrums, in both the visible and the invisible bandwidths.

Still another temperature-tied relationship awaits us now. How do blackbody surface temperatures influence the actual magnitude or levels of the peak-power locations in the spectrum of the resulting radiation?

The answer here is somewhat more tedious to give, but none the less important. In the case of the spectrum of the Sun, careful measurements indicate that per square meter of its surface it radiates about 21.3 watt within a small interval around its maximum, or peak-power, wavelength of 4.8×10^{-7} m.

But what is the physical meaning of "within a small interval" in that sentence? It is not possible for any measuring device applied to the Sun's spectrum to measure *just* the power emitted at precisely 4.8×10^{-7} m wavelength, while totally disregarding the power emitted at wavelengths a tiny bit longer or shorter than that.

The result is that every valid measure of spectral power distribution—that is, of the power around or near some particular wavelength or frequency—must include some specification of the bandwidth or interval on which the measurement is based.

Typical of such unit combinations are these: the watt per square meter per meter ($W/m^2 \cdot m$), which is ill suited for measurements having to do with the wavelengths of light or infrared radiation, but fairly well suited for measurements of many radio radiations; and the watt per square meter per angstrom (W/m²·Å), the angstrom (Å) being equal to only 10^{-10} meter. One angstrom is an excruciatingly tiny wavelength difference. Hence we find also such unit combinations as the watt per square meter per 100 angstrom (W/m²·100 Å). The measurement cited for the Sun—21.3 W/m²· 100 Å—was based on this particular unit combination.

This power peak at or close to the 4.8×10^{-7} m wavelength in the Sun's spectrum is quite sharp. It is even somewhat sharper that it would be if the Sun were actually a perfect blackbody radiator. In order to make the length relationships even clearer, we can note that the peak itself can be expressed in the angstrom unit, for 4.8×10^{-7} meter is 4,800 Å. What our measure means, then, is that a total power of 21.3 W per square meter of the Sun's surface is indicated within the bandwidth between 4,750 Å and 4,850 Å. This is a bandwidth of just 100 Å centered on the peak level of 4,800 Å. The bandwidth, in this case, represents just about $100/4,800$, or 2.1 percent of the central wavelength at which the peak is located.

There are also unit combinations in which the bandwidth as well as the position of the peak power are expressed in frequency terms rather than in wavelength terms. For example: watt per meter squared per hertz. Or watt per square meter per wave number. The wave number is the number of times the wavelength of the peak power radiation goes into 1 meter. Thus, with a peak power at 4.8×10^{-7} meter, the corresponding wave number is 2.083×10^6 per meter. Clearly, the greater the wave number, the higher the frequency of the corresponding radiation, and the lower or less the wavelength.

These unit combinations for analyses of spectral power distributions present difficulties not only to laymen but even to those whose work requires their use. However, these difficulties are trivial compared with the confusions or errors that might result were the unit combinations to omit some suitable and definable bandwidth measure in the denominator.

How high are the peaks?—We have still to deal with the quantitative issue. Assume that we have made measurements of the spectral power at the peak wavelength or frequency for some particular blackbody temperature. How will the corresponding power levels of the peaks change as the temperature of the radiating surface is raised? We know that doubling the surface temperature of the Sun, for example, would double the frequency (halve the wavelength) of the peak power in the spectrum of the resulting star, radiating at 12,000 K. But how would it alter the actual *power level* of that doubled frequency?

The answer is that the actual comparable spectral power levels increase as the *fifth power* of the thermodynamic temperature (θ^5)! A star with surface temperature of 12,000 K—and there are such—would have a spectrum in which its peak power, centered at half of 4.8×10^{-7} m, or 2.4×10^{-7} m—would show 32 times the 21.3 watt per square meter per 100 angstrom measured for the Sun. In short, each square meter of the surface of such a star would radiate about 682 watt in the bandwidth of 100 angstrom, centered around the 2,400 angstrom wavelength. That wavelength is well into the ultraviolet, far beyond the visible wavelengths in which the Sun pours out its peak radiation powers.

The enormous effect of surface temperature on such outpourings of power accounts for the fact that the so-called ultraviolet stars are believed to burn through their internal thermonuclear fuel supplies within a relatively few million years, while a comparatively "cool" star like the Sun, with only 20 million K of core temperature, remains stable after an estimated 4.5 to 5 billion years of radiating at much the same levels as today.

As these lines are written, a great astronomical satellite called Copernicus is about to be placed in orbit by the United States. Six different telescopes are included in its total weight of nearly 2.5 tons. Its intended orbit, some five hundred miles high, will keep it above the atmosphere in order, it is hoped, to permit its telescopes to make readings of both ultraviolet and X-ray radiations for a period of as long as five years aloft.

Its largest telescope, with a mirror diameter of some 32 inches, is designed to gather the ultraviolet radiations that the Earth's atmosphere does not allow to reach the great telescopes in surface observatories. Its five smaller X-ray telescopes, based on designs developed at University College, London, are intended to gather data on even more energetic (shorter wavelength, higher frequency) radiations believed to pour out from such objects as pulsars and quasars.

The enormous effort and expense of the Copernicus orbiting observatory is justified by the present certainty that there are numerous energy sources in the universe—hundreds, in fact—which behave as if their core temperatures were as high as 50 million K, or 2.5 times that of our Sun.

The screening effects of the Earth's atmosphere has deprived the measurers of the skies of data needed to complete the great spectral analyses. Enormous blank spots in the spectral levels beyond visible light remain to be filled in. The basic concepts and the tools of measurement outlined here will apply, however, to the findings from the space-based telescopes that Copernicus-type satellites aim into space.

Whether the object of interest is a relatively old, slow, steady star like the Sun, or a hot young ultraviolet wastrel such as those that Copernicus is designed to deal with, the basic concepts of radiation measurement must remain the same.

34 THE RANGES OF RADIATION

We are no other than a moving row
Of Magic Shadow-shapes that come and go
  Round with the Sun-illumin'd Lantern held
In Midnight by the Master of the Show . . .
      —EDWARD FITZGERALD,
        Rubáiyát of Omar Khayyám (1859)

Most so-called "measurements" of radiation emission are actually long-distance deductions or extrapolations. They are based on the distribution and the relative magnitude of radiations absorbed at the receiving end. Such reception or absorption takes place amidst temperatures necessarily far lower than those prevalent where the radiation is emitted. This truth, or truism, applies when we refer to "measurement" of the Sun's radiative emission, which actually takes place 149 million kilometers distant from the Earth. It applies also to "measurement" of the radiation of a carbon arc light or a tungsten filament in a bulb safely operated in some laboratory on Earth.

There is no way to apply a thermometer to the surface of the Sun or other shining star. Every thermometric substance we use on Earth—mercury, platinum, or any other—would be vaporized before it could be placed at that 6,000 K solar surface.

Thus "measurement" of radiation is an operation at a distance from the radiating surface. What is received and measured is only a small sampling of the total radiation emitted by the surface. The process of ascribing radiation levels to the source necessarily involves estimates of how large the received sample is, relative to the entire emission by that source.

Thus, the reception, or absorption, measurements (that precede the transmission, or emission, estimates) must make use of physical variables which relate amounts of received radiation power to the sizes of the areas that have received or collected them.

In the case of the all-important solar radiation reaching the Earth, total power at all receivable frequencies appears to average about 1,395 watt per square meter. The word *average* is used here because the Sun's radiation is found to be not totally uniform. It is moderately manic-depressive in its behavior. During its manic or hyperactive phases the Sun may radiate up to about 1 percent more than its average; and in its depressive or quiet cycles, about 1 percent less.

Consider now an actual radiation receiver exposed to the full power of the Sun as that power reaches a high mountain or an orbiting satellite. This receiver may be a strip of well-blackened platinum with a total area of only 1 square centimeter, or a mere $\frac{1}{10,000}$ of a square meter. Hence it intercepts and converts into surface heat only 0.139 5 watt of radiated power. The interpretation of its temperature behavior cannot be complete without taking into account its physical size and radiation-collecting capacity.

Now, consider what happens in a spectral analysis of the Sun's radiation. The visible spectrum includes the well-known rainbow sequence of colors from the lowest frequency, red, on up through orange, yellow, green, blue and violet. Below the red lies the invisible infrared. If we place a tiny bolometer across a band spectrum and slide it from the red toward the violet end, its changing surface temperature will show that increasing amounts of radiant power reach it, attaining a maximum when it is centered at the peak wavelength of 4.8×10^{-7} meter (4,800 angstrom) in the green region.

Suppose further that our bolometer has a width corresponding to just 100 angstrom. The entire visible spectrum has a bandwidth of about 3,700 angstrom—from 3,900 Å in the lowest visible red to 7,600 Å in the highest visible violet. Hence 100 Å represents a bandwidth of about 2.7 percent of the total *visible* spectrum.

When the bolometer is receiving radiation in that peak bandwidth centered on 4,800 Å, it records power representing only a tiny fraction—about 1 part in 3 million—of the total power it records when it is exposed to the full solar radiation.

The interpretation is that, though each square meter of the Sun's surface radiates a total of some 64.17 million watt, only about 21.3 watt of that total lies within the bandwidth of 100 angstrom centered on the peak-power wavelength. Even though this is the peak segment of radiation power, it is so thin a sliver compared with the entire bandwidth of solar radiation (invisible as well as visible) that it contains but a tiny segment of the total radiated power.

After all, though wavelengths below about 3,900 Å are not visible, the Sun does radiate, at diminishing power levels, all the way down to wavelengths as long as 10 cm or longer. A 10 cm wavelength is identical with 10^9 (1 billion) angstrom. Hence it is reasonable that the slender sliver of peak power in the visual green should include so small a fraction of the over-all total power received from the Sun.

Concerning unit combinations—The complexities of radiation measurement and estimation can be dealt with best if great care is used in selecting the unit combinations used to express them.

Total radiation powers are best expressed in terms of the watt, as with the estimate of 4×10^{26} watt of total power radiated by the Sun into space. That is the radiative emission of the Sun as a whole.

The term *emissivity* is often used when the radiative power is related to the area from which it comes. Thus, the Sun has a total or integral emissivity of about 64.17 million watt per square meter of its surface. The opposite concept, for the receiving end, would be *receptivity*, though that term is not used. Rather something like *radiation density* is employed, as when we say that solar radiation, at the distance of the Earth, has a radiation density of about 1,395 watt per square meter. (That means one square meter of area measured perpendicular to the direction of the center of the Sun.) Emissivity and radiation density are often measured or estimated in such non-SI units as calories per second per square centimeter (cal/s·cm²), and so forth.

Next come the corresponding measurements and estimates in which only a specific segment or spectrum fraction of radiation is singled out. These are the measures of *spectral emissivity,* such as that which estimates 21.3 watt of emissive power per square meter of the Sun's surface, within the particular bandwidth of 100 angstrom centered on the peak-power point of 4,800 angstrom wavelength.

When the bandwidth is expressed in terms of frequency rather than of wavelength, we find such unit combinations as the following: watt per square meter per hertz (W/m²·H). Here each hertz is a bandwidth difference of just 1 cycle per second. At frequencies such as those of light and infrared radiations, this is an extremely thin bandwidth. For example, in the case of green light it represents only about 1 part in 6×10^{14} parts! If the radiation singled out is at radio frequencies, however, 1 hertz represents a far larger ratio. For example, in the standard broadcast bandwidth between about 5 and 15 kilohertz, it represents roughly 1 part in 10,000.

No measurement of spectral emissivity can be properly compared with another unless both are expressed in terms of the same bandwidths in their unit denominators. Thus, it would be fallacious to compare a spectral emissivity based on a per-meter bandwidth with one based on a per-hertz bandwidth.

In radio astronomy, and in radio reception in general, spectral measures of received radiations are constantly used. Radio astronomers measure the strengths of the signals they receive in terms of what they call the *flux unit*. The physical variable that it has been tailored to fit is commonly called the *flux density* of reception from astronomical radio sources. That same flux unit serves also to evaluate the relative strengths of signals sent back to Earth from manmade satellites launched into space.

The standard flux unit, abbreviated "fu," is 10^{-26} watt per square meter per hertz (10^{-26} W/m²Hz). The ordinary unit combination in the International System (SI) would be just 1 W/m²Hz, but such a measure is trillions of times too large for the realities of radio astronomy.

In this widely used flux unit, the area element (m²) does not refer to a unit area of the distant star or mass of matter from which radio waves are received on Earth. Rather, it relates to the area of the great radio "mirror" on Earth that collects and focuses those radio waves for the receiving devices. It is the area of the "dish" that serves the same purpose in radio astronomy that the great mirror of a reflecting telescope serves in optical astronomy.

The entire geometrical area of such a radio dish cannot be included in these calculations. Rather, complicated antenna theory shows that the effective areas of these paraboloid collectors amount to about half of their actual geometrical areas.

For example, suppose we are receiving radio signals from the Sun by means of a metal paraboloid with a radius of 10 meter. Its geometric area is thus about 314.2 m², and its effective area in the neighborhood of 150 m². The radio waves it collects are fed to a receiver tuned to respond to a bandwidth of 10^7 hertz (10 MHz), centered on a wavelength of 1 meter, equivalent to a frequency of about 300 MHz. Thus, the radio power to which the receiver responds lies in the frequency bandwidth from 295 MHz to 305 MHz.

In this frequency range, radio astronomers expect to receive radio waves from the Sun with an intensity of about 5×10^7 (50 million) flux units, equal to 5×10^{-19} W/m²Hz. In our example, with a bandwidth of 10 MHz and an effective area of the receiving antenna of 150 m², the total radio power passed along to the receiver is about 7.5×10^{-10} watt, seemingly very small indeed.

Yet it is actually extremely powerful in comparison with the powers with which radio astronomers must work when their great dishes are aimed at distant stars and quasistellar (quasar) objects. Neighboring stars in our galaxy are roughly 3×10^5 times as distant as is the Sun from Earth. Should such a star emit radio waves no more powerful than those of the Sun, they would be received on Earth at a level of about 0.001 flux unit. This would not be considered a radio object worth mention, for its received power per hertz of bandwidth is simply too insignificant.

A striking and even sensational instance of the use of the flux unit in radio astronomy emerged while this section of the book was being prepared. During the 1960s a fantastically powerful radio source was discovered in the constellation of Cygnus (the Swan), part of the Milky Way of our own galaxy. It was first identified as a strong X-ray source by instruments carried into orbit by a research satellite designated by the imposing name of UHURU. This source could not have been detected from Earth's surface, because the atmosphere screens out X-ray radiations from the heavenly bodies.

This newly found radiator was given the name Cygnus X-3. Before long the great dishes of the radio telescopes were turned in its direction, and it was found to be also a radio source, though only a rather weak one, as such sources go. Its radio power as received on Earth was measured at about 0.1 flux units.

The intensity of its high-energy X-ray emissions plus its simultaneous radio emissions has led astrophysicists to speculate that Cygnus X-3 (which has never been detected optically) might be one of those fantastic processes now called a "black hole." This is a totally collapsed star of such mass and density that no light or other radiation can leave it, but into which other matter can be drawn by its intense and enduring gravitational field.

The continual outpouring of powerful X-rays suggested that matter, sucked into that black hole, might be converted swiftly into radiation far more intensively than matter can be converted into radiation by the thermonuclear processes in the cores of still-shining stars such as our Sun.

Early in September 1972, the radio telescope of the Canadian National Research Council, at Algonquin Park, about 150 miles west of Ottawa, was being used by Dr. P. C. Gregory. He intended to make measurements of radio emissions from the well-known star Algol, but conditions in that direction proved unfavorable. He decided, therefore, without previous plan, to turn his instrument toward Cygnus X-3, which he had observed only a few nights before, emitting radio waves at its usual moderate level of 0.1 flux units.

This time, however, Dr. Gregory was astounded. The signal strength was fantastically greater than before. In fact, his receiver was being "saturated." After various precautionary checks and tests, he became convinced that there was no sudden defect in his equipment, and he managed to measure the apparent new level of the radio emissions from Cygnus X-3. This worked out to about 22 flux units, more than 200 times the measurements made previously!

Dr. Gregory lost little time in spreading the sensational news, seeking confirmation or correction. He called his colleagues at the well-known Greenbank radio observatory in the United States, and found that they almost "dropped through the floor" when they heard his report. Within about an hour their own observations and measurements had confirmed his.

The word flashed from center to center, around the world. So large and sudden a change in the "dynamic output" of a radio object had never before been observed.

The truly mysterious case of the great upward leap by Cygnus X-3 shows that, though the actual radiation powers measured by the radio astronomers' flux units are extraordinarily tiny, the sophisticated equipment and methods of today permit the kinds of comparisons that can detect such mysterious events in the heavens.

In coldly quantitative terms, the Canadian radio telescope had, at first, collected from Cygnus X-3 only about 10^{-27} W of power per hertz of bandwidth, on each square meter of the effective area of the receiving dish. When, a few days later, that 10^{-27} W was found to have jumped up to about 2.2×10^{-25} W, it was a surprise of historic significance. We cannot here pause to speculate on the physical events on or near Cygnus X-3 that may have caused this leap in its power output.

It suffices that, as Dr. Gregory noted, there was no precedent in all the annals of radio astronomical measurements for any radio object to increase its radiated power so sharply in so short a time.

The S of flux density—Radio astronomers commonly use the symbol S to stand for flux density, measured in flux units. This is by no means to be confused with the S used to symbolize the very different physical variable called entropy.

Measurement of radio signals received from various heavenly objects involves certain complications and special situations. At these relatively long wavelengths (or low frequencies), a true, or full, blackbody radiator emits power that diminishes just about in proportion to the square of the wavelength on which the receiver is centered. Conversely, the power could be said, in that general range, to change proportionately to the square of the frequency on which measurement is centered.

The Sun and stars are by no means perfect blackbodies in their radiation behavior, but the relationship just mentioned is close to the effects that measurement of received radio signals reveals. The result is that radio astronomers have good reason to work at the highest possible frequencies, or shortest possible wavelengths. Thus, radio astronomy is largely committed to careful measurement in the realm of microwaves, not much longer in wavelength than the longest infrared radiations.

One more unavoidable consideration now enters. The physical variable of flux density, or S, is really appropriate only when the emitting body appears so small that it is like a geometrical point. If it has any substantial disk or area as seen from Earth, then a further refinement or complication is indicated, in order to take into account the geometrical realities. This refinement became unavoidable as the receiving antennas that were developed for radio astronomy narrowed the beam or width within which they received incoming radio signals. They became ever more focused and spatially selective in their reception.

Hence the variable of flux density S had to be related to the solid angle of the beam accepted or received by the particular radio telescope or telescope combination. Solid angle, we have seen, is measured by the *steradian* (sr), of which 4π, or 12.57, constitute a full sphere. Thus if the measure is made in terms that mean S/sr, or flux density per steradian of the receiving antenna, then a new physical variable is at hand. It has received the name of *brightness*. This is symbolized by "b" and refers to the radio brightness or intensity of the particular portion focused on in the emitting body, such as the Sun or some great radio-emitting cloud in space.

A brightness measurement is always made in terms of some particular frequency or wavelength. It is not a composite or cumulative measurement for the full spectral range of radiations of the body aimed at, or even for those parts of its radiations that could be called radiations at radio frequencies.

Once a brightness, b, has been measured, and the wavelength, λ, at which it is measured is noted, then a simple relationship can be assumed. The temperature T_b of a blackbody that would radiate with that brightness at that wavelength is proportional to the square of the wavelength times the brightness—or $T_b = \lambda^2 b$.

So-called "brightness temperatures" are constantly computed and stated for large heavenly objects or for particular portions of them. In many cases, the measure of brightness temperature is given, even though it is obvious that conditions in the body do not permit the existence of any ordinary blackbody-equilibrium radiation. This applies especially to nebular and gaseous bodies whose radiations are quite apparently due rather to magnetic or electrostatic forces than to the usual thermal situations.

Thus, the analysis of radiations, and especially of radio radiations, leads to the extension of the variable called *temperature* into situations for which it was not originally intended.

One striking example helps to illustrate this. Measurements of radio emissions from the Sun show that its corona—the thin but far-flung gaseous envelope far beyond its visible surface—has a brightness temperature of about 1 million K, at wavelengths of about 3.7 meter, even though the Sun's surface below has a thermo-dynamic temperature only of 6,000 K.

The physical meaning is almost certainly that atoms and particles in the corona attain speeds, and hence kinetic energies, equivalent to 1 million K. This means a kinetic energy per corona particle about 5 percent of that of a typical particle in the inmost thermo-nuclear core of the Sun.

Science has benefited from this seeming paradox. Radio tele-scopes, using the longer wavelengths of radiation, "see" a much larger Sun than we can see with our eyes. The radio Sun, at 3.7 m wavelength, has nearly 2.5 times the diameter of our visual Sun. The difference is caused by the corona, and it is the coronal radia-tions, resulting from extremely fast particle movements through its thin gas, that the radio telescope reads as if those motions meant 1 million K temperature in the corona.

Thanks to the selective effect of radio receptions at various wavelengths (frequencies), it has become possible to know some-thing of conditions at various levels of the Sun, ranging from some-what below its apparent visual surface to far out in space, 1 solar radius or more beyond that surface.

Absorption as well as emission is a fundamental part of the meas-urement of radiation events in the universe. Absorption usually is expressed as a ratio of radiation absorbed to radiation received. Thus we have seen that the Earth reflects away about 30 percent of the Sun's radiation and absorbs, in one manner or another, about 70 percent. Absorptions take place also in transmission through a section of substance. There is absorption of radiation in the atmo-sphere, accounting in particular for our surface inability to receive ultraviolet and X-ray information from the hottest stars in the heavens. Absorption and "absorptivity" vary at different frequen-cies and wavelengths. Thus, we know that the Earth's atmosphere is relatively transparent to the visual frequencies while behaving as a fairly effective screen against the ultraviolet and X-ray emissions that shower down upon it.

Radiations inside the stars—Radiation is usually presented as if it were solely or mainly a transporting of energy through space, from body to body or from particle to particle. Yet it is constantly at work on a massive scale *within* the largest bodies in the universe—the shining stars themselves.

All but an insignificant portion of the enormous thermonuclear energy generated in the core of the Sun makes its way to the surface in the form of incessant radiations and absorptions and reradiations by the particles and atoms composing the Sun. Ordinary thermal conduction of heat carries but a minor fraction of the power of the 4×10^{26} watt that pour out from the Sun.

The sheer pressure of this flow of radiation, from center to surface, is also responsible for the continuing size of the Sun itself. A body of hot gas exerts pressure because of the incessant bombardment by its flying and bouncing molecules. A substance through which great streams of radiation pour, from a high-temperature furnace (such as the Sun's thermonuclear core) toward the outside space, is sustained also by separate *radiation pressure,* over and above the *kinetic pressure* of the moving particles.

The pressure bearing down on matter in the central core of the Sun has been estimated to be more than 4×10^{10} times the standard atmospheric pressure at the surface of the Earth! Why does not the Sun collapse internally under this gigantic pressure? Because of the two counterpressures mentioned. The radiation pressure is estimated at from 14 to 17 percent of the entire pressure. The rest is the "normal" pressure of a body of gas raised to about 20 million K temperature.

This means that radiation alone—if it can be thought of apart from the movements of the particles and atoms—exerts in the heart of the Sun a pressure of about 6 billion times that with which air presses down on Earth's surface at sea level.

Other computations indicate that in stars whose total masses are four or five times the mass of our Sun, radiation supports about 30 percent of the total pressure, or nearly double its apparent proportion in the Sun.

Like total radiation power, pressure from radiation rises in proportion to the fourth power of the thermodynamic temperature of the substance. Radiation pressure indeed sets a limit on the mass that a star can attain and maintain. No single star has been found whose apparent mass is greater than about 2×10^{32} kg, or 100 times the mass of our Sun. In a star so massive, internal temperatures rise to such a level that the radiation pressure makes the star unstable, prone to blow apart.

At intervals, existing stars appear to go wild and flare out with vastly increased radiation power. These are the nova or, in extreme cases, the supernova stars. Typically they spew into the space around them vast quantities of matter. These eruptions of mass are propelled outward at fantastic velocities by the radiation pressure of the star whose thermonuclear processes have run wild.

Even our relatively tame and subdued Sun exerts undeniable radiation pressure in the space around it. The tails of comets that approach the Sun always point away from it. These tails are formed by particles and vapors driven out of the cometary substance by the Sun's insistent radiation pressure.

Whether through the outer space outside the Sun and stars or through the particle-crowded space within them, radiation always travels at the same ultimate velocity of 3×10^8 m/s, known as c. That is the only velocity at which electromagnetic radiation can travel in space.

Nevertheless, though the transits of radiation from particle to particle and atom to atom are so swift, there are so many such exchanges that the estimates of the average length of time required for a single bit of energy to work its way from the core of the Sun to the surface vary from hundreds to millions of years! Trillions-times-trillions of radiation-absorption-radiation sequences must take place in this strange flow of power from the thermonuclear furnace at the core to the radiating surface and coronal layers of the Sun.

During this enormous chain of radiation-absorption interchanges, significant alteration takes place in the nature of the radiation. Speaking from the point of view of radiation as transfer of energy by bundles of photons, we can say that the photons typical of the core processes are far more energetic than those typical of the final surface emissions. Since the amount of power that streams out from the Sun's surface is in balance with the amount of power released at the thermonuclear center, it is clear that as the great radiation wind streams out from the center, ever more and smaller photons are emitted and absorbed—smaller, that is, in terms of the energy that each carries, not in terms of anything so concrete as physical size or volume.

The ratio of typical core-photon energy to typical surface-photon energy is about 3,000 or even 3,500 to 1, for the typical photon of radiation in the core corresponds to X-ray frequencies, whereas we know that the most representative photons from the surface into space are at green-light frequencies, about midway in the visible spectrum.

In actual frequencies, we can say that the typical thermonuclear radiations of the core correspond to about 1.8 to 2.0×10^{18} hertz, whereas those of the surface correspond to 6.25×10^{14} hertz.

Taking a photon census is impossible, of course. But the great sequences of energy interchanges from center to surface increase by thousands of times the numbers of photons radiated, and diminish by corresponding thousands of times the average energy carried per photon.

This is another way of saying that a sort of energy degradation takes place in the flow of radiative wind from the solar center to its surface. It is reflected also in the scaling-down of the effective temperature from some 20 million K in the thermonuclear heart to about 6,000 K at the surface. As the energy bits work their way outward in endless encounters, their typical energies and frequencies diminish, their typical or corresponding wavelengths increase.

It is often assumed that the Sun radiates as it does, with a total power of 4×10^{26} watt, because its surface temperature is maintained at 6,000 K. That is a reversal of the physical facts and sequence. It would be more meaningful to say that because a power of 4×10^{26} watt is constantly pouring outward from center to surface of our Sun, its surface temperature reaches and maintains an average level of some 6,000 K! The surface-temperature pattern is the consequence, not the cause, of the mighty flow of radiative power from the burning heart of our central star.

Indeed, it is sometimes difficult to realize that the power source of the Sun is relatively restricted in terms of volume. Most of the Sun, in terms of its total volume, is passively reacting to the flood of radiative energy emanating from the small inner volume where the thermonuclear reactions take place. They cannot take place in those vast volumes where the average surrounding temperature is far below the 20 million K level.

A rough estimate can be attempted of the total energy content of the Sun at any given instant. This is the kinetic and radiative energy combined within the Sun. It does not include the vast rest-mass energy corresponding to the fact that each of some 2×10^{30} kg of mass in the Sun is equivalent, by Einstein's equation $E = Mc^2$, to 9×10^{16} joule of "massergy." Our search here is solely for the combined energy of particle motions (thermal energy) and radiative energy within the Sun.

The astrophysicist Martin Schwarzschild, in his *Structure and Evolution of the Stars,* rates the Sun's net energy store as equal to the energy that the Sun radiates from its surface during some 10^{15} second of shining at its present average rate. This is a time period of nearly 32 million years, and corresponds to a total store of internal energy of approximately 4×10^{41} joule, all of which is said to be available for radiation when it reaches the surface of the Sun.

This reserve is additional to the not-yet-tapped reserves of nuclear energy within the Sun. What Schwarzschild calls the "net energy store" is often regarded as synonymous with the Sun's total thermal energy; however, it must be recalled that it includes also the radiation quanta or photons ceaselessly passing from particle to particle within the volume of the Sun itself.

This grand total of about 4×10^{41} joule of net energy store probably represents between 0.5 (five tenths) and 0.7 (seven tenths) of 1 percent of all the energy the Sun has radiated into space during the past 4½ to 5 billion years since it began to shine at more or less its present rate. It is indeed a vast amount of energy "in transit" through the substance of the Sun. Yet it is equivalent to less than 2 parts in 1 million of the "massergy" represented by the Sun's remaining mass, which is equivalent to a total of about 1.8×10^{47} joule.

Estimating solar heating of Earth—How much warmer are we on Earth than we would be if the Sun were not at its present rather convenient and necessary distance from us? In other words, what temperature level can we assign to interstellar space where there is no nearby Sun, but only the scattered stars that we see elsewhere in the heavens?

During the 1920s, an eminent astronomer and theorist, Sir Arthur Eddington, ventured a fascinating computation. He estimated that apart from the Sun, the Earth receives from all the stars about the same radiation energy as if there were in the heavens around us only some 2,000 first-magnitude stars.

Each star of the absolute first magnitude radiates with a power in the visual range of about 36.3 times that of the Sun. Also, when any star is at a distance of 3.1×10^{17} m, its absolute magnitude equals its apparent magnitude. In other words, the Earth receives about the same total stellar radiation as if there were scattered more or less uniformly around us 72,600 stars like the Sun, each at a distance of 3.1×10^{17} m.

The combined effect is that each square meter of the Earth's surface receives about 2.3×10^{-5} watt of radiation power from the stars. Since radiation moves through space always at 3×10^8 m/s, this means an average density of star-emitted radiation energy of about 7.67×10^{-14} joule per cubic meter in space around the Earth. Even a cubic kilometer of space contains only 7.67×10^{-5} joule of energy of stellar radiation.

If a blackbody, or a thermometer with blackened bulb, is placed in space subject to such a radiation level, it reaches a temperature of just about 3.18 K. At this very low temperature its own radiation emissions reach the same low level as its own absorption of the stellar radiations. That is its blackbody-equilibrium temperature.

And that, accordingly, is the approximate temperature of interstellar space. Cold enough, by any standards that we heat-conditioned creatures can comprehend!

Earlier we mentioned measurements showing that the density of the Sun's radiation as it reaches the Earth's upper atmosphere is about 1,395 watt per square meter of cross section. Hence we can conclude that the radiation we receive from the Sun is about 60 million times stronger than that which we receive from all the stars.

The energy density of the Sun's radiation in space near the Earth is about 4.6×10^{-6} joule per cubic meter. And per cubic kilometer of space it is about 4,600 joule, which is a rather respectable amount of energy, even for so large a volume as a cubic kilometer.

These relative figures for the power of starlight versus that of sunlight at Earth match rather well with the observed average temperatures on the surface of Earth.

The average energy density of the radiations of bodies at ordinary "room temperatures" around us is about 1.4×10^{-6} joule per cubic meter, which is about 30 percent of the energy density of the radiation of the Sun alone by the time it reaches space near the Earth.

The Sun's radiation packs more energy into a given volume of space. However, Earth's total absorption of energy from the Sun must be spread out over a spherical surface that is four times the area of the cross section that the Earth shows to the Sun. Also the Earth, as noted, is a less perfect radiation absorber than the ideal blackbody.

The result of these and other factors is that the over-all average temperature of the Earth's entire surface is somewhat lower than what is commonly considered to be "room" temperature.

We note the overwhelming role of radiation in conveying energy from the thermonuclear furnaces through the substances of stars to their surfaces. Yet it is not only in such glowing masses of plasma and gas as the Sun and other stars that we need to realize that radiation is internal to matter as well as common to empty space between bodies of matter. Generally speaking, we can regard the radiational relations within a mass of matter as approximately those of a blackbody or perfect radiator-absorber maintained at the temperature of that matter.

Indeed, radiation is nowhere absent in our universe. It is utterly and unendingly ubiquitous, and any system of measurement units and concepts must be prepared to cope with it, however complex the actual situations.

The pervasiveness of radiation in space as well as in time is dramatically, or even melodramatically, manifested by a "new" kind of radiation, discovered within the past dozen years. The quotation marks around "new" are justified, for it is widely believed that this radiation has been kicking around our universe, so to speak, since its beginning—since the occasion of the supposed "Big Bang," when an incredibly compacted amount of "massergy" exploded, with the result that the universe now around us evolved.

The names by which this peculiarly pervasive and low-level energy is known in science include "general background radiation" and "cosmic microwave radiation." It is associated with a thermodynamic temperature of about 2.75 K, which is close to, but still less than, the 3.18 K that Eddington estimated as the radiation temperature typical of interstellar space in general.

This general background radiation is scattered quite uniformly, traveling in every direction equally. Hence it cannot be linked with a pattern of emission from the Milky Way, or the plane of our galaxy. Nor does it show any indication of originating from any other particular direction relative to the galactic plane.

It was first noticed during radio-wave measurements that were being made in search of microwave "noise" originating in the atmosphere of Earth itself. At first it appeared during receptions at 7 cm wavelength, equivalent to about 4.3×10^{10} hertz. The experimenters noted not only that it seemed to come from every which way, but also that its power level appeared to increase as reception was tuned to shorter wavelengths (higher frequencies).

Generally speaking, the rise in power followed a pattern indicating a peak at about 1 mm wavelength (about 3×10^{11} hertz), which would correspond to a blackbody radiating at the thermodynamic temperature of 2.75 K. The spectrum of this all-pervasive microwave radiation is still being studied. It is indeed rather close to the blackbody radiation pattern, but perhaps not quite close enough.

The proponents of the Big Bang hypothesis of the origin of the universe have taken much heart from this evidence of a cosmic background radiation. Indeed, the late George Gamow, a gifted science writer as well as an imaginative physicist, pointed out, some time previous to the discovery, that the Big Bang hypothesis called for universal surviving microwave radiation, representing the cooled-off and frequency-diminished result of the intensely hot radiation that filled the expanding universe during the first instants after the Big Bang was triggered.

As the universe expanded under the enormous radiation pressures, the radiation had to do much work—driving matter apart at high velocities. For these and other reasons, this radiation could be expected to cool down, to become longer in wavelength and lower in frequency. Gamow indicated that its present peak level might indeed correspond to a very few degrees kelvin of blackbody radiation.

It is not important for our purposes here whether or not cosmologists and astrophysicists of the future regard the undeniable prevalence of this 2.75 K radiation as a surviving souvenir of the Big Bang birth of the cosmos. Sufficient that the possibility of such a cosmic process is taken seriously and made the object of extraordinarily careful and complete measurements, at levels of power intensity so tiny that they would have been regarded as utterly undetectable a few years ago.

Entropy and the hierarchy of energies—The concept of entropy can be applied meaningfully to measurements of radiative as well as conductive transfers of energy between and within masses of matter. In fact, all the major processes of energy interchange can be compared on a kind of scale or sequence, an "order of merit" arrangement.

This pattern we owe to the theoretical physicist Freeman J. Dyson, author of "Energy in the Universe," a revealing paper published in *Scientific American* in September 1971. Dyson compares the entropy, if any, of each kind of energy interaction, with the unit energy developed. In other words, the revealing unit combination is the entropy per joule of energy developed or unfolded in each of the various processes and interactions. The higher that entropy-to-energy ratio, the greater the likelihood that change tends in that direction, and the less likelihood that it will go in the other direction.

The pecking order or processional direction in this hierarchy, so to speak, is from less-entropy-per-energy unit to more-entropy-per-energy unit, not the other way about.

Three kinds of purely mechanical-energy relationships head the list. For each of them the entropy is zero, because there is no heat necessarily involved in their processes.

The first is gravitational energy, being essentially the potential energy of bits of matter widely scattered about in space, and their increasing kinetic energy as they begin to "fall together" under mutual attraction—but before they have actually clumped together with accompanying heat effects.

The next is rotational energy, as of a spinning planet, star or galaxy.

The third is orbital energy, as of the Earth around the Sun, the

Moon around the Earth, or binary stars around a common center of gravity.

In the light of what follows, we will do well here to note that the magnitudes of orbital and especially rotational energies can indeed equal or even surpass thermonuclear energies on stellar and cosmic scales. A brief and very contemporary commentary here will help to demonstrate this.

Extraordinarily large rotational energies are believed now to characterize myriads of aged, shrunken stars—especially the pulsars, whose existence was not even suspected until relatively few years ago. As this is written, about seventy radio-emitting pulsars have been located and additional members of the strange family are still being found. Their pulsed or intermittent radio signals show them to be spinning at rates ranging between about 1 revolution per second to 30 or 40 revolutions per second.

Only a star or ex-star shrunken to tiny size and enormous density could spin at such rates without being torn apart and destroyed by centrifugal forces. Pulsars are widely believed to have shrunk to densities close to those of nuclear particles, such as neutrons. They are sometimes called "neutron stars." The total spinning mass probably remains in the range between one and three times the present mass of our Sun.

That Sun, with mass of nearly 2×10^{30} kg, has a radius of about 695 million meter, and revolves slowly on its axis, completing a turn in about 1 month. Its rate of rotation in SI terms is thus only some 2.4×10^{-6} radian per second, or 2.4 microradian per second. The Sun's rotational energy is thus negligible compared with its content of thermal energy, constantly maintained at about 4×10^{41} joule by the thermonuclear furnace in its innermost core.

If the Sun were to shrink to a radius of only 10 km (10^4m) it would rotate about 1,000 times per second, or 2.6 million times its present rate, and would possess about 10^{43} joule of rotational energy. This is more energy than can be supplied by the Sun's present reserves of thermonuclear fuel. In fact, "a spinning neutron star begins its life with more kinetic energy of rotation than its parent star could derive from nuclear fusion," in the words of Dr. Herbert Friedman, chief scientist of the Hulburt Center for Space Research, Washington, D.C.

What is the source of such stupendous rotational energies? They result from the drastic gravitational shrinkages, releasing vast quantities of energy, some of which is converted into the swift spinning of the neutronized star. In the imaginary case of the Sun shrinking to 10 km radius, the release of gravitational energy amounts to about 1.5×10^{46} joule, which is about 1,500 times the expected resulting rotational energy of 10^{43} joule.

Thus, there is no doubt that radical shrinkages of burned-out stars can provide power more than sufficient to account for stupendous rotational energies such as those envisaged by Dr. Friedman. Such shrinkages not only accelerate the spin rate enormously but also intensify the surface magnetic field that was present in the preshrunken star. Our Sun shows a typical magnetic flux density of about 1 gauss, equal to 10^{-4} tesla (T) in the corresponding SI unit. This would be increased to 10^{10} gauss, or 10^6 T, on the surface of a neutron star of 10 km radius.

The combination of extremely swift spinning and extremely intense magnetic fields on the surfaces is believed to account for the undeniable fact that pulsars generate powerful, periodic, and readily recordable radio pulses that reach the Earth. These are the pulses for which pulsars were named. They mean that the pulsar's vast rotational energy is being gradually converted into radio-wave emission. Hence, it is hardly surprising that the spin rates of pulsars, at first believed to be quite constant, have lately been found to diminish very gradually. Thus, the gravitational (shrinkage) energy that became rotational energy is now gradually being leaked away as energy of radio-wave radiations.

Gravitational, rotational and orbital energies are associated with zero entropy, in the great cosmic rating scale devised by Dyson. The next level is that of thermonuclear energy, in which the entropy is quite small in relation to the energy emitted. Dyson's scale is set up in such a way that the sunlight radiated from the surface of our present Sun is arbitrarily considered to rate as 1.0 in its entropy-to-energy ratio. On that scale, thermonuclear processes receive a rating of only 10^{-6}, or 1 part in 1 million of sunlight's rating.

The next higher level is reached in the transfers of energy which take place within the Sun and the stars themselves, from the thermonuclear furnaces in the cores to the radiating surfaces. Here the entropy-to-energy index rises to 1 part in 1,000, or 1,000 times greater than the index for the original thermonuclear process in the core.

Finally comes the surface launching of radiative energy into space—resulting, for the Sun, in what we call *sunlight,* though so much of the radiation is invisible to us. Here, by Dyson's reckoning, the index rises again a thousandfold, to just 1.0.

Still another realm of energy interchange is found, especially on Earth, in chemical reactions, including the burning of fuels and other processes based on the chemical bonds in matter. The index for chemical interactions ranges from 1 all the way to 10, since the amount of added entropy per joule of energy continues to rise.

Yet another process, already mentioned here, shows a still greater magnitude in its index. This is the re-emission into space of longwave radiation from the Earth, which was warmed in the first place by the shorter-wave radiations from the Sun. For this process the necessary index values lie between 10 and 100, so large is the entropy relative to unfolded or transmitted energy.

Is this the end of the road? Or is there some farther final point in the increasing downhill slope of energy toward irreversible finality? According to Dyson, that ultimate lies in the background microwave radiation of the universe itself. This pattern of radiation, roughly that of a blackbody radiator at 2.75 K, is the ultimate sink or discard of energy. Its index is 1,000—compared with the 1 for sunlight.

Once energy has been so reduced or degraded, it cannot by any now describable process be converted from the microwave background radiations into the other levels in the hierarchy. Also, in the words of the paper mentioned, "no way is known in which this energy could be further degraded or converted into any other form."

Degradation here is not a term of reproach or disapproval.

Thus, we see that the basic or cosmic progress along the great continuum or ladder of radiations is primarily a descent. At each successive level, the energy-changing events in matter correspond with radiation packets or photons, or with electromagnetic frequencies, somewhat farther down the spectrum.

First in the thermonuclear furnaces of the stars we find photons with energies of gamma rays and X-rays. Then, as the energy transport moves outward through the stellar substance, we find photons with energies of ultraviolet radiations. At the surface of stars in the Sun's temperature level, we note the preponderance of the emergence of photons of visible light.

Then, arrived on Earth, these energies are converted into the long-wave heat radiations or are stored in plants that are later chemically combined (burned) to emit mainly such heat wavelengths.

The frequencies of radiation at these low levels correspond to such material events as the rotation of molecules or the vibration of atoms.

With every level of radiational energy—with each spectral segment in the great continuum—are associated different kinds of energy transactions in matter. The most energetic involve the fusing-together or bursting-asunder of the nuclei of atoms. Next come the events involving the alterations of electron orbits around atomic nuclei. Then come events or situations involving the mutual motions of entire atoms within molecules, or of molecules themselves with respect to the entire mass of matter that contains them.

In all the broad world of measurement there is no broader or more challenging realm than this—the application of measures of energy and power and temperature to radiative emissions and absorptions on the one hand, and, on the other, to the analysis of the specific changes in the bodies of matter in which those emissions and absorptions take place.

V / The Body Electric

I sing the body electric. . . .
  —WALT WHITMAN,
    from poem of the same title (1855)

Is it a fact—or have I dreamed it—that by means of electricity the world of matter has become a great nerve, vibrating thousands of miles in a breathless point of time? Rather, the round globe is a vast head, a brain instinct with intelligence . . .
  —NATHANIEL HAWTHORNE,
    The House of the Seven Gables (1851)

. . . Through Space and Times fused in a chant,
  and the flowering eternal identity,
To Nature encompassing these, encompassing God—
  to the joyous electric all. . . .
  —WALT WHITMAN,
    As They Draw to a Close (1871)

35 ARRIVAL AT THE AMPERE, BASIC UNIT OF ELECTRIC CURRENT

Electricity is of two kinds, positive and negative. The difference is, I presume, that one comes a little more expensive, but is more durable; the other is a cheaper thing, but the moths get into it.
—STEPHEN LEACOCK,
Literary Lapses (1910)

The present International System of Units (SI) is basically an outgrowth of an earlier, less complete unit system that was commonly identified by the initials MKSA or mksa. They stood for *meter-kilogram-second-ampere,* and today the ampere remains one of the six units adopted "to serve as a practical base for a system of measures for international relations." Besides the four basic units named in the preceding lines, the other two, already presented here, are the kelvin, unit of temperature, and the candela, unit of luminous intensity.

Thus, the ampere is a basic unit of the SI, not a derived one. It is intended to be treated as no less basic than the meter of length, the kilogram of mass, or the second of time.

Yet its definition sounds in many ways as if the ampere were derived, not basic, for that definition states: "The ampere, unit of electric current, is the constant current which, if maintained in two parallel conductors of infinite length, of negligible cross section, and placed 1 meter apart in a vacuum, will produce between these conductors a force equal to 2×10^{-7} newton per meter of length."

Let us see why an electrical unit was elevated to the position of one of the basic units from which all the nonbasic units must be derived in the SI structure. Through many years and in many different unit systems, no electrical unit was, in fact, considered basic.

Instead, electrical units were derived from the three basic mechanical units of length, mass and time, meaning either the centimeter-gram-second or the meter-kilogram-second combination.

If the ampere were not established as basic, then the definition just quoted would have the following meaning in dimensional terms: the current of 1 ampere in each of the two parallel wires is a current times a current, or a current squared. It is equal to—that is, it produces—a specified force (10^{-7} newton) per unit length of the wires, when the two wires are separated by just 1 unit of length.

Current is usually symbolized by I. Hence I^2 is equal to a force that has the dimensions of length times mass per time squared (since a force is a mass times an acceleration). In symbols: $I^2 = L\ M/T^2$, and thus I must equal the square root of $L\ M/T^2$, which is $L^{1/2}\ M^{1/2}\ T^{-1}$.

Actually, the correct statement is that the current I is proportional to, rather than equal to, the square root of a length, times the square root of a mass, divided by a time. This is the "dimensionality" of electric current if we insist on regarding length (L), mass (M) and time (T) as primary and every other physical variable as necessarily derivative from those three. However, dimensional recipes of this kind are crabbed and confusing. Nobody can get a clear physical concept of the square root of a length or of a mass.

In order to avoid such dimensional confusions and make at least one physical variable of electricity a primary one, the ampere of current was chosen and elevated to that rank of primacy. The choice might have been any one of several other electric variables, such as charge (measured by the coulomb), or electric potential difference (measured by the volt), or even electric resistance (measured by the ohm). However, the ampere was picked, and it does very well to disentangle the dimensional mixups that plagued more than one generation of students of electricity and science.

Custom has helped to create the belief that there is something more essentially "basic" about the physical variables of length, mass and time than about a physical variable such as electric current. From a strictly theoretical point of view this appears to be an unjustified prejudice. From a practical point of view it appears to be a needless obstacle to efficiency in making and interpreting measurements.

It would be quite possible to create a complete and consistent system of units to measure all important physical variables, using as basic or primary only length, mass and current, and deriving the unit of time from those basics. That would be an MKA (meter-kilogram-ampere) system, rather than the MKSA (meter-kilogram-second-ampere) system, the category in which the SI is often said to belong.

The full truth is that the SI now makes use of six, not four, basic units, for the kelvin (K) of thermodynamic temperature and the mole (mol) of molecular or atomic or particle mass have been accorded equal standing with the other four basic units. This leaves outside the inner circle only the candela (cd). True, it is called part of the practical base of the SI, but what it is designed to measure belongs in the category of psychophysical or sense-conditioned effects.

Autonomy and equality for the ampere—Here, at any rate, the ampere, unit of electric current, will be considered as truly basic. Such recognition reflects the stubborn physical fact that electric, or rather electromagnetic, interactions are not simply reducible to equivalent mechanical interactions. The inclusion of the one necessary electromagnetic unit among the inner circle of basics enables us to give up the procrustean efforts to force electromagnetic interactions into molds made to fit mechanical interactions. The latter were known first, but the former are no less deserving of autonomy and primacy on that account.

Electromagnetism developed relatively recently as a large body of measurable and comparable data. These measurements and comparisons are best made when one basic unit, such as the ampere, is accorded independent validity and not regarded as merely derivative from the traditional trinity of length, mass and time.

The SI definition of the ampere reflects the fact, first noted by André Marie Ampère (1775–1836), that currents flowing through conductors produce magnetic effects in the space outside those conductors. Thus, if currents in two parallel conductors flow in the same direction, their magnetic fields interact in such a way that the conductors are drawn closer together. If the currents flow in opposite directions, however, the resulting force tends to separate them further. Two parallel wires carrying opposite currents thus generate a *repulsive* force; two wires carrying currents of like direction generate an *attractive* force.

It is not theoretically essential that the currents flowing in each of the two parallel conductors have the same strength, though it is experimentally far more convenient when that is the case.

Actual measurements of this kind are not made by setting up parallel conductors of "infinite length." In fact, concentric coils of very fine wire of closely measured dimensions are used. The devices commonly employed for the most detailed current measurements in the great standards institutions of the world are called "current balances." They use systems of coils suspended on both sides of an extremely sensitive balance scale, the coils containing a total of many hundreds of turns of fine bare copper wire.

The force of attraction or repulsion between the coils can be measured, as the equilibrium of the scale beam is restored by compensatory weights. Such a current balance, operated by patient experts, can measure current to a few parts in 10^6 (1 million). But it is no simple matter. On one occasion, at the National Physical Laboratory at Teddington, near London, nearly 1,500 measurements of coil diameters and 4,500 of the axial position of the coils were made in connection with just one set of precision current determinations.

Measurements of forces between conductors are not the only procedures that have been used to quantify electric currents. In fact, for a long time an earlier form of the ampere, known as the *international ampere,* was defined in an electrochemical fashion, as the "unvarying current which in 1 second deposits 0.001 118 gram of silver from an aqueous solution of silver nitrate [$AgNO_3$]."

No mention occurs there of force or length, though the definition does specify an interval of *time* and a *mass* of a particular substance. Accompanying that definition were detailed instructions as to how such measurements must be carried out in order to be truly comparable. The silver nitrate was to constitute not less than 15 nor more than 20 percent of the weight of the distilled water in which it was mixed. A solution was to be used but once, and only so long as not more than 30 percent of the silver in the solution had been deposited by the current. The anode substance must be silver, the cathode substance platinum. Also, the current density at the anode must be no more than 0.2 ampere per square centimeter;

that at the cathode not more than 0.02 ampere per square centimeter. And so on and on . . .

The present ampere, known as the *absolute ampere,* is very slightly larger than the onetime international ampere. The difference is just 165 parts in 1 million. Hence, if an electrochemical definition of the present absolute ampere were to be given, it would have to require that the amount of silver deposited in each second (per ampere of current) be 0.001 119 845 gram!

Today the absolute ampere is the only universally standard unit of electric current. However, it is not the only current-measurement unit that can be found in textbooks and technical papers still used in libraries and research institutions of the world.

In the so-called cgs-emu (centimeter-gram-second electromagnetic-unit) system, the unit of current is the *abampere,* equal in magnitude to 10 absolute ampere. Another name sometimes used for the abampere was the *biot,* after J. B. Biot (1774–1862), a French physicist, who first provided mathematical expressions for the laws of heat convection.

The abampere, or biot, was defined in terms much like those for the ampere in the SI, except that the parallel conductors were to be placed just 1 cm (instead of 1 m) apart, and the resulting force between them was to be 2 dyne per centimeter of length (instead of 1 newton per meter of length). The dyne was the force unit in the centimeter-gram-second-unit system, and 1 newton equals just 10^5 dyne.

Still another system of electrical units was used alongside the cgs-emu. This rival system was designated as the cgs-esu (centimeter-gram-second electrostatic-unit) system. Its particular and rather monstrous unit of electric current was called the *statampere,* and its magnitude was almost incredibly tiny; for 1 absolute ampere equals 2.998×10^9 statampere. That number may look familiar, and should, for it is just ten times the number that represents the speed of light, c, expressed in meter-per-second units.

Little need appears at this stage for an analysis of the theoretical principles underlying the *emu* and *esu* varieties of the *cgs* unit system. They served well the needs of eras of swift and intensive electrical development, but beyond doubt the SI is rapidly replacing them, and for very good reasons of practicality and international understanding.

Still another unit of current that may be encountered from time to time is the so-called Faraday per second. The absolute ampere is equal to a flow of electric charge at the rate of just 1 coulomb (C) per second. That coulomb is indeed a large charge unit for all practical purposes of charge measurement. However, the Faraday charge unit is still larger, for 1 F equals 96,500 C. Thus, 1 Faraday/second equals 9.65×10^4 A.

Relating current to charge—Current is related to charge in the same way that power is related to energy. This is illustrated when we note that 1 ampere equals 1 coulomb per second, while 1 watt equals 1 joule per second. Conversely, a direct current of 1 A delivers a total of 1 C in 1 second; and a power of 1 W delivers an energy of 1 J in 1 second.

In the case of electric current, what actually is "delivered" by the conductor? What is the nature of the flow in the wire through which measurement may show that a current of just 1 ampere is passing? It is a flow or passing-along of the ultimate unit of negative charge, the electron. A conductor is like a pipe through which electrons flow, and every electric battery or direct-current generator is like a pump, maintaining such flows (when the current is passing).

Scientists today have measured quite precisely both the mass and the actual charge of the electron. Hence we can state rather accurately just how many electrons are moving past a given point or plane in a wire when a current of 1 ampere flows through it: each second 1 coulomb of charge passes, and 1 C is equal to the charge of $6.241\ 45 \times 10^{18}$ electrons. Very roughly that is $6\frac{1}{4}$ million million million electrons per second, constituting just 1 ampere of current!

The physical picture is not that of individual electrons rushing at relativistic speeds from one end of a wire to the other. The electrons themselves are sometimes said to drift through a conductor. Yet there is no doubt that when a current of 1 ampere deposits the stated amount of silver out of a silver nitrate solution, it does so because it has supplied to the silver nitrate atoms

$$6,241,450,000,000,000,000 \text{ electrons}$$

per second.

Each electron is the smallest possible negative charge. Hence it is from the negative terminal of an electric battery that the electrons are emerging, and it is to the positive terminal that electrons return. The same is true of the negative and positive terminals in a direct-current generator.

Our household and industrial electric power supplies are in the form of alternating current, with the myriads of electrons surging back and forth a large number of times each second.

Mass is not an important consideration in connection with electric current, yet it is interesting to consider the equivalent in electron mass of the flow we call 1 ampere. The mass of an electron at rest is very nearly 9.11×10^{-31} kilogram. Thus the total electron flow per second of 1 ampere current represents a mass of less than 5.7×10^{-12} kilogram. A steady current flow of 1 ampere would have to continue roughly 100,000 years before the electron mass that passed a single position in the wire amounted to 1 kilogram.

Electric current through a conductor, of course, does not deplete the metallic atoms in the conductor of all their electrons. New electrons are always arriving to replace those that have been shifted onward in the direction of the current. It is indeed tempting, but risky, to press too far the concrete pictures that we derive from our macroscopic experiences, into the submicroscopic realm of electrons and other elementary particles.

The choice of the name ampere for the unit of current was a deserved honor for a remarkable experimenter and interpreter of measurements. During one magnificent burst of creative work in a single week of 1820, A. M. Ampère measured and analyzed mathematically the major magnetic effects produced by uniform electric currents flowing through conductors.

Ampère had learned of the discovery by the Danish savant Hans Christian Oersted (1777–1851) that passing an electric current through a wire would cause a nearby magnetic compass to turn until it pointed at right angles to the wire. In an incredibly short time, Ampère had leaped far into the future—he found the forces between parallel current-carrying wires; discovered the powerful magnetic effects of a long current-carrying coil of wire, called a solenoid; and even reached the stage of understanding that enabled him to suggest that all magnetism might be due to electric currents of some kind. Oersted had talked of "conflict" in the wire that moved the nearby compass needle. Ampère, however, clearly conceived currents and their field effects external to the conductor through which they moved.

A strange, religious, often tormented personality, Ampère in more than superficial ways earned the great tribute given to his memory by James Clerk Maxwell, who called him "the Newton of electricity."

Concerning the coulomb, unit of electric charge—The coulomb, SI unit of electric charge, suffers from one drawback that cannot be called small, but certainly should be called noncritical. It is very big in relation to the effects whose measurements actually engage or interest scientists and technicians. Hence it is more likely than most of the other SI electromagnetic units to appear, in practice, with such submultiple prefixes as *micro-* (one millionth), *nano-* (one billionth), *pico-* (10^{-12}), and so on.

The earliest electrical events that humans noticed had to do with charge. Some substances, when rubbed with other substances, became "electrified" and would attract bits of dust or fiber. In cold, dry weather, a person might become so "charged" by shuffling across a carpet that a painful spark would actually leap between his hand and a metal doorknob or latch that he was about to grasp.

Amber was one of the substances which, when properly rubbed, produced such effects; hence the Greek word for amber, *electron,* became the progenitor of the entire enormous linguistic family that includes *electricity, electronics,* and even *electron* (the ultimate unit of negative charge).

We have already displayed the electron particle as a charge unit and seen how, in combination with the unit of electric potential difference, the volt, it becomes a measure of the tiny energies associated with particle and atomic events—the electron volt (eV).

The corresponding energy-measuring unit combination with the coulomb would be the coulomb volt (CV). One coulomb of charge, responding to a potential difference of 1 volt, would gain just 1 joule of energy. Thus, we can say that 1 CV equals 1 J. And it is no less true that 1 V equals 1 J/C. In other words, 1 volt is the electric potential in response to which 1 coulomb of charge gains 1 joule of energy.

In order for a body to be charged negatively to the extent of 1 coulomb, it would have to accumulate about 6.24×10^{18} surplus electrons. That could not include the electrons within the complete atoms making up the normal substance of the body, for every complete atom is electrically neutral, having as many positive charges (protons) in its nucleus as it has negative charges (electrons) bound to that nucleus.

However, it is difficult for a relatively small body to acquire and hold so many surplus or free electrons, and for a very basic reason: like charges repel each other; unlike charges attract each other. Thus the more than 6 million million million electrons required for a full coulomb of negative charge display a most powerful mutual "dislike." They tend to spread and scatter.

If two tiny or pointlike objects, each with 1 C of negative charge, could be placed 1 meter apart, they would repel each other with enormous force. However, each tiny electron in each of these two bodies would be repelled also by the others in its own collection. Hence concentration of charges as great as 1 C on small bodies is difficult. This does not invalidate the coulomb as a unit, however.

The definition of the ampere in the International System led us to note that parallel wires attract each other when they carry currents moving in the same direction, but repel each other when their currents move in opposite directions. In the case of charged objects, if both are negatively charged, they repel each other; if both are positively charged, they repel each other; but if one is positively and one negatively charged, then they attract each other.

This inevitably has led to romantic analogies about temperamental opposites attracting and like natures repelling.

Since we describe a negatively charged body as one with a surplus of electrons, can we describe a positively charged body as one with a deficiency of electrons? We can and must. In the various procedures for generating "static" electricity in small objects, when the rubbing or stroking action tends to withdraw electrons from the atoms in the surface of an object, then that object becomes "positively" charged. When the rubbing or striking action tends to convey to the body additional or surplus electrons, then that body becomes negatively charged. Charge sign—plus or minus—is a relative matter. Compared with a negatively charged body, a neutral or uncharged body appears to be positively charged; and vice versa. Thus we see that either a negatively charged or a positively charged object may attract light bits of matter which themselves show no sign of charge, one way or another—or at least did not do so until the charged object was brought near them.

Static electrical interactions are fascinating and often utterly surprising. Endless tricks and special effects can be performed with their aid. They not only provide amusement but also serve important industrial and technical needs.

True measurement entered electrostatics thanks to the efforts of Charles Augustin de Coulomb (1736–1806). This French army officer, who spent years overseeing the erection of fortifications in the island of Martinique, was the first investigator to apply mathematics to electrical and magnetic interactions.

He was not only a good mathematician but an avid and able experimentalist. He had to be, for the forces he worked with were not only tiny but all too often temporary; the electrostatic attractions and repulsions had an infernal way of vanishing, because of inadequate electrical insulation.

In order to make quantitative measures of tiny forces, he developed the idea of the torsion balance, allowing fine wires or fibers to be twisted and using the angle of their twist as an indicator of the force that had caused it.

Using methods so subtle and yet simple, Coulomb was able to complete the measurements on which he based his inverse-square law in electrostatics. It was a striking analogue to Newton's law of gravitational forces. Stated most simply, if Q_1 is the charge on one very small body, and Q_2 the charge on another very small body, with D the distance between them, then F, the force with which they either attract or repel each other, must be proportional to Q_1Q_2/D^2.

Newton had shown that if M_1 is the mass of one (spherical) body, M_2 that of another, and D the distance between their centers of mass, then F, the force with which each body attracts the other, must be proportional to M_1M_2/D^2. In the gravitational interaction the force is always an attractive one; in the electrostatic interaction it may be either attractive or repulsive.

Coulomb's law, as it is properly called, is valid only when the two bodies are quite small relative to the distance between them. It works fully, in other words, with imaginary "point charges." Any substantial diameters dull the fine edge of its mathematical precision.

In both the Newtonian and the Coulombian laws, as simply summarized here, the words "proportional to" mean that there is a constant which, when multiplied by the stated fraction, makes it precisely equal to the observed force.

The constant of proportionality in the case of Newton's law is the famous gravitational constant G. The constant of proportionality in the case of the Coulombian relationship is no less exact and well known. It differs in value according to the system of units being employed. We shall not explore it specifically here, however.

Other units for measuring electric charge—The coulomb is not the only unit that has been widely employed to measure charge. In the electromagnetic-unit (cgs-emu) system, charge is measured by the *abcoulomb,* whose magnitude is just 10 times that of the coulomb. In the electrostatic-unit (cgs-esu) system, on the other hand, charge is measured by the *statcoulomb*. Like the statampere it has an extremely small magnitude. In fact, 2.998×10^9 statcoulomb are equal to 1 coulomb.

Still another system of electrical units, the Gaussian system, has been used widely for work in electromagnetics. It combines elements of both the esu and emu systems. In the Gaussian system, the unit of charge is also that same very small statcoulomb. So far as the International System is concerned, the statcoulomb and abcoulomb retain historical interest only.

The faraday unit (with or without a capital F) was used to some extent in the past to define a charge far larger even than the present coulomb. Several different values were used for the faraday, a situation repeating the kinds of confusion we have noted in the case of the calorie, the Btu, and even the horsepower. There was the "physical" faraday with a value of 96,421.9 coulomb; the "chemical" faraday with 96,495.7 C; and the "carbon 12" faraday, with 96,487 C!

Thus, when the faraday per second was mentioned earlier, it should have been specified which of the three possible faradays was meant. These are typical of the confusions and duplications which the SI has managed to eliminate.

Both the ampere of electric current and the coulomb of electric charge are linked with the meter, unit of length, in combinations designed to measure important physical variables in the realm of electricity.

Current density, for example, relates the magnitude of a uniform current through a conductor to the cross-sectional area of that conductor transverse (perpendicular) to the direction of the current flow. In the usual case of a wire carrying a current, the ampere measure of the current is divided by the cross-sectional area of the wire. The SI unit combination for current density is the ampere per square meter (A/m^2).

It is true that conductors, even the heaviest bus bars in power generating stations, ordinarily have cross-sectional areas amounting to no more than a small fraction of a square meter. However, the unit combination does satisfactory service in its present form. For example, if a wire with uniform cross section of 1 square centimeter is carrying a steady current of 5 amperes, the current density all through that wire is just 5 times 10,000, or 50,000 A/m^2, since 1 m^2 contains 10^4 cm^2.

In many important instances, electric currents travel through very thin films or through metallic layers whose thickness is either not measured or is too variable for any single consistent thickness to be recorded. The interest of the measurers is in the width rather than in the total cross section of such ultrathin conductors. They seek what is commonly called the *linear* current density, rather than the cross-sectional current density just described. The ampere per meter (A/m) is the unit combination in which linear current density is measured. The denominator of that fraction is the width, transverse to the direction in which the current flows, of the thin film or sheet. For example, if a current of 1 mA (milliampere) flows through a thin film just 0.1 mm (millimeter) wide, then the linear current density is 10 ampere per meter (A/m). The linearity, it must always be remembered, is measured at right angles to the direction of current flow.

In the case of the coulomb, important physical variables are measured in relation to volume (length cubed), to area (length squared), and to distance (simple length). Volumetric charge density is the name usually given to the total charge contained within a unit volume. The unit combination is thus coulomb per cubic meter (C/m^3).

If we think of the surface of any three-dimensional shape enclosing a volume containing electric charge, then each bit of such surface area can be credited with the effects of a portion of the total enclosed charge. The simplest case, of course, is that of a perfectly spherical surface enclosing a volume of just 1 cubic meter, at the center of which is a point charge with magnitude of just 1 coulomb. The volumetric charge density is thus just 1 C/m^3.

However, a sphere containing just 1 m^3 has a surface area of about 3.35 m^2, so each m^2 of its surface area would correspond to just $\frac{1}{3.35}$, or about 30 percent of the 1 coulomb charge within. In order to get a proper unit relating internal charge to surrounding surface area, we need a sphere whose surface area is just 1 m^2, surrounding a central charge of just 1 coulomb. Then the total electric-charge effect, called "the electric flux," as measured at the sphere's surface is just 1 coulomb per square meter (1 C/m^2).

Electric *flux* may seem a rather unfortunate term here, though it is the one constantly used. A flux of heat is a flow or motion of heat; a flux of particles is a flow or motion of particles. But an electrostatic flux reveals no physical motion. It is simply the field or force lines radiating out from the charge or charges. Similarly, we find that magnetic flux, to be presented later, represents the magnetic field, rather than a physical *flow*.

The unit combination C/m^2 is used in two different, yet related, situations. The first, already described, has to do with the electric flux density (per unit area) surrounding central charges. The second is the situation in which the surface itself, such as a metal sphere, contains the charge. In either case, the total charge, measured by the coulomb unit, is divided by the area. In the first case it is the area *through which* the effects of an internal charge become apparent; in the second it is the area *on which* the charge rests.

One name for the first situation is "electric bias," described as the electric flux across a transverse cross section of unit area. A common name for the second situation is "surface electric charge density" or "surface density of electric charge." In either case, the unit combination used in the SI is the C/m^2.

Electric flux itself is measured simply by the coulomb unit. If a small or point charge of just 1 coulomb magnitude is surrounded by a surface of any size that completely encloses it, then the total electric flux on that entire surface must be just 1 coulomb also. The electric bias, or electric flux per unit area (C/m^2), at various parts of that surface will vary, however, depending on the distance of the portion of the surface from the point charge within.

If a point charge of 1 C is enclosed by a plastic sphere with a radius of 1 meter, then the 1 C of electric flux is distributed over that sphere's total surface area, which is 12.57 m^2, resulting in an average electric flux of $\frac{1}{12.57}$ or 0.079 6 C/m^2 at that surface. If the same charge is enclosed by a sphere with a radius of 2 meter, and a resulting area 4 times as great, or 50.28 m^2, then the average flux at that surface becomes $\frac{1}{4}$ as great, or about 0.02 C/m^2. The total flux in either instance is identical with the charge at the center —just 1 C.

Electric charge and dipole measurements—Finally, charge is related to simple length or distance in the measuring of so-called "electrical dipole moment." A charge dipole consists of positive and negative charges connected together at a fixed distance. If the two charges, one negative, one positive, have a total charge difference of 1 C, and if the distance between them is 1 m, then the electric dipole moment is 1 coulomb-meter (1 C·m).

We are surrounded by tiny electric dipoles. Almost every molecule is a somewhat lopsided configuration, in which charge differences are found at definite distances apart. The charge difference is multiplied by the distance separating its two components. Thus, the greater the charge difference between the two ends, the greater the electric dipole moment for any given separation. And the greater the separation, the greater the electric dipole moment for any given charge difference.

If in the laboratory we construct an electric dipole with 1 microcoulomb (10^{-6} C) of charge separated by a distance of 1 cm (10^{-2} meter), then its electric dipole moment is just 10^{-8} C·m. If then we make another dipole with a 2 microcoulomb charge and a 2 cm separation, its electric dipole moment is 4 times as great, or 4×10^{-8} C·m.

Vast numbers of detailed measurements have been made to

determine the molecular dipole moments of a great variety of substances important in the chemical technology of our era. Indeed, to most chemists and physicists, "dipole moment" seems essentially a molecular concept. The SI unit combination of coulomb times meter (C·m) is extravagantly huge for the purposes of such molecular measurements or estimates. It could conceivably be cut down to more convenient size by the use of such a submultiple prefix as the *atto*coulomb (aC), which is just 10^{-18} C. However, in actual practice the typical unit for molecular dipole moment measurements has been one often called the *debye* (D).

One coulomb equals $2.997\ 925 \times 10^9$ statcoulomb, the charge unit in the electrostatic, or cgs-esu, system. The debye unit equals just 10^{-8} statcoulomb-centimeters. Hence 1 coulomb-meter equals about 3×10^{19} debye, and 1 debye equals about 3.33×10^{-20} coulomb-meter. Those differences truly can be termed astronomical. The result, however, is that the debye unit has a size rather neatly suited to its uses.

Many substances show zero dipole moment, indicating that their molecules are formed by in-line and symmetrically placed atoms. However, hundreds of other substances of great interest reveal definite dipole moments. Thus, water (H_2O) has a moment of 1.85 debye (D); ammonia (NH_3) of about 1.47 D. Some substances with exceptionally large moments include potassium fluoride (KF) with 7.33 D; cesium chloride (CeCl) with 10.42 D; and cesium iodide (CeI) with 10.2 D.

Measures of molecular dipole moments are incomplete unless there is an indication of the temperature or temperature range that prevailed when the determination was made.

The debye is named for the eminent physical chemist Peter Debye (1884–1966). In extensive investigations culminating in 1912 he recognized two distinguishable responses of atoms and molecules to electrical fields acting upon them. All atoms, he noted, possess a type of polarizability, known as *displacement* polarizability. Roughly, it means that the atom's electron or electrons, being negatively charged, are pulled in one direction by an electric field, while the nuclear proton or protons, being positively charged, are pulled in the opposite direction by the field.

However, over and above this, Debye found that some molecules, because their own structure produces spatial separation of unlike charges, form permanent dipoles, and thus show additional and separate response to electric fields.

For example, hydrogen chloride (HCl), well known as hydrochloric acid, consists of molecules each formed of two oppositely charged ions; one is a hydrogen ion (H^+), which is in effect just the nuclear proton, lacking its planetary electron; the other is the chlorine ion (Cl^-), which has acquired that surplus electron, and so shows negative charge.

Such a molecule is in effect composed of two opposite charges separated by a tiny but precisely determinable distance. The magnitudes of the charges are perfectly definite; that of the hydrogen ion is positive $1.602\ 192 \times 10^{-19}$ coulomb; that of the chlorine ion is minus the same submultiple of the coulomb. The separation of the two is the same for each and every HCl molecule.

By measuring such molecular dipole moments physical chemists have learned an amazing amount about the shapes and sizes of molecules. The fact that carbon dioxide (CO_2) has no detectable dipole moment shows that its three atoms are ranged in line symmetrically. The three atoms of water molecules for long were thought to be in-line and symmetrical also. However, dipole measurement and other measurements have shown water molecules to have a triangular configuration, with the oxygen atom at the vertex, so to speak, and the directions from it to the two hydrogen atoms forming an angle of just $104° 31'$. Not $103°$ or $105°$, mind you— but just the cited precise angle, an impressive evidence of exactitude.

Ammonia (NH_3), on the other hand, is found to have a pyramidal form. The three hydrogen atoms form a sort of triangular base, while the nitrogen atom is at the vertex "above" that base. The "above" is necessarily a figure of speech, for in fact the nitrogen atom is alternately on one side and then on the other of that triangular base, as the molecule vibrates in the way that molecules will and must under the influences of heat and radiations.

By comparing dipole moment measurements made at various temperature levels, chemists have been able to make close estimates of the comparative magnitudes of (1) the permanent molecular dipole moment, and (2) the displacement polarization produced by an applied electric field.

Some substances show only the latter effect, having no detectable dipole moment. Some, such as HCl, SO_2 (sulfur dioxide), and ammonia, show well-marked dipole moments combined with relatively small displacement polarizations. Thus, at room temperature, ammonia's total response is only 16 percent attributable to displacement polarization, and 84 percent to dipole moment. In ethyl ether, on the other hand, the dipole moment contributes only 14 percent, as against 86 percent from the displacement polarization.

The myriads of measurements of molecular dipole moments show emphatically that the standard units of electrical variables are by no means limited in application to the gross and macroscopic events around us. They are perhaps even more revealing when applied to submicroscopic interactions, for they disclose the essentially electrical structure of matter and the fact that the chemical bond can be comprehended solely in terms of charges and charge interactions.

The intercharge distances involved in molecular dipole moment measurements are far too small ever to be measured by any ordinary length-determining procedures. Nevertheless physical chemistry is able to assert with confidence what must be the distances and the angles involved in the structures and configurations of important and complex molecules.

The world of measurements includes many mansions. That of the ultrasmall atoms and molecules is the tiniest, but by no means the least important among them.

36 THE VOLT OF ELECTROMOTIVE FORCE, THE OHM OF RESISTANCE, AND OTHERS

Volta is full of knowledge and the ability to display it . . . where electricity was concerned, he saw with the eyes of a Newton. First and foremost he is an explicator, a valid thinker and, with all that, a fine, charming person. He is good-looking and the focus of attention with the ladies, always cheerful when the occasion demands, and with fine ability to express himself. He argues with violence, curses when his experiments run wrong, and smiles angelically when they work well.
—GEORG CHRISTOPHER LICHTENBERG,
German physicist and philosopher, describing a visit
by Volta to Göttingen in 1784

The sequence in which we display the many other electromagnetic units in the International System is a matter of convenience rather than mandate or logic. They form, indeed, a sort of seamless web, each related to the rest, and all relying on the one basic electrical unit, the ampere, as their link to the firm SI foundation with its septet of fundamentals: the meter of length; the kilogram and the mole, both units of mass; the second of time; the kelvin of thermodynamic temperature; the candela of luminous intensity; and the ampere of electric current.

The volt (V) is probably the derived unit best suited to consider after the ampere. It is variously called the unit of (1) electric tension, (2) potential difference, (3) electromotive force, and (4) voltage, pure and simple. Each of these alternative titles is justifiable; each singles out some important aspect of the interactions measured in terms of the volt.

In the standard table of supplementary and derived units included in the International System, a small parenthetical fraction appears alongside the abbreviation V (for volt); it is "(W/A)." This parenthetic fraction offers almost infinite riches in a little space, for it shows that the volt can be recognized as the watt per ampere, or the amount of power delivered per unit of current flowing.

To illustrate this relationship: if a current of just 1 ampere is delivering 1 watt of power, we know that it flows in response to an electromotive force of 1 volt. However, if 100 watt of power is being delivered by a current of 1 ampere, we know that an electromotive force of 100 volt is operating.

This, in a slightly different form, is the old and useful rule of thumb: "Watts equals volts times amps." The electric-power delivery is the product of the electromotive force (in volt) times the current (in ampere). Thus the volt is adequately defined by the watt, the universal derived power unit of the SI, plus the ampere, the basic electrical unit of that system.

The actual SI definition of the volt sounds a bit stuffy, but is admirably careful in its formulation:

> The volt, unit of electric potential and electromotive force, is the difference of electric potential between two points of a conducting wire carrying a constant current of 1 ampere, when the power dissipated between these points is equal to 1 watt.

This definition shows that we may measure the voltage drop or voltage difference between various power-consuming components placed in series in one single circuit. For example, suppose we have a circuit supplied with current by a battery whose voltage we do not know. We do measure the current flow at 10 ampere—a flow that must be identical all through the circuit. Three small heating elements (resistances) are strung out at intervals in this circuit. We apply a voltmeter across the first one, and find a potential drop or difference of 5 volt, indicating an electric-power consumption there of 10×5, or 50 watt. The next resistance also shows a voltage drop of 5 volt, indicating added power consumption there of 50 watt. The third shows a voltage drop of 2 volt, indicating a power consumption of 20 watt.

Thus we have total power consumption in the circuit of 120 watt, and total voltage drops of 12 volt. Obviously the voltage supplied by the battery is 12 volt. Why did we not apply the volt-meter to the battery's terminals directly? Either it was inaccessible, being part of a modern overpowered automobile; or we decided not to, in order to make the example more informative. In either case, we note that if the three heating elements were bunched together they would show a combined voltage drop of 12 volt, a combined power consumption of 120 watt, and the same invariable current of 10 ampere.

The volt, from our present point of view, is the promising off-spring of current on the one hand and power on the other. Even more graphically: current times voltage is the measure of the power consumed or supplied electrically.

In an earlier chapter we suggested the analogies between mechanical force, temperature difference, and voltage difference. A mechanical force times the speed at which it is applied measures a power—as we saw in the case of James Watt's derivation of his original horsepower from the rotary processionals of the plodding workhorses of the mines of Britain. An electromotive force times a current measures a power also. Thus, current measured by the ampere is the electrical analogue of velocity measured by the meter-per-second unit combination.

We may extend the analogy by saying that current is to velocity as charge is to length; and that mechanical force is to electromotive force (in volt) as mechanical power is to electrical power (both measured by the watt), or as mechanical energy (measured by the joule) is to electrical energy (also measured by the joule).

The various earlier systems of electromagnetic units that the SI is rapidly replacing, provided their own peculiar units for measuring this same physical variable of electromotive force. Thus the electromagnetic (cgs-emu) system had its abvolt, 10 million of which equaled 1 volt in the SI. The electrostatic (cgs-esu) system had its statvolt, equal to 299.796 of the SI volt (also called "absolute" volt).

Various preliminary values for the volt were formerly in effect, differing slightly from the present worldwide value. In the United States prior to 1911, for example, a so-called "international volt" was recognized; its value was 0.999 16 that of the present absolute volt. England changed the values of its official and legal volt four times between 1905 and 1911.

Volumes many times larger than this could be devoted solely to the proposals, debates, modifications and alterations that affected nearly all the principal electromagnetic units. These shifts were not the results of whim or vacillation. The forging of a commonly acceptable and practical pattern for measurement procedures in this new and rapidly growing area of technology was no easy matter. If duplication and confusion ran rampant for so long in such ancient and relatively easy operations as measuring lengths or areas or volumes, was it not inevitable that general agreement would be difficult of attainment in the great new worlds of electricity and magnetism?

Around charged bodies voltage differences show themselves. Enormous voltage differences are generated in every television tube in order to accelerate the electron beam from its source toward the picture screen. One of the principal needs in the quantitative analysis of such situations is to have a meaningful measure of electric field strength. This is provided by computing a gradient: the average change—gain or loss—in voltage per meter of distance.

The unit combination for this is simply the volt per meter (V/m). For example, if, within a vacuum tube, electrons emerge from an "electron gun" at one end and just 0.25 m distant strike a metal target maintained at a 15,000-volt potential above that of the electron gun, then they have traversed an electric field with a strength of 60,000 volt per meter. That electric field strength is, accordingly, greater than that in an evacuated chamber in which, across a distance of 10 meters, we find a voltage difference of 100,000 V—for in this latter case the electric field gradient is 10,000 V/m. Electric field strength is thus much like a gradient or slope. Its steepness is what is being measured.

Ohm's law in action—Perhaps the best-known relationship studied in elementary electricity is that called *Ohm's law*. It states that the magnitude of a current flowing in response to a given potential difference will be inversely proportional to the resistance in the circuit through which it flows.

In terms of units this can be stated as follows: volt divided by ohm (of resistance) equals ampere. Or it can be stated this way:

volt divided by ampere equals ohm (of resistance). Finally, it can also be stated as volt equals ampere times ohm of resistance. These forms are equally valid, and the choice depends solely on convenience.

We have thus led the way to the definition of the ohm, universally recognized unit of electrical resistance. In brief, the ohm, whose abbreviation is the Greek letter *omega* (Ω), can be represented also as V/A, the volt per ampere. The actual language of the SI definition for this resistance unit is:

> The ohm, unit of electric resistance, is the electric resistance between two points of a conductor when a constant difference of 1 volt, applied between these two points, produces in this conductor a current of 1 ampere . . .

For safety's sake one final proviso is added:

> . . . this conductor not being the source of any electromotive force.

In other words, no additional voltage should be supplied by any part of the circuit between the two points mentioned.

The present absolute ohm, or volt per ampere, of the SI is the culmination and, almost certainly, the termination of a varied assortment of resistance units that have emerged during the past century, more or less. The most enormous in magnitude was the so-called *statohm,* the resistance unit of the electrostatic (cgs-esu) system of units. Each statohm was equivalent to $8.987\ 76 \times 10^{11}$ of the present SI absolute ohm.

Just prior to the establishment of the value of the absolute ohm, the prevalent worldwide unit was the "international ohm," whose value was 1.000 495 times that of the absolute ohm, which replaced it after 1948. The international ohm was defined as the resistance of a uniform column of mercury containing just 14.452 1 gram of that liquid in a column 1.063 meter long, and maintained at $0°$ C. Such a column was commonly cylindrical in shape, but did not have to be, so long as its cross section was uniform throughout the specified length.

In France prior to 1911 a different "international" ohm was used, with a resistance equal to that of 1.003 395 present absolute ohm. In England in 1884, on the other hand, the legal ohm had a resistance equal to 0.997 67 of the present SI absolute ohm. Still another resistance unit, called the siemens, had considerable use also. Its actual resistance was about 0.941 2 of the present SI absolute ohm. The best evidence indicates that it was named after the important electrical engineer and theorist William Siemens (1823–83). He had been born Wilhelm von Siemens in Germany, but most of his working life was spent in Britain. He is, incidentally, to be distinguished from two other German engineer-inventors—his brothers Ernst Werner von Siemens (1816–92) and Friedrich von Siemens (1826–1904).

Some confusion threatens the siemens resistance unit, for the name siemens was also proposed for a unit of energy, equal to 1 watt-hour, which would be the same as 3,600 joule. Later the siemens name was proposed for the unit of conductance, which is the inverse or reciprocal of resistance. This siemens conductance unit is now preferred to a unit called the *mho*, which is the word *ohm* spelled backwards—one of measurement's prize novelties.

Authorities have disputed which Siemens the conductance unit was meant to honor. The author of the present book leans to the view that it was William Siemens, but the other side holds that it was his older brother, the Ernst Werner von Siemens who developed self-acting dynamos and made important contributions to electrical telegraphy.

The Siemens saga, taking the family as a whole, is one of the richest in electrical technology. William Siemens invented the alternating-current dynamometer for measuring power in alternating-current circuits. In 1879 he was the first to apply the electric arc to fusion, and in connection with this he developed a more satisfactory process for cremation of human remains—a process not used, incidentally, by the cremation furnaces so active in the extermination centers operated by Hitler's Germany. William Siemens was also a pioneer in application of electricity to rapid transit. He was responsible for one of the earliest electric tramways, known to the Yanks across the sea as a *streetcar* system.

The present importance of the siemens of resistance, like its inverse, the siemens of conductance, is purely historical. Even the ohm, so well established, had to win out against various possible competitors.

The tiniest resistance unit in the entire zoo is the *abohm,* a part of the electromagnetic (cgs-emu) system. The abohm equaled just 1×10^{-9} absolute ohm. In other words, 1 billion (American style) abohm made the equivalent of 1 present absolute ohm in the SI.

Resistance and resistivity—Conductance has already been defined as the inverse or reciprocal of electrical resistance. Thus, a circuit with total resistance of 5 ohm has a conductance of 0.2 mho; while one with 100 ohm resistance has a conductance of 0.01 mho; and so on. The common symbol for the physical variable of electrical resistance is R. That for electrical conductance is G. Their mutual relationship can be stated either as R equals $1/G$, or as G equals $1/R$.

The three convenient variations of the famous Ohm's law can be formed using the siemens of conductance rather than the ohm of resistance: (1) amperes divided by volts equals siemens; (2) amperes divided by siemens equals volts; and (3) volts times siemens equals amperes.

Resistance of a wire or rod of known dimensions and composition may be easily calculated in advance with the help of a measure called the electrical *resistivity* of the substance.

Some metals, such as silver and copper, have very low resistivity, which means that they are correspondingly high in electrical conductivity, the reciprocal of resistivity. Other metals, such as iron, show considerably more resistivity. The highest resistivity of all belongs to the substances commonly called nonconductors; and if their resistivity is high enough they can function as electrical insulators.

In the SI, the unit of resistivity is the ohm-meter squared per meter ($\Omega m^2/m$). This means, in the numerator, a cross section of 1 m² and in the denominator a length in the direction of the current of 1 meter. Unit resistivity would be shown by a substance that, with 1 m² cross section and 1 meter of length, developed a total resistance of just 1 ohm. Tables of resistivities in SI units refer to the unit combination as the "ohm-meter" (Ω m), because the m^2 in the numerator cancels the m in the denominator, leaving only m in the numerator.

Actual measurements of resistivities of materials are, of course, not made with monstrous cubes, 1 meter on an edge. More typical is the use of a wire of some known cross section, such as 1 cm^2, and some known length, such as 10 m. The resistance then found is converted by means of simple arithmetic into the resistance for such a 1 m^3 cube of the same substance, assuming current were flowing from one face to the face opposite through the intervening 1 m of distance.

Typical resistivities for metals, expressed in SI units, may be around 10^{-7} ohm-meter. Since the reciprocal of resistivity is conductivity, a substance with a resistivity of 10^{-7} ohm-meter has a conductivity of 10^7 mho per meter (mho/m).

Let us suppose that we have decided to form an electrical resistance out of an alloy whose resistivity has been measured and found to be 5×10^{-7} ohm-meter. We intend to use wire with a cross section of 0.1 square centimeter (equal to 10^{-5} square meter), and our problem is what length of that wire needs to be used in order to make a resistance totaling just 0.3 ohm.

The recipe for a total resistance, R, is the resistivity times the length, divided by the cross-sectional area, all expressed in consistent units (here, as before, we shall use SI units). The necessary length L must equal the cross-sectional area of the wire (A) times the desired total resistance, divided by the resistivity of its material.

In this case we have $0.3 \times 10^{-5}/5 \times 10^{-7}$ which tells us that the required length of this particular wire is 6 meter. If a wire were used with just half that resistivity, the needed length would be doubled—to 12 meter.

Computations using either measured resistivity or measured conductivity (reciprocal of resistivity) can be made, with the help of the physical dimensions of the materials to be used, to arrive at the resulting resistances or conductances respectively.

Tables of conductivities commonly give measurements for solids or solutions in such unit combinations as the mho, or "reciprocal ohm," per cm or per m. If measurements are stated in units of the mho per cm, then the corresponding reading in mho per meter must be just $\frac{1}{100}$ of that magnitude, for conductivity measures the "willingness" of a particular configuration of matter to allow electron flow (current), and a hundredfold increase of distance (from 1 cm to 1 m) must mean that such willingness is diminished to one one-hundredth of its former amount.

Conductivity and resistivity of all substances, solid or liquid, must be stated with specification of the temperature at which the measurements are made. Conductivities of metals typically diminish with rising temperature. Conductivities of various standard solutions of acids in water, such as the sulfuric acid used in most automobile storage batteries, tend to increase as their temperatures rise. For example, a typical sulfuric-acid solution at 30° C shows about two thirds more conductivity than at 0° C.

The confusion that at one time beset comparisons of conductivity and resistivity, of conductance and resistance, can be substantially reduced, thanks to the greater clarity and uniformity of the International System units.

Some inconsistencies of nomenclature may still be encountered, however. Thus a recent translated summary of the very complete system of the standards of physical variables approved for use in scientific work in the U.S.S.R. names a unit called the siemens (S) as the unit of electrical *conductivity* and defines it as the electrical conductivity of a conductor whose resistance is just 1 ohm. Strictly speaking, such a siemens would be a unit of *conductance,* not of *conductivity.*

Another unit combination used in the U.S.S.R., the siemens per meter (S/m), is indeed a measure of electrical *conductivity,* for it is defined as the conductivity exhibited by a cylindrical conductor with a cross-sectional area of just 1 m² and a length of 1 m (in the direction of the current), the electrical conductance (not conductivity!) of which is just 1 S.

As with the thermal variables in an earlier chapter, the suffix *-ivity* in words such as conductivity and resistivity should signal us to look for the relevant dimensions of the material through which the effect (heat or electricity) is to flow. These relevant dimensions are the length, in the direction of the flow, and the cross section, transverse to that flow.

The greater the cross section, the greater the resulting conductance and the less the resulting resistance. The greater the length, the less the resulting conductance and the greater the resistance. Coupled with these important relationships of physical size are the particular characteristics of the substance shaped into the conductor of electricity or heat, as the case may be. Some substances are very reluctant to convey electricity or heat; they have high resistivity, low conductivity. Others are remarkably willing to convey electricity or heat (as the case may be). Such substances show low resistivity, high conductivity.

Specific resistivities of metals used to conduct electricity in practice are usually presented in terms of the metric unit combination symbolized by $\Omega mm^2/m$, which stands for the resistance of a wire just 1 m long with a cross section of just 1 mm^2. The specific conductance is correspondingly presented in terms of the reciprocal, which can be symbolized by $S \cdot m/mm^2$, where S stands for the siemens unit or the mho (inverse ohm).

Here are some sample measurements, all made at 20° C (293 K):

SUBSTANCE	RESISTIVITY ($\Omega \cdot mm^2/m$)	CONDUCTIVITY ($S \cdot m/mm^2$)
silver	0.016	63
copper (pure)	0.0172	58.1
copper (ordinary)	0.0178	56
aluminum	0.0286	35
zinc	0.059	16
nickel	0.070	14
iron	0.098	10
platinum	0.105	9.5
quicksilver (mercury)	0.96	1.04
ALLOYS		
nickelin	0.43	2.3
konstantin	0.50	2.0
chrome nickel	1.1	0.91

Nickelin is an alloy of about 54 parts copper, 26 nickel and 20 zinc. Konstantin is one of the names for an alloy with about 54 parts copper, 45 nickel and 1 manganese. It has a very desirable characteristic: its resistivity changes very little as its temperature changes. Resistivities of most metals rise troublesomely as they become hotter.

Iron, the constant companion of civilization, reveals an exceptionally great resistance response to changes in temperature. Thus, at about 25° C its resistivity gains about 0.5 percent with each rise of 1° C; by 850° C this rate of increase has grown to about 1.8 percent; then at still higher temperature it declines, so that at 1,000° C it is once again about 0.5 percent per 1° C. Temperature changes thus make drastic or even dangerous alterations in the resistivity of iron that is used to carry currents.

Alloys such as konstantin, however, show changes in resistivity as small as 4 or 5 parts per million for each added 1° C at the temperature levels prevailing in most laboratories or electrical devices.

Extreme nonconductors = insulators—When we turn to the extreme nonconductors commonly called "insulators," we find that a slightly different metric unit combination typically is used to express their very high resistivities. It is called the ohm-centimeter and is abbreviated as $\Omega \cdot cm$. Actually it represents the resistance of a 1 cm cube of the substance, in which whatever current passes must flow a distance of 1 cm through a cross section of 1 cm^2.

In terms of this unit, the specific resistance of quartz has been found to be 10^{19} $\Omega \cdot cm$; that of amber, the same; of paraffin, 10^{16} $\Omega \cdot cm$; of hard rubbers, between 10^{12} and 10^{18}; of mica, 10^{14} to 10^{17}; all in the same unit combination.

Completely pure water is found to have a resistivity of about 2.3×10^7 $\Omega \cdot cm$, which is markedly decreased by even the slightest mineral content or contamination. Thus, even well-distilled water may show as "little" as 5×10^5 $\Omega \cdot cm$.

The enormous range of comparative conductivities and resistivities can be illustrated dramatically. A "wire" of quartz, just 1 m long and with a uniform cross section of 1 mm^2, has resistance about 6×10^{21} times that of a pure silver wire of the same dimensions. In short, silver's conductance is about 1.6×10^{22} times that of quartz, one of the better insulators.

A unit to measure some significant physical variable serves its purpose best when two great requirements are fulfilled. First, it should be consistent with, or easily convertible into, units used to measure other, but related, physical variables. This kind of harmony is usually called the coherence of the units in a system.

Second, and perhaps even more important for the everyday rough-and-tumble of measurements, a unit serves best when it is fairly well suited in size to the kinds of effects and interactions that it will be employed with. For example, for daily use on Earth around us, a length unit equal to the distance called 1 light-year ($9.463\ 7 \times 10^{15}$ m) would hardly prove appropriate. Nor would a unit of time equal to 1 millennium (about $3.155\ 69 \times 10^{10}$ seconds).

Rated according to their size suitabilities, these first electrical units show up rather well.

The volt unit is admirably suited in size to its practical applications. The great majority of actual measurements are made in the range between 1 millivolt (1 mV or 10^{-3}V) and 1 megavolt (1 MV or 10^6 V). This is a ratio range of 10^9 times, for 1MV = 10^9 mV. It has been pointed out by a keen metrological commentator, M. Danloux-Dumesnils, that the volt itself lies in the lower third of this ratio range, "which is the most convenient position."

The size of the ohm also proves to be sufficiently convenient. Not many measurements are made to magnitudes much less than 1 ohm, but a very great many are made between there and as much as 10,000 megohm (10^4 MΩ). Resistors calibrated in megohms are produced commercially and constitute essential components of many common electronic devices.

The physical variable of electric resistance was one of the first in the entire electromagnetic family for which practical units were sought and established. As early as 1861, when the Committee on Electrical Standards (appointed by the British Association) began its historic work, the search for a suitable unit of resistance was the first project to occupy its attention. This resulted in part from the fact that physical standards of resistance—resistors, in short—are considerably easier to design, make and maintain than are standards of electric current or of voltage, for that matter.

The concept of electric resistance prevalent in the mid-1800s was a peculiar one. It was identified with velocity. This came about because the important early system of electrical units proposed by Wilhelm Weber (1804–91) was so structured that resistance had the same "dimensional" recipe as velocity—a distance divided by a time (as in m/s).

In fact, in the middle of the nineteenth century, William Siemens had already introduced and popularized to some extent in Britain a resistance unit that was dimensionally equivalent to 10^7 m/s or 10^9 cm/s, as it was then stated. The Committee on Electrical Standards accordingly chose that 10^9 cm/s resistance for its recommended unit, calling it the *ohmad*. That Arabic-flavored title did not long endure, however. The *ad* was subtracted, and it became simply the Teutonic monosyllable, ohm—a sound that seems to cause some sensitive ears to react distressfully indeed. (Yet there are specialists in esoteric psychophilosophical exercises who find in the similar sound *oo-oo-oom* an ineffable and unfailing consolation.)

The decision to symbolize the ohm by the Greek omega (Ω) persists as an unfortunate convention of modern metrology. It does prevent the ohm from being mistaken for a fat zero, however. On the other hand, one wonders what would be the effect if some most unusual measurement problem called for results in terms of microohms, whose approved symbol combination would be all Greek: $\mu\Omega$!

The unmistakably convenient sizes of the volt and the ohm help to support satisfaction with the ampere unit's size also, for a potential of 1 V will always push through a resistance of 1 Ω a current of just 1 A.

What, however, of the size of the ampere-second, more properly known as the coulomb, unit of electrical quantity or charge? This C, which has been part of the electrical-unit family since 1881, has a rather schizophrenic relationship with the uses to which it is actually applied. It is excessively large for almost all the needs that arise in the course of measuring *electrostatic* interactions. Yet it is by no means large—if anything, it is a bit small—when applied to measuring interactions arising from electric *currents*.

For example, in the enormous field of electrochemistry, the electrolytic depositing or precipitation of metals out of solutions plays a great role. It is known that to deposit a 1 mole mass of some substance with single valency requires a total quantity of electricity of very nearly 9.65×10^4 coulomb. Thus, if the current had a value of 100 ampere, it would have to flow for 965 second to deposit that 1 mole mass.

In industrial measurements a larger unit of quantity of electricity is often used in preference to the coulomb. It is the ampere-hour, equal to 3,600 C. The depositing of that 1 mole of a single-valence metal would thus require 26.8 A-hr, equal to the 96,500 C.

The ampere-hour is a construction of the somewhat regressive or topside-to character as the all-too-common energy unit called the kilowatt-hour (kw hr), equal to just 3.6×10^6 (3.6 million) joule.

Count Volta and Herr Ohm—The masters of measurement whose respective lifeworks are commemorated by the volt and the ohm could hardly have been more different in their backgrounds.

Count Alessandro Volta (1745–1827) was an aristocrat, a man of culture and distinction, and an eminent savant of northern Italy. He was not only the founding father of what is now called electrochemistry, but an innovator in several other directions also. His great and epochal development was the first electrochemical battery or "voltaic pile," a source of uniform and steady current, units of which could be connected in series, so as to build up the electrical pressure that we now commonly call "voltage."

Though a fine communicator, a writer of distinction and a man of the world, he left behind him no clear picture of the thought processes that led him to the first of his current-providing piles. Never before had men disposed of a reliable and controllable source of electric currents. The dynamo, or mechanical "electron pump," was not known or even suspected when Volta assembled his first simple batteries.

His revelation, just at the turn of the nineteenth century, probably did as much as any other one step to make that the era of rapid and far-ranging development of the knowledge and the measurability of electrical and magnetic interactions.

Georg Simon Ohm (1787–1854) was Bavarian-born, son of a locksmith. Much of his outstanding work, however, was accomplished during a decade when he taught high school at Cologne in the Rhineland. It was then that he made the measurements on which he established Ohm's law, as the great tripartite relationship of current, voltage and resistance is now known.

Authorities around him remained indifferent or even mildly contemptuous toward his work. He spent years without recognition or official support, and was fifty before he was granted a rather minor position in the Polytechnic Institute at Nuremberg, and well over sixty before he was placed in a university teaching post at Munich.

His is by no means the only story of stubborn and stupid neglect of major contributions to science. The resistance that Ohm's measurements showed the voltage must overcome in order to send current through a circuit was as nothing compared with the social and academic resistance that confronted his revelations.

37 THE FARAD OF CAPACITANCE, THE HENRY OF INDUCTANCE, AND SOME MORE

. . . the electric chain wherewith we are darkly bound.
—GEORGE GORDON BYRON,
Childe Harold (1817)

Place two conducting surfaces parallel to each other, but separated by a nonconducting substance or space. Then apply a voltage, or electromotive force, so that one conductor is "at" a certain number of volts more than the other. A typical and important process at once takes place. Electric charge flows into this simple device, and simply "sits" there, so long as the voltage difference is maintained and the nonconducting separator between the two charged conductors is not breached.

One of these conducting plates, which we may designate as N, is negatively charged with respect to the other. This means that N has a surplus of surface electrons confronting the other plate, P, which is positively charged with respect to N. In other words, the surface of P, facing N, has a deficiency of electrons. The resulting voltage difference between N and P may be maintained by an electric battery, of the sort first developed by Volta. Or it may be produced by an electrostatic machine or even by a direct-current dynamo.

This simple setup is commonly called a capacitor, but more colloquially known as a condenser. That name survives from the days when it was believed to condense electric charge on its confronting conducting surfaces—which is not a bad simile for what indeed it does. Large glass capacitors—called Leyden jars after the Dutch city where they were believed to have originated—were used in many a striking and spark-studded demonstration during

459

the seventeenth and eighteenth centuries. They are still used widely. Every radio and television set makes use of many capacitors, some fixed as to their capacitance, others variable by means of a turning knob or a sliding control.

The word *capacitor* suggests capacity. The electric variable created to summarize what capacitors do is called capacitance, and just as a bottle or flask has a capacity for gases or liquids, so a capacitor has a capacity for electric charge. This capacity is related not only to the dimensions of the capacitor but also to the electromotive force applied in order to pack the electric charge into it.

The worldwide unit for measuring capacitance is the *farad,* abbreviated as F. It is so called in honor of Michael Faraday, a magnificent scientist and exemplary human being, whose life and attainments are summarized later in these pages. Faraday's contributions to electromagnetic insights were so varied and basic that almost any one of the units for electromagnetic physical variables could have been named in his honor. The farad is not a poor choice, however, for it presents peculiarly close links between essential electrical interactions and familiar physical processes that were known before electricity.

The most obvious link is a figurative one: between the geometrical concept of room or capacity for storing substances, and the concept of electrical room or space in which it is possible to store electric charge, the physical variable measured by the unit of the coulomb (C). When the basic 1 ampere current flows for 1 second it delivers a total charge of just 1 coulomb.

Now, if 1 coulomb of charge is stored or placed in a capacitor by an electromotive force of 1 volt, then that capacitor has a capacitance of exactly 1 farad. But, if 1 volt stores in a capacitor a total of 0.01 coulomb of charge, that capacitor has 0.01 farad of capacitance; and so on.

With any fixed capacitor, the stored charge is directly proportional to the applied voltage. The constant of proportionality that indicates how many coulomb of charge will go into the capacitor for each volt of applied electromotive force is another name for the capacitance of that particular capacitor.

The simple recipe for the relationship between these three essential electric variables is this: quantity of electricity equals electromotive force times capacitance. Or, in terms of the units, coulomb equal volt times farad. Conversely, farad equal coulomb divided by volt.

Thus, if a particular capacitor stores 0.000 02 coulomb per volt of applied electromotive force, then its capacitance is just 0.02 millifarad (mF), or 20 microfarad (μF). Now, if we apply to its two terminals a voltage difference of 200 V, it will accept and store 4×10^{-3} F, or 0.004 F, also known as 4 mF.

Capacitors can be compared to cylindrical pots whose cross-sectional areas we know. Pot P has a cross section of just 1 m². Hence, when we pour into it 1 m³ of liquid we know that liquid will stand just 1 m high. Pot Q, on the other hand, has a cross section of just 0.2 m². Hence, when we pour 1 m³ of liquid into it, that liquid will stand 5 m high. The cross-sectional area of the pot is analogous to the capacitance of a capacitor, the volume of liquid to the quantity of electricity (charge) stored in the capacitor, and the height of the liquid in the pot to the voltage across the plates of the capacitor.

A pot, beaker or test tube may have so little cross-sectional area that 1 m³ of liquid rises high enough in it to shatter it by the resulting pressure. Just so, the insulation in an overloaded capacitor may break down, permitting charge to leak from one plate to the other. Sparking and arcing are the spectacular evidences of such condenser overloads.

Every electrical thunderstorm on Earth reveals such a breakdown in the insulation between two strange charged surfaces— that of the Earth beneath, and that of a cloud above. Clouds accumulate much charge, either from contact with layers of air that are electrically unlike them or from the falling of charged raindrops. Enormous voltages are generated until, with a blinding flash followed by reverberating roars of thunder, the great release of lightning carries charge from one side to the other. The charge actually oscillates rapidly back and forth between cloud and earth, much as a harpstring or guitar cord that has been plucked vibrates until its energy is dissipated.

Billions of capacitors of various sizes are fabricated and put into service each year. A capacitor is an essential element in the ignition system of every automobile, except for some particularly sophisticated "solid-state" systems of very recent design.

The farad (F) can be identified either as the coulomb per volt (C/V) or as the ampere-second per volt (As/V). The non-conducting substance or arrangement between the conducting plates of capacitors varies a good deal. Ordinary dry air will do. Other dielectrics used in capacitors include mica, hard rubber, various plastics and, of course, glass.

Calculating capacitance—The capacitance of a capacitor depends on (1) the total area of its opposing or confronting plates; (2) the separation between them; and (3) the dielectric effect or efficacy of the substance, or the absence of substance, that occupies this separation.

The recipe is rather simple. Suppose a capacitor has just two plates, of identical area A (measured in m^2), and that these are separated by the distance D (measured in m). The substance occupying that separation is characterized by the dielectric constant ε, the small Greek epsilon. This constant is measured in terms of the unit combination farad per meter (F/m). Then the capacitance of the capacitor, measured in farad units, must be:

$$\text{Capacitance} = \varepsilon\, A/D$$

The very lowest ε is that of a total vacuum. In the International System (SI) it has its own symbol ε_0, and a very important value: ε_0 equals 8.854×10^{-12} farad per meter (F/m) or coulomb per volt-meter (C/V·m). The dielectric constants of substances are commonly given as multiples of this ε_0. Thus, the dielectric constant of dry air at normal or standard atmospheric pressure is about 1.000 594 times ε_0. This is only about 6 parts in 10,000 greater than that for empty space or total vacuum.

For amber or hard rubber, on the other hand, the dielectric constants are about 2.8 times ε_0. Various kinds of quartz have dielectric constants ranging between 3.8 and 5.0 times ε_0. For mica the dielectric constant rises to between 7.1 and 7.7 times ε_0.

With information of this kind it is possible to calculate in advance of building a capacitor the capacitance it should have. For example, suppose the capacitor is to have two parallel plates with 25 cm^2 area and a separation of 1 mm, that separation to be filled with hard rubber. Its capacitance must then be 0.002 5 m^2/0.001 m times 2.8 times 8.854×10^{-2} F/m. The product of this is

6.2×10^{-10} F, or 0.62 nF. (The nanofarad, or nF, equals 10^{-9} farad.)

If the hard-rubber separator were removed and a vacuum were substituted between the plates, the capacitance would be reduced to 0.22 nF. On the other hand, if a good grade of mica, with 7.5 times ε_0 as its dielectric constant, were substituted for the hard-rubber separator, the capacitance might be increased to 1.66 nF. If only air were left between the plates, and their separation were increased from 1 mm to 2 cm, an increase of 20 times, the capacitance would be reduced to 0.011 nF, or 11 pF (picofarad).

Compared with the capacitances actually measured in laboratories and workshops, the farad (F) is a very large unit. One seldom encounters cases in which the result of measurement is as large as one or more full farad. Typical measurements are expressed in the far more common submultiples: the mF (millifarad), the μF (microfarad), as well as the nF (nanofarad) and even the pF (picofarad).

Capacitance, charge and electrical energy—The physical variable called capacitance is related in a significant way to both the variable of energy, measured by the unit of the joule (J), and the variable of charge or quantity of electricity, measured by the unit of the coulomb (C). In the International System (SI) the farad, unit of capacitance, is identical with the coulomb squared per joule:

$$1 \text{ F} = 1 \text{ C}^2/\text{J}$$

This fraction can be looked upon as the ratio of the square of the charge pushed into the capacitor, to the energy that has been expended in that operation.

This equality makes clear that when 1 coulomb of electric charge is stored in a 1 farad capacitor, an energy of 1 joule is expended in that electric operation. Likewise, when that 1 coulomb charge is allowed to escape from such a capacitor, that departure releases 1 joule of energy.

In the same way, if 1 joule of energy has been used in compressing a spring and that spring is released and allowed to go back to its unstressed state, it can do 1 joule of work on the way.

We disregard here the slight molecular and atomic friction that (within the spring or within the capacitor) converts into heat a small amount of the energy.

What happens when we apply to a 1 farad capacitor sufficient voltage to store in it 2 coulomb of charge (rather than just 1)? The energy involved is now 4 joule (not 2). And if we increase the voltage enough to store 3 coulomb of charge in that 1 farad capacitor, the energy involved is 9 joule (not 3).

Since the energy that is needed to move charge into a capacitor increases as the square of the amount of that charge, capacitors can store large amounts of energy electrically. For example, if a capacitor of 1 farad capacitance, in response to an applied electromotive force of 100 volt, stores a charge of 100 coulomb, the energy required is 100,000 joule. And when this charged capacitor is discharged, it releases an energy of 100,000 joule—less what has been scattered or dissipated as heat during the charging and the discharging processes.

We noted earlier the old rule of thumb that "amp times volt equal watt." Parallel to it is the rule that does for the joule of energy what that first rule did for the watt of power; and the new rule is "coulomb times volt equal joule."

In the case of a 1 farad capacitor, the number of coulomb of charge stored is equal to the number of volt of electromotive force that did the charging. Hence, the energy of such a charging operation must be proportional to the square of the applied voltage, for if the number of coulomb stored increases in direct proportion to the voltage applied, then the product of that number of coulomb times the voltage must be proportional to the square of that voltage.

The process of charging a large capacitor or a series of such capacitors may be a slow one, but the process of discharging one or more is a rapid one, for it is driven by the electromotive force or voltage that exists across the terminals of the capacitor. It is much like the sudden release of a compressed spring or the release of a tuning fork that has been bent and held under tension.

When such a tuning fork is let go, it vibrates for a while, gradually dying away. The force of the released spring causes it to overshoot the mark, first one way, then the other. Its vibrations diminish as the energy is converted into internal heat among the molecules of the elastic tuning-fork blades.

When a capacitor is discharged, by a contact from its negative to its positive plates, that discharge is oscillatory also. The negative plate has stored a surplus of electrons. When a path suddenly opens up for these, they rush in myriads across to the positive plate. But more rush over than are needed to establish electrical balance or neutrality. For an instant the plate that had been positive becomes negative with respect to the other. Then the surplus electrons rush back again, though not quite in the same number or with the same total energy as before. Dozens, scores, hundreds of these diminishing oscillations may be detected in the discharge of a capacitor. The analogy to elastic bodies that have been stressed and released is very close indeed.

The compressed-spring analogy helps further to understand the fact that the stored energy increases as the square of the charge that is stored in a capacitor by an applied voltage.

We have seen that when a uniform force is applied in a particular direction, the extent of the motion (distance moved) in that direction times the force gives the amount of work done. Hence, we say that 1 newton of force times 1 meter of distance moved equals 1 joule of work done on the way. However, when a force is exerted to compress a spring, each tiny additional distance moved against that spring's elasticity requires an *increase* in the applied force. The farther the spring is compressed the more it pushes back against the compressing force. The result is that, with a truly elastic spring working well within the range of its elasticity, the amount of work required to compress it increases as the *square* of the distance moved. If 1 joule of energy is required to compress a particular spring by a distance of 1 cm, then 9 joule will be required to compress it by a distance of 3 cm; 16 joule for 4 cm compression; and so on.

The applied mechanical force is measured in newton units; the applied electromotive force is measured in volts. The resulting "give" or change in dimension of the spring is measured in meter units; the "give" or compliance of the capacitor is measured in farad units. Just as the energy required to compress a spring is proportional to the square of the distance by which it is compressed, so is the energy required to charge a capacitor proportional to the square of the charge stored in it.

Since electromotive force (in volt) is the analogue of mechanical force (in newton) and charge (in C) is the analogue of distance moved (in m), what then is the mechanical analogue of capacitance (in F)? It is the physical variable called *elasticity*, measured in units of the force required per unit length moved against that force and in its direction. The SI unit combination for expressing the elasticity of a spring is newton per meter (N/m).

There is an inverse or reciprocal of elasticity, called *compliance*. The less stiff a spring is, the more compliant it is, and vice versa. Mechanical compliance is measured by the SI unit combination of the meter per newton (m/N).

Similarly, there is an inverse or reciprocal of capacitance, called *elastance*. The unit for measuring elastance is 1/farad, sometimes called the *daraf* ("farad" spelled backward), just as the mho is the ohm spelled backward. A capacitor with very low capacitance has high elastance. In fact, a capacitor of 1 mF capacitance necessarily has an elastance of 1,000 daraf; while a capacitor of 1 picofarad (pF) capacitance has 1 billion daraf of elastance.

The concept of stiffness (as in a spring) helps us comprehend more clearly the meaning of capacitance. The stiffer the spring, the greater the energy required to compress it by a given distance. The larger the capacitance of a capacitor, the greater the energy required to charge it with a given electrical charge (measured in coulomb units).

Springlike capacitors—In still other significant ways, capacitors behave much as springs do. Suppose we have three identical springs—identical in size and stiffness (measured in units of N/m). Now we attach them in a series, like links in a chain. The new spring shows just one third of the stiffness of any one of the three separate springs. That is, a given force will stretch the triple sequence of springs three times as far as it would stretch any one of the three.

Next we take three identical capacitors. For simplicity we use 1 farad capacitance for each. Now we connect them in series, and we find that the chain of three exhibits a capacitance of just ⅓ farad.

Its electrical "stiffness" has been reduced to a third of the value of that for any one of the triplets.

Back to the three springs again. This time we connect them in parallel. To one bar we attach one end of each spring, and the other ends to another bar. Now we test the force required to pull the two bars apart. We find that the triple spring is now trebly stiff. The force which would have extended a single spring by one unit of length now suffices to increase the separation between the two bars by only one third of a unit of length.

Our three identical capacitors are now connected in parallel, and we find that they behave like a single capacitor with three times the capacitance of any one of the three elements. The electrical stiffness has been increased; and, of course, the elastance has been divided by three. The capacitance of the three in parallel is 3 farad; thus the elastance of the three is reduced to $\frac{1}{3}$ daraf.

These effects do not require the use of capacitors of identical capacitance. Suppose we have four capacitors, each with a different capacitance: C_1, C_2, C_3 and C_4. If we connect them in parallel, we get a new and larger capacitance, C_n, which is equal to the sum of C_1 plus C_2 plus C_3 plus C_4.

On the other hand, if we connect the same four capacitors in series, instead of in parallel, we find that the relationship becomes this series of fractions:

$$1/C_n = 1/C_1 \text{ plus } 1/C_2 \text{ plus } 1/C_3 \text{ plus } 1/C_4.$$

In these respects, of course, the capacitors behave quite like springs in mechanical analogies. In this second example (series connection) C_n is *less* than C_1 or C_2 or C_3 or C_4.

A simple laboratory demonstration can reveal one possibly unexpected characteristic of capacitors. We begin with two metal plates, placed just 1 cm apart, and find that when a 10 volt electromotive force is applied they store a charge of 10^{-8} coulomb, indicating a capacitance of 1 pF (10^{-9} farad).

Now, without allowing this rather small condenser to discharge, we slowly pull the plates apart until the distance between them is increased to 10 cm. This must mean that the capacitance has become just $\frac{1}{10}$ of what it was before, or 0.1 pF. However, the stored charge has remained the same, at 10^{-8} C. What, then, has happened to the voltage? If we have kept a voltmeter connected to the two plates, we will find that the voltage has increased tenfold, from 10 V to 100 V, as the plates were pulled apart.

This corresponds to the basic recipe for every capacitor: coulomb equal farad times volt (C = FV), or farad equal coulomb per volt (F = C/V).

Note that the energy stored by the capacitor has increased tenfold during this process of separation. At the beginning, with the plates just 1 cm apart, its stored electrical energy was 10^{-7} joule. Afterward, with the plates 10 cm apart, the stored electrical energy was 10^{-6} joule, or 1 microjoule. The stored energy has been increased 10 times.

Where has this additional energy come from, since we have neither added nor taken away charge from the two plates? It has come from the work done in pulling the plates apart. The plates of a charged capacitor attract each other, for each one has an opposite charge (as "seen by the other"). Unlike charges attract each other. It is more work to pull apart the plates of a charged condenser than to pull apart uncharged pieces of metal of the same size and shape.

Capacitors provide useful reservoirs for electrical energy that may be needed in sudden great bursts. Capacitor banks are charged and then used to supply huge surges of current in flashlamps, producing brilliant peaks of illumination, for photographic and other needs. Solid-state lasers are energized by such flashlamps, powered by capacitors.

It is typical that the charging period for a great capacitor bank is thousands or millions of times the duration of the actual discharge time.

Introducing electrical inductance—If an electric current flows through a coil with a large number of turns, that coil behaves like a magnet. Indeed, it is called an electromagnet, and if properly suspended it will even line itself up with the Earth's magnetic field, just as does a well-mounted compass needle of magnetized iron.

If the current through such a coil is changed swiftly—as when it is turned off and turned on—the coil produces a strange effect on the behavior of the flow of electricity. That effect is quite different from the effect of a capacitor in the circuit.

Because of the multiple turns of the coil the electric flow seems to have acquired a kind of inertia: if it is flowing, it seems to want to continue flowing, even after being turned off. When it is finally off, it seems not to want to get started again. In much the same way, a heavy object, though on a level surface and with freely turning wheels supporting it, seems reluctant to be put into motion and, once in motion, seems reluctant to come to a halt again. This is the effect of mass. And the coil has given the electric current that moves through it a kind of "massy" behavior.

This is an important effect, and it is referred to as *self-induction,* a physical variable measured by the SI unit called the *henry* (H). The definition of the henry is: ". . . the inductance of a closed circuit in which an electromotive force of 1 volt is produced when the electric current in the circuit varies uniformly at the rate of 1 ampere per second."

In terms of the electrical units already presented, the henry is described also as the volt-second per ampere (V·s/A), which simply restates the relationship given in the definition: the generation of 1 volt per change of 1 ampere per second in current.

A basic rule of electric currents and circuits is that the direction of the electromotive force provided by self-inductance is such as to *oppose* the change taking place in the current. For example, if the current has been *off* and is switched *on,* the voltage generated in the inductance coil is such as to retard or oppose the current that otherwise would begin to flow instantly and freely in that circuit.

On the other hand, when the current has been flowing uniformly, if it is cut off by a switch, the induction coil generates an electromotive force in a direction tending to *continue* the current flow in the direction it had been going before being switched off.

Above all, induction has a slowing effect, for the countervoltages that are generated are proportional to the rate at which the basic current flow is changing. When the switch is turned on, the change is, of course, an increasing current flow; when it is later turned off, the change is a decreasing current flow. Each of these changes—the increase and the decrease—is opposed and delayed by self-inductance in a circuit.

Measuring mutual inductance—Though the International System defines the henry in terms of self-induction wherein the inductance is part of a single circuit in which current is being increased or decreased, there are other important electrical interactions called *mutual inductance,* for which the henry also serves as the unit of measurement.

Mutual inductance takes place when the current in which the variations—increases or decreases—originate is electrically separate from the current which is "induced" or generated by those changes. The wires through which the primary current flows are not connected with the wires in which the induced current results.

The only connection between those wires is afforded by the winding of one circuit in a coil around a coil contained in the other circuit. Those coils are not in electrical contact, however. They are connected solely by the inductance effect, which we shall not explain in greater detail here.

The relationship of importance in mutual inductance can be symbolized. If we represent by A/s, the rate of change of the primary current, in ampere per second; and by e_s the electromotive force induced in the secondary (in volts); then the mutual inductance (M), in henry units, becomes

$$M = \frac{-e_s}{A/s}$$

The minus sign ahead of the numerator is not mysterious. It simply recognizes the fact that the direction of the voltage in the secondary is such that it tends to produce a current opposing the change taking place in the primary coil. When the current in the primary is increasing, the change A/s in the denominator is a positive one; when it is decreasing, that change is a negative one. The secondary electromotive force (voltage) in the numerator, being opposed in effect to the change in the primary, must be given the opposite sign.

Both self-inductance and mutual inductance fall within the generalized relationship of the henry to the units for voltage, time (the second), and current (the ampere): $1\ H = V \cdot s / A$.

Let us apply these principles to a concrete example of mutual inductance. Certain measurements or items of information are necessary. We need the total number of turns of wire in the primary coil, symbolized here by N_p, and the total length of the primary coil, symbolized by L_p. Also the total number of turns of wire in the secondary, symbolized by N_s. Finally, we need the cross-sectional area A of the primary coil.

For the sake of simplicity we make our primary coil rather long—0.6 meter—with 500 turns of wire evenly spaced along that length. The radius of that cylinder of wire is 3 cm. Hence its cross section A is just 0.002 83 m². The secondary, consisting of 1,500 turns, is wound over the middle section of the primary, directly on top of it, but electrically insulated from it. We wish to compute the mutual inductance, M, produced when a unit current, 1 ampere, is flowing through the primary.

One more measurement is needed. It has to do with the so-called *permeability* of whatever is inside the concentric coils. The symbol *mu* (μ) is the accepted one for permeability, which in the SI is measured by the unit combination henry per meter (H/m). By far the most important permeability is that of empty space, or total vacuum. This is symbolized by μ_o, called "*mu* sub zero," and the permeabilities of all substances are compared with it as a standard. The actual value of μ_o in SI units is 1.257×10^{-6} H/m.

The permeability of ordinary air at standard pressure is so very little different from μ_o that the latter is commonly used for approximate calculations when inductances are wound around air rather than around perfect vacuum. In our example we assume this to be the case.

Hence, the complete equation for the mutual inductance is given by a fraction whose numerator is the product of four measurements and whose denominator is just the one measurement of the length of the primary coil. This is the equation

$$M = \mu_o N_p N_s A / L_p$$

The resulting numerator is thus the product of 1.257×10^{-6} H/m times 500 (primary turns) times 1,500 (secondary turns) times 0.002 83 m², all of which equals 0.002 7 henry-meter (H·m). The denominator of that fraction is 0.6 m. Hence, the result is a mutual inductance of 0.004 5 henry, also expressible as 4.5 mH (millihenry).

The equation for the self-inductance of a single coil is quite similar to the one just given. Such self-inductance, often called the "magnetic flux linkage" of a coil, is symbolized by the letter L. If N is the number of turns in the coil, while its length, or distance from end to end, is represented by D, and its cross-sectional area is A, then the equation has three magnitudes in its numerator, divided by the one D in the denominator:

$$L = \mu_0 N^2 A / D$$

We have, as before, ordinary air or perhaps a vacuum inside this single coil. Let us assume that it has the same dimensions as the primary coil in the previous example: 500 turns spaced out over a distance of 0.6 m. Then the numerator is the product of 1.257×10^{-6} H/m times $1,500^2$ (square of number of turns) times $0.002\ 83$ m², which equals 0.008 henry-meter. Since the denominator is 0.6 meter, the result is L = 0.013 H, or 13 milli-henry.

In this simplified example we have neglected the effect of the ends of the solenoid coil. Since it is rather long relative to its radius and number of turns, those end effects would not be large, but for high precision work they could not be totally neglected.

Comparing permeabilities—The permeability of any substance is greater than that of empty space (μ_0). If we use μ_0 as our standard, then the μ for ordinary air is nearly 1.000 004, or only some 4 parts per million higher. For pure water the μ is about 81 times μ_0. This means that if we were to fill a solenoid coil with pure water and place charged bodies in the water, the magnetic forces on those charges produced by a current flowing through the coil would be about 81 times smaller than if those charges were in the same place but surrounded by vacuum instead of water.

We have seen that every coil through which a current flows steadily behaves like a magnet. The kind of magnetic field that forms within such a coil, and also around it, depends heavily on the *permeability* of the substance, if any, within the coil.

The really large permeabilities are those for various kinds of iron and nickel. For iron the permeability is typically about 10,000 times μ_0. Thus, if a solenoid coil is filled with soft iron, its self-inductance, as measured in henry units, is enormously increased.

The mutual inductance of separate coils wound one around the other is also increased enormously when the space within them is filled by soft iron or alloys of even higher permeability.

The apparatus that is thus arranged is a *transformer*. Our entire system for the distribution of alternating-current power from its source in generating stations to its delivery to our homes and factories is dependent on transformers, whose mutual inductance is raised to high values (measured in the henry unit) by extremely permeable "fillings" for their coils.

Self-inductance and mutual inductance, so closely related, are electrical effects, depending on variations in current flowing through coils. Yet they are also, as we begin to see, magnetic effects. The henry unit, also known as the volt-second per ampere (V·s/A), thus provides a sort of bridge from the electrical to the magnetic variables in the great and complex world of electromagnetic measurements.

The farad is named for Michael Faraday (1791–1867), an experimenter of extraordinary achievements, thanks to persistence, patience and an inspired imagination. Faraday was a self-educated man of much sweetness, modesty and religiosity. In many ways he was one of the saintliest of the great scientists. His contributions to electromagnetics are almost unbelievably broad and basic. Only a few highlights can be singled out here.

Faraday was a poor young bookbinder's apprentice when his scientific enthusiasm and devotion attracted the attention of Humphrey Davy, the brilliant but somewhat erratic chemist chosen as the first head of the Royal Institution in London. Faraday became Davy's rather underpaid and underprivileged laboratory and personal aide. When Davy was obliged to retire somewhat prematurely, Faraday replaced him.

While carrying forward valuable researches in chemistry, Faraday became engrossed with electrical problems and tackled several prime areas head on. First he managed to construct the ancestor of all electrical motors, in which current was transformed steadily into mechanical motion and power. It was only a current-carrying wire that revolved continually around a permanent magnet pole, but its progeny fills the world today.

A decade of unremitting labor went by while Faraday vainly sought ways in which to convert magnetism into electric current.

Then he discovered electromagnetic induction and built the first mechanical generator of electric current. This was, in truth, a dynamo, the inverse of that first primitive electric motor.

Though not an accomplished mathematician, Faraday had a concrete and brilliantly pictorial imagination for physical effects. He first pictured the action-at-a-distance of electric currents and magnets as being caused by invisible "lines of force," the totality of which constituted what he called a "field" in space. Later, his field concept was completed, codified and mathematicized by James Clerk Maxwell, but it was Faraday's great idea to begin with.

Faraday also discovered the laws of electrolysis, thus taking his place with Volta and Davy as a founding father of electrochemistry.

His personal modesty and preference for experimentation over prestige and glory led him to decline the sought-after presidency of the Royal Society. He likewise refused knighthood.

Tributes to the greatness of Faraday have been many. One of the most discerning was that of the British science writer J. G. Crowther, who called him the greatest physicist of the nineteenth century, as well as the greatest of *all* experimenters, well worthy to be ranked with such supreme scientists as Archimedes, Galileo, Newton, and very few others.

Albert Einstein, whose capacity for appreciating the contributions of others was almost as marked as his own theoretical insight, once stated that in the history of science two pairs or couples of equal magnitude had appeared: Galileo and Newton; and then Faraday and J. C. Maxwell.

Crowther commented that even the "wonderful" Maxwell could not be considered as great in science as Newton. Hence, if Einstein's assessment of the pairs on a par was valid, Faraday "must be accounted" as greater even than Galileo. No unit for measuring or comparing scientific greatness has been created, not even one called the Nobel (prize). Ratings of relative greatness, however intriguing, are probably misleading and fallacious at best. Yet how else can one do justice to the extraordinary qualities and attainments of this most modest genius, Michael Faraday?

The American Joseph Henry (1797–1878) in many ways paralleled the interests and attainments of Faraday in electromagnetic effects. His academic career began as a teacher at Albany Academy in New York State, and it was there that he made his first researches into magnetism and electrical induction, whose unit, the henry, was named in his honor.

He was in his middle thirties when he was appointed professor of "natural philosophy" at Princeton University, New Jersey. The range of his teaching there appears almost incredible in our era of narrow specialization. His classes included physics, chemistry, geology, mineralogy, astronomy, mathematics, and even architecture!

Working largely alone, Henry in important respects foreshadowed the future. He demonstrated the principles of electric telegraphy, constructed a crude but operable electric motor and proved for the first time that the discharge of a capacitor (such as a Leyden jar) is oscillatory, not just a single rush of current from the negative to the positive plate. When later on Heinrich Hertz demonstrated in his German laboratory the existence of the electromagnetic oscillations we call radio waves, it was such oscillatory discharges that were employed in his first crude "transmitters."

By 1846, Henry was led—chiefly by a sense of obligation, it appears—to become the first director of the Smithsonian Institution in Washington, D.C. As its head he contributed vastly to the expansion of scientific work in the United States; but, as a result of his involvement in endless administrative labors, his own important work in physics was virtually terminated.

Among achievements attributable largely to his influence from the Smithsonian were the establishment of the National Museum and the launching of the United States Weather Bureau. Henry's services to the Union during the Civil War were especially important.

During his long life as discoverer and administrator he repeatedly refused personal profit from his own innovations. Thus, the electric telegraph, which he well might have claimed as his invention, was credited in large part to a mediocre painter and anti-Semite named Samuel F. B. Morse.

The use of the designation "henry" for the unit of induction is well justified. Joseph Henry doubtless produced and recognized self-induction even before Michael Faraday in London, though the latter, a careful and accomplished communicator of scientific results, published his findings first.

It is a relief to be able to report that no quarrel about priorities ever tainted the relationships of these two pioneers, Henry and Faraday. Indeed, when Faraday finally learned about Henry's early experiments, he was delighted, and exclaimed, "Hurrah for the Yankee scientist!"

Henry's "firsts" included even the prototype of that indispensable device based on mutual inductance—the electric transformer. All the multimillions of transformers of various sizes in which alternating currents are either stepped up or stepped down in voltage are thus the progeny of this part of his pioneering.

Beyond doubt his happiest and most rewarding years were those when he experimented rather than administered in science. He was a devoted and successful metrologist, devising measuring instruments and using them with fine results. Thus by means of his own thermoelectric galvanometers he was able to show that the peculiar blemishes on the face of the Sun, called "sunspots" today, radiate less heat than do the surrounding portions of the solar countenance.

Henry investigated by careful measurement important aspects of what is now called the solid state, and also the liquid or fluid state, of matter. He demonstrated that, in general, the extent of cohesion in each of these states was similar, even though "common sense" might seem to require that the solid state show far more cohesion than does the fluid state.

Henry's integrity, devotion, and unflagging fidelity to what he regarded as his duty make him one of the truly great personalities of science in American history.

38 MAGNETIC MATES: THE WEBER OF MAGNETIC FLUX, THE TESLA OF MAGNETIC FLUX DENSITY, AND FURTHER ITEMS

I guess I am mainly sensitive to the wonderfulness & perhaps spirituality of things in their physical & concrete expressions.
—WALT WHITMAN,
letter written in 1888

. . . *[physical] laws are only as good as the accuracy of our measurements.*
—MILTON A. ROTHMAN,
The Laws of Physics (1963)

Calculations of inductance often deal with a unit combination called simply the ampere per meter (A/m). It actually means the ampere-turn per meter, but the turn, or loop, or 360-degree swing-around, of the conducting wire is usually not now accorded the status of a unit on its own account.

In the International System, the ampere-turn per meter (At/m) is the unit combination tailored to measure the physical variable often called *magnetic field strength*.

Let us take again the example of the primary coil described in the preceding chapter, with 500 turns of wire evenly spaced along a cylindrical length of 0.6 m. Such a coil has an average of 833 turns per meter, and if a current of 1 ampere is sent through it, the resulting magnetic field strength obviously is 833 ampere-turn per meter. If, on the other hand, the current is increased to 3 ampere, the magnetic field strength increases proportionally, to 2,499 At/m.

These are measures of the magnetizing force inside the coil or solenoid. However, even when we know the magnetizing force (in At/m) we do not yet know the magnitude of the magnetic flux that force will produce inside the coil. Two additional measurements are needed: (1) the cross-sectional area of the coil, and (2) the magnetic permeability (μ) of whatever occupies the interior of the coil—vacuum, air or some other substance.

To borrow an analogy from the electrical variables—we may know the electromotive force (voltage) applied to a conductor, but we cannot predict what current will flow through it until we know (1) the cross-sectional area and length of that conductor, and (2) its resistivity. Then these two will give us the resistance in that circuit, and the resulting current in amperes will equal the volts of electromotive force divided by the ohms of resistance.

Magnetic flux is thus analogous to electric current flow. In a current, something is actually moving—the electrons through the conducting wire. In a magnetic flux nothing tangible is actually "flowing" or moving from place to place, but the lines of magnetic force are shaped much like the flow lines in a fluid moving through pipes.

Our sample coil of 0.6 m length and 833 ampere-turn per meter is wound in cylindrical form with a radius of 3 cm, and a resulting cross-sectional area of 0.002 83 m². Since only air is inside it, the μ (permeability) is 1.257×10^{-6} H/m. When the current flowing through its turns has a magnitude of 3 ampere, then the resulting magnetic flux inside the coil is the product of 2,499 At/m times 0.002 83 m² times 1.257×10^{-6} H/m. This works out to about 8.9×10^{-6} ampere-henry (AH).

However, the unit combination of the ampere times the henry (AH) has been converted into a single SI unit, called the weber (Wb). It is equal to 1 ampere times 1 henry and is the appropriate measure for magnetic flux. Our sample coil thus has a magnetic flux of 8.9×10^{-6} weber. The unit abbreviation is usually Wb, so that it shall not be confused with the watt, whose abbreviation is simply W. Thus 8.9×10^{-6} Wb may be expressed also as 8.9 μWb (microweber).

The weber, in fact, is equivalent also to 1 volt-second, meaning 1 volt times 1 second. It may seem odd that the unit designed to measure the flux or seeming flow of magnetism could be compounded from the volt of electromotive force and the second, unit of time. However, this relationship is made quite clear in the SI definition for the weber unit:

> the weber . . . is the magnetic flux which, linking a circuit of one turn, produces in it an electromotive force of one volt as it is reduced to zero at a uniform rate in one second.

The first "it" in that definition refers to the circuit of just one turn. The second "it," however, refers to the magnetic flux, or magnetic state (if we prefer to call it that), which acts upon that hypothetical coil of one single turn.

When a magnetic state with a magnitude of 1 Wb has existed, but is reduced within just 1 second to a magnitude of 0 (no magnetic flux), then the effect of this magnetic change on the single turn of coil is to generate in it an electromotive force of just 1 V.

If the coil had 10 turns rather than 1, but was subjected to this same vanishing of 1 weber of magnetic flux within 1 second, then an electromotive force of 10 V would be generated in that coil, and so on. The vanishing or dying-away of the magnetic flux causes magnetic lines of force to move past the conductor (wire) of the coil, and when magnetic lines of force move past a conductor, or a conductor moves past magnetic lines of force, then electromotive force is generated in that conductor, as a sensitive voltmeter will show.

The great basic interaction in electricity is that whenever a charge moves, it produces a magnetic field, or "magnetism." Conversely, the movements of a magnetic field produce, or can produce, the movement of electric charges. The dying-away of a magnetic field and the increase or growth of a magnetic field are both "movements" of the lines of force composing such a field.

The relationship is a symmetrical and reciprocal one: moving charges create magnetic effects; moving or changing magnetic effects set charges into motion. No wonder that this great area of the world of measurements is commonly called *electromagnetic,* rather than either electrical or magnetic.

Total magnetic flux versus flux density—The weber (Wb) measures the total flux of a magnetic field without regard to how densely packed that flux may be, relative to the area through which it passes. Hence the International System supplies another unit, equal to the weber per square meter (Wb/m^2) and given the single unit name of *tesla* (T). It measures the important physical variable of magnetic flux *density*.

We have mentioned the analogy between electric current measured in ampere and magnetic flux measured in weber. Current density is often measured in ampere per square meter. For example, a wire with a cross section of just 1 cm^2, if carrying just 1 ampere, has a current density of 10,000 A/m^2; whereas a conducting bar with a cross section of 100 cm^2 has a current density of just 5,000 A/m^2 when carrying a current of 50 A.

Magnetic flux density too is measured "per square meter." In the example of the coil previously used, the magnetic flux density inside the coil is just 0.003 14 Wb/m^2 or tesla (T).

The italic capital H is almost universally used to symbolize the physical variable of magnetic field strength, which in the SI is measured by the unit combination of ampere-turn per meter (At/m). Likewise the symbol B is used to symbolize magnetic flux density, which in the SI is measured by units of the tesla (T) or the weber per square meter.

Earlier unit systems provided other units to measure these two crucial physical variables, B, the magnetic flux density, and H, the magnetic field strength. Thus, magnetic field strength, H, in the cgs-emu (electromagnetic) system was measured by the *oersted* (Oe) unit, and 0.012 57 Oe equaled 1 At/m. In the cgs-esu (electrostatic) system, on the other hand, the corresponding unit was the *statoersted*, about 3.77×10^8 of which equaled 1 At/m.

Some other unit combinations, such as the ampere-turn per centimeter, and even the ampere-turn per inch, have been used to measure this same physical variable of H, or magnetic field strength. The At/cm equals 100 At/m; and the At/in equals 39.37 At/m.

On the other hand, B, magnetic flux density, was measured in the cgs-emu (electromagnetic) system by the *maxwell per* cm^2 (Mx/cm^2), 10,000 of which equal 1 tesla. In the cgs-esu (electrostatic) system there was a strange unit called the *statmaxwell per* cm^2. It was huge, equal to nearly 3 million tesla and to 30,000 million maxwell per cm^2. The SI tesla, accordingly, equals about 3.336×10^{-7} statmaxwell per cm^2, a unit now obsolescent but still in need of clear definition.

Magnetic flux density, the B variable, has also one special unit called the *gauss*, in the special Gaussian unit system which combines features of both the cgs-emu and the cgs-esu systems. Ten thousand gauss equal 1 tesla.

The variable of magnetic flux, for which the SI provides the weber unit, has been measured in a variety of other units also. Indeed, it has been called by names other than magnetic flux; among them are "flux of magnetic induction," and even "magnetic pole strength," or simply "pole strength."

In keeping with Faraday's concept that the magnetic field is made up of invisible curved lines of magnetic force, a unit of magnetic flux called the "line" has been used, equal to just 10^{-8} weber. Another name for this line unit was the *maxwell* (Mx). Since the line, or maxwell, was very small relative to the effects it was called upon to measure, there was use also of the kiloline and the megaline, equal respectively to 10^3 and 10^6 line.

In the cgs-emu (electromagnetic) system, the unit for measuring magnetic flux was sometimes called the unit pole. It was equal to just 4π (or 12.57) times the line, or maxwell. In the cgs-esu (electrostatic) system, on the other hand, the magnetic flux unit was staggeringly large, being equal to very nearly 3×10^{10} maxwell or line units, and 3×10^2 weber units.

Such listing of units and unit combinations which are *not* now included in the SI becomes tedious at times. However, since these older units still appear in much literature, they warrant mention. Also the rationale behind the simplifications and streamlinings of the International System becomes more apparent in the light of the duplication and confusion that it seeks to eliminate.

One of the stickiest of the complications has to do with past name changes. The name "oersted" was given only in 1930 to the unit for magnetic field strength (H) in the cgs-emu (electromagnetic) and Gaussian systems. Previously that same unit had been known by the name "gauss," and the name "oersted" had been used prior to 1930 for a unit that measured a quite different physical variable called reluctance, one which we fortunately feel safe in omitting here.

It was the International Electrotechnical Committee that decided on these name changes. Having freed the name "gauss" from the magnetic field strength variable, the Committee gave that name to the unit for magnetic flux density (B), a unit equal to 1 maxwell per square centimeter, and so 10,000 times smaller than the tesla unit in the SI.

Today this gauss unit for magnetic flux density remains alive, well, and widely used. Indeed, it is still encountered more often than the tesla, though that situation probably will change in time. Coupled with the gauss and used to measure very small magnetic fields is still another unit, called the *gamma,* equal to $\frac{1}{100,000}$ gauss. Thus, we have these clear decimal relationships: 1 tesla equals 10^4 gauss or 10^9 gamma; while 1 gamma equals 10^{-5} gauss and 10^{-9} tesla.

The gauss is widely used to measure what is mistakenly called the power or strength of the field of an electromagnet. Our Earth, a rather weak electromagnet, has a field that, at the surface, shows values somewhat less than 1 gauss. Between the poles of a typical toy horseshoe magnet the field is probably some hundreds of gauss. Powerful magnets designed for industrial and technical purposes may show about 1,000 gauss. Large laboratory electromagnets produce fields as high as 20 to 30 thousand gauss, which means 2 or 3 tesla.

Two particularly powerful superconducting electromagnets made for the Lawrence Radiation Laboratory of the University of California at Berkeley were described in 1973 publications. One, a dipole magnet designed to bend a beam of heavy ions, developed within its bore of 20 cm a strength of 4 tesla. The other, a quadruple doublet magnet designed to focus the beam, developed a magnetic gradient of 25 tesla per meter within its 20 cm bore. Thus, at the center of that small passageway the magnetic flux density differed by about 2.5 tesla from that prevailing at the outside or circumference of the bore.

Such high magnetic-field strengths control quite precisely the directions followed by the swiftly moving positive-charged heavy ions, which have been accelerated by the Bevatron and are on their way to locations where experiments are being conducted.

A measured magnetic flux density of 4 tesla means, of course, 40,000 gauss in the older and more common unit. This suffices to emphasize that the tesla is indeed a rather large unit for the effects that it is at present called upon to deal with on this planet.

All measurements expressed in terms of the tesla deal with—or should deal with—the important physical variable symbolized by B, rather than with that symbolized by H. The latter variable is properly measured in terms of the rather odd unit combination called the ampere-turn per meter (At/m, or sometimes written simply as A/m).

Reviewing a parade of electromagnetic units—Our progress has taken us step by step through more than half a dozen important units for electrical and magnetic measurements, including (1) the ampere of current; (2) the coulomb of charge; (3) the volt of electromotive force or potential difference; (4) the ohm of electrical resistance, and its inverse, the mho of conductance; and (5) the henry of inductance, both self-inductance and mutual inductance.

Next, in the realm of outright magnetic measurements, we may assign a sort of priority to (6) the weber, or unit of magnetic flux, which is in effect an energy-per-unit-of-current, or joule-per-ampere. Following it, we may place (7) the tesla, unit of flux density or induction, which is the number of webers per unit cross section (m^2) through which they pass. The tesla, in effect, is the joule per ampere per square meter of cross section (J/Am^2).

Now we come to a physical variable not previously mentioned: (8) *magnetic force,* sometimes called magnetic potential. Its unit combination in the SI is simply the ampere-turn (At), the number of turns times the magnitude of the current flowing through them. A single unit At in the International System is a coil of 1 turn through which just 1 ampere flows. In the earlier cgs-emu (electromagnetic unit) system a unit named the *gilbert* measured the same physical variable; 1 gilbert was equal to just 0.795 8 ampere-turn in the SI. In the cgs-esu (electrostatic unit) system, on the other hand, the corresponding unit was called the abampere-turn, and equaled exactly 10 ampere-turn in the SI.

The name and concept "turn" seem clear enough. However, as mentioned, it is not generally looked upon as a full-fledged unit. The proposal has been made to accord it full dimensional status, and even to give it a unit name, such as the "spat." This has no connection with a minor private quarrel, but is derived from the Latin *spatium,* meaning space. Should the "spat" ever take hold, it would be a sort of alternative or even a replacement for the SI unit of plane angle, the radian. In fact, 1 spat (or full turn) equals just 2π radian, or 6.283 2 rad.

The SI unit for measuring the variable of magnetomotive force might thus become the ampere-spat rather than the ampere-turn. But its magnitude and its basic physical meaning would not thereby be altered.

Following this crucial physical variable of magnetomotive force we come logically to the SI unit of *ampere-turn per meter* (or *ampere-spat per meter*), which measures the variable of magnetic field strength symbolized by H.

Earlier we saw that electrical field strength is measured by relating changes in electromotive force (measured by the volt) to distance (measured by the meter)—thus giving us the volt per meter (V/m). Intensive electrical fields, such as those that build up inside the massive clouds called thunderheads, reach levels of 300,000 volt per meter, or more.

In a similar way, magnetomotive force, which is measured by the ampere-turn, is related to distance and becomes the ampere-turn per meter, the unit combination that measures magnetic field strength. When a move of 1 meter distance brings a change (either gain or loss) of just 1 ampere-turn in magnetomotive force, then the field has a strength of 1 ampere-turn per meter. If the change is 2 ampere-turns, then the field strength is 2 At/m, and so on.

The concept of field strength, either electrical or magnetic, includes the idea of a test body that is subjected to the field at the particular point or region being measured. In the case of electrical field strength, if the test body is a unit charge of 1 coulomb, then in a field of 1 V/m it will be subjected to a force of just 1 newton. And if that 1 C charge moves a distance of 1 meter in response to that field-generated force, then work or energy of just 1 joule has been expended on it.

Similarly, the unit test object for a magnetic field would be the so-called "unit pole," with just 1 weber of magnetic flux. When it is acted on by a magnetic field with strength of 1 At/m, then it is pushed (or pulled) with a force of 1 newton, and if it moves 1 meter in the direction of that force, then 1 joule of work or energy has been expended on it.

Actually, the concept of a "unit pole" is rather faulty. It is not possible to saw a magnet in half in order to get a "north pole" and a separate "south pole." Each half will develop its own north and south poles. Thus the situation is not completely like that of electric charge, which can exist in purely negative form, as in the case of the electron particle, and in purely positive form, as in the case of the far heavier proton particle.

In spite of these differences, the analogies between the principal electrical and magnetic variables are both numerous and interesting in the study of units and dimensions.

The importance of permittivity and permeability—Particular importance attaches to the physical variables called *permittivity,* symbolized by Greek *epsilon* (ε), and *permeability,* symbolized by Greek *mu* (μ). The former, measured in units of farad per meter (F/m), is on the electrical side; the latter, measured in units of henry per meter (H/m), is on the magnetic side.

The actual capacitance, in farad units, of a condenser or capacitor is proportional to the permittivity of the substance or lack of substance (space) between its oppositely charged conducting plates. The greater the permittivity of this intervening situation or substance, the greater the capacitance of the capacitor.

We have seen, too, that the self-inductance of a coil or the mutual inductance of separate concentric coils is proportional to the permeability of the substance or space located within the spiral windings. The greater the permeability of that substance or space, the greater the inductance, measured in terms of the henry unit.

The fact that permeability is measurable by the unit combination henry per meter (H/m) reveals that it is identical with the ratio of B, the magnetic flux, to H, the magnetic force or field strength in response to which the flux has formed. In short, $\mu = B/H$. If Wb/m^2, the unit combination that measures magnetic flux B, is divided by At/m, the unit combination that measures magnetic force or field strength H, the result is equal to henry per meter (H/m), the unit combination that measures permeability, μ.

Magnetic permeability μ thus has to magnetic force H a relationship like that which electrical conductance (measured by the mho) has to electromotive force (measured by the volt). The product of μ times H is the magnetic flux density B. The product of conductance times the voltage applied is the current that flows, a flux of electrical charge, measured by the ampere.

Another and more concrete interpretation of permeability, μ, for any substance and situation is this: It is the ratio of (1) the magnetic flux density that actually develops in the substance, to (2) the flux density that existed in the same volume before that substance or, any other substance was placed there.

We have noted that in the International System, the flux density of empty space is called μ_o and is measured at 1.257×10^{-6} henry per meter. Almost every material substance has a higher μ than this. Hence those permeabilities are often expressed as multiples of μ_o. Using the measured permeabilities of tens of thousands of different substances as a guide, we can now make certain generalizations about the varied magnetic behavior of the kinds of matter all around us. When placed in the midst of magnetic flux, all substances behave in one of three basic fashions. (1) Some, called *diamagnetic,* take in even fewer lines of magnetic flux than would be present in the vacuum. They are even less permeable, magnetically speaking, than "nothingness." Hence their μ is less than μ_o, and so less than 1.0. (2) Other substances, called *paramagnetic,* show slightly more permeability than does empty space. Finally, there is (3) the rather small group of substances that show enormously high permeabilities, and these are called the *ferromagnetic* substances.

Diamagnetic behavior is shown by the inert gas nitrogen (N_2) and by some other gases that form molecules rather than single atoms. Of all the metals, bismuth (Bi) is the only one that shows it markedly. Paramagnetism is shown by atoms of sodium and potassium, also by molecules of oxygen gas (O_2), to name but one instance.

In a uniform magnetic field, a diamagnetic substance sets its long axis at right angles to the lines of force in the field, whereas a paramagnetic or ferromagnetic substance sets its long axis parallel to those lines of force (as we see with iron compasses). In a divergent

magnetic field, a diamagnetic substance is repelled, whereas paramagnetic and ferromagnetic substances are attracted.

The magnitudes of diamagnetic effects are extremely small.
Their permeabilities are about of the order of 1 part per million less
than the permeability of empty space. The paramagnetic effects too
are small. Their magnitudes typically are of the order of from 1
part in 100,000 to 1 part in 10,000.

On the other hand, the effects of the ferromagnetic substances
are truly enormous. On the same scale they range from about 10
times the permeability of empty space to 10,000 times that of μ_o.

Actually, only three solid metals belong in the ferromagnetic
group of giant effects: iron, cobalt and nickel. The compound
Fe_3O_4, an iron oxide, is also ferromagnetic. So too are various
alloys composed of differing proportions of nickel, copper and
chromium. Ferromagnetism is so large in its responses to magnetic
fields that it was the first kind of magnetic behavior observed (as
in lodestones) and was for long the only sort of magnetic behavior
that was recognized. Other increasing delicacy or precision of measurement revealed the faint diamagnetic and paramagnetic behaviors in other substances.

Measurements of the permeabilities of ferromagnetic substances
make much use of what is called the "initial" or starting permeability. This is about 150 μ_o for iron, 110 for nickel, and 70 for cobalt. Some of the special alloys far exceed these figures for "initial"
permeability, however. For example, permalloy, a mix of some 78
parts nickel and 22 of iron, shows about 9,000 μ_o at the outset. The
initial permeabilities of other special alloys, based on various metallic mixtures, have been measured as about 20,000 times μ_o!

These permeabilities do not persist, however, as the magnetic
field strength (H) is raised to ever higher values. There comes a
point at which any ferromagnetic element or alloy seems "fed up,"
to have taken in all the magnetic flux that its atoms and molecules
can cope with. Then, further increases in H are not matched by
corresponding increases in B, and the fraction B/H, which defines
the permeability for the substance, tends gradually downward toward 1.0.

The substance is said to have become magnetically "saturated."
In order to escape these limitations, some magnetic experimenters

in the effort to attain the very highest possible B, or magnetic flux density, have deliberately worked with empty coils—small in diameter, carrying large current in amperes, and squeezing together myriads of lines of force in the pencillike interior.

What of the close relationship suggested between permeability, μ, and permittivity, ε? The former, we know, is measured by the unit combination henry per meter (H/m). The latter is measured by the analagous unit combination farad per meter (F/m), and the ε_0, or permittivity of empty space, has been measured at the value of 8.855×10^{-12}F/m.

The mathematical analyses of J. C. Maxwell showed that there were certain situations in which it became necessary to multiply the permeability and the permittivity of empty space. The product of $\mu_0 \varepsilon_0$, he found, was a very interesting fraction:

$$\frac{1}{8.98 \times 10^{16} \text{ meter}^2 \text{ per second}^2}$$

In other words, its denominator included the square of a velocity, because m/s measures velocity and m^2/s^2 measures velocity squared. But what velocity could correspond to the square root of that telltale magnitude 8.98×10^{16}? That square root, in round numbers, is just 3×10^8; and 3×10^8 meter per second corresponds, within a tiny fraction, to the velocity of light in empty space.

This and other important considerations led Maxwell to propose in a very positive way that light was identical with certain frequencies of electromagnetic oscillations. He indicated that there must be also other frequencies of such oscillations that we do not see as light but which can somehow be emitted and received.

Not very long afterward, the physicist Heinrich Hertz in his laboratory in Germany produced the so-called "hertzian waves" by purely electrical means, and detected them at a distance, also by electrical procedures. We call them today radio waves, or microwaves, and know that they too travel through space with that universal velocity of very nearly 3×10^8 m/s.

Thus, to the two great constants of the void or vacuum, μ_0 and ε_0, is added a third, c, the universal velocity of electromagnetic radiation everywhere in the absence of intervening matter. Together the three form two particularly elegant and meaningful relationships:

$$(1)\ c^2\,\mu_0\varepsilon_0 = 1$$

and

$$(2)\ c = 1/(\mu_0\varepsilon_0)^{\frac{1}{2}}$$

or

$$c\,\mu_0^{\frac{1}{2}}\varepsilon_0^{\frac{1}{2}} = 1$$

If we attach a known mass to a spring of known stiffness, the inertia of the mass and the elasticity of the spring will combine to produce a certain frequency—the frequency at which that mass will vibrate up and down when it is displaced from its equilibrium point and then let go. We have noted the close analogy between capacitance, measured by the farad unit, and the stiffness of a spring; also the analogy of inductance, measured by the henry unit, and the inertia of a mass of matter.

Now, if we combine in one circuit a capacitor of known capacitance (in farad units) and an inductance of known value (in henry units), what is the result? For simplicity let us suppose they have the large values, respectively, of 1 farad and 1 henry. What is the product, dimensionally speaking, of 1 farad times 1 henry? It proves to equal just s^2—that is, 1 second squared.

What is the time or duration to which this squared second might refer? It is the natural or inevitable period of oscillation (electrical) in the circuit formed by the combination of that capacitance and that inductance.

If the combination had a 1 millifarad capacitor and a 1 millihenry inductance, then the product would be 10^{-6} s^2, or $s^2/10^6$. Extracting the square root of the fraction, we get $\frac{1}{1,000}$ s, or 1 millisecond, which would be the natural oscillation period for the electrical circuit thus assembled.

The combination of suitable capacitance and inductance makes possible the tuning of radio and television receiving sets, and the transmitting of signals at precise frequencies by broadcasting stations. The construction of continuously tunable circuits for transmission and reception thus depends on the magnitudes of such physical variables as those measured by the farad, the henry, and the rest of the unit hierarchy.

Weber, pioneer in electromagnetic metrology—Wilhelm Weber (1804–90), for whom the SI unit of magnetic flux is named, is preeminent among those who first labored to organize consistent and complete systems of electromagnetic units and measurements. His labors in these fields were long and intensive.

Weber was not yet thirty when he was appointed to the chair of physics at the University of Göttingen, Germany. There he worked closely with the great genius Karl Friedrich Gauss (1777–1855), a prodigious mathematician, astronomer and metrologist. Together, Gauss and Weber formulated the leading ideas that have determined the future course—or courses—of all major electrical measurements.

Weber's systematizing was not abstract. He himself made absolute measurements of a wide range of electrical physical variables, including that of electromotive force or potential difference (now measured by the volt), of current (ampere), of resistance (ohm), of capacitance (farad), and of inductance (henry). The parenthesized units were not those that Weber used, for most of them had not yet been evolved; but the physical variables that they now measure were those that Weber found ways to measure.

He devised measuring instruments of his own when he could find none at hand suited to his needs. One of his devices is known as the dynamometer, another as the earth inductor. Actually it was Weber, rather than J. C. Maxwell in Britain, who first realized that in any complete and self-consistent structure of electromagnetic units, the velocity of light, c, was an inescapable, though often hidden, ingredient.

Not surprisingly, Weber supported the principle advocated by Gauss that the units chosen to measure electromagnetic variables be built up from the three basic mechanical variables of length, mass and time—those that the SI measures by means of the meter, the kilogram and the second, respectively. The Weber-Gauss principle did not permit of a separate basic electrical unit, such as the SI today uses in the form of the ampere of current.

Weber, for example, based his absolute measurement of electric current on the force exerted by that current on what Gauss had called a unit magnet pole, placed at a unit distance from the current-carrying conductor (wire). It would be equally true to say that it was based on the force that such a unit magnet pole exerted on a current-carrying conductor located at unit distance from the pole. In either case, current strength is measured by, or is in proportion to, the mechanical force resulting from that current flow under certain specified conditions.

This approach is still used in the SI definition of the ampere as the current that, flowing through two parallel wires, causes each to exert on the other a force of 1 newton per meter of wire length. However, the SI takes care to elevate the ampere, thus defined, to dimensional equality with the meter of length, the kilogram of mass, and the second of time. The Weber-Gauss approach was to interpret the ampere and all other electromagnetic variables as being complicated derivatives from the basic units of length, mass, and time only.

Of all scientists whose names have been given to units for electromagnetic measurement, Weber was the first-born among those whose lives were passed entirely in the nineteenth century. He was in his upper seventies when, in 1881 in Paris, the international gathering was held at which names for the electrical units were chosen. Both Weber and Gauss were at that time omitted from the list of honored individuals, much to the indignation of many Germans. They did not fail to insist that it was Gauss and Weber who had first offered the basic pattern of electrical units that the Paris congress had adopted for worldwide use.

For some time afterward, many German scientists and science writers deliberately used the name "weber" rather than "ampere" for the new unit of electric current. It was their unilateral protest against the wrong they felt had been done. Ampère's name, however, survived for the basic current unit, and later the name of Weber was given to the SI unit of magnetic flux, the same unit that is equivalent to 1 volt-second (1V·s).

Weber's renown in the academic and intellectual world was not limited to his great achievements in the metrology of electromagnetic interactions. He made history also as a staunch advocate of civil liberties and academic freedom. When the king of Hannover, whose principality included the University of Göttingen, suppressed the country's constitution, Weber was one of some seven members of the university faculty who dared to protest publicly against this action. They were all summarily dismissed, for divine right of rulers was more potent there than traditions of academic freedom, and there was no secure tenure, even for men of Weber's eminence.

He survived for several years as a private tutor. Meanwhile, he managed to travel a good deal, covering great distances on foot, for he was a passionate pedestrian. Indeed, one of his most painstaking and complex series of measurements, made in cooperation with one of his brothers, was an analysis of the mechanisms and motions by which humans walk. Eduard Friedrich Weber was the name of that collaborating brother, and the book they produced bore the down-to-earth title of *Mechanik der menschlichen Gewerkzeuge,* approximately equivalent in English to *Mechanics of Human Apparatuses.*

Wilhelm Weber collaborated with another brother, Ernst Heinrich by name, in an admirable pioneer study of how surface waves form and move on water. Their work was both experimental and mathematically analytical.

By 1843, half a dozen years after his ouster from the Göttingen faculty, Wilhelm Weber found an academic position at the University of Leipzig, Saxony. By 1849 he returned to the university at Göttingen again, a move that honored the institution more than the man himself.

Among many basic electromagnetic measurements by Weber and Gauss was their analysis of the actual magnetic field of the Earth. For these tremendously important terrestrial measurements Weber devised the instrument called the magnetometer; and also a variation of the common magnetic compass, called the "declination needle," for measuring the angle of dip of the Earth's magnetic field from any point on its surface.

In search of significant ratios—In 1840, while still "between" university positions, Weber began making the measurements deliberately designed to reveal the ratios between units in the two separate self-consistent systems—the electrostatic (cgs-esu) and the electromagnetic (cgs-emu).

The essential point of departure of the former (esu) system is that it sets the value of the permittivity of empty space (ε_0) to be unity (1) and dimensionless. One of the results of this is that the magnetic permeability of space (μ_0) has both a numerical value and dimensions; in fact, in the cgs-esu system, the permeability of space acquires this constant value: $1.112\ 6 \times 10^{-21}\ s^2/cm^2$. The units of second squared per centimeter squared suggest that this permeability constant is dimensionally equivalent to the inverse or reciprocal of the square of a velocity.

In the electromagnetic (cgs-emu) system, on the other hand, it is the permeability of empty space (μ_0) which is set at unity (1) without dimensions. And in this system, the permittivity of space (ε_0) acquires that same constant value: $1.112\ 6 \times 10^{-21}\ s^2/cm^2$. Thus, the permittivity constant becomes dimensionally equivalent to the reciprocal of the square of a velocity.

The numerical value $1.112\ 6 \times 10^{-21}\ s^2/cm^2$ equals this fraction: $1/8.988 \times 10^{20}\ cm^2/s^2$. And that fraction, not by any means accidentally, is equal to $1/c^2$. Thus, the reciprocal of the square of the velocity of light in space, $(1/c^2)$, enters into each of those two basic systems, the esu and the emu.

A third system, previously mentioned, also was widely used—the so-called Gaussian unit system, a sort of compromise between the esu and emu systems. It went all the way and assigned values of unity (1) without dimensions to *both* the permittivity (ε_0) and the permeability (μ_0) of empty space. The result is that the Gaussian system has the same units as the electrostatic (cgs-esu) system for measuring the electrical variables of current, charge, electromotive force and potential difference, resistance and capacitance. At the same time it has the same units as the electromagnetic (cgs-emu) system for measuring inductance, magnetic field strength (H) and magnetic flux density (B).

The Gaussian unit for magnetic flux density (B) is called the gauss; the corresponding measure in the cgs-emu is symbolized by a unit combination (maxwell per square centimeter). However, their values are identical, for 10,000 gauss and 10,000 Mx/cm^2 are alike equivalent to the SI magnetic-flux-density unit, the tesla (T), also known as Wb/m^2.

Both Weber and Gauss are indissolubly connected with the origins and interpretations of the world of electromagnetic measurements. Though the International System, for reasons of simplicity and practicality, departed from their principles, the difference is rather an improvement over, than a rejection of, their work.

The tesla of magnetic flux density—The tesla (T), unit of magnetic flux density in the SI, is the latest of all the eponymous units chosen for electromagnetic variables in that system of rapidly increasing international acceptance. To many an electrotechnician or scientist, its equivalent, the weber per square meter (Wb/m^2), may seem more familiar. And still others may find themselves somewhat unsure just what the tesla measures, if they are not first reminded that it is equal to 10,000 gauss, and as such is the SI measure for the variable generally symbolized by B.

The tesla is named for Nikola Tesla (1857–1943), born in Croatia, now part of Yugoslavia. He lived and worked in the United States during most of the years of his long and unusual life. By his ancestry and ethnic background, Tesla is the only Slavic scientist or inventor included among the eponymous unit names in the electromagnetic area.

The son of a Serbian Orthodox priest, Tesla received his early engineering training in Prague, then the metropolis of the Bohemian portion of the strangely assorted Austro-Hungarian Empire. His first engineering assignments took him to Budapest, the Hungarian capital, and to Paris. Tesla, a multilinguist, possessed of an almost photographic memory and a self-consuming energy, soon looked for new worlds that might afford greater scope for his abundant ideas and ambitions.

He was twenty-seven when he arrived in the United States, bearing a letter of recommendation to the great inventor Thomas Alva Edison.

Tesla's major concentration was on better methods for generating and delivering electric power. He was a strong advocate of the use of alternating rather than direct current for such power supply. Within a few years after his arrival in America, he was working in association with George Westinghouse of air-brake fame. They represented the pro-alternating-current team in a keen competition with the Edison interests, who were just as feverishly trying to put through their direct-current systems.

Tesla patented a long series of electrical devices and techniques.

Most important, from today's points of view, were his concept and contribution of the so-called "induction" type of electric motor, dependent for its rotating magnetic field on alternating-current supply. Modifications of this rotating-magnetic-field principle became basic to the great generators that now convert hydraulic energy into electric power at Niagara and so many other installations. Other Tesla inventions included spectacularly high-tension (high-voltage) induction coils, novel and improved capacitors, transformers, generators, arc lights, and so on.

As he grew older, Tesla became increasingly compulsive and peculiar in his personal habits and interests. He has been called one of the most eccentric persons of whom there is any record. His hang-ups included a veritable phobia about microbes, a deep devotion to the care and feeding of pigeons in the park, and other quirks that his fellow engineers and the public found puzzling or bizarre.

During his last years Tesla was quoted often in the press in comments or predictions on matters of general interest. He made more than one weird statement about "death rays" and mysterious forces of unimaginable potency. Yet he asserted that "atomic power is an illusion." But death overtook him just as nuclear energies were being fashioned into weapons horribly and unprecedentedly lethal.

Tesla's years of apparent decline do not negate, however, the many bold contributions and inventions of his brilliant youth and prime.

39 ELECTROMAGNETIC MEASUREMENTS: A FINAL SURVEY

If one may measure small things by great . . .
—VERGIL,
Georgics IV (1st cent. B.C.E.)

Many an ambiguity has beset the physical variables and the specific units used in magnetic measurements. Even with the aid of the International System units, it is sometimes impossible to make clarity prevail, but it is worth some effort.

One source of uncertainty is the loose use of the term *field* with reference to "the" magnetic field. At least three different but formally distinct uses of *field* can be encountered.

1. Sometimes the word is used to mean any space that contains magnetic flux, the physical variable measured by the SI unit of the weber (Wb). The symbol for the variable of magnetic flux is the Greek capital letter *phi* (Φ).

2. Sometimes *field* is used to mean the magnetizing or inducing field, the physical variable that is symbolized by H and measured by the SI unit combination of ampere-turns per meter (At/m), or by the unit of the oersted (Oe) in the cgs-emu (electromagnetic) and Gaussian unit systems. This is the variable called magnetic field strength.

3. Finally, and most properly, *field* is used to mean the magnetic flux density, which is symbolized by B and measured by the SI unit tesla or by the unit combination of the maxwell per square centimeter (Mx/cm^2) in the cgs-emu (electromagnetic) system and the gauss (Gs) unit in the Gaussian unit system. This B variable is the product of the H variable times the magnetic permeability (μ) of

496

the substance that is responding to the H, or magnetizing-field, effect.

The safest course is to restrict "magnetic field" to this last use, and to avoid it with the other two. In short, *magnetic field* should refer to the actual flux density, B, and not to H, the magnetizing force which has produced that flux density.

In the SI pattern, a unit magnetic field is that flux density (B) which exerts a force of 1 newton on each 1 meter length of a conductor (wire) carrying a current of 1 ampere. Thus, 1 tesla is equivalent to 1 newton per ampere-meter; and the relationship can be stated either as 1 T = 1 N/A·m, or as 1 T·A·m = 1 N.

In contrast, the variable H means a "force field," which is to say, a force per distance. The force is that exerted by one ampere-turn —a 1 A current flowing through a single loop or turn. The distance is 1 m, and thus H, or magnetic field strength, is measured by At/m, where the "t" simply means a full turn.

The variable of permeability, symbolized by μ, is measured by the fraction B/H for whatever substance is involved. Thus μ can be measured by the SI unit combination of the weber per meter-ampere-turn (Wb/m·At). Since 1 Wb equals also 1 joule per ampere (1 J/A), permeability can also be measured by the unit combination of the joule per ampere squared meter (J/A²·m).

The most important permeability, that of empty space, is equal in the SI units to either 12.57×10^{-7} J/A²·m, or to 12.57×10^{-7} Wb/m·At, or even to 12.57×10^{-7} H/m (henry per meter). Each of these three is an alternative unit combination with the same physical meaning.

Another physical variable called *reluctivity* sometimes appears. It is the inverse or reciprocal of permeability, much as conductance (in mho units) is the reciprocal of resistance (in ohm units). Thus reluctivity is equivalent to H/B and may be expressed in such unit combinations as the A²·m/J, the At·m/Wb, or the m/H (meter per henry).

Reluctivity must be distinguished from *reluctance,* which is symbolized by the script capital letter R and measured in the SI unit combination of ampere-turn per weber (At/Wb). Reluctance is the magnetic analogy to electrical resistance, measured by the ohm unit. Reluctivity is defined as reluctance per unit length. Hence, reluctivity times unit length equals reluctance, and reluctance may be symbolized by $H{\cdot}L/B$, where "L" stands for the variable of length, measured by the SI meter unit.

The correspondence or analogy of electrical resistance and magnetic reluctance is part of a pattern of parallels that are both striking and, sometimes if not often, helpful in making the magnetic side of things seem less alien to understanding. Beginners in electrical studies learn Ohm's law, which says, roughly speaking, that current (in ampere) equals electromotive force (volt) divided by resistance (ohm) in an electrical circuit: $A = V/\Omega$, and $A\Omega = V$. The magnetic counterpart of that is Rowland's law, which states that magnetic flux (in weber) equals magnetomotive force (in ampere-turn) divided by reluctance (in ampere-turn per weber).

No single unit name has ever been given to the At/Wb unit combination, but if someday it were to be named, for example, the rowland, one could summarize Rowland's law as neatly as Ohm's law, and in this way: $Wb = At/R$, and $Wb\,R = At$.

Analogies again—The analogies between electrical and magnetic measurement situations are far-reaching. For each of these two categories or directions, there is a physical variable corresponding to a flow or flux; another corresponding to the force that establishes the flow or flux; and still another corresponding to the opposition or obstacle (resistance) that the force overcomes in establishing that flow or flux.

This is shown most vividly by setting side by side the two corresponding lists. Where non-SI units have been included for completeness, they are distinguished by an asterisk. (See next page.)

The rel, a non-SI reluctance unit, is rather unimportant, but has been included for good measure.

Two other meaningful parallels can now be added. Each system has a physical variable representing the intensity of the force—that is, the increase (or decrease) of the force per unit distance moved in a volume in which it is effective. And each system has a physical variable that represents the density or crowding of the resulting flow or flux: the magnitude of the flow or flux per unit volume.

Intensity of electric force is called electric field strength (E) and

Parallels in measurement of electrical and magnetic circuits (Suitable units of measurement are placed in parentheses below the name of the physical variable and its most common symbol.)

GENERAL DESCRIPTION	IN ELECTRIC CIRCUITS	IN MAGNETIC CIRCUITS OR SITUATIONS
FLOW or FLUX	Electric current, I (ampere)	Magnetic flux, Φ (weber or maxwell*)
FORCE	Electromotive force, F (volt)	Magnetomotive force, F (ampere-turn, or gilbert*)
OPPOSITION	Resistance, R (ohm)	Reluctance, R (ampere-turn per weber, or rel*)
Basic law	Ohm's law: $I = F/R$	Rowland's law: $\Phi = F/R$

is measured in the unit combination volt per meter (V/m). Intensity of magnetic force is called magnetic field strength (H) and is measured in the unit combination of ampere-turn per meter (At/m, often abbreviated simply to A/m). In non-SI units it is measured also by the oersted (Oe).

Density of flow or flux in electrical situations is called current density (J) and measured by the unit combination ampere per square meter (A/m²). Density of flux in magnetic situations is called magnetic flux density (B) and measured by the SI unit of the tesla (T), or by the non-SI units of the gauss (Gs) or the line per centimeter (li/cm).

It is this B variable which—as emphasized here—constitutes the real and legitimate "magnetic field."

Thus, no fewer than half a dozen important analogies exist between the electrical and the magnetic variables. We can proceed a few steps further in tracing parallels. The magnetic situations are in certain instances comparable to two different electrical situations: (1) the electrostatic, and (2) the electrokinetic. The electrostatic situations are those that emphasize the interactions of electric charges, rather than of electric currents; the electrokinetic, however, emphasize the current rather than the static-charge effects.

Since energy, or work, measured always in the SI by the joule unit, is so basically important, let us see what variables in each of the three categories can be combined to yield energy magnitudes.

The magnetic combination that yields an energy product is symbolized by $m \times F$. The m stands for a variable sometimes called magnetic mass, but also known by such diverse designations as magnetic flux, magnetic-induction flux, or strength of magnetic pole. It is measured by the weber (Wb) unit, which helps to pin it down more precisely and to remind us that it can also be measured by the volt-second (V·s), which equals the weber.

The variable F has already been shown as magnetomotive force, measured by the SI ampere-turn (At).

The relationship can be expressed compactly by unit abbreviations: Wb $\times$ At = J. A magnetomotive force of 1 At in conjunction with a magnetic-induction flux of 1 Wb represents an energy of 1 joule.

The electrostatic combination that equals an energy is symbolized by $Q \times F$. Here Q is the quantity or charge of electricity, measured by the coulomb unit in the SI; and F is the electromotive force in units of the volt. Thus, a charge of 1 C acted on by a force of 1 V represents an energy or work of 1 joule.

The electrokinetic combination that equals an energy is symbolized by $\Phi \times I$. Here Φ is again the variable of magnetic flux or magnetic-induction flux, measured by the SI weber unit; while I is electric current, measured by the ampere. The relationship in abbreviated units is just Wb $\times$ A = J (weber times ampere equals joule of energy or work). Clearly this compact equation is closely similar to that for the magnetic situation, in which the t was added to the A for ampere as a reminder that the situation typically involves current flowing through coils of many turns.

Another important kind of measurement is that which relates energies to the volumes of space in which they act or appear. The SI unit combination for energy density is joule per cubic meter (J/m^3).

The magnetic pair whose product is such an energy-per-volume is symbolized by $B \times H$. These are the familiar "twin" variables: B is the magnetic induction or flux density, measured by the SI unit of the tesla (T). H is the magnetic field strength, or magnetic intensity, measured by the SI unit combination of the ampere-turn per meter (At/m).

In concrete terms, if an H of 1 At/m results in a B of 1 T, then the energy represented is just 1 joule. It requires this much energy, we may say, to produce the 1 T of flux density in the particular substance concerned, by subjecting it to the effect of magnetic intensity of 1 At/m.

Earlier we examined B and H in a very different relationship, for B/H is the magnetic permeability (μ) of the substance; whereas H/B is its reluctivity, the inverse of the permeability. Here, however, we multiply, rather than divide, B by H.

The electrostatic combination whose product is an energy density is symbolized by $D \times E$, where D stands for surface density of electric charge, measured in the unit combination of the coulomb per square meter (C/m^2), and E is the electric field strength, which is measured in units of volt per meter (V/m). The relationship can be abbreviated this way:

$$C/m^2 \times V/m = J/m^3.$$

That is, 1 coulomb per square meter in conjunction with 1 volt per meter represents an energy density of 1 joule per cubic meter.

The electrokinetic combination whose product is an energy density is $J \times \Phi/L$. J is the electrical current density, measured in ampere per square meter (A/m^2), while Φ, as before, is the magnetic-induction flux, measured by the weber unit. The symbol L stands for length, measured by the meter. The product of this expression is an energy density measured in joule per cubic meter. If the current density is 1 A/m^2 and the other (fraction) is 1 Wb/m, then the resulting energy density will be 1 joule per cubic meter of space.

Combinations of variables equivalent to power—Power, the time rate of energy conversion, is measured by the watt in the SI and, together with energy, has broad and basic importance, whether we deal with mechanical or thermal or electromagnetic interactions.

Two magnetic variables when divided by time have power as their product: $m \times F/T$. Again m is the same magnetic "mass" or amount of magnetization, measured by the weber unit; while F is magnetomotive force, measured by the unit of the ampere-turn

(At); and T stands for time, measured by the unit of the second. In units, the recipe is: 1 weber ampere-turn per second equals 1 watt; or, abbreviated, 1 Wb At/s = 1 W.

An electrostatic combination that is equivalent to a power is represented by $F \times Q/T$. Here Q is charge, measured in the coulomb unit, while F is electromotive force, measured by the volt, and T stands for time, measured by the second. Thus, the unit relationship is 1 V·C/s = 1 W.

Now, 1 coulomb per second (1 C/s) is identical with 1 ampere (1 A). Hence the combination can also be presented as $I \times F$ equals power. I stands for current, measured by the ampere unit. This recipe is the familiar "ampere times volt equal watt."

That is, in effect, an electrokinetic combination, for I represents the variable of *current*, not that of static charge, represented by Q. Another electrokinetic combination that yields power is symbolized by $I \times \Phi/T$. Here I again is current, measured by the ampere unit; while Φ again stands for magnetic-induction flux, measured by the weber unit, which is equal also to the volt-second combination. T represents time, measured by the unit of the second. Specifically, a current of 1 ampere associated with a magnetic-induction flux of 1 weber represents an energy of 1 joule, and such an energy per second is equivalent to 1 watt.

These comparable combinations yielding powers, energy densities, and energies reveal the similarities that can be stressed between the adjacent yet distinguishable areas of electromagnetics, electrostatics, and electrokinetics. Thanks to the superior simplicity and consistency of SI units, it is possible to analyze these analogies with greater ease than in earlier unit systems.

As a further aid to simplification and certitude, we offer here a listing of the principal electromagnetic variables that may be encountered in reading and working in the field. They are arranged alphabetically in the order of the symbols most commonly used to represent them—first the symbols derived from the alphabet in English, then those derived from the Greek.

Some duplications will be found, for there are a good many more physical variables to symbolize these days than there are symbols available for them. The list represents things as they are, not as they would be if rigorous logic had always prevailed in metrology. In each case, the SI unit or units appropriate to the measurement of the indicated variable are given, together with some relationships of interest between those units and others. Units not marked by asterisks are part of the International System; those with asterisks are not part of that system, but are important enough to include for completeness.

Tools of electromagnetic measurements

SYMBOL AND COMMON NAME FOR THE PHYSICAL VARIABLE		DESCRIPTION OF THE VARIABLE BY MEANS OF APPROPRIATE UNITS OR UNIT COMBINATIONS
B	susceptance	mho (reciprocal of the ohm) See also the variable of conductance.
B	magnetic-flux density or magnetic induction	tesla (T), equal to Wb/m². Also measurable in gauss* (Gs) (see also J, magnetic polarization)
C	capacitance	farad (F), equal to C/V
D	electric displacement or charge density	coulomb per square meter (C/m²)
E	electric field strength or intensity	volt per meter (V/m), equal to W/A·m
E	electromotive force or induced emf	volt (V), equal to W/A
F	magnetomotive force	ampere-turn (At)
G	conductance	mho, the reciprocal of the ohm of resistance (see also susceptance B above)
H	magnetic field strength or intensity	ampere-turn per meter (At/m) (also measurable in the oersted* Oe)
I	current	ampere (A), equal to 1 C/s
J	current density	amperes per square meter (A/m²)
J	magnetic polarization (corresponds to magnetic-flux density B)	tesla (T), equal to Wb/m²
L	inductance or self-inductance	henry (H), equal to 1 V·s/A
m	magnetic mass, or amount of magnetization, or magnetic flux	weber (Wb), equal to 1 V·s

SYMBOL AND COMMON NAME FOR THE PHYSICAL VARIABLE		DESCRIPTION OF THE VARIABLE BY MEANS OF APPROPRIATE UNITS OR UNIT COMBINATIONS
N	electric induction flux	coulomb (C); see also Q or quantity of electricity.
p_m	magnetic moment	joule-meter per ampere (J·m/A)
	magnetic dipole moment (no standard symbol)	ampere per square meter (A/m²) Also measurable in maxwell per cm (Mx/cm)
P	permeance	weber per ampere (Wb/A) equal to volt-second per ampere (V·s/A)
P	power (electrical or mechanical	watt (W), equal to 1 J/s
Q	charge or quantity of electricity	coulomb (C), equal to 1 A·s
	reluctivity (no standard symbol)	ampere-turn per weber-meter (At/Wb·m) equal to reluctance per unit length.
R	resistance	ohm (Ω), equal to V/A
R	reluctance	ampere-turn per weber (At/Wb) equal to ampere-turn per volt-second (At/V·s)
U_m	magnetic potential difference	ampere-turn (At) (see also F magnetomotive force)
U	sometimes symbol for electromotive force, potential, or voltage; see V below	
V	voltage, potential difference, or electromotive force	volt (V), equal to W/A
Y	admittance (See also B, susceptance)	mho, reciprocal of the ohm
W	energy or work	joule (J), equal to the watt-second
X	reactance	ohm (Ω), equal to V/A, and also to W/A²
Z	impedance	ohm (same as above)

Variables symbolized by Greek letters

γ	(gamma) conductivity	see symbol sigma σ below

SYMBOL AND COMMON NAME FOR THE PHYSICAL VARIABLE	DESCRIPTION OF THE VARIABLE BY MEANS OF APPROPRIATE UNITS OR UNIT COMBINATIONS
ε (epsilon) permittivity, or dielectric permittivity	farad per meter (F/m), equal to A·s/V·m, derived from variables B/H
λ (lambda) permeance	see symbol L for inductance, above
μ (mu) permeability, or magnetic permeability	henry per meter (H/m) equal to V·s/A·m
ρ (rho) resistivity, or volume resistivity	ohm-meter (Ω·m)
ρ (rho) also used for volume density of charge	coulomb per cubic meter (C/m³)
σ (sigma) surface density of charge	coulomb per square meter (C/m²)
σ (sigma) may also stand for conductivity, or volume conductivity	mho per meter, equal to 1/ohm· meter
Φ (phi) magnetic flux, or magnetic induction flux	weber (Wb), equal to V·s, also equal to W·s/A, and to J/A
Ψ (psi) flux linkage, or magnetic flux	weber (Wb). See above equivalents.
Ψ (psi) electric flux	coulomb (C). See also Q, charge or quantity of electricity.

VI / Problems of Pressures, Densities, Strong Drinks, Sound, and Flow

Galileo showed men of science that weighing and measuring are worthwhile. Newton convinced a large proportion of them that weighing and measuring are the only investigations that are worthwhile.
—CHARLES SINGER,
A Short History of Medicine (1944)

40 MEASUREMENTS AT THE BOTTOM OF THE OCEAN OF AIR

The man who undertakes to solve a scientific question
without the help of mathematics undertakes the impossible.
We must measure what is measurable and make measur-
able what cannot be measured.
—GALILEO
during his period in Padua, before
moving to Florence in 1610

. . . the beginning of modern science in the Renaissance
consisted in a new philosophy, which considered systematic
experiment to be the main source of knowledge.
—MAX BORN,
Experiment and Theory in Physics (1943)

The International System (SI) has a single unit for pressure, called the *pascal* (Pa), defined as the pressure resulting when a force of 1 newton acts uniformly over an area of 1 square meter. Thus 1 Pa = 1 N/m², and the combination "newton per square meter" will often be found in writings that do not use the pascal itself as a single unit.

The *tor* also has been used as the name for that unit of one newton per square meter, but it is now officially superseded by the pascal. Still very widely used is yet another pressure unit known as the *torr* (two *r*'s), which is substantially larger than the pascal. Thus 1 torr equals 133.322 pascal; and 1 pascal equals 0.007 5 torr.

Still other pressure units and unit combinations remain in use or in the literature. They are expressed in lengths of a column of mercury or a column of water (at some specified temperature). Or they relate a gravitational force, such as 1 pound of weight or 1 kgf, to a specified area over which the force is distributed. Among these, of course, are the familiar "pounds per square inch" (psi) used on the pressure gauges that test the tires on our arrogant motor chariots. The psi is much larger than the pascal, for 1 psi equals 6,894.757 2 Pa; and 1 Pa equals 0.000 145 psi.

These and other oddly assorted measures for pressure could be disposed of with perfunctory statements that the "tor" and the "torr" (both from the surname of Evangelista Torricelli [1608–47]) and the SI "pascal" (for Blaise Pascal [1623–62]) entered the terminology of metrology. However, the events, step by step, that led to understanding of pressure and its measurement are highly meaningful for all of metrology; hence some of these highlights of history are surveyed in the following pages.

It was Galileo Galilei (1564–1642) who first grappled directly with the problems of the measurement of pressure and with those of its opposite, absence of pressure or degree of vacuum. In this area, as in that of temperature measurement, he was only partly successful, but he pointed in directions that his disciples and successors were able to follow. Indeed, even the failures of Galileo seem more seminal in their influences than do the successes of many a less bold and basic investigator of nature.

Galileo was seventy-two when he wrote his final masterpiece, the *Dialogues Concerning Two New Sciences*. He was about seventy-four when it was at last published in Holland. He was by this time in part a recluse both because of age and the orders of the Inquisition, which confined him to a villa in Arcetri, near Florence. Yet the pages of that book are full of practical problems, of situations involving the labors and concerns of the people around him, for Galileo had closely watched the world and taken to heart the needs and the difficulties of its workers, craftsmen, builders and engineers. With a persistence that reminds at times of his so-different predecessor, Leonard da Vinci (1452–1519), he examined actuality—not only the world of nature untouched by human intervention, but also the technical needs and hang-ups of carpenters, masons, shipwrights and miners.

One of the most teasing technical difficulties of that time involved the efforts to drain the ever-deepening mines from which metallic ores were being taken in increasing amounts for smelting and fabrication. Various primitive drainage devices were in use, including scoops or buckets mounted on looped chains. By means of crude gearing these were powered from the surface, either by horses or oxen treading round a capstan, or in some instances by water wheels, turned by flowing streams.

A far more convenient drainage device, however, was the suction pump. It needed only a pipe reaching into the sump from which waste water was to be sucked up to the surface. The pump itself could be placed at the source of power. However, a baffling limitation soon enough became apparent: no suction pump, however carefully built or fully powered, could suck up water that lay more than about 10 meter below the pump itself.

Puzzled, miners came to Galileo with their problem. He did not find this beneath his dignity and status. Before long he was satisfied that some basic physical relationship imposed this distance or height limit. But what was its nature?

He was still grappling with that question when, in his vigorous and effective Italian prose, he wrote his *Two New Sciences*. The relevant passages reveal that, like the waste water from the mine drains, he had gone in the right direction—but not far enough. Some historians of science, however, have given Galileo less credit than he deserves in this respect.

Galileo had a clear enough idea of what a suction pump did when pulling water upward by less than that strange 10-meter-height limit. The pump first pulled some of the air from the intake pipe, then the water rose in that pipe to the pump level. Indeed it was no great mystery, since every Italian youngster who used a hollow straw to suck up water from a fresh pond was doing the same thing on a smaller scale.

But if removing air from the pipe could raise water by 8 meter, what prevented the process from working when the height difference was, for example, 12 meter? Galileo's efforts toward a tenable answer involved some remarkable early measurements. He set out deliberately to determine whether air indeed has weight and density and, as nearly as possible, how much of each.

Even the asking of such physical questions was iconoclastic in terms of establishment natural philosophy as derived from the works of the great authority Aristotle and his scholastic interpreters. The prevalent view was that air, like fire, was not possessed of gravity or weight. Indeed, air and fire were in this respect regarded as kinds of antimatter, possessed of levity, the opposite of gravity. Earthy substances, such as the ores taken from the mines, had a natural tendency to settle down to rest as close as possible to the interior of the Earth. Thus, solids and liquids had gravity, or were "grave" in their behavior, whereas air and fire had a natural tendency to rise and, so, had "levity." To speak of air having *weight* was a contradiction of the principle that air was naturally levitative and upward-tending.

Galileo weighs air itself—To determine whether air has weight Galileo used a great glass flask. Then with a syringe or pressure pump he squeezed into it as much extra air as he could, sealed its top, and weighed the flask (a "pressurized" flask, we would probably call it today). That was weight A. After allowing the excess air (pressure) to escape, he weighed it again, and found weight B. The difference (A minus B) corresponded to the downward force (weight) of the additional air that had been allowed to leak out.

In modern terms, we can suppose that a large flask with a capacity of just 1 cubic meter weighs 10 kg empty (that is, filled only with air at normal atmospheric pressure). Then, with a pump similar to those used to pump up bicycle tires, we push into the flask an additional 101,325 pascal of pressure, and seal off the flask before weighing it. (The 101,325 Pa is picked because that represents an additional 1 standard atmosphere of air pressure.)

The pressurized flask shows a weight of about 11.3 kg, but when that extra pressure has been allowed to escape, its weight drops back to 10 kg. The additional air forced into the flask must have had a weight of some 1.3 kg. This is substantial evidence that air is by no means a mere nothing, and that it cannot possess the inborn antigravity, or "levity," attributed to it by the Aristotelian dogmatists.

But does air have density? To have a measurable density there must be a magnitude of mass per unit volume under definite conditions. In the days of Galileo density was conceived rather as a weight (downward force) of such a mass per unit volume.

Galileo once again questioned nature by means of a large flask of Italian glass. First he weighed it empty (filled with air at surrounding pressure). The result was weight A. Next he put water into the flask in such a way that though the water entered, none of the air already inside was allowed to escape. The result was that water filled about three quarters of the flask's volume, while the air was squeezed into the remaining one quarter. Again he weighed the flask, finding weight B. Now he opened a valve in the flask, allowing the excess pressure of the trapped air to escape. Again he weighed the flask, finding weight C.

The following arithmetic is easy. Weight B minus weight C equals the weight of the air that had filled three quarters of the flask at the beginning—the same three quarters later occupied by water. Weight C less weight A equals the weight of the water that filled three quarters of the flask.

Galileo compared the magnitude of that water weight (C less A) with that air weight (B less C). The former weight was obviously much greater. Galileo's measurements led him to conclude that the air weight, and hence the air density, was about $\frac{1}{400}$, or 0.25 percent, of the corresponding amounts for water. In more modern terms, he found that the air around us has a specific gravity of 0.002 5, water being 1.0.

This measurement was almost double that revealed by modern techniques. Air at standard pressure has a specific gravity of about 0.001 3, or just a little more than half that indicated by Galileo's pioneer measurement.

Considering the crudity of the devices at his disposal, Galileo's finding was an extraordinary achievement. Even more extraordinary was the fact that he had dared to ask the question and had been able to devise a positive, step-by-step procedure for supplying an answer.

Now, having demonstrated that air—the atmosphere around us—has both weight and density, Galileo sought also to find a way to measure the force or attractive effect exerted by an absence of the usual amount or pressure of air. (Such a partial reduction of pressure takes place when a suction pump goes to work.)

Galileo used for this concept a phrase that now seems somewhat unfortunate or inverted, "the force of a vacuum." Today we would insist that a vacuum exerts no force, for the force comes the other way—from the pressure which is not counteracted by the vacuum.

In his *Two New Sciences* Galileo pictured a rather neat experimental device that he had invented for measuring this force of a vacuum. It consisted of a cylinder, closed at one end only, and of a wooden piston, "perfectly fitting" to that cylinder, and containing a valve that could be opened or closed by means of a metal lever extending to the outside.

The procedure he indicated was this: The cylinder was placed on its closed end and partly filled with water. Then the piston was carefully fitted into the open end of the cylinder and forced downward, with its valve open, allowing the escape of all the air between the water and the bottom of the piston. Next the valve was closed, and the cylinder carefully inverted, so its closed end was uppermost and the piston below.

The piston remained in place, even though it now supported the weight of the water. It was held in place by "the force of the vacuum," as Galileo called it.

But how great was this force? Attached to the bottom of the piston was a metal hook and fastened into this was a bucket into which could be placed "sand or any heavy material" in such quantity as to eventually pull the piston away from the water and cylinder above it. That total weight of bucket and contents, of piston, and of water resting on the piston, would indicate the "force of the vacuum" above a piston of that particular area.

We cannot know now, and probably shall never know, whether Galileo actually performed the experiment here indicated. It may well have been a project he outlined for future execution.

There is no doubt, however, what the approximate results would be if such a test were conducted carefully. If the area of the piston is just 1 square meter, then a weight of more than 10,000 kgf, or more than 10 tons, would be required to pull the piston out of the cylinder.

What Galileo had indicated was a demonstration of surrounding atmospheric pressure such as was carried out in 1654, a dozen years after Galileo's death, by Otto Guericke (1602–86), the mayor of Magdeburg, in Saxony, credited with having invented the first true air pump in 1650. Guericke's Magdeburg "hemispheres," made of copper, were held together solely by the "force of the vacuum" that his pump had created inside by evacuating air. Thirty horses, fifteen pulling each way, were unable to separate the hemispheres, while Emperor Ferdinand III and his retinue looked on in amazement. But when Guericke admitted air to the interior through a valve, the hemispheres simply fell apart of their own weight.

Galileo, however, was on the track of water and its response or lack of response to suction from above. His interpretation of what happened, or should happen, with his cylinder and piston was complicated by his ideas of the cohesiveness of the column of water above the piston or in the intake pipe of a suction pump. It was his error to assume that the vacuum exerted its "force" through such columns of water, and if the columns that could be lifted were limited in length, it must be because the water failed, through rupture or parting company with itself, to transmit that force from above.

This interpretation is supported by the following significant comment in his *Two New Sciences:* "Whenever a column of water is subjected to a pull and resists the separation of its parts, this can be ascribed to no cause other than the resistance of the vacuum."

Under ordinary conditions, water can be parted with ease. Every Italian boy who jumped into a river for a swim demonstrated this. However, Galileo believed that the "force of the vacuum" changed this—giving water a degree of cohesiveness, just enough to allow it to be lifted about 10 meter, but not more.

Galileo was here applying to water the same concepts of tensile strength that he applied to metals, stone and wood in his pioneer studies of the strengths of materials. He was much impressed by the idea that if a rod or column of any substance were long enough, it would break of its own weight when it was hung from the top. Or if its own weight was not sufficient to bring about such rupture, then hanging additional weight to its bottom, as in a bucket or basket, would do the trick.

Galileo conceived the column of water in the intake pipe of a suction pump as a sort of rod, hung from its top by the "force of the vacuum." Today we understand that such a column of water is supported *from the bottom* by the pressure of the atmosphere which acts on the water, and the rise of the column in the suction pipe toward the pump is caused by the fact that the pump removes most of the corresponding atmospheric pressure from the top of that column in the pipe.

The limit on the height of a column of water thus sustained is 10.333 m when the water is at its maximum-density temperature of about 4° C and when the surrounding atmospheric pressure is at the standard value of 1 atmosphere, or 101,325 pascal. The water cannot rise higher, because the atmospheric pressure of 101,325 pascal corresponds to the pressure at the bottom of such a column of water of just 10.333 m, no more and no less. If the surrounding air pressure were to be reduced by, say, 1 percent during the course of some change in the weather, then the maximum height of the water column would likewise be reduced by 1 percent, so long as the "force of the vacuum," or suction from the top, was used to raise that water.

Galileo's admirable errors—Correcting the mistaken ideas of Galileo is easy and insignificant today, thanks to other insights contributed by him and by thousands of scientists who followed along the trails he blazed. In singling out tensile strength of water and other substances for study and speculation, Galileo was actually employing the measurement relationship that underlies all pressure units or unit combinations—that is, he was relating a force to an area on which or through which it is exerted.

In the International System, that force is always the newton (N), and the area to which it is related is the square meter, thus resulting in the pascal unit, equal to 1 N/m^2. However, rupture tests on metal or wooden "rods" with a cross section of 1 square meter are unwieldy, if not impossible. Hence, engineers have often measured tensile strengths of rods or wires with cross sections as small as 1 square centimeter, or the 10,000th part of 1 m^2. And because of the prevalence of weights as units of force, these tensile-strength measurements have commonly been stated in terms of the number of kilograms of weight (kgf) required to rupture such a 1 cm^2 sample of the substance.

Thus, tin has in these terms a tensile strength of 280 kfg/cm^2, which corresponds to about 2.75×10^7 Pa, since 1 kgf/cm^2 equals 98,066.5 Pa. Steel wire, on the other hand, shows tensile strength from 10 to 12 times greater than that of tin. The tensile strengths of most metals are between about 1×10^7 Pa and 33×10^7 Pa.

Tensile strength, however, is a measurement concept almost totally inapplicable to water, whether or not it is subject to "the force of a vacuum." The molecules of water slide past one another so freely—water is, indeed, so *fluid*—that a column of the substance shows virtually no coherence and cannot be hung from its top at all. If the water is chilled until it freezes to ice, that is another story, which we need not examine here.

If there had been validity to Galileo's notion that suction pumps reached their height limit because of the limited tensile strength of water columns, then we should be obliged to assign to water a tensile strength of about 10^5 Pa, or about $\frac{1}{274}$ of the tensile strength of tin and about $\frac{1}{3,000}$ that of steel. The unquestionable fact that it was an invalid idea does not rob it of interest, nor—as we shall see—did Galileo's authority and influence so dominate his closest disciples that they hesitated to depart from and improve on his speculations.

Perhaps this was one of the greatest merits of Galileo as thinker and stimulator of thought; he did not seek to set himself up as an omniscient authority, or a counter-Aristotle. Indeed, some of the richest comments in his *Two New Sciences* have to do with this very aspect. In particular, he noted in conclusion that as he wrote, new ways and means had been opened up, "of which my work is merely the beginning." He predicted that by using these new ways and means "other minds sharper than mine" will explore "the most distant corners" of the new sciences.

Newton, who was born in the year in which Galileo died, more than once stated his awareness of the limitations of his own enormous contributions—even though he did not commonly preserve friendly feelings toward those who challenged his conclusions or contested his claims to priority. In a superb essay on the occasion of the 200th anniversary, in 1927, of the death of Newton, Albert Einstein stressed that Newton was himself more fully aware of certain weaknesses in the structure of his thought than was the generation of scientists who followed him. "This fact," Einstein added, "has always excited my reverent admiration . . ."

In the last stage of his long and laborious life, following the publication of *Two New Sciences,* Galileo, like Moses on Mount Nebo, was looking down on a promised land that others, not he, would enter and explore.

One of its great areas was that of pressure and pressure's opposite, or inverse, vacuum. Despite his inability to supply the missing elements to the problem of the limits of suction pumps, Galileo had effectively cleared out of the way of his followers one of the most stubborn and devoutly believed myths of natural science; it was known by the Latin name of the *horror vacui*. This was the doctrine that Nature abhorred a vacuum, did not tolerate it, and would go to extravagant lengths to prevent its appearance.

The deep-rooted nature of this doctrine was such that it was part of the standard or academic explanation of how solid objects, such as arrows, bullets or thrown stones, could travel some distance through the atmosphere. The explanation was necessary, for part of the orthodox Aristotelian interpretation was the thesis that the "natural state" of grave matter was motionlessness.

Why, then, did not an arrow come to a stop as soon as it left the bowstring? Why did not a bullet halt after it emerged from the barrel of the gun? What kept a stone flying after it had left the hand of the thrower? The original impetus of bowstring, gunpowder or throwing arm—so went the explanation—caused behind the flying object a vacuum or, rather, the threat of one. The surrounding air, to forestall any such obscene outrage against Nature, rushed into this space with such vehemence that the continued forward motion of the object was sustained.

According to this picturesque version of dynamics, far from having a retarding effect on objects thrown through it, the atmosphere was what kept them in continued motion. Even motion itself would be impossible in the void. Existence itself was alien to nothingness. Vacuum was regarded as the antithesis of all that was proper, ordained or allowable in nature. It was a sort of physical subversion —utterly intolerable and banned by the powers that be.

What about the endless motions of the Sun, the Moon, the planets, and the "fixed" stars themselves about the central Earth? They did not move in any abhorrent vacuum, according to this long-prevalent doctrine. Either they were mounted in crystalline invisible spheres which moved one over the other to the accompaniment of their own music, or they were conveyed and convoyed by flights of angels assigned to those endless and glorious tasks.

Moreover, the rules and relationships that prevailed in the terrestrial environment were basically different from those in the celestial. One could not conclude that what prevailed in the distant sky was of a piece with what happened on Earth, or vice versa . . .

It was against such backgrounds that the new physics of Galileo and his followers had to be established and developed.

Torricelli, the evangelist of vacuum—Evangelista Torricelli, who gave to Earth the abhorred vacuum, was more than forty years younger than Galileo. He had done well as a student in Faenza, and at nineteen he went to Rome for study at the well-named Collegio de Sapienza. His mathematics professor there was Benedetto Costelli, a friend and former student of Galileo. It was from Costelli that Torricelli gained an introduction to Galileo's revolutionary contributions to the understanding of mechanics and motion.

Soon after the publication of *Two New Sciences,* in Holland, Torricelli read it with deepening understanding and enthusiasm. It turned him increasingly toward experimentation and applied science and, within a few years, was instrumental in taking him to the side of Galileo himself.

Torricelli originally worked as what we might now call a "research mathematician." His investigations included the conic sections, the cycloids, and the so-called "acute hyperbolic solid." He developed further Galileo's demonstration that trajectories of cannon balls, bullets and thrown stones followed curves of the class called parabolas, one of the basic conic sections. One of Torricelli's works, included in his *Opera Geometrica,* was called *De Motu* (Concerning Motion).

It was not until 1641, the year before Galileo's death, that Torricelli, still in his early thirties, took up residence at Arcetri as an assistant and amanuensis to Galileo, now blind but still irrepressibly full of activity and projects. It was not to the aid of a spent or senile has-been that Torricelli came. The final years of Galileo are among the wonders and the glories of the history of science.

Something of their significance can be glimpsed from the visit a few years earlier of another young man, drawn by the renown of the greatest natural philosopher of his time. That young man was a well-to-do, cultured, earnest Englishman named John Milton, making his grand tour at age thirty. He was able to communicate in Italian, Latin, Greek or Hebrew, as well as English.

The old scientist made a deep impression on the poet from Albion. This appears powerfully in a passage published about six years later as part of Milton's famous prose polemic, *Areopagitica,* urging Parliament to vote against the required licensing (censorship) of books.

Recalling his travels, Milton wrote, "I could recount what I have seen and heard in other countries where this kind of inquisition tyrannizes . . ." Censorship, he said, "had damped the glory of Italian wits" to such an extent that "now these many years" nothing "but flattery and fustian" had been published in Italy.

Then, with the roll and thunder of characteristically Miltonic rhetoric, he recalled how there in Italy "I found and visited the famous Galileo grown old, a prisoner of the Inquisition, for thinking in astronomy other than the Franciscan and Dominican licensers [censors] thought."

The paths of science and literature crisscross throughout the broad realm of human culture. Milton himself, like Galileo, was to live out his final and perhaps most creative years in blindness. He dictated not only his two epics, *Paradise Lost* and *Paradise Regained,* but also the final and most intensely personal masterpiece, *Samson Agonistes,* a tragedy in the classical style with choral poetry of stupendous sweep and richness. More than likely, when he dictated the *Samson* so many years after his visit to Galileo, the recollection of that blind giant of thought was intermingled in Milton's mind with his own plight, under the Restoration regime.

Quite near the end of Galileo's life both Torricelli and a still younger man, Vincenzo Viviani (1622–1703), came to serve him and learn from him. During the first several months after Torricelli joined the household, Galileo was occupied principally with developing new ideas on impact—the accelerations produced when large forces act on bodies for relatively short periods of time. His impact concepts, though still unperfected, were in general directed toward the force concepts that were later to be systematized by the English genius Isaac Newton, as yet unborn. Galileo was in fact dictating to his disciples material on the impact problem when he became ill with a fever that led, rather rapidly, to his death just about a week after the beginning of 1642.

With the master gone, Torricelli received recognition in Florence and the state of Tuscany, of which it was the famed metropolis. He was appointed mathematician and professor to Grand Duke

Ferdinand of Tuscany, to this extent being recognized as Galileo's successor.

His real heritage from Galileo, however, was not a title or even notes and outlines. It was the bold, direct way of looking at the real world; of questioning nature by means of experiment; of measuring to determine magnitudes that could be analyzed and interrelated mathematically. It was, above all, the unintimidated but disciplined and tireless exercise of the imagination.

Torricelli takes up a Galilean problem—Galileo himself had drawn Torricelli's attention to the problem of the curiously limited length of the water column that could be sustained by suction-pump action. Torricelli, assisted by Viviani, continued these investigations.

Three factors were thought to play a role, according to the ideas that Torricelli developed: (1) the density of the water itself (its weight per unit volume); (2) the weight of the atmosphere that pressed down on the surface of the water that was being pumped, or whose pumping was desired; and (3) the removal, by means of the suction-pump action, of some of the usual atmospheric pressure from the upper part of the suction pipe.

To Torricelli it seemed that the best test of the validity of the first factor would be the use of a liquid with density far greater than that of water. There was but one obvious choice—the strange silvery liquid called *Argento Vivo* (Living Silver, or quicksilver, now known best as mercury), a metallic element whose specific gravity is about 13.5 times that of water.

The historic experiment itself, performed by Viviani pursuant to Torricelli's ideas, seems astoundingly simple. A long glass tube, sealed at one end, was filled to the other end with mercury. Then, with a thumb carefully closing off the open end, it was inverted into a bowl partly filled with additional mercury.

The tube was held upright. Almost at once the level of the mercury in it dropped from the upper (sealed) end and reached a level which, when measured, proved to be about 0.76 meter (76 cm) above the mercury in the bowl below. There the level stood in the tube, seemingly neither falling nor rising.

The two experimenters tried changing the shape of the tube. They tried the same operation with an arrangement that had a large bulblike expansion of the tube at its upper, sealed end. This did not alter the uniform height of about 76 cm above the level of mercury in the bowl.

What now filled the space above the top of the mercury column in the tube? Torricelli gave a bold and confident answer: *Nothing!* He was certain that the experiment had been performed in a way that prevented the entry of air or any other substance into that seemingly empty portion of the tube. The mercury itself in the lower part of the tube effectively sealed off the upper part. Hence this emptiness above the mercury in the tube must be that impossible and abhorred something, or nothing, the *vacuum* which Nature supposedly would not tolerate anywhere. Ever since that time, vacuums created by similar means have been known as Torricellian vacuums.

The detailed analysis supplied by Torricelli left no doubt except in the minds of skeptics. He showed that the actual height of the mercury in the tube had the same ratio to the maximum height that suction pumps could lift water as the density of water had to the density of mercury. In other words, in tubes of equal cross section, this new situation (vacuum above, ordinary air pressure below) would support equal weights of mercury on one hand and water on the other.

Torricelli cut the Gordian knot by indicating the pressure of the Earth's atmosphere as the supporting factor, not the "force" or "attraction" of the vacuum above the liquid in the tube. His interpretation is validated easily. If a small valve, inserted into the top of the tube, is opened after the Torricellian vacuum has been established, the level of mercury in the tube gradually sinks as more and more outside air pressure is admitted. Before long the level of liquid in the tube has sunk to that of the liquid in the bowl around it.

In a paper dated 1644 Torricelli offered interpretations to his contemporaries. He insisted that the force holding up the column of mercury in the tube was external to that tube; it was not a "something" or an effect residing in the vacuum at the top of the tube. In fact, in words of admirable clarity, he concluded that "the quicksilver rises to the point at which it comes into equilibrium with the weight of the outside air pressing down on it."

Finally, in words picturesque enough to survive to our own time and beyond, he reminded his fellow mortals that all of us "live submerged at the bottom of an ocean of the element air." This element, he concluded, is "known by experiments beyond question to possess weight."

From his own careful experiments he was able to report that many learned natural philosophers had been quite wrong about the character of a vacuum. Light passes through it freely. So do heat radiations. It does not interfere with the action of a lodestone (magnet). In fact, nature and nature's processes seem quite at home in and around this supposedly subversive and horrible vacuum.

Before his death at the early age of thirty-nine, hardly half a dozen years after his master and mentor Galileo, Torricelli made other discoveries and advances that cannot be detailed here. Had his life been somewhat longer, it is probable that the knowledge and influence of his epochal demonstration of the action of air pressure would have spread throughout Europe more rapidly than it did.

The torr, the bar, and the barye—Today the *torr,* named for him, is in wide use, especially for the measurement and comparison of extremely small pressures—the pressures associated with near-vacuums, in fact.

The torr is based directly on the length of the mercury column that corresponds to standard atmospheric pressure, commonly called for short just "1 atmosphere." At this pressure, the mercury column, when maintained at the temperature of $0°$ C, has a length of 760 mm; and 1 torr is the pressure that sustains just 1 mm of mercury, or $\frac{1}{760}$ of the pressure of 1 atmosphere. Even 1 torr is equal to 133.322 pascal; and 0.01 torr to 1.333 22 pascal.

For the full definition of 1 torr, we must add that the prevailing gravitational acceleration should be just the standard (g_0). If it is less than g_0, the mercury column will stand higher, whereas if the acceleration is greater than g_0, the mercury will stand lower than at g_0. Adjustments are made from prepared tables to compensate for deviations from the standard value of the gravitational acceleration and of the temperature to which the column of mercury is subjected.

A century or two ago a pressure of just 10^{-2} torr would have been regarded as a fairly "good" vacuum. Today, high-vacuum technology is far advanced and still advancing further. Makers and users of high-vacuum apparatus deal constantly with pressures as low as 10^{-7} to 10^{-11} torr!

Perhaps it sounds paradoxical to speak of so infinitesimal a measure as 10^{-9} torr as a "pressure," yet the presence of even so tiny an amount of gas as that indicates can drastically alter the results of a critical experiment involving particles and their interactions.

The torr was first proposed as a pressure unit shortly before the outbreak of World War I, in 1914. For a number of years it was used, largely in German-speaking countries; but in 1958 it was adopted by the British Standards Institution and is today a seemingly permanent part of the metrological picture.

Another pressure unit, rather widely used for high-pressure work, is the *bar,* its name being derived from the same root that has given us barometer, barograph, and so on. Its relationship with the approved pascal unit of the SI is a decimal one, for 1 bar equals 100,000 (10^5) pascal.

In fairly common use is the submultiple, millibar, equal to 10^2 pascal. A unit whose name is deceptively similar to that of the bar is the *barye,* its magnitude being just 0.1 pascal. The barye is a pressure defined as equal to a force of just 1 dyne per square centimeter. (The dyne was the force unit in the centimeter-gram-second [cgs] system, 1 dyne equaling 10^{-5} newton.) Since 1 cm^2 equals 10^{-4} m^2, it follows that 10 barye are identical with 1 pascal; and 0.1 pascal with 1 barye.

In high-pressure technology, pressures are often rated in kilobar units, and the pressures in stellar interiors may be rated in multiples such as the megabar or even more.

The following chapter makes clear something of the spread and extension of the great discovery by Torricelli and Viviani, and the justification for assigning the name of the pascal to the one approved pressure unit in the International System.

41 PILGRIMAGE FOR PASCAL

*Our dreadful marches [changed] to delightful
measures.*
 —SHAKESPEARE,
  Richard III

*The problem of physics is how the actual phenomena, as
observed with the help of our sense organs aided by instruments, can be reduced to simple notions which are suited
for precise measurement and used for the formulation of
quantitative laws.*
 —MAX BORN,
  Experiment and Theory in Physics (1943)

From Paris in November 1647 an intense and insatiable young
genius named Blaise Pascal wrote a letter to his brother-in-law,
François Périer at Clermont in the Auvergne. The letter, urgent
and ardent like its writer, proposed that Périer climb a mountain
in the interests of science, and with as little delay as possible.

Périer, a respectable counselor in the Court of Aids, complied
in due time. However, it was ten months before the report of the
resulting historic pilgrimage could reach Pascal in Paris. The
expedition, a strange mingling of stately formality and unprece-
dented experiment, had ascended the Puy de Dôme, an extinct
volcano that stands high over the countryside where Pascal himself
had been born.

That pilgrimage established, just as Pascal had expected and
intended it should, that with increasing altitude the atmospheric
pressure diminishes, and thus the Torricellian barometer became
also what we now call an altimeter, a device to measure heights
above sea level.

The pilgrimage up the Puy de Dôme, like many another idea of
brilliant import, was born in the highly original and independent

mind of Blaise Pascal. Yet the stimulus had come all the way from Florence, Italy, in one of those international exchanges that were becoming ever more common, as science became a boundary-vaulter and a language of its own among its adherents and enthusiasts.

Word of the mercury-column demonstrations by Torricelli and Viviani had reached one of Galileo's leading French disciples, the Abbé Marin Mersenne (1588–1648), a member of the order of Minim Friars, resident at the Convent of the Annonciades in Paris. Stimulated by Galileo and others, Mersenne had increasingly devoted himself to science and to mathematics, in which he made notable contributions, especially in its application to musical theory.

As a younger man, Mersenne had been a classmate of the great French philosopher, mathematician and methodologist René Descartes (1596–1650), and Descartes when in Paris was one of the group of savants who visited Mersenne to discuss the new ideas and insights that were bursting out in this era of accelerated breakaways from past patterns of thought.

The informal group in touch with Mersenne included also such thinkers as Pierre Gassendi (1592–1655); Pierre Fermat (1601–65), a mathematician of notable attainments; and now this Pascal (1623–62), not only the youngest but in many respects the most dazzling and unpredictable of them all. At the age of twenty-four when he proposed the Puy de Dôme expedition, he was already a leading creative mathematician, inventor, and writer of memorably beautiful French prose.

Not long before, Mersenne, carrying forward the work of Galileo, had carefully measured the length of the "seconds pendulum"—that is, the pendulum which, at Paris in this case, completes a cycle of to-and-fro oscillation in just 2 second. Precise and readily reproducible measurement of significant physical effects—this was part of the new spirit animating this group. Pascal, perhaps more than any of his older associates, was stimulated by these ideas.

He knew at this time only the bare facts about Torricelli's mercury columns, and he had not yet studied Torricelli's important analytical explanation of the phenomena in terms of atmospheric pressure. However, Pascal on his own quickly comprehended these essentials. Indeed, he drew from them conclusions that went beyond those of Torricelli himself.

A mind that amazes—The great leaps of Pascal's insights seem astounding, even after so long a time. The critic Walter Pater, writing more than two centuries later, said of him: "Physics, mathematics, were with him largely matters of intuition, anticipation, precocious discovery, short cuts, super-guessing. It was the inventive element in his work, and his way of painting things that surprised most of those able to judge."

These rare gifts had been marked when Pascal, as a mere boy, had independently discovered or derived difficult mathematical principles that had never been taught him by tutor or textbook. Pater, again, declared, "He might have discovered the mathematical sciences for himself, it is alleged, had his father, as he once had a mind to, withheld him from instruction in them." Before he was eleven, Pascal had taught himself most of geometry as then known. He used charcoal to draw the diagrams on the tiles of his playroom. Not being acquainted with the traditional names of the principal geometric curves, he had devised his own designations for them, thus originating his own nomenclature in addition to the principles underlying them.

Even before Pascal received from Périer the report on the pilgrimage up the Puy de Dôme, he had been certain what those results *should* be. Certain as he was that air had weight, he knew also that measurable differences in its pressure must show themselves as measurements were made at ever greater heights above sea level.

Pascal was determined to do away, once and for all, with the old Aristotelian dogmas of the natural levity of air and fire, as opposed to the natural gravity of water, earth and other solid matter.

"It is commonly known," he wrote, "that a balloon when inflated weighs more than it does when empty." However, "if the air possessed levity, then the more the balloon was inflated, the lighter it would be, for it would contain more air."

The defenders of Aristotelian physical orthodoxy had tried to explain this by saying that the apparent weight of air was the result of impurities and vapors that it contained (smog being then blessedly unknown). Genuine pure air, they insisted, should exhibit levity, not gravity; it should seek ever to ascend, and the more of it there was, the less would be its apparent weight.

Pascal's enduring fame as a polemicist was well deserved. He made short work of such *ad hoc* objections. "I am not familiar with *pure* air," he wrote with a fine razor-edged scorn, "and I believe it may be very hard to discover. I am dealing here only with the sort of air we breathe, no matter what may be its ingredients." And, he continued, "it is the entity that I call 'air,' and *that*, I declare, does have weight."

The letter that Périer wrote in September 1648 at last brought to Pascal word of the memorable pilgrimage he had instigated. Pascal's instructions had been followed. First, at the foot of the Puy de Dôme, Périer had carefully measured the height of the mercury column in a Torricellian barometer. It stood at an apparent height that, in the future metric system, was to be designated as about 663 mm. Then the strange procession began to move up the slope.

Périer had not undertaken to conduct so crucial an investigation alone; he had assembled for the occasion "a distinguished group of professional and ecclesiastical gentlemen," who could lend validation as well as prestige to the outcome. When the group had climbed to a level nearly a thousand meter above their starting point, the Torricellian mercury column was set up again. This time it stood only some 586 mm above the surface of the mercury in the bowl below.

There had been a definite loss of some 77 mm, or about 11 or 12 percent, in its length. The pressure of the atmosphere at the bottom of this famous mountain was obviously able to sustain a substantially greater weight of mercury than was the atmospheric pressure near the top. Quite approximately, for each 100 m of ascent, the mercury column had dropped an average of 7½ to 8 mm.

Typically for his talent and temperament, Pascal was ready with an explanation before the inevitable questions could properly be asked. "The air's weight and pressure is the only cause," he announced.

Then he dismissed with his characteristically sharp irony the long-cherished doctrine that nature abhors a vacuum. The cause of what happened on the Puy de Dôme could not be "the abhorrence of the vacuum, for it is totally certain that more air pressed on the foot of the mountain than on its summit, and one cannot well say that nature has a greater abhorrence for a vacuum at the foot of the mountain than at its top."

(Pascal is one of the gifted few in science history whose felicity of expression makes ordinary science literature seem hopelessly inept and limping . . .)

Today, the city of Clermont-Ferrand serves as the capital of the Department of Puy de Dôme, named for that same mountain. This historic town of the lovely Auvergne has become a manufacturing center, sometimes referred to as the Akron of France. Its factories produce tires, rubber goods, plastics, textiles, chemicals, and electronic and mechanical products.

Amidst modern traffic, vapors, fumes and odors, the community of Clermont-Ferrand counts as its glory the fact that here was born one of the most memorable geniuses of French culture and of Western thought. Modern atlases, even in brief notations about Clermont-Ferrand, mention that it was the birthplace of Blaise Pascal.

His influence persists, in ways often overlooked. For example, in the great industrial plants of that region and elsewhere in the modern world, huge hydraulic presses exert forces of tens of thousands of kilograms, or multimillions of newtons, actuated by relatively small hydraulic pumps. They put into practice the principles of transmission of pressure in liquids, first enunciated clearly by Pascal. Those pressures, indeed, can be and should be measured in units of the pascal (Pa)!

Tourists, rushing through the region to visit its Roman ruins and lovely countryside, depend for their safety on the hydraulic-brake systems in their automobiles—systems operating by means of the same transmissions of pressures through liquids that Pascal studied and explained.

Meantime, in the schools of Clermont-Ferrand, and indeed in those of all France and elsewhere where the French tongue is studied and appreciated, the writings of Pascal still stir and excite lovers of language. His is the brilliance of clarity and pithy precision, not the pseudo profundity of long-winded obscurity.

It was this Pascal, student of pressure, atmospheric and liquid, who was so aware of compression of expression that he asked to be pardoned for having written a long letter because "I did not have time to write a short one." This is but one of his winged expressions that soar above the centuries of humdrum communication.

Pascal is most often identified with his *Pensées* (*Thoughts*), a strange, incomplete collection of fragments and jottings, intended originally to become a definitive defense of Christianity. Among those fragments are scattered such enduring gems as these:

"The heart has its own reasons, which are quite unknown to the head."

"Man is not more than a reed, the frailest in nature—but he is a *thinking* reed."

"If Cleopatra's nose had been shorter, the whole face of the earth would have been changed."

"The more intelligence one has, the more people one finds to be original. Commonplace people see no differences between men."

And also one of those pithy profundities to which the present author, and so many others, must pay tribute: "The last thing one discovers in writing a book is what to put first."

Enduring principles and premature conjectures—Every weather forecast by a modern meteorologist makes use of principles that Pascal first clearly called to the attention of his world. Indeed the concept of a describable structure and spatial differentiation of the Earth's atmosphere can be said to begin with him. He even extrapolated his conceptions and assumptions to the point of venturing an estimate of the total weight of the Earth's atmosphere. In terms of modern SI units, he calculated that this atmosphere must exert a total downward force of about 1.73×10^{14} newton. This corresponds to 1.76×10^{13} kgf (kilograms of force, or weight), and to about 3.9×10^{13} lbf (pounds of force, or weight).

The total area of the Earth's surface is about 5.1×10^{14} square meter. By Pascal's estimate of a total atmospheric force of 1.73×10^{14} N over this area, the average atmospheric pressure would be about 0.34 pascal (newton per square meter). However, standard atmospheric pressure is actually 101,325 pascal.

It would seem that Pascal's bold estimate was about 300,000 times too small! Indeed, twentieth-century calculations indicate that the total mass (not weight) of the Earth's atmosphere is about 5.2×10^{18} kilogram. This includes the mass of the denser atmosphere at sea level and that of the less dense air at great altitudes, all the way up to heights at which the air is so thin as to be virtually negligible. The mass of the entire Earth is more than 1 million times the mass of its atmosphere.

Pascal's early estimate of the "weight" of all the air was far from what is revealed by the measurements that have been made since his time. Its importance today lies in the fact that he could conceive of such a total weight and the further fact that even his enormous intuitive brilliance in approaching natural phenomena was not sufficient, in the absence of adequate measurement data, to come close to the realities.

Within a few years after publication of the measurements made in the strange processional for Pascal up the slopes of Puy de Dôme, great numbers of measurements of mercury columns were coordinated with the known altitudes at which those measurements had been made.

These new barometers responded not only to changed altitudes, but also to continual and unmistakable variations in atmospheric pressure resulting from changes in "the weather." Today, the standard or "normal" air pressure at sea level is considered to be 101,325 pascal, or 760 mm of mercury, or 760 torr. This is a sort of agreed reference midpoint. When the weather produces "an area of low pressure," the mercury column may fall by 1 percent or more, even though the barometer remains located just where it was.

The standard atmospheric pressure of 760 mm of mercury (at 0° C) corresponds to 29.92 inches of mercury at the same temperature. Barometers sold for general use in the United States are usually calibrated in these inches-of-mercury units. At the moment of writing these lines, the author's least unreliable barometer indicates almost exactly 29.9 inches of air pressure, at an elevation of not more than about 15 feet above the mean sea level of the nearby Pacific Ocean. The dial of that barometer is calibrated all the way down to 26 inches to deal with extraordinarily low pressures, and up to about 32 inches, to deal with unusually high pressures.

Thus, the instrument is at least theoretically capable of responding to air pressures down to some 15 percent below the agreed standard, or up to some 7 percent above that standard (29.92 inches).

The pilgrims of the Puy de Dôme processional were not at sea level when they made their initial reading of the length of the mercury column. They were, in fact, about 500 m above sea level. This (together with the state of the weather) accounts for the fact that their initial reading was only some 663 mm, or between 12 and 13 percent less than the presently agreed "standard" air pressure at sea level.

By the time they had climbed to their highest point on the old volcano, they were at an altitude of about 1,500 meter above sea level, and the mercury column was correspondingly shorter than at the start of their climb.

Pascal in Paris was stimulated further by the confirmation that came to him with the report from his brother-in-law Périer. He lost no time in attempting some even more difficult measurements on his own. No proper mountain was conveniently near, and the Eiffel Tower was some centuries ahead, in a future that not even Pascal could foresee. However, he found near Paris the tower named for St. Jacques de la Boucherie, standing nearly 50 meter in height.

He measured a mercury column at its base, then climbed with it to the top of the tower and measured again. According to his report the column had lost something more than 2.5 mm in length as a result of that 50-meter climb.

Never given to halfway or leisurely activity, Pascal then applied his new altimeter to private houses of much more moderate height than the tower. Between ground level and the top story, "about ninety steps high," he reported in one instance he had noted a difference of about $\frac{1}{3}$ mm in the length of the mercury column. Moreover, he declared that these results were in perfect agreement with the measurements reported by his brother-in-law from the Puy de Dôme.

Historians of science may be forgiven if they suspect that Pascal, in an excess of excitement and zeal, somewhat "doctored" these almost infinitesimal measurements. Differences of a fraction of a millimeter were hardly to be taken seriously under the conditions then prevalent with the Torricellian tubes of mercury. However, the accelerating technology of the nineteenth and twentieth centuries has justified Pascal to this extent: it is possible now to register beyond serious doubt even the tiny diminution in atmospheric pressure that results from an ascent of "about ninety steps."

Pascal was aware that atmospheric pressure does not have a straight-line correlation with altitude increase. An ascent from sea level to 1,000 m altitude results in a greater loss in the length of the mercury column (and hence a greater drop in pressure), than does the ascent from 1,000 to 2,000 m altitude, and so on.

Today's abundant and precise measurements of atmospheric pressures enable us to sketch the approximate pattern in some detail. As a rocket climbs from sea level to the altitude of 10 km, the air pressure drops from about 760 to 210 mm of mercury—a loss of some 72 percent of the initial pressure. Between 10 and 20 km altitude, a further loss of 178 mm of mercury is noted—or about 23 percent of the initial sea-level pressure. At 30 km the pressure is down to about 5 percent of that considered standard at sea level.

At 40 km the pressure has fallen to 9.5 mm of mercury, or just 1.25 percent of the sea-level standard. Finally, at 50 km altitude, the remaining pressure is only 0.75 mm of mercury, or less than $\frac{1}{10}$ of 1 percent of its sea-level value.

In terms of total pressure drop, the first 10 km of ascent above sea level brings a change about 3 times that of the second; and the second about 5.5 times that of the third; the third about 4.5 times that of the fourth.

Thus the pattern of atmospheric-pressure decrease with altitude increase is not a simple one. However, it has been carefully studied, especially in those altitudes in which most air traffic now moves.

42 THE PASCAL AND OTHER UNITS FOR SQUEEZING, STRETCHING, AND TWISTING

Cedar that resists and reed that bends . . .
—VICTOR HUGO,
Eviradnus

. . . *heart with strings of steel*
Be soft as sinews of the new-born babe.
—SHAKESPEARE,
Hamlet

Pressure, such as that exerted by the atmosphere on the surface of the Earth or by a liquid on the walls of the vessel that contains it, is only one of the physical variables constantly measured by scientists and technicians in terms of a force and the area over which it is exerted. Indeed, the pascal (Pa) and the other, non-SI units by which pressures are measured are also the units used to measure a variety of squeezing, stretching, compressing and twisting effects, highly important in the technology of our times.

How do bodies of matter respond to forces that are applied to them? These responses can be grouped into three principal forms or patterns, depending on the direction of the applied forces relative to the surfaces of the body they act on.

First place here is given to compression, or squeezing, acting on all surfaces of a body. The atmosphere has been shown to press down everywhere on Earth's surface with an average pressure of about 101,325 pascal, or newton per square meter. This is equivalent to 14.7 pounds per square inch (psi) in the antiquated units still prevalent in the United States.

If an object is carried down under the surface of the sea, the pressure exerted all over it rises rapidly, at a rate of about 10,000 pascal additional per meter of descent under the surface. At a depth of about 10 meter, the pressure has risen to 2 atmospheres; in other words, has doubled. By about 1,400 meter depth the pressure is about 140 atmospheres, or 1.4×10^8 pascal—roughly 1 ton per square inch.

If a solid object is exposed to such considerable increases in ambient pressure, it may be expected to compress and so to occupy a smaller volume than above the surface of the sea. The question is, by what proportion will the former volume be reduced by some specific increase in the external pressure? Some substances obviously are compressed far more easily than others. For instance, certain kinds of soft rubber. Others, such as steel, seem highly resistant to compression.

Such differences in compressibilities can be expressed quantitatively with the help of the pascal unit. Suppose, for example, that a measured pressure rise of 100,000 pascal is found to reduce the volume of a body by 2 percent. The force per unit area is regarded as the *stress* on the body, and the decrease of 2 percent in volume is a distortion or yielding referred to as the *strain* exhibited by the body.

Strain measurements are free of unit tags. They are, accordingly, called "dimensionless." The 2 percent reduction in volume is found whether the measurements are made in cubic centimeters, meters, inches, feet or even miles. The numerical measure of a stress, however, varies according to the units used. Thus the increase of 100,000 pascal in the example becomes an increase of 14.5 psi, or of 0.986 923 atmosphere, and so on.

Now, what is the "compressibility" of the substance in the example just given? It is found by dividing the strain (distortion) by the stress which causes it. In this case, since 10^5 pascal produced a strain of 0.02, the fraction is $0.02/100,000$, or 2×10^{-7} per pascal. In physical terms it means that each additional pascal of pressure further reduces the volume of this substance by 2 parts in 10 million.

Certain cautions need to be noted. This compressibility measure implies that an increase of ambient pressure by 5 million times would result in the reduction of volume by 10 million parts in 10 million—that is, by the volume occupied by the substance being reduced to *zero*. This, of course, does not take place. As the molecules and atoms are crowded ever closer together, they "fight back" harder, and most substances can be expected to show decreased compressibilities under extremely high pressures. But within the pressure ranges readily attainable in factories and most laboratories, such changes are seldom likely to be important.

The compressibility of 2×10^{-7} per pascal for the substance in the example just given is substantially greater than that measured for many familiar materials. For example, it is about 25,000 times greater than that typically shown by steel, which is about 8×10^{-12} per pascal.

The meaning of a little measure—Engineers and scientists generally use the inverse or reciprocal of compressibility, and so obtain a measure which in this case they call *the bulk modulus, the volume modulus,* or *the modulus of volume elasticity.* In each instance, modulus is derived from a word which meant "a little measure." The usual symbol for bulk modulus is either K or B.

Every bulk modulus is derived from dividing the strain (distortion) into the stress that causes it. In the example already given, the fraction is $100,000\%_{0.02}$ or 5×10^6 pascal. This is the reciprocal of the compressibility of 2×10^{-7} per pascal.

Here are typical bulk modulus (B) and compressibility $(1/B)$ figures for some significant liquids and solids.

SUBSTANCE	BULK MODULUS (B)	COMPRESSIBILITY $(1/B)$
ether	1.3×10^7 pascal	7.7×10^{-8} per pascal
alcohol	1.13×10^9 "	8.8×10^{-10} " "
water	2.13×10^9 "	4.7×10^{-10} " "
lead	8.0×10^9 "	1.2×10^{-10} " "
mercury	2.76×10^{10} "	3.6×10^{-11} " "
glass	4.0×10^{10} "	2.5×10^{-11} " "
brass	6.1×10^{10} "	1.6×10^{-11} " "
aluminum	7.0×10^{10} "	1.4×10^{-11} " "
iron (cast)	9.4×10^{10} "	1.1×10^{-11} " "
copper	1.2×10^{11} "	8.3×10^{-12} " "
steel	1.6 to 1.3×10^{11} "	6.2 to 8.0×10^{-12} " "

Water is often regarded as virtually incompressible. However, the table shows that it is more than 10 times as compressible as mercury and about 19 times as compressible as glass. The great specialist in high-pressure physics, Percy W. Bridgman, by applying 2×10^9 pascal (20,000 atmospheres) pressure was able to compress a sample of water to 25 percent less than its normal volume. This result indicates a compressibility of about 1.23×10^{-10} per pascal for water, whereas the table indicates nearly four times as much. This seeming decrease in compressibility or rise in the bulk modulus is probably the result of the crowding together of water molecules under the extremely high pressure. Water under the pressures prevalent around us shows the table-indicated compressibility.

Some substances simply are crushed or disintegrated when they undergo pressure well below that equivalent to their measured bulk modulus (B). Here are typical limits of resistance to crushing for four familiar building materials: (1) concrete, between 6 and 26×10^6 pascal; (2) limestone, between 41 and 170×10^6 pascal; marble, between 5 and 140×10^6 pascal; and granite, between 67 and 230×10^6 pascal.

Once crushed or disrupted, of course, such a substance does not resume its former volume and shape when the excess pressure is removed. Thus B, the bulk modulus of a substance, usually remains useful only in the ranges of pressure from which that substance can return elastically to its former volume and shape.

The most-used modulus of all—Similar cautions apply in the case of another elastic modulus, even better known than B. It is commonly called *Young's modulus* and is symbolized by either Y or E. It is designed to measure the responses of materials to forces that tend to stretch them in one particular direction, or to compress them in one direction. Most typically Y is measured for columns, pillars, rods, beams and wires.

More precisely, Young's modulus is a measure of the longitudinal force required to produce a particular uniaxial extension (or compression) in a substance. As such, Y is the reciprocal of what could be called the *stretchability* (or the *linear compressibility*) of that substance.

Each measured Young's modulus results from a fraction whose numerator is the strain produced by stretching (or compressing) a wire, rod, or bar of the substance, and whose denominator is the stress, or force per unit cross section (measured in a pressure unit such as the pascal) by which that strain was produced.

For example, if a pull or tension of 10,000 newton is exerted on a metal rod or bar with cross section of 10^{-4} square meter, then the stress is $10,000\!/_{10}{}^{-4}$, or 10^8 pascal. Now, if this stress is found to extend (elongate) the rod by 0.1 percent, or 1 part in 1,000 of its former length, then the linear stretchability of that metal is $10^{-3}/10^8$, or 10^{-11} per pascal, and its Young's modulus is the reciprocal of that, or 10^{11} pascal.

Here are typical Young's modulus (Y) figures and linear stretchability-compressibility $(1/Y)$ figures for some familiar substances.

SUBSTANCE	YOUNG'S MODULUS (Y)	STRETCHABILITY-COMPRESSIBILITY (1/Y)
oak	1.0×10^{10} pascal	1.0×10^{-10} per pascal
lead	1.7×10^{10} "	5.9×10^{-11} " "
brick or concrete	2.0×10^{10} "	5.0×10^{-11} " "
glass	5.0×10^{10} "	2.0×10^{-11} " "
granite	5.5×10^{10} "	1.8×10^{-11} " "
aluminum	6.8×10^{10} "	1.5×10^{-11} " "
brass	9.2×10^{10} "	1.1×10^{-11} " "
copper	1.1×10^{11} "	9.0×10^{-12} " "
iron (cast)	1.2×10^{11} "	8.3×10^{-12} " "
steel (mild)	2.1×10^{11} "	4.8×10^{-12} " "

Here again certain cautions must be noted. The Young's modulus retains its meaning and usefulness so long as the stress does not exceed the substance's *limit of proportionality* (L.P.). Below that limit the relative lengthening or shortening (strain) changes approximately in proportion to relative change in the exerted stress. Above that limit, the strain begins to increase more rapidly than the stress. This continues until the stress attains the *elastic limit* (E.L.), the maximum from which that substance can return elastically, when once the stress is removed. When the elastic limit is passed, the body remains permanently altered or deformed.

Still higher in the scale of increasing stress lies the *yield point* (Y.P.). When this is passed the deformed body continues to grow longer, even when the stress on it does not grow.

Finally the stress attains the point of *ultimate strength* (U.T.) or tensile strength (T.S.), which causes the bar or wire to tear apart and thus destroys it.

Steel, with a Y of about 2.1×10^{11} pascal, attains its elastic limit at about 2×10^8 pascal. This means that it can be stretched by no more than about 1 part in 1,000, and thereafter return elastically to its original size and shape. Even special heat-treated spring steel can be stretched no more than about 1 part in 200 without permanent deformation.

Some substances exhibit relatively high elasticity but relatively low ultimate strength. Glass at room temperature is one such. Other substances show relatively low elasticity but relatively high ultimate strengths. These include certain plastics and synthetics.

Engineers and technicians often make crucial comparisons in terms of the pascal and other pressure units. For example, rolled manganese bronze, with a Young's modulus of 1.1×10^{11} pascal, reaches its elastic limit at 2.4×10^8 pascal, or when it has been stretched by only some 2 parts in 1,000. Finally, it arrives at its ultimate strength and ruptures at 5.9×10^8 pascal, a stress which is only some 2.5 times the stress that brought it to its elastic limit.

Young's modulus is named in honor of the amazing Dr. Thomas Young (1773–1839), possibly the most versatile and widely gifted of scientists who have worked in Britain. Several chapters would be required to list even briefly his principal contributions to the physical and life sciences, to say nothing of scholarship and linguistics.

G, the modulus of torsion, twist, and shear—A third important modulus is related to the twistability of substances. It is commonly called the *shear modulus* and is symbolized by G. It is the reciprocal of the very common physical variable of *twistability*. Its measurement can be visualized rather simply. Picture a cube of the substance, of unit length (in this case a cube 1 meter on each edge). One side, which can be called the bottom, is rigidly anchored. To the opposite side, the top, is now applied a force, directed along the plane of that side and parallel to the two opposite sides of the cube (hence at right angles to the face of the cube toward which the force is directed).

Suppose this to be a force of 10,000 newton. Hence, the top of the cube is subjected to a stress of 10,000/1 m², which equals 10,000 n/m², or 10,000 pascal. Had the same force been applied to a cube only 1 centimeter on edge, and hence with sides having only 10^{-4} m² area, then the stress would be $10,000/10^{-4}$ m², or 100,000,000 pascal.

Such stresses produce twisting, or torsion, effects of various extents, depending on the substance in question. The strain, or change of shape in response to such stresses, is measured in terms of the angle produced on the sides of the cube that are parallel to the direction of the exerted force and perpendicular to the top and bottom surfaces. The angle of tilt produced by the stress is measured in the SI angular unit, the radian, equal to nearly 57.3° of angle.

If the 10,000 newton force on the top is found to produce a tilt of 0.01 radian in the cube of 1 cubic meter volume, then the resulting shear modulus, or G, for the substance forming the cube must be 10,000 P/0.01, or 1,000,000 (10^6) pascal. That is the G figure. The twistability ($1/G$) figure is, of course, the reciprocal, or 10^{-6} per pascal, for that substance.

Sometimes the name *modulus of rigidity* is used instead of shear modulus, but the symbol for the variable is almost always given as G.

Here are the approximate G and $1/G$ magnitudes as measured for a number of familiar substances.

SUBSTANCE	SHEAR MODULUS OR MODULUS OF RIGIDITY (G)	TWISTABILITY (1/G)
lead	7.0×10^9 pascal	1.4×10^{-10} per pascal
glass	2.0×10^{10} "	5.0×10^{-11} " "
aluminum	2.6×10^{10} "	3.8×10^{-11} " "
brass	3.7×10^{10} "	2.7×10^{-11} " "
iron (cast)	4.9×10^{10} "	2.0×10^{-11} " "
steel (mild)	8.0×10^{10} "	1.3×10^{-11} " "

The G, shear modulus or rigidity modulus, and the Y, Young's modulus or stretchability modulus, are not totally unconnected and independent for individual substances. For many of the most familiar and important construction metals, the G has less than half the magnitude of the Y. In fact, it is typically found within the range of 37 to 42 percent of the Y figure. Thus the percentage equivalent of the G measure divided by the Y measure is 27 percent for rolled or drawn copper, 37 percent for aluminum, 38 percent for manganese bronze, 39 percent for brass, 40 percent for high carbon steel, and 41 percent for cast iron.

The B, or bulk modulus, for solids ranges in value anywhere from less than 50 percent of Y to 110 percent of Y. Obviously, in liquids or gases there are no Y or G measures, though they do have B, bulk modulus, measures.

For most solids, one strange-appearing relationship exists, approximately, among the three elastic moduli.

$$Y \approx \frac{9\,B{\cdot}G}{3\,B + G}$$

This means that Y is approximately equal to a fraction whose numerator is 9 times bulk modulus times shear modulus, and denominator is 3 times the bulk modulus plus the shear modulus. In some instances, the fraction may be as much as 5 or 6 percent greater than Y, in others as much as 5 or 6 percent less than Y; but mostly it is found to be quite close to Y.

Elastic moduli at work—The measured elastic moduli of basic construction materials are used by engineers and technicians to calculate in advance the capacities and responses of beams and structural members of various sizes and shapes. Thus they solve problems that plagued builders and designers since long before the days of Galileo. Indeed, it was Galileo who may be said to have made the solutions to such intensely practical problems part of the legitimate and respectable scope of science.

He was intensely concerned with problems of structural strength and safety, and he brought to them the best of his mathematical and analytical insights.

Let us illustrate the use of measured moduli of materials under several typical and common conditions. The first is that of a rigid beam, supported on one end and bearing a load on the other—a cantilevered beam, in short. Let us call the length of the beam L, its depth D, and its width W. Also we know the downward force of the load F at one end. Finally, we know Young's modulus Y of the substance from which we propose to make the beam.

Our question is: How much deflection downward will take place at the load-bearing end of that cantilevered beam? The recipe or formula to use in this case is

$$\text{Deflection} = \frac{F}{Y} \times \frac{L^3}{WD^3}$$

At once we see that the amount of the deflection will be inversely proportional to the Y of the substance composing the beam. Also that the deflection increases as the cube of the length from the point of support to the load-bearing end. Likewise it *decreases* as the cube of the depth of that beam. This mathematical recipe contains volumes of information explaining the favored shapes and cross sections of structural members used when strength is essential to safety.

The next question: How much deflection takes place in a rectangular beam supported at both ends and bearing its load in the center, halfway between those ends? This time the recipe is

$$\text{Deflection} = \frac{F}{4Y} \times \frac{L^3}{WD^3}$$

Yet another common situation is that of a horizontal beam bearing no load except its own weight, which is here symbolized by F, as the total earthward force of the beam. This beam too is supported at both ends only:

$$\text{Deflection} = \frac{5}{32} \times \frac{F}{Y} \times \frac{L^3}{WD^3}$$

These useful measures can be employed also when analyzing behavior of beams that are not to be mounted horizontally. Young's modulus Y also helps to predict how much load a vertical pillar or rod can support before it will begin to buckle. Once again F stands for the force, or weight, of that load. L now stands for the height (length) of the column, supported from below. The other two dimensions, W and D, refer to its cross section, assumed here to be oblong. The recipe here makes use of the familiar π, the ratio of the circumference of a circle to its diameter (3.141-59 . . .):

$$\text{Maximum F} = \frac{\pi Y}{12} \times \frac{WD^3}{L^2}$$

If the cross section of such a vertical beam or rod is square, with a side of size S, then the recipe changes somewhat:

$$\text{Maximum } F = \frac{\pi Y}{12} \times \frac{S^4}{L^2}$$

One final aspect of Y, the important Young's modulus, may offer an element of surprise. What is the physical significance of Y? The fact is that it represents for any substance the tensile (stretching) stress which theoretically should *double* its unstressed length . . . Actually, for all building materials of importance, such doubling is in fact impossible. Long before it could take place the material has reached elastic limit and also its ultimate strength limit.

Thus, the basic moduli, indispensable to the mechanics of materials, must be used with constant reference to the realities. Here again metrology is revealed as the area of science and technical endeavor in which theory and practice merge and interact most fruitfully.

43 DEALINGS WITH DENSITIES

The physical variable called density offers few obstacles at the outset. Being simply the mass per unit volume, it is measured in the International System by the basic combination of kilogram per cubic meter (kg/m^3), and in non-SI units by such combinations as pounds per cubic foot.

Indeed, it was the density of water, at its maximum density temperature of about $4°$ C, that originally served in the attempt to derive the basic metric mass unit from the length unit, the meter. The gram, first unit of mass, was to be the mass of a cube of maximum-density water with a volume of just 1 cubic centimeter. And so nearly was that intent realized that, for all but the most precise computations today, each cubic centimeter of pure water near that $4°$ C level can be regarded as having just 1 gram mass.

Because the SI uses the kilogram rather than the gram as the basic mass unit, and the meter rather than the centimeter for all unitary areas and volumes, it can be stipulated that 1 m^3 of pure water, since it contains 1 million cc, represents a mass of 1,000 kg. Thus, the SI figure for the density of water is, for all ordinary purposes, 1,000 kg/m^3.

Ever since the days of Archimedes, whose detective work with the aid of density measurements has been told and retold and need not be repeated here, the density of water provided the common and almost inevitable standard of reference for other densities.

Thus arose the physical variable now known as *specific gravity,* which expresses densities of other substances as multiples or submultiples of the density of water.

Even more practically stated, specific gravity is the *ratio* of the weight of a given volume of the tested substance to the weight of a like volume of pure water. The word *gravity* may obscure the fact that this is really a measure of density.

Besides density, or mass-per-unit-volume, its reciprocal is sometimes used—that is, the volume-per-unit-mass. For any given substance or situation this is the volume, in cubic meters, that contains just 1 kilogram of mass. In the case of water, this specific volume is 0.001 m³. Originally it was intended that the liter, metric measure of capacity, should have just that volume. However, unintended obstacles led to a very minor deviation from this goal, and the liter is now known to be 27 parts per million larger than just 0.001 m³. In truth it has a volume of 0.001 000 027 m³!

Considerations of this kind had not arisen when, in the 1670s, the important British natural philosopher Robert Boyle (1627–91) made public the details of the first hydrometer, a simple device for determining densities, either absolute or comparative, of liquids.

Boyle's original instrument had the essentials of the millions of hydrometers that have been constructed and used since then. A weighted bottom, the weight being placed in a lesser bulb; a float above, in the form of a large bulb; and above that a stem, the extent of whose projection above the surface of the liquid being tested indicates by means of some suitable scale what that liquid's density is. By the same principle that causes bathers to float higher in the Great Salt Lake than in Lake Tahoe, the higher the stem stands above the surface of the liquid, the greater is that liquid's density.

However, the stems of most hydrometers are made uniform in cross section, or as nearly so as possible. The result is that a difference of 1 cm of stem immersion at the upper end represents a different change of density in the liquid than does a difference of 1 cm at the lower end of the stem, the end closer to the bulb that supplies the buoyancy.

This rather simple but undeniable physical fact has combined with various historical accidents, diverse national habits and enthusiasms, and considerable accumulations of human illogic. The result of this mélange has been an extraordinary number of diverse hydrometer scales and tables based upon them, associated with a variety of names and inconsistencies. Some of the names are compactly monosyllabic, such as Bates, Beck, Brix and Sikes. Others, retaining greater present importance, happen to be bisyllabic—Baumé (Bé) and Twaddell (Tw).

The annals of densitometry are still instructive, not because they present a pattern of how metrological progress should be made, but rather because they are so intertwined with familiar human habits, vices and misfortunes. The habits and the vices are apparent enough, when we note that a very large part of work on densitometry has been in the category of alcoholometry, the measurement of the relative alcohol content of spirituous beverages by means of their densities. The unfortunate aspect arises from the fact that such measurements were demanded, not for the health or well-being of the imbibers, but for the benefit of the excisemen who were about to tax those spirits.

Densitometry flourished in part also because of its relation to the extraction of sugar, in which aspect it bears the name of saccharimetry. That area, somehow, lacks the flavor and aroma of the alcoholic efforts.

Two types of alcoholic beverages were being made commercially and consumed in increasing quantities in not-always-merry England: those in which sugars were transformed into alcohol by fermentation, as in the beers and ales; and those in which alcohol, already formed by such fermentation, was concentrated by distillation, like whiskies, gins, brandies, and so on.

The latter became known as spirits because of that same Aristotelian dogma that what flew upward, as fire and air and vapors, had levity, not gravity. It was not grave or heavy or downward-tending. What was distilled out of the fermented mash did not belong to Earth, but rather to the upper and ethereal realm of the *spirits*. The temptation to make modern comments on the connections between spirituous beverages and levity is not easy to resist, but it can be done . . .

The taxation of both fermented and distilled liquors was a continuing source of conflict and concern. So too was the pricing problem. A certain amount of consumer protection did exist, and the law said that malt brew might be sold either as penny ale or two-penny ale. But to qualify for the higher price it had to be stronger and have more body than the cheaper brew.

Concerning the curious conners of ale—How could this be established to the satisfaction of customers? In an era prior to proper densitometric devices, the answer required the services of one of the most extraordinary group of professional measurers in recorded history. They came to be called the ale-conners, a name doubtless derived from the verb meaning *to know*. ("D'ye ken John Steele?" And in modern German a *Kenner* is one who knows his specialty.) It was *not* linked to the con in con man.

When an alehouse keeper had a batch of new brew, he would hang a bush out of his first-floor window, as a sort of summons. (Hence the widespread occurrence of "Bush" in the names of pubs in Britain.)

Thus invited, the conners would enter the establishment, wearing the leather shorts or abbreviated trousers that were part of their measuring equipment. The ale in question was poured onto a bench, allowed to form a thin film, and then a conner would sit down on it, waiting a few moments before he tried to get up again. If then he found that he tended to stick to the bench, his verdict would be that this was fit for the twopenny rate. If, on the other hand, the breakaway was easy and effortless, the ale was declared to qualify for no more than the penny price.

Taxation as well as pricing problems led to strange makeshifts and devices. Many of the most important questions and solutions of density measurements arose directly from the complicated levies devised to derive revenue from brewing and from distilling. Densities have played essential roles in the fiscal destinies of the fermented and the distilled beverages of the great nations for more than three centuries.

In a survey of the highlights of each, the brewing of beers and ales will be examined first. This brewing typically involved measurements of densities significantly greater than that of water (that is, with specific gravity greater than 1.0). The succeeding chapter will do much the same for the story of the densities of distilled spirits, which are typically less than those of ordinary pure water.

Though the "spirits" are stronger, alcoholically speaking, than the brews, fermentation comes first. Before there can be any extraction and concentration of its alcoholic products by distillation, fermentation must take place.

Basically, beer is a drink produced by alcoholic fermentation of a cereal that has first been crushed or ground and otherwise prepared to invite and encourage the organisms (yeasts) that supply the enzymes that convert sugars and starches to ethyl alcohol, humankind's longtime companion and consolation, or curse—depending on which aspects of an enormously ramified cultural phenomenon one decides outweighs the others.

The variety of starchy plants from which beers can be brewed is truly astounding. Most familiar in our own culture has been the use of barley, maize, millet, rice, rye and even wheat. Hops are well-known additives, but they serve flavoring and preservative functions, rather than that of supplying food for the "wee beasties" or "yeasties" on whose enzymes the basic chemical transformations depend.

The grain ingredients of beer are often induced deliberately to germinate or sprout to some extent before the fermentation process is launched. In this way the starchy stores of the seeds are converted partly into sugars, intended for the nourishment of the tender sprout that is being prepared. When the sprouting, or *malting,* process has been carried as far as the brewmasters choose, it is halted, usually by roasting and grinding the grain.

The ground-up, mashed-up, cooked-up mixture of grains and water, when it is ready for the great fermentation, is known by the traditional though rather inelegant name of *wort.* The density, or specific gravity, of the many different types of wort became important and has remained so.

The parallels between breadmaking and beermaking have more than once been pointed out. Even the use of yeast as a leaven for bread is analogous to the use of yeast as a converter of the sugar-starch infusion, or wort, into alcohol and other biochemicals. However, had breads been the sole goal of humans in working with grains and yeast and waters, it is doubtful whether density measurements would have received the keen attention that history reveals.

When the grains have been prepared and ground, the subsequent process of combining their ingredients with water is, somewhat like teamaking, the production of an infusion or suspension of solids in water. As with the first tea made from a teabag and the second or third use of the same leaves, there were distinct differences between the potency and desirability (from the drinkers' point of view) in the beers made from initial straining, and those made from later or final washings of the mash. The former, because of the far heavier content of sugars and starches, resulted in what was admiringly called "strong" beer. The latter produced the also-ran beverage denigrated variously as "weak" or "small" beer.

In terms of modern density measurements, it can be said that the wort suitable for fermentation into a truly strong beer may well show a density as much as 10 percent greater than that of pure water, whereas the wort destined to become a small or mild beer may be not more than 3 or 4 percent denser than pure water.

The actual process of fermentation, with the breaking-down and rearranging of sugars, starches, and water into alcohol, characteristically produces further decrease in the liquid density. By the time the brew is completed, filtered and ready to be imbibed, the excess of its density over that of pure water may have diminished to a third or even a fourth of what it was before the yeast first went to work.

Dwindling densities and a rough rule—It is interesting to report that a rough rule of thumb can be offered—unofficially, so to speak—indicating how the dwindling of this excess density during the brewing process is likely to be related to the resulting percentage of alcohol by weight in the finished brew. Very nearly, the percentage loss of density between the stage of the original wort and that of the completed and potable beer or ale equals the percentage of alcohol in the beverage itself.

Exceptions can be found, especially in unusual or esoteric brewing techniques. But for the kind of brewing that has become the almost universal pattern in the great industrial nations, this rule of thumb provides a useful guide. Following are several examples, taken from British brewing records and arranged in the order of increasing percentage of alcohol in the finished brews. In each case, a measurement has been added: the percent by weight of the *extractives*—that is, the dissolved or suspended solids—remaining in the finished brews ready to be imbibed. These are solids that, rather obviously, have *not* been fermented (converted into alcohol). In so-called full-bodied or heavy beers these solids may represent double the weight of the alcohol in the brew. In light beers, on the other hand, they may weigh no more than, or even a bit less than, the associated alcoholic content.

Specific gravity (sp gr) means, quite simply, the density on a scale that takes the density of pure water as 1.0.

1. Wort of 1.032 specific gravity loses 0.026 of that amount to become 1.006 sp gr *very mild ale*, with 2.6 percent alcohol content and about 2.8 percent remaining solids.

2. Wort of 1.040 sp gr loses 0.032 of that amount to become 1.007 sp gr *light bitter*, with 3.28 percent alcohol content and about 3.1 percent remaining solids.

3. Wort of 1.046 sp gr loses 0.031 of that amount to become 1.014 sp gr *porter*, with 3.2 percent alcohol content and about 5 percent remaining solids.

4. Wort of 1.048 sp gr loses 0.036 of that amount to become 1.011 sp gr *draught pale ale*, with 3.7 percent alcohol content and about 4.2 percent remaining solids.

5. Wort of 1.055 sp gr loses 0.043 of that amount to become 1.010 sp gr *bottled pale ale*, with 4.4 percent alcohol content and about 4.2 percent remaining solids.

6. Wort of 1.060 sp gr loses 0.039 of that amount to become 1.019 sp gr *strong stout*, with 4.8 percent alcohol content and about 7 percent remaining solids.

7. Wort of 1.080 sp gr loses 0.052 of that amount to become 1.028 sp gr *strong ale*, with 5.2 percent alcohol content and about 9.6 percent remaining solids.

The preceding rather informal series of examples uses the names —not always quite consistent—by which British pub patrons are likely to ask for the respective brews. Those names differ almost unnervingly from one big brewery to another. However, the indicated densities of the brews, their alcohol content and their approximate content of solid matter provide more positive identification than do the names themselves.

It can be noted that 2, the light bitter, with a relatively low alcohol content, contains even less solid matter, by weight, than alcohol. It is clear, too, that basic differences are bound to exist between such a beer with barely more than 3 percent remaining solids and the strong ale, 7, with nearly 10 percent of such solids.

It is not necessary to explore traditional or technical distinctions between beers and ales, especially since they are often inconsistent or even contradictory. Nor is it essential to delve into the fascinating biochemical and gustatory differences between the characteristic *lager* style of bottom fermentation at lower temperature, and the characteristically British type of top fermentation at higher temperature levels. To be avoided here also, though rather regretfully, are discussions of the causes and natures of differences in flavor and bouquet, as between the superb hop-informed and hence slightly bitter masterpieces of Pilsener brewing; the rather sweetish and malt-suggesting beer associated with Munich; the moderately noncommittal but pleasant brew of Dortmund; and so on and on . . .

The focus here is rather on areas in which the tax collectors—especially in Britain, where it all started somewhat earlier—clashed with the brewmasters and finally worked out adjustments and compromises.

From fermentation to taxation—The first formal fiscal onslaught on beer production took place in the Restoration, under Charles II in 1660. From the beginning it was a dual taxation, for "small," or weak, beer was taxed 6 pence per barrel, while the stronger stuff was required to pay 6 times as much.

Typical of beer taxation were its repeated and rather rapid increases. Yet for a long time they followed this pattern of two levels: a stiffer levy against the strong, a lesser against the weak beer.

Yet, who was to say whether a new batch belonged in the former or the latter category? Obviously the fiscal officers would not let the brewmasters or proprietors make such a decision. And there is ample evidence that the latter did all they could to confuse or impede the former, seeking, wherever possible, to get the strong called weak.

The test methods were for long almost absurdly crude and

direct. From an act of Parliament we can learn that the officer was accustomed, or even expected, to be "dipping his finger into the wort, tasting it, etc." Indeed, this was called his "only means of discrimination."

It has been pointed out, in a donnish aside by an esteemed British consulting chemist in the *Encylopaedia Britannica,* that today we do not know just what was implied by that ambiguous "etc."!

Near the end of the eighteenth century, the first saccharimeters (sometimes also called saccharometers) were developed. These were not the more sophisticated optical devices which measure the extent by which solutions of sugar in water rotate the polarization of light transmitted through them. These were, rather, essentially hydrometers, or liquid-density indicators. They were so calibrated that—on the assumption that the sole important solid substance dissolved in the solution was sugar (or were *sugars*)—the device would indicate very nearly the proportion of solid sugar contained in that solution.

Thus, if a saccharimeter revealed that a ready-to-ferment wort had a density of 9 or 10 or 11 percent greater than that of plain water, the fiscal officer did not need to dip in his fingers and taste the stuff; he knew that it was properly taxable at the strong-beer rate, not the weak.

Beer taxation moved in other directions too. Beginning in 1697 under King William III, new taxes were levied against the ingredients used by brewers. At the start, a tax of 6 pence a bushel was levied on malt, and this was boosted by stages till it reached 31 pence per bushel.

By this time, when a barrel of beer was finally ready to be imbibed, the total taxes against it had amounted to about 164 pence (more than 13½ shillings). This averaged, in fact, no more than 4½ pence per gallon. Yet it galled many an ale-loving Englishman to recall that his ancestors, not many generations before, had been able to *buy* small beer at 1 penny per gallon and strong beer at 4 pennies!

In 1711 still another levy was added, in the form of a tax on hops. This continued, with various vicissitudes, through the next century and a half, but was finally abolished, because its yields proved too erratic and irregular.

A new fiscal phase on the favorite fermented beverages of Britain began in 1830. The tax on beer itself was abolished, but that on malt was continued at 54 pence per bushel for another quarter century, a period in which the hops tax also persisted. New brewing procedures were rapidly developing as the industry became ever more concentrated in a few hands. In 1847 the use of sugar, as such, was permitted in the brewing process. The sugar, too, was taxed, beginning in 1850 at 16 pence per hundredweight, then rising to 78 pence per hundredweight. When the general tax on sugar (nonbrewing as well as brewing sugar) was revoked in 1874, a special levy of 138 pence per hundredweight was introduced for the sugar that went into the making of beers and ales.

It was 1880, nearly a century ago, before a major turning point came in the battles of beer taxation in Britain. All the ingredient taxes were eliminated. In their place came a single density-determined tax on the wort itself. Following a pattern already prevalent in the taxation of distilled spirits, this fiscal measure set up a sort of standard wort density, namely, 5.7 percent greater than that of water, or 1.057 sp gr. The rate was 75 pence tax on each 36-gallon barrel of wort of this "normal" density. If, however, density measurements showed the wort actually to be substantially greater, or less, than this norm, then the 75 pence rate was adjusted either upward or downward to conform approximately with the difference.

As an example, such a barrel of 1.035 sp gr wort, representing 98 percent of the standard density, would be taxed at 98 percent of 75 pennies, or 73½ pennies. On the other hand, a barrel of 1.080 sp gr wort, being about 2.2 percent over the density standard, would be taxed at about 76.6 pennies.

The standard wort density was slightly reduced in 1889 to 1.055 sp gr, but the same principle remained operative. Meantime, with that fiscal inevitability that scarcely needs explication, the basic rate was constantly raised. It became 81 pennies by 1884, 93 by 1900. Then came World War I (1914–18) and the lid was off. The rate rapidly was boosted to 1,200 pennies (100 shillings) per barrel of wort, still based on the standard 1.055 sp gr.

The 1880 legislation that set all this density-determined taxing into motion had appeared under the memorable title of the Free Mash Tun Act, four peculiarly awkward monosyllables marching all in a row. That title may not have qualified for a worldwide euphony award, but the statute it headed did move beer-taxing effectively from the realms of wort-tasting or leather-pant-sticking procedures.

Though wort is by no means equivalent to a solution of pure sugar mixed with water, it is valid to observe that the 1.055 sp gr standard when applied to sucrose-water solutions represents a mixture of 14 percent sugar ($C_{12}H_{22}O_{11}$) with water. Its sugar content amounts to 147.7 gram per liter. And on another density scale, known as the Baumé (Bé), such a solution is measured at 7.5.

One may wonder at the use of the density of the wort, or preparatory solution. Why not the density of the finished beer, ready for the imbiber or guzzler? There is a logic to the choice of the wort density rather than that of the final product that emerges from it after the fermentation. The strength, or rather the density, of a wort is a good indicator of its potentials.

The desirability, the quality, and largely the price to the consumer of beers and ales are—or should be—determined to a great extent by two factors. One is the alcoholic content; the other is the flavor, body, or palate-pleasing effect. Each of these two is produced by utilization of the starchy-sugary solids contained in the original infusion, the wort. The solids that are fermented into alcohol do not, in a general way, remain intact to produce flavor and body. The solids which do have not been fermented to produce greater alcoholic content.

Alcoholic content of beer now produced and sold for public consumption seldom reaches a level greater than 5 or 6 percent by weight. The exceptions are fascinating but unimportant for our present purposes.

Why has taxation played so large a role in the brewing industry not only in Britain but also elsewhere in the so-called civilized world? Basically, because it is a relatively easy and dependable source of revenue for the state. The moralistic motives—to discourage a deleterious indulgence, to encourage the workman to take home his inadequate pay, not push it across the counter of the neighborhood pub—these factors appear to have been operative only occasionally and to minor extents.

The old brew, and the new—There was a time, before tea and coffee were introduced from faraway places on earth, when beer and ale were virtually the only beverages available to the masses, other than clear cold water from a stream or pond. As other beverages were increasingly consumed, and as distilled spirits came into wider use, the trend in beer and ale consumption shifted noticeably toward the lighter, brighter, less alcoholic brews.

Beyond doubt this trend was promoted, and consciously so, by the increasingly powerful brewery interests. By the close of the nineteenth century and the opening of the twentieth, the voices of the beer barons and their aides were often raised in praise of these trends. Thus we find that a member of the Council of the British Institute of Brewing, Dr. Philip Schidrowitz, writing during the first decade of the present century, greeted with unconcealed satisfaction the product then emerging from the vast, industrialized breweries of Britain and compared it favorably with "the muddy, hopped, and alcoholized beverages to which our ancestors were accustomed." This new, more rapidly brewed product was clearer, lighter, brighter, less hopped, and distinctly less alcoholized.

Had British beer taxation been tied by some formula to the specific gravity of the finished brew, the rate of returns per barrel or gallon produced would have shown an unwelcome decline, for the brews were becoming lighter and less alcoholic.

Today, with American commercial beers so lamentably alike in flavor, weight, and other characteristics, one is almost compelled to cite types and categories from the British area. There are still four fairly broad groups, into which the many rather inconsistent names can be fitted.

The first, with least alcoholic content by weight, includes the light ales and bitters, distinguished by relatively high ratio of solids to their actual alcoholic content. Early in this century, this group was typified by alcoholic content ranging from 4 to 4.5 percent and solid content between 3 and 4 percent, all by weight.

Next are the mild ales, with somewhat higher alcoholic content

556 / The World of Measurements

(between 4.5 and 5 percent), and with percentages of solids anywhere from $\frac{1}{5}$ to $\frac{2}{5}$ greater than their alcoholic percentages. Such ales typically emerge from worts whose initial densities show specific gravities between 1.055 to 1.072.

The third large group may be called the pale and stock ales, sold largely in bottles rather than via the cask. They typically have from 5.6 to 6 percent alcohol content, with solid content somewhat higher than the mild ales. This group usually results from worts of specific gravities between 1.06 and 1.08. The use of the adjective "pale," it is apparent, is in no way synonymous with weak or feeble, so far as these beers are concerned.

In the fourth group are the stouts and porters (the latter so called because this was the characteristic choice of the market porters of London). Their alcohol content tends to lie in the same range as that of the pale and stock ales. However, they are typically marked by far higher solid contents than the latter beverages. The stouts and porters are generally the end products of worts whose specific gravities ranged between 1.07 and 1.09.

In taxing the "makings" of beers in proportion to densities prior to the fermentation process, the British levy observed an unchallengeable logic, for the suspended or dissolved solids in the wort provided the possibilities for both kinds of desiderata in the brew when it was ready to be dispensed: (1) the alcoholic percentage that it would contain, and (2) its residue of flavor-producing and body-building "extractives," or solids.

(Needless to say, the phrase "body-building" refers to the beer's body or general impression on the palate of the imbiber; not to the kind of addition to the imbiber's body which is so graphically referred to as the *Bierbauch* in German or the beer belly in English!)

The theme here is density and its typical uses in the service of human needs and wants, not the drinking habits of humanity. At the prefermentation stage, the wort density provides a fairly simple index to the possibilities among which the brewmaster and his employers may choose. The two possibilities, however, work in opposite directions so far as their effects on the alteration of the density are concerned. The higher the percentage of alcohol in the readied beer, the lower its density will be, for any given percentage of the solids or "extractives" remaining in the completed beer. On the other hand, the higher the percentage of those remaining solids, the higher the density of the beer, for any given percentage of alcoholic content.

Here is a hypothetical example of the brewing process at work. A suitable wort is prepared with specific gravity of about 1.046 and placed in a container of 1 cubic meter capacity. This 1,046 kg mass of wort is composed of 121 kg of solids—mostly starches and sugars—and 925 kg of water.

After the yeast is added, fermentation gradually begins and then accelerates. It produces a total of 28.3 kg of alcohol (C_2H_5OH) and 24 kg of carbon dioxide (CO_2). Most of the CO_2 escapes from the fermenting mass, but about 10 kg of it remains dissolved in the mixture.

When fermentation has gone far enough, the yeasty sludge is drawn off and filtered out, thus discarding about 47.5 kg or 39 percent of the original solids, but leaving 21.4 kg of unfermented solids in the beer as flavoring and "extractives." A total of about 26.2 kg of water is added during the various steps in these delicate operations.

The concluding beer, ready for bottling or casking, comes to 1,011 kg, composed (by weight) of 2.8 percent alcohol, 2.1 percent solids, nearly 1 percent carbon-dioxide gas, and the remaining 94 percent of water, with its small mineral content.

By volume, rather than weight, this beer contains 3.5 percent of alcohol—just about standard for United States brewing practices today. Its specific gravity is about 1.011. The original wort was about 4.6 percent denser than water; the completed beer is 1.1 percent denser. Thus, the completed beer has lost 3.5 percent of its density in relation to water, and this percentage conforms almost exactly to the alcoholic percentage by volume.

A mass of about 61.6 kg of solid and gaseous matter has been rejected—not retained—in the finished beer. The brew in this hypothetical case would be a very light one, containing only 2.1 percent of extractives or solids by weight.

The dissolved carbon-dioxide gas gives beer its bubbles, sparkle and frothy "head." Carbon dioxide, no less than ethyl alcohol, is the product of the enzyme action that the yeast plants bring to bear on the sugars of the wort. The typical biochemical reaction of alcoholic fermentation, in fact, is the conversion of a molecule of glucose ($C_6H_{12}O_6$) into 2 molecules of ethyl alcohol (C_2H_5OH) and 2 molecules of carbon dioxide (CO_2). In terms of mass, about 51 percent of the converted sugar becomes alcohol, and about 49 percent becomes carbon dioxide.

Translating brewer talk—Brewers of Britain developed their own way of talking about relative densities of worts and beers. They spoke of "gravity," instead of specific gravity. And where an engineer or physicist might speak of specific gravity of 1.055, meaning 1.055 times the density of the water standard, they said simply "55 degrees of gravity," understanding that this meant 55 *thousandths* greater than the standard water density.

To convert such a "gravity" degree into the corresponding percentage of solid by weight, the "gravity" degree was divided by the factor 3.86. Thus, the standard wort of 55 degrees (1.055 sp gr) was equivalent to $55/3.86$, or 14.25 percent of solid by weight. The remaining 85.75 percent was, of course, the water with which the wort had been made.

Some stouts and porters began as worts with specific gravities as high as 1.082. Almost as high is the density of the wort from which is brewed the famous and formidable *Bockbier,* whose consumption is an annual rite in Germany.

Early in the present century some breweries in the eastern United States produced "top fermentation" ales on the British model, and not to be outdone, worked with wort as high as 1.084 sp gr (nearly 22 percent solids and 78 percent water, by weight!). The ale resulting from the subsequent fermentation contained about 6.5 percent of alcohol by weight and 5.7 percent of remaining solids or extractives.

The Teutonic *Bockbier* typically retains about 10 percent of solid matter by weight. On the other extreme is the bright and sprightly *Weissbier,* popular in the Berlin region. It shows only some 3 percent of solid matter by weight, less than a third that of the battering *Bockbier.* (The *Weissbier* is actually *Weizenbier;* that is, "wheat beer" rather than "white beer.")

The massive hangovers typical of overindulgence at the *Bockbierfest* are attributable not to the alcoholic content, which is usually around 4.5 percent by weight, but rather to the heavy load of relatively unassimilable solid matter in this particularly potent special brew.

The ideal hydrometer for measuring densities of beer worts is one calibrated to show just 1,000 on its stem when it is allowed to float in pure water, and graduated down to about 1,100, each of the 100 intervening intervals representing one British "degree of gravity."

Such a scale differs from the scales used on the so-called saccharimeters on the one hand or on the alcoholometers on the other. The Brix or Balling saccharimeter shows directly the percentage of sugar (sucrose) by weight in the water solutions in which it is floated. It is customarily calibrated to register correctly when the solution is at 17.5° C. Thus, the degree Brix is identical with the percentage point of sugar by weight. The principle is the familiar one of the hydrometer, the stem rising higher above the surface in a dense solution than in a less dense one.

The alcoholometer, on the other hand, is designed especially for testing solutions of alcohol and water. Such solutions are lower in density than pure water, whereas the sugar-water solutions are higher in density than pure water. Alcoholometer instruments are typically calibrated to show directly on their scales the percentage of alcohol *by volume* (not weight) in the solution in which they are floated.

Neither saccharimeters nor alcoholometers are completely logical measuring devices for beers and ales, but the saccharimeter could be employed in a pinch by someone careful to make proper conversion of its readings. The reason is that all but the most exceptional brews have densities higher than that of water. In this they differ basically from the distilled spirits, which form the focus of the next chapter.

Beer is sometimes looked upon as the budget-minded American's proper tipple. It is assuredly not the best buy for those unfortunate individuals seeking rapid intoxication. A sequence of measurements and arithmetical operations can readily show why.

A typical can of beer as purveyed by the multimillions to Americans contains 12 fluid ounces, equal to very nearly 341 cubic centimeters of the brew. If the beer within has a typical specific gravity of about 1.008, then the mass of beer in the container is about 343.7 gram. If its alcohol content is 3.5 percent by volume, it is 2.8 percent by weight, so the can contains 9.62 gram of ethyl alcohol, and if the purchaser has paid 20 cents for that can, the price per gram of alcohol has been about 2.08 cents. At that rate, 1 kilo of alcohol costs 2,080 cents or $20.80.

On the other hand, the unfortunate alcohol craver who buys a bottle of run-of-the-mill vodka for, say, $3.00 a fifth (of a gallon), is likely to find that its label specifies its contents as "80 proof." This means that the vodka is 40 percent alcohol by volume, not weight.

The bottle contains a volume of alcohol equal to 8 percent of the volume of a gallon, or 363.7 cubic centimeters of alcohol. The rate per cc of alcohol is thus about 0.82 cent, and since the specific gravity of alcohol is very nearly 0.8 times that of water, the rate per gram of alcohol proves to be about 1.03 cents—equivalent to $10.30 per kilogram.

Thus, for the distorted vision that sees alcohol as the only justification for imbibing beer or spirituous beverages, the beer-borne alcohol proves to cost about twice as much as does the alcohol in a familiar distilled form.

44 THE LIGHTER SPIRITS
AND DENSITOMETRY

Be large in mirth; anon we'll drink a measure
The table round.
 —SHAKESPEARE,
 Macbeth

Whatever cannot be Taxed, is unlawful and witchcraft.
Spirits are Lawful, but not Ghosts; especially Royal Gin
is Lawful Spirit. No smuggling real British Spirit and Truth!
 —WILLIAM BLAKE,
 Annotations to Dr. Thornton's
 New Translation of the Lord's Prayer (1827)

As was true of the brewing of ale and beer, fiscal and physical facts are also inextricably intertwined in regard to the measurements of spirits, those stronger alcoholic beverages obtained by distillation of previously fermented farinaceous ingredients.

From the first efforts to extract state revenue from the whiskies and other distillates, difficulties beset the tax collectors and tax-payers because of the lack of any objective and reproducible way to measure alcoholic content. And it was always the alcoholic content to which the taxes were tied.

An early and spectacular but unsatisfactory test in Britain consisted in the officers' taking a sample of the batch to be taxed, mixing it with a small amount of explosive powder, then trying to kindle the combination. If it took fire and burned briskly, it was deemed to be "above proof." If, on the other hand, it would not support combustion despite this encouragement, it was pronounced "below proof." Above proof paid more than the basic tax rate; below proof paid less. Apparently a slow burn corresponded to the proof state itself.

The American Indian term "firewater," so descriptive of one of the blessings the white men brought them, seems to fit this burn-or-no-burn test. Yet such trial by fire was crude, erratic and unsatisfactory. At best it could divide the taxable spirits into two broad categories, but could hardly differentiate degrees within either of them.

The term *proof,* now so well established, goes back to these efforts of the sixteenth century. The sense of the word in this connection was that of test or demonstration, as in the true meaning of "The proof of the pudding lies in the eating thereof." Or as in that other saying, so often misrepresented: "The exception proves the rule." (It tests it, and if the exception stands, the rule falls; it does not mean that a rule is validated or supported by an exception!)

Measurements in one area of technology and science usually provide material for measurements in other, and even unsuspectedly distant areas. During the last decade of the seventeenth century a practical basis for alcoholometry was supplied. It took the form of multiple measurements, published as tables by the young Royal Society, showing the specific gravities of hundreds of different mixtures of alcohol and water.

Those tables made it possible to answer the burning question *How much of each?* without burning gunpowder or the tax collector's long nose.

Ethyl alcohol—or "spirits of wine," as it was commonly called —was known to be substantially less dense than water. Measurement revealed the relationship precisely: depending on the temperatures at which water and ethyl alcohol, respectively, were tested, the specific gravity of the latter ranged from about 0.787 4 (as at 25° C) to 0.794 3 (at 15° C).

Also, it was recognized that so far as the whiskies, brandies and other distilled spirits were concerned, their densities were determined essentially by their relative proportions of alcohol and water content, for the density influence of their flavoring substances or trace ingredients was negligible. They might mean much to the palate of the connoisseur, but the metrologist seeking solely to attain accurate density determination could disregard them.

From the period of Queen Elizabeth I onward, the rulers of Britain had overseen and increasingly regulated and taxed both production and purchase of distilled spirits. Royal licensees under Elizabeth were commissioned to report on the state of the nation with respect to the liquors then called "aqua vitae," "aqua composta," and "usquebaugh." The last is a Gaelic word, which by a series of mispronunciations and misspellings has become either "whisky" or "whiskey."

In the early 1640s, excise duties—taxes, for short—were levied against spirits. Before the 1650s had ended, the Parliament under Cromwell greatly increased the excise officers' controls over both distillers and dealers in "the stuff." Nevertheless for the purposes of the taxation operations, the measurement tools and methods available to the excise officers were crude and questionable, as we have seen.

Not until about 1730 was there produced in Britain a hydrometer intended and suited to indicate the approximate alcoholic content of a distilled liquor as a function of its density. The early device, called the Clarke, was complicated and inconvenient. Each instrument came equipped with more than forty separate weights. Most of these were hung onto the hydrometer in order to reset it for various ranges of specific gravity. Others were called "weather weights" and were used to compensate for changes in the prevailing temperature at which the test was conducted.

The lower the temperature of the liquor, the greater its density. Hence a mixture that might be taxed at the full rate when a summer heat of, say, 80° F prevailed, might seem lower in alcohol content when the thermometer showed only 40° F. To prevent such fiscal loss, one of eleven weather weights was added for each 5° F fall in temperature. Thus the Clarke provided, rather roughly, a compensation covering a range of 55 degrees F.

These intervals were, however, too coarse. In warm weather, for example, a batch of liquor might seem, according to the Clarke, to contain more than 10 percent over the "proof" standard, and thus call for an increase of 10 percent over the basic levy. Rather than face this, the distillers would often choose to dilute the liquor until it was down to the "proof" level at that temperature.

Despite these and other deficiencies, the Clarke instruments persisted as the excise alcoholometers of Britain for some three generations. The difficulties were not due solely to design problems. No hydrometer can provide more information about alcoholic content of tested solutions than is contained in the tables of prior measurements with which it must be used. Those tables must be both detailed enough and dependable enough to cover the full range of varying proportions (alcohol to water) and the full range of temperatures at which the solutions are to be measured in actual fact.

The concept of a standard or normative "proof" strength was operative in Britain from the early days of the burn-or-no-burn tests. It paralleled the normative concepts applied to brewing—first the wort standard set at 1.057 specific gravity, then the modification to 1.055 sp gr.

Why was such an arbitrary standard sought out and set by legislative action? The reasons were probably more legalistic and fiscal than physical. If a tax rate was decreed for the realm, it was proper that it should apply to some standard or definable situation, from which all deviations would result in an alteration or adjustment of the levy. But there had to be some stable pole in the flux of phenomena. The strength of spirits was too variable, apparently, to pin down in any other way.

The law formalizing what is now known as British proof strength for distilled spirits dates back only to the time of George III. However, it was based on procedures and concepts that long antedated it. Today it seems curiously complicated and arbitrary. Proof spirit was declared to be that whose density was just $12/13$ (or about 0.923 077) times that of pure distilled water measured at the same temperature.

But what was this establishing temperature to be? The first determinations of this George III definition were made at or very near 51° F. Later the level was raised to 60° F. Even this 9 degree increase brought perceptible changes. At 60° F, proof density was reduced from the former 0.923 077 sp gr to 0.919 76 times the density of water at the 60° F temperature—a reduction of 0.033 17 in specific gravity.

This prevailing 60° F proof density, when measured most carefully, is found to correspond to a proportion (by mass or weight) of 49.28 percent ethyl alcohol, or 57.1 percent *by volume*. The remaining 50.72 percent of the mixture (by mass) or 42.9 percent of the mixture (by volume) is water.

However, the well-established and famous proof strength of Britain is not habitually referred to in terms of its actual specific gravity, nor by the percentages of its alcohol content, either by mass or by volume. Rather it is restated on a degree scale. Indeed, it has become involved with two diverse degree scales!

When we deal with the spirit proof, pure and simple, we can call it 100 degree proof. The word is *degree,* but it can be considered to mean percent—100 percent of established proof. Proof percentage obviously can rise far above this 100 percent; in fact, 175.35 degrees (or percent) of proof is the upper limit—representing pure alcohol, without water admixture.

Likewise, proof percentage can drop far below that 100 percent; in fact, all the way down to 100 degrees below it, or to zero percent, another name for pure water without alcoholic admixture.

The practical principle on which all this operates is the following. The number of degrees "over proof" tells how many additional batches of spirit *at* proof could be made from the batch being tested. Thus if 1,000 gallons are designated to be "67 above proof" or o.p. (for "over proof"), we know at once that they can be converted by appropriate dilution to 1,000 plus 670, or 1,670 gallons of proof spirits.

Similarly, if 100 gallons of spirit is found to be 10 degrees below proof, it means that they contain only the amount of alcohol that would be found in 100 minus 10, or 90 gallons of proof spirits.

The o.p. and u.p. abbreviations ("over proof" and "under proof") have been common and quite understandable to those who are *au courant* with the British proofing system. The word "degree" is often omitted, as in "30 over proof" rather than "30 *degrees* over proof," and so on.

From Clarke to Sikes—The crudities of the Clarke hydrometer, or alcoholometer, became intolerable toward the end of the eighteenth century. It was then that the prestigious Royal Society completed an important alcoholometric assignment for the British government, detailed measurements of the actual specific gravities of a wide variety of alcohol-water mixtures at different typical temperatures. The actual measurements, abundant and onerous, were carried out by the principal staffers of the Society—Charles Blagden, its secretary, and George Gilpin, its clerk. The resulting tables were published in the famed *Philosophical Transactions* of the Royal Society.

The government, bent on sharpening its tools for taxing the

products of the distilleries, had advertised its desire for a new alcohol-oriented hydrometer and associated scale. Thus it was that in 1794 Bartholomew Sikes combined the new tables and his own practical ideas and produced the small Sikes hydrometer. It was far superior to anything in the field before it; and more than a score of years after the first model was demonstrated, it became, in 1816 under George III, the only standard for the fiscal measurement of spirits.

The Sikes design did not do away entirely with the use of hang-on weights in addition to the permanent hydrometer shape. But it made them simple and easy to keep track of. It had the usual three parts: beginning at the bottom with the ballast weight for stability and uprightness; then the hollow bulb for buoyancy; and finally, like a mast above a yacht, the uniform, slender, hollow stem with graduated markings so that the measurer might note accurately the depth to which it sank in the alcoholic beverage.

The Sikes instruments were usually made of brass, to resist the corrosive effect of the solutions into which they were placed to sink or swim. In the protective box provided to transport each, nine graduated weights or *poises* were placed. These were shaped so they could be attached easily to the bottom of the bulb. Their numbers, from 10 through 90, indicated their relative weights and the extent to which they tended to pull the instrument lower in the liquid that sustained it. The stem on top was divided into ten major intervals, readily visible. Each of these was subdivided into five others.

The most remarkable aspect of the Sikes device, however, was not its mechanical construction, but the arbitrary degree scale created for it, and for which it had been designed. This scale ranged from 0 to 100 degrees Sikes. A reading of 0 Sikes meant that the liquid supporting the device was at "66.7 over proof." On the other hand, a reading of 100 Sikes meant that the liquid was pure water, containing no alcohol at all.

Since pure water is substantially denser than a 66.7-over-proof mixture, the assignment of the higher degree number to the former was justified. However, between the Sikes degree and the degree over proof or under proof, a curious and unfortunate confusion may arise, and should at all costs be avoided.

The actual reading of the Sikes degree was fairly simple. It consisted in adding the poise weights required to bring the stem as close as possible to its zero mark in the liquid and, if necessary, adjusting that figure according to the number of marked intervals by which that zero mark was submerged below the surface.

For example, if the Sikes device was found to sink to the zero mark without any poise weight attached, this gave a reading of 0 degree Sikes and indicated the liquor to be 66.7 over proof.

On the other hand, if the stem required poise weights 60 and 40 (totaling 100) to pull it down to the zero mark, this indicated 100 Sikes degrees and showed the liquid to be all water, no alcohol.

Or, if just the poise labeled 10 was needed to bring the zero stem mark level with the surface, this indicated 10 degree Sikes. Consultation with the detailed tables supplied with each instrument showed this to correspond to a liquid 58.4 over proof.

Possibly it might be found, however, that the 10 poise brought the stem only to the mark labeled 2 on the stem. This obviously meant the liquid to be denser than 10 degree Sikes by 2 degrees— giving it a reading of 12 degree Sikes, which the table would readily convert into the equivalent "over proof" figure.

Used with each Sikes alcoholometer was at least one reliable thermometer, for the tables supplied the equivalents in one-degree-Fahrenheit intervals, all the way from 30° F to 100° F.

Even the limitations of this basic Sikes design were not permitted to restrict the progress of advancing alcoholometry. Two other special Sikes models, designated as A and B, were added. The first was designed to cover the range of strengths between 66.7 over proof and 73.5 over proof, at 60° F. The second was intended to measure proof strengths even greater than that.

From the arbitrary Sikes degrees, tied to the particular and rather peculiar instrument, the carefully measured tables led back to the degrees over or under proof and so, for those who wished to pursue the matter that far, to the actual specific gravities and densities of the spirituous beverages tested.

American readers caught up in the Sikes degrees and the degrees over or under proof may be tempted to exclaim, "For Land's Sikes!" and withdraw into a fallacious and unjustified smugness with respect to their own nation's way of dealing with the "proofing" of distilled spirits.

The truth is that at the time of this writing, the measures of alcohol content applied in the leading nations of the globe are diverse and sorely in need of a unifying and simplifying change. All these measures have dual purpose: to help the tax collector extract revenue for the state and to let the purchaser of spirits know something of what he is getting. But these aims could be better served in our shrinking world by a consolidated and uniform system that would be simple in concept and application. Obviously, such a system would expedite the importing and exporting of spirits.

On the way to suggesting such a system, let us note again that the venerable British Sikes scheme did have some advantages for the fiscal authorities. For example, a basic tax rate was levied against each gallon of proof spirit (spirit at its proof equivalent). Now, if a batch was measured and found to be 10 degrees under proof, that meant it contained only so much alcohol as 0.9 times as much spirit *at* proof. Hence, the basic tax rate was cut by one tenth, as applied to the actual number of gallons. Or—amounting to the same thing—the tax rate was left unchanged, and the number of gallons was reduced by one tenth. When the Sikes device revealed an over-proof situation, the opposite changes were made.

Alcohol has long been distilled from fermented substances for technical and industrial uses as well as for human drinking. There are various special distilling processes and compromises that may make the procedures thus far described inapplicable, for the product may not be carried all the way to the stage of a possible potability.

Three density methods to determine the tax—The fiscal authorities of Britain were able to deal also with these details. They devised two other methods for taxing the distillation industry and used the simple rule that whichever of the three methods resulted in the highest tax payment was the one to be applied.

These alternatives too depended on accurate density measurements. The first consisted in assessing the tax chargeable for one gallon of proof spirit to each 100 gallons of wort (the fermentable mixture), and then adding a like charge for each "5 degrees" of density decrease that took place during the fermentation process—

a change we have noted carefully in connection with the brewing of beers and ales.

For example, if the wort, when readied for fermentation, showed a density 20 percent greater than that of water, while the resulting "wash" or liquid result of fermentation was down to only 5 percent above the density of water, then the decrease or "attentuation" came to 15 percent or degrees, and three times the basic proof tax would be added for each 100 gallons concerned, thus bringing the total tax to four times the basic proof rate, per 100 gallons.

The second method was aimed at an intermediate stage in the distillation processes and based its charges on the actual quantities —at proof strength, as revealed by density—of the early distillation of the "wash." These early distillation products are known in the trade as "low wines."

The third approach, of course, was that previously set forth, and it dealt with taxation of the ultimate and high-alcoholic products which distillers are wont to call by such intriguing titles as "feints" and "spirits."

Each of these three possible approaches to deriving revenue from the fermentation-distillation combination depended basically on reliable measurements of densities. On the way to the fermentation-ready wort, the specific gravity of the mixture increases beyond the 1.0 magnitude corresponding to the density of pure water. During the fermentation process, and even more during the subsequent distillation operation, the densities of interest show significant decreases. The final outcome, as in the case of a vodka rated at 100 proof in the United States system, may be a specific gravity of only 0.934 4, more than 6½ percent below the density of water at a like temperature.

An observer not caught up in the problems of taxation nor enslaved by alcoholic addiction might view the alcoholometric scene with a certain levity, and conclude that feint spirits never dilute the dense heart of the avid fiscus, as he or she distills state revenues from the unceasing reduplication of Demon Rum!

We have begun with British designs for using density measurements for the excision of revenues, but only because they had a certain priority in time and provided the standards that stimulated other governments to develop deviant systems of their own. The variety and range of the resulting schemes are almost enough to drive the layman to drink. Yet each is at work influencing the lives and welfares of hundreds of millions of humans today, whether or not they are teetotal in their personal habits.

In France, J. L. Gay Lussac (1778–1850), a chemist-physicist notable for his meticulous, precise and persistent style of work, published in 1810 and 1815 papers making substantial contributions to the understanding of the complex processes of alcoholic fermentation. Such processes were basic to the wine industry, though we have hardly mentioned it here, since the brewing of beers and ales offers an easier access to our prime subjects.

In 1824, as one of his many services to the French nation and its industrial development, Gay Lussac unveiled his so-called "centesimal" alcoholometer. The quoted adjective meant merely that it was based on a range of 100 degrees—actually percents—ranging from a zero that meant all-water-no-alcohol to a 100 percent that meant the unlikely fluid composed solely of alcohol, without water admixture.

This densitometer-alcoholometer won for Gay Lussac particular praise from the French Institute. It survives today as the scale in which French and Belgian spirits are measured and marked. It is based on the percent of alcohol content by volume, not by weight. The temperature at which Gay Lussac determinations are made is 15° C, equal to 59° F. Thus a 30-degree indication on a bottle of spirits in France or Belgium means that by volume it contains 30 percent alcohol, 70 percent water or water equivalents, when measured at 15° C.

In terms of the Gay Lussac scale, the British proof strength (100 UK proof, as it is often called) is equal to 56.9 degrees.

In the United States, extensive studies on the densities of various alcohol-water solutions at many different temperatures were undertaken, first in 1848, and again, with more sophisticated methods, in 1913 and 1928. Among the results has been the distinctive and unduplicated pattern of proof that prevails in this country. This pattern is laid down in a rather orotund and legalistic pronouncement: "Proof spirit shall be held to be that alcoholic liquor which

contains one-half its volume of alcohol of a specific gravity of
0.793 9 at 60° Fahrenheit."

The specific gravity in this statement applies, to be sure, not to
the proof spirit itself, but to the pure alcohol, which shall constitute one half of its volume. The specific gravity of pure ethyl alcohol at 60° F (15.6° C) is, of course, just 0.793 9. However, when
the mixture is composed of half alcohol, half water, by weight, then
the specific gravity at this temperature is 0.934 4, which is 0.140 5
more than the specific gravity of the pure alcohol, but only 0.065 6
less than the specific gravity of pure water (which is 1.0, by definition).

In short, 100 proof in the United States system is equivalent to
a specific gravity of 0.934 4. In the U.S. system the British proof
spirit is classified as 114.2 proof. Thus British proof is 14.2 percent stronger than U.S. proof; while the latter is about 12.4 percent
weaker than the former.

The U.S. 100 proof is very nearly 50 degrees in the Franco-
Belgian Gay Lussac system, while the common U.S. 80-proof
strength of whiskies, vodkas, and so on, is equivalent to 40 degrees
Gay Lussac.

For quick determination of equivalents in these three leading
systems, a table is used in which, except for a few entries, whole
numbers appear rather than decimals. Hence, though accurate
enough for its common uses, it is approximate rather than totally
precise:

U.K. (BRITISH)	U.S. PROOF	GAY LUSSAC
35°	40%	20°
50°	57%	28.5°
60°	68%	34°
70°	80%	40°
75°	86%	43°
83°	95%	47.5°
87.6°	100%	50°
100°	114%	57°

The basic British, or United Kingdom, proof, 100° on this table,
actually is equivalent to 114.2% U.S. proof and to 56.9° Gay

Lussac, rather than to 114% and 57° respectively. However, these are trivial deviations in terms of the requirements of customs officials.

Further alcoholometric oddities—In 1865 Dmitri Mendeleev (1834–1907), the great Russian chemist who had been born in Siberia, turned his formidable talents to the study of alcoholometry also. Today, however, in the U.S.S.R. as well as in Austria and Italy the official measures of alcoholic content follow the Tralles system, which originated in Prussia shortly before the beginning of World War I.

Like the Gay Lussac and U.S. systems, the Tralles is based on volume, not weight, at a determining temperature of 60° F, or very nearly 15.6° C. Tralles readings do not differ much from those on the Gay Lussac system; yet the differences are sufficient to cause minor complications. Thus the strength of British proof spirit, rated 56.9° by Gay Lussac, is 57.1 percent on the Tralles scale.

In Germany today the Tralles scale is not in official use. Instead, the Windisch scale, based on alcohol weight, not volume, prevails. British proof spirit strength, for example, is measured at 49.3 percent Windisch.

Finally in this tour of the Babel of disparities of modern alcoholometry, one comes to the most "different" system of all, still adhered to by Spain. It is the Cartier system, dating back to 1800, and it assigns its 10-degree level to zero alcohol content (meaning pure water), while it assigns its 44-degree level to 100 percent alcohol content (meaning no water)! Thus, the 34 Cartier degrees between 10 and 44 cover the entire range of 0 to 100 percent alcohol, based on volume, not weight.

The Cartier scale is numerically askew in terms of the scales mentioned earlier, as the following facts indicate: The 10 degrees Cartier between its 10-degree and 20-degree points are equivalent to an interval of 52.5 degrees Gay Lussac. The next 10 degrees Cartier (from 20- to 30-degree level) represent a 25.9-degree Gay Lussac interval. The following 10 degrees Cartier (from 30- to 40-degree level) represent a 17-degree Gay Lussac interval, which is an average of 1.7 degree Gay Lussac per 1 degree Cartier. Then, in the final step from 40 to 44 degrees Cartier, the Gay Lussac equivalent is only 4.69 degrees, which is an average of 1.15 degree Gay Lussac per 1 degree Cartier!

Beyond reasonable doubt, the most rational and defensible of these diverse systems is the Gay Lussac. However, the recommendation of the International Office for Legal Metrology, headquartered in Paris, is that the nations all agree on and adopt a slight modification of Gay Lussac, in the form of a volume percentage system based on a determining temperature of 20° C rather than 15° C as in the Gay Lussac.

Even the seemingly small difference of 5 degrees C does make perceptible changes in alcoholometric measurements. Thus, if we make measurements on a spirit that has been rated at 80 proof in the United States system, we find that its actual density at 20° C is about $\frac{3}{10}$ of 1 percent less than at 15° C.

Density as the key to sugar content—Similar basic considerations apply to the measurements of the densities of sugar solutions by the hydrometer devices known as saccharimeters and identified with the names of Bates and Brix. Solutions of sucrose (cane sugar) in water reveal a far larger range of specific gravities than do the wort solutions prepared for fermentation into beer and ale, or for distillation after fermentation.

A modern table based on the comparison of sugar solutions measured at 20° C with pure water at its maximum-density temperature of 4° C ranges all the way from a shade less than 1.0 for pure water sans sugar, to 1.473 sp gr for a solution containing 89 percent sugar ($C_{12}H_{22}O_{11}$). Such a solution is so loaded with the sweet stuff that when 1 liter of it is evaporated to eliminate the water portion, there remains more than 1.31 kg of sugar!

Aqueous sugar solutions, as well as a great many other solutions of organic and inorganic compounds in water, are commonly measured by a famous hydrometer scale named the Baumé, abbreviated as Bé. Together with another persistent scale named the Twaddell, abbreviated as Tw, it forms a pair that has somehow not dropped entirely out of use.

The Twaddell scale was designed primarily to deal with liquids denser than water. It takes as its zero point water at the maximum-density temperature of 4° C. For each 1 percent increase in that density, the Twaddell scale rises by 2 degrees. Thus 20° Tw represents specific gravity of 1.1, while 100° Tw represents 1.5 sp gr. Accordingly, conversion from °Tw to specific gravity is quite easy.

The Baumé scale or scales, on the other hand, appear more arbitrary and complicated than the Twaddell, yet are more widely known. It may seem strange that when they were first proposed in 1768 by the eminent French chemist Antoine Baumé, they were generally greeted as improvements on the hydrometer systems then in use.

Baumé created two scale systems. One was for his hydrometer designed to measure solutions with densities greater than that of simple water; the other was for liquids less dense than water. The first scale showed its 0° point in pure water brought to a temperature of 10° Réaumur; and its 15° point was set at the density of a 15 percent salt-water solution. The scale between these two determining points was divided into equal intervals, and also extrapolated upwards, beyond the 15° mark.

On the other hand, Baumé's scale for liquids less dense than water was calibrated to show 0° when floating in a 10 percent salt-water solution, and to show 10° when floating in pure water. Here again the in-between intervals were equal, and the scale was extrapolated beyond the 10° point to take care of densities still less than that of pure water.

It is the first, or denser-than-water, Baumé scale that is more likely to be encountered today in tables and reference works. In using the Baumé scale measurements, one is just about obliged to double-check to be certain which of the two scales is actually in operation, and what particular formula ties that scale to the actual measure of specific gravity.

In many tables, the formula that relates the Baumé measurement of solutions denser than water to their actual specific gravities is Bé $= 145 - 145/\text{sp gr}$. Or if we know the Bé measure and wish to derive the specific gravity from it, we can use this equivalent equation: sp gr $= 145/(145 - \text{Bé})$. For example, 1.2 sp gr is equivalent to very nearly 24.2° Bé.

Another equation, for the Baumé scale which measures solutions less dense than water, looks like this: Bé $= 140/\text{sp gr} - 130$. This also can be expressed as sp gr $= 140/(\text{Bé} + 130)$. For example, in a table of the Baumé degree measurements and the corresponding specific gravities of solutions of ammonia in water (aqua ammonia), we find that 22 Bé corresponds to 0.921 1 sp gr; and 29 Bé to 0.880 5 sp gr.

One of the conveniences offered by the Baumé scales when they were first introduced was the fact that on these hydrometers the scale subdivisions were all equally spaced. We are accustomed to equal-interval spacings on our thermometers and on our yardsticks too. Such scales are not only easier to calibrate but also seem easier to read.

However, the fact is that Bé degrees at different points in the scale do not stand for the same density differences. The Bé degrees toward the lower-density ends of the scales represent larger density differences than those at the other ends. For example, in the case of the Bé scale for the denser-than-water solutions, we find that between 0° and 10° Bé, specific gravity rises by 0.074 1, but between 50° and 60° Bé it rises by 0.179 6, or more than twice as much.

Some typical Bé denser-than-water equivalents follow: 20° Bé equals 1.16 sp gr; 40° Bé equals 1.381 sp gr; and 60° Bé equals 1.750 9 sp gr.

45 NOW HEAR THIS!—ESSENTIALS OF ACOUSTIC MEASUREMENTS

Why cannot the Ear be closed to its own destruction? . . .
Why an Ear, a whirlpool fierce to draw creations in?
—WILLIAM BLAKE,
Book of Thel (1789)

The youthful rock band calling itself The Shade was signed to supply music for a Saturday night dance at Fullerton High School in southern California, during the autumn of 1972. A strangely significant clause was included in the agreement: the music was to have a volume, or loudness, not over 92 decibels. Payment of the fee was made contingent on that safety measure.

Richard Jackson, assistant principal of the school, reported later on what the standard "decibel meter" had indicated when used under his supervision to measure the music output. The reading, it is true, had risen sometimes as high as 104 to 106 decibel, but Mr. Jackson concluded that the spirit of the agreement had been observed, for the benefit of all concerned.

Bob Dire, identified as the only over-21 member of The Shade, admitted there was justification for that decibel clause—". . . they've got to have levels. I've been some places so loud it goes past music." His group prided itself on making music, not—like some "acid rock" players—blasting so loud that "it drives people."

The Fullerton limit on loudness was not the result of an administrator's private whim. The year before, at nearby Anaheim Union High, bands had been so enormously amplified that monitors, armed with decibel meters, had threatened to disconnect the power from the amplifying systems.

Increasingly in an age of acute and accelerating noise pollution the public is becoming aware of the unit called the *decibel* as a measure of the loudness and the lethal threats of sound, both as regards intended music or unwanted noise.

The Fullerton school authorities felt they were reasonable in designating the 92-decibel limit, for it corresponded to the loudness of a modern jet plane in full flight only five-hundred feet overhead. Abundant medical evidence shows that extended exposure to such loudness levels impairs human hearing.

Even 85-decibel sound intensity can be harmful to hearing. Modern sound-level meters offer three scales to fit our hearing perceptions and different intensity levels. The A-scale is weighted for intensities around 40 decibels, the B-scale for the 70-decibel intensity region, and the C-scale for even greater intensities. The most suitable scale is selected by switching.

Most frequently used is the A-scale, which emphasizes high-frequency tones. These tend to be more harmful than the low frequencies, even at similar intensity levels. Frequencies up to about 20,000 hertz and down to about 18 or 20 hertz are perceived as sound by young listeners with normal hearing. As we age, this range becomes narrower. Aural response is greatest in the 3-octave bandwidth between 500 and 4,000 hertz. Below 500 and above 4,000 hearing responses fall off noticeably.

In our most responsive frequency range lie the 1,000-hertz tones that acousticians have agreed to use for their standard in setting decibel ratings for the sounds we perceive and compare. The effort is always to find how our subjective sensations of loudness and pitch correspond to or distort the physical factors of sound intensities and vibrational frequencies.

The procedure is to use a panel or group of persons with normal hearing in both ears and let them listen to pure 1,000-hertz tones of gradually increasing intensity. At first the sound is too weak to be perceived by any listener. Then one after another perceives it. When the sound is strong enough that a substantial number of the group hear it (but just barely), then that strength is identified as threshold sound, to which the unit value of 0 dB (zero decibel) is assigned.

The term "strength" is an ambiguous one here, and deliberately so. There are two different ways in which such "strength" can be measured by objective units. But they differ significantly in behavior. One way is to measure the intensity, or power per unit area, of the 0 dB sound, using, for example, the SI unit combination of the watt per square meter (W/m^2). The other way is to measure the pressure—or, more precisely, the pressure *difference*—within the waves of the 1,000 hertz sound as they reach the ears of the listening judges. That can be measured in the SI pressure unit, the pascal (Pa), equal to 1 newton per square meter (N/m^2).

The pressure measurement has gained a certain precedence over the intensity measurement in acoustics, but both are important. Based on tests, and by general international agreement, the pressure associated with the 0 dB reference sound wave is 2×10^{-5} Pa, which can also be stated as 2×10 μPa (micropascal). On the other hand, by equally broad agreement, the intensity or power per unit area associated with that same sound is just 10^{-12} watt per square meter, which can also be stated as 1 pW/m^2, using the submultiple of the picowatt.

This means that the density of the sound power that reaches the listener is 1 pW/m^2, or 10^{-12} W/m^2. A human eardrum has an area of very nearly 1 cm^2, or $1/10,000$ of the area of 1 m^2. Hence, at 10^{-12} W/m^2 intensity, each ear receives about 10^{-16} W of power, or 10^{-16} joule of energy per second, an extraordinarily tiny amount indeed!

Tests such as these are not tests of the acuity of hearing of the listening panel members. Such determinations, in fact, work the other way around. They use the performance-tested sensitivities of a sufficient number of healthy, unimpaired human hearings to set a threshold sound with which all other sounds can be compared, paying always suitable regard to the frequencies of those other sounds, insofar as those frequencies differ from the standard test frequency of 1,000 hertz.

Today the physical characteristics of the international standard "0 dB" reference sound are no longer in doubt. So long as existing usage continues, the reference sound is an undistorted wave of 1,000-hertz frequency, driven by pressure changes of 2×10^{-5} pascal, sending power of 10^{-12} watt through each square meter of area as it reaches ears of human listeners.

Many a mature person who by no means suffers from "deafness" or even unusual hearing impairment is bound to find that this standard least-audible sound lies below, not above, her or his level of perception. Hearing acuities differ even in the young; and it is incontrovertible and probably unavoidable that the sensitivity of hearing perception declines with advancing age. For example, normal listeners in their middle forties are likely to find that the intensity of the standard 0 dB sound must be multiplied about 3 times before they begin to perceive it. (The intensity, mind you, in W/m^2, not the "loudness"!) And equally normal listeners in their middle fifties are likely to find they do not honestly begin to hear a 1,000-hertz tone until its intensity is about nine or ten times that of the 0 dB international standard.

Nevertheless, and without the slightest intentional disrespect to the richer judgment that is supposed to be attained with greater maturity, the fact remains that the 0 dB or standard reference sound is established, and all other sounds or sound combinations are compared with it, by methods that are unfortunately rather complex. This complexity arises not by choice but because of the complexities of the relationships between subjective human perceptions and the measurable objective characteristics of the sounds that call forth those perceptions in the human ear-mind network.

In search of the essentials of sound—At this point it is high time to explore the nature of sound. What is it? Sound consists of waves of greater and less compression that pass through matter, gaseous, liquid or solid. Many different kinds of disturbances, deliberate or inadvertent, initiate audible sound: the vibrations of human vocal cords in speech, the vibrations of the sounding board of a violin or guitar, the impact of a hammer against the head of a nail, the titanic pressure pulses from a jet engine as a plane takes off, and so on.

These waves of alternately more and less compression move at typical speeds through different sorts of matter. Through dry air at standard atmospheric pressure (0° C) they move at about 331.5 m/s; through water at 25° C they move at about 1,500 m/s; through steel, at about 5,000 m/s; and through cool glass, which is still more rigid, at about 5,500 m/s (meaning meter per second).

Generally speaking, the velocity of sound through whatever medium does not depend on the frequency of its waves. The lengths of those waves, however, are inversely proportional to the frequency of the sound, for any given velocity of movement. Thus, the standard 1,000-hertz wave shows a length in standard atmospheres of just about $\frac{1}{3}$ meter. That is the distance from one peak of pressure to the next following peak. The distance from a peak of pressure to the trough of less pressure that follows it is, of course, just half that, or $\frac{1}{6}$ meter.

A sound of 10,000 hertz forms air waves of $\frac{1}{10}$ the length of the 1,000-hertz sound, or $\frac{1}{30}$ meter. And a sound of 20,000-hertz frequency, at or just beyond the frequency-perception limit of a great many human hearers, consists of air waves $\frac{1}{60}$ m, or about 17 mm, long.

The frequency of any sound wave is a time measurement. Thus, the standard 1,000-hertz frequency means that within 1 millisecond the eardrum responds to one full cycle of movement: the higher-pressure peak of the wave pushes the eardrum in a little; the lower-pressure trough of the wave draws the eardrum out a little. Thus the eardrum and its linked bones and the liquid in the semicircular canals which they actuate all execute two opposite motions—a motion "to" and a motion "fro"—during each hertz, or cycle per second. With a 1,000-hertz sound, each such motion must be completed within $\frac{1}{2,000}$ second, or 0.5 millisecond.

We have seen that healthy young people perceive as sound air vibrations of 1,000-hertz pitch with intensities of only 10^{-12} watt per square meter. In ordinary conversation the air pressures in front of a speaker's mouth vary by only one-millionth (10^{-6}) of normal atmospheric pressure. At a distance from the speaker, these physical changes are even tinier. Such ultra-tiny or minimal physical processes are constantly perceived and interpreted by intact human hearing. Our ear systems are superbly sensitive when not abused by excessive loudness or diminished by ageing.

A rather simple relationship exists between the pressure-difference (P) that drives a sound through the air, the velocity (V) with which that sound moves, and the density (D) of the air, on the one hand, and the intensity (I) of that sound, on the other—

$$I = P^2/DV$$

Most important, this equation tells us that the intensity, or the power per unit area, of the resulting sound must increase as the *square* of the pressure difference that exists between the peaks and troughs of sound waves.

In the case of standard atmosphere, we know the velocity V to be about 331.45 m/s; the density D to be about 1.293 kg/m³; and the pressure P, as already noted, to be about 2×10^{-5} pascal. When these values are inserted into their proper places in the equation, calculation shows that the resulting intensity, I, is very nearly 10^{-12} watt per square meter, the amount already mentioned as the agreed standard for the 0 dB reference sound.

Notice that if the pressure P is increased 10 times, the resulting intensity I is increased 10×10, or 100 times, and so on. Thus the ratio of a change in sound intensity is the *square* of the ratio in the change of the sound pressure that has produced it. Obviously, when our measurements are focused on the varying intensities of sound, measured in W/m² units, we are confronted by far greater ratios than when our measurements are focused on the varying pressures in sounds, measured in units of the Pa (pascal).

Just one example provides a small preview. The pressure differences in the horribly powerful sounds near jet aircraft as they take off are found to be about 10^6 (1 million) times greater than the pressure difference in the reference sound, the threshold-of-perception wave designated as 0 dB. This means that the intensity, in W/m² unit, of the former sound must be 10^6 times 10^6, or 10^{12} (1 million million) times the latter.

These are enormous ratios. Since the sounds we listen to, or are forced to hear, have so great a dynamic range, some simple and compact way is needed to express these ratios. A reference sound, such as that which we call 0 dB, has little value unless we have some sort of unit or measure to express the resulting ratio comparisons.

How the bel was born—That is the situation which brought about the invention and first application to sound of the units called the *bel* and the *decibel,* the tenth part of the bel. They were devised in the 1920s by research engineers of the Bell telephone system. The purpose at that time was not related directly to sound as such; a method was being sought to measure more conveniently the weakening or "attenuation" which takes place in the electrical power of telephone signals transmitted over long lines.

The method chosen was much like the convenient "power of 10" that we have used in this book, with 10^3 representing one thousand, 10^6 standing for one million, 10^{-3} standing for 1 one-thousandth (0.001), 10^{-6} for 1 one-millionth (0.000 001), and so forth. The bel and decibel were based on logarithms, the "common" logarithms, in which the base number is 10. Any table of logarithms will show that log 10 is 1, log 1,000 is 3, log 1,000,000 is 6, and so on. Likewise, log 0.1 is -1, log 0.001 is -3, log 0.000-001 is -6, and so on.

Now, if the original electrical power in a telephone line is P_1 and the attenuated or reduced power is P_2, then the ratio of original to attenuated power is P_1/P_2. To express these typically huge ratios handily, they were converted into logarithms: log P_1/P_2, and the result was treated as a unit and given the name *bel*. It was derived from the last name of the inventor of the telephone, Alexander Graham Bell (1847–1922), who died just the year before the bel was named.

Thus, a power attenuation or diminution by a ratio of 100,000 to 1 could be expressed simply as a loss of 5 bel, while one of 10 million to 1 would be a loss of just 7 bel, and so on. The reverse held true too. If amplifiers were installed in a telephone line with a resulting increase of 10,000 times in the power of the signals, this could be called simply a gain of 4 bel; while a hundredfold amplification was a gain of 2 bel; and so on.

Amplification and attenuation ratios do not all end in neat strings of zeros. A ratio of 1,000 is represented by 3 bel, one of 10,000 by 4 bel, but a ratio of 2,000 works out to 3.3 bel, and a ratio of 20,000 to 4.3 bel. Likewise a ratio of 6,000 is 3.6 bel, and one of 60,000 is 4.6 bel. With the help of a table of common logarithms just about any large ratio can be converted into a compact and handy "bel" unit equivalent.

It was soon found that the bel was rather large and coarse for the ratios that were actually encountered in practice. Hence, its place was taken by the tenth-of-a-bel, or *deci*-bel, now known worldwide as the decibel and symbolized by dB, or sometimes by just db, which is less respectful to the memory of Dr. Bell. It is this decibel, devised originally to deal with the attenuations and amplifications of electrical power in telephone circuits, that acoustics experts now employ so widely and persistently.

Decibels for forces versus decibels for powers—When the decibel is used to compare powers or energies, it is defined simply as 10 times the common logarithm of the ratio of the powers or the energies that are being compared. But an important difference must be noted when the decibel is used to compare forces or force-like physical variables. In the case of electricity, electromotive force, measured in units of the volt, is such a variable. The power that flows through any circuit of fixed resistance is proportional to the *square* of the voltage that is at work. Hence, when the decibel is used to summarize the ratios of voltages, it is defined as *20* times the common logarithm of such a ratio. For example, an amplifying system that converts an input voltage difference of 1 microvolt (10^{-6} V) into an output voltage difference of 1 volt has effected a gain ratio of 1 million or 10^6 times. However, this becomes not 60 dB, but rather 120 dB, because of the rule just given.

By using the factor 20 when dealing with voltage gains or losses, the decibel changes in that variable are kept in step with the decibel changes in the variable of power, measured by the watt. A millionfold (10^6) increase in voltage corresponds to a 10^{12} times increase in power.

Now, in sound, pressure is the analogue to voltage in electricity. Other things remaining equal, the changes in the power of a sound will increase as the *square* of the changes in the pressure difference found in that sound.

Thus, when we deal with sound pressure, we can and must say that every gain of 20 dB means a tenfold increase in that pressure; or every loss of 20 dB means a reduction to one tenth of what that pressure had been before the loss. But when we deal with intensity or power per square meter of a sound, then we can and must say that every gain of 10 dB means a tenfold increase in intensity, and every loss of 10 dB means a reduction to one tenth of what that intensity had been before the loss.

Most of the decibel ratings of sound will be found to be based on *pressure* differences, as related to that standard 0 dB pressure.

When pressure is the physical variable being compared, then we can make the following approximate conversions of changes of 1 dB at a time: a gain of 1 dB indicates a pressure increase of 1.12 times; 2 dB, 1.26 times; 3 dB, 1.41 times; 4 dB, 1.59 times; 5 dB, 1.78 times; 6 dB, 2.0 times; 7 dB, 2.24 times; 8 dB, 2.43 times; 9 dB, 2.82 times; 10 dB, 3.16 times; 11 dB, 3.55 times; 12 dB, 4.0 times; 13 dB, 4.47 times; 14 dB, 5.01 times; 15 dB, 5.62 times; 16 dB, 6.31 times; 17 dB, 7.08 times; 18 dB, 7.95 times; 19 dB, 8.91 times; and 20 dB, of course, 10 times.

Thus each 20 dB gain represents a multiplication by 10; each 20 dB loss, a reduction to one tenth of the former pressure level.

On the other hand, if intensity, or power per unit area, of sound is the variable being compared, then the decibel relation to the ratios is quite different. Each 10 dB gain represents a tenfold multiplication in intensity; each 10 dB loss, a reduction to one tenth of previous intensity. And these are the approximate conversions 1 dB at a time: 1 dB indicates an intensity gain of 1.26 times; 2 dB, 1.59 times; 3 dB, 2.0 times; 4 dB, 2.43 times; 5 dB, 3.16 times; 6 dB, 4.0 times; 7 dB, 5.01 times; 8 dB, 6.31 times; 9 dB, 7.95 times; and 10 dB, of course, just 10 times.

Rivals of the decibel—The decibel is not the only unit devised to represent such ratios compactly and conveniently. Another ratio unit called the *neper,* and abbreviated Np or Nep, has been used mostly in Europe. It is based on the same principle, but uses the so-called "natural" logarithms, based on a number other than 10, the base of the common logarithms. We need not explore the number here, but it will suffice to say that there is a constant-proportion relationship between the dB and the Np units: 1 dB equals 0.23 Np; and 1 Np equals 4.34 dB. Likewise, 1 bel equals 2.3 Np; and 1 Np equals 0.434 bel.

At this stage we are still dealing with physical variables, rather than with psychophysical perceptions and sensations applied to the measurement of sounds. Evidence of this physical situation is the fact that, using the variables we have already been dealing with, we can even compute what must be the average motion back and forth of the molecules of air as these sound waves pass through them—and hence what might be the typical response, or "excursion," of the human eardrum when it reacts to these oscillating changes in pressure of the air that impinges on it.

If we represent the frequency of the sound wave by F, measured in hertz, and use the other variables of velocity (V), density (D) and intensity (I), which were previously defined, then we can make use of the following equation to find L, the average length of the back-and-forth molecular motion as the sound wave goes through the molecules of the air:

$$L^2 = I / 2\pi^2 \, V \, D \, F^2$$

The group $2\pi^2$ is a constant equal to about 19.74. The frequency F, since we deal primarily with the standard reference wave, is 1,000 hertz. The intensity I of that reference sound is, as already noted, 10^{-12} W/m². When we take the square root of both sides of the above equation, it comes out:

$$L = (I / 19.74 \, V \, D \, F^2)^{1/2}$$

The superscript $\frac{1}{2}$ at the right indicates the square root of the value of the fraction inside the parentheses.

Completion of the calculation for standard atmosphere shows that L must equal just about 1.1×10^{-12} meter. This means that 0 dB sound, the threshold or least-audible level, is associated with average back-and-forth motions of only about 1 millionth of a millionth of a meter in the molecules of the air, as a result of the sound waves passing through them.

This is extraordinary, for it is a length of about one fifth of the radius of a hydrogen atom! It is one more evidence of the extraordinary sensitivity of healthy young ears. The fact is that such healthy and unspoiled hearing is so very sensitive that if it were but a relatively small amount more sensitive, humans would be in danger of hearing as sounds the constant collisions and bombardments of the molecules of the air, resulting from their unceasing thermal motions. Such acuteness in hearing would be neither restful nor serviceable to us.

At the other end of the dynamic scale, we find that these ultrasensitive ears are somehow able to tolerate for brief times sounds whose pressure differences are a million times greater than those associated with this seemingly infinitesimal back-and-forth of our eardrums. They are sounds whose intensities, in terms of watts of power impinging on each unit area, are 10^{12} times that of the zero decibel sound.

A major and growing problem of our era is the noise pollution amidst which increasing scores of millions of urban dwellers exist. Using the decibel of comparative sound pressures as a guide, let us compare typical situations all the way from the least-audible to the highest-tolerable or even beyond that danger point.

The following decibel ladder is arranged with the least first and the greatest last. It represents a sound-pressure range of 140 dB, equal to 14 bels, and, since 20 dB means a ratio of 10 times when we deal with pressures rather than powers, the pressure range here is as 1 to 10^7 (10 million). The corresponding intensity range, were we dealing with watt-per-square-meter measures, would be as 1 to 10^{14} (100 million million).

This decibel "ladder" lists intensities like those shown on the appropriate scales of standard decibel meters, matching our responses to intensities at both high and low frequencies as well as in the central frequency bands.

Perceived loudness diminishes as distances increase between our ears and the sound sources. Accordingly in several situations distances are indicated (in meters), to increase the consistency between the objective physical factors determining the intensity of air vibrations, and the resulting perceptions of loudness that are reported by our ear-mind equipment.

    0 dB: minimum audibility for intact hearings.
  10 dB: just audible.
  20 dB: well-designed broadcasting studio with nothing going on in it.
  30 dB: a soft whisper at distance of about 5 meter.
  35 dB: interior of an orderly library.
  40 dB: typical bedroom, without TV or conversation.
  50 dB: "very quiet" condition.
  55 dB: light auto traffic, 15 m distant.

60 dB: air conditioning unit operating at about 6 m distance.

65 dB: average conversation—not an argument!

70 dB: typical light freeway traffic.

75 dB: freight train passing at 15 m distance ("conversation has by this point become noticeably difficult").

80 dB: "annoying" sound level.

85 dB: pneumatic drill at 15 m distance.

90 dB: heavy truck passing at 15 m distance.

95 dB: N.Y. City subway station ("very annoying").

100 dB: loud shout at 15 m distance.

105 dB: jet plane takeoff at 600 m distance.

110 dB: riveting gun not far away.

115 dB: maximum vocal loudness without amplification.

117 dB: discotheque at full blast.

120 dB: jet takeoff at 60 m.

130 dB: limit of amplified speech.

135 dB: ("painfully loud").

140 dB: jet operations on an aircraft-carrier deck.

There is no need here to add more extreme cases. Such possibilities exist, chiefly as nightmare visions of sound so lethal that it might serve as a deliberate weapon against some captive foe.

The super-Babel of sounds typical of various rungs in this acoustic ladder are by no means all, or even mostly, sounds at the arbitrarily chosen reference frequency of 1,000 hertz. However, by acousticians' agreement, that frequency has been used as a standard, since some such standard there must be, in view of the ear's uneven sensitivities to sounds at various frequencies up and down the complete audible spectrum.

From the objective to the subjective in sound measurement—At last we are prepared to approach an important transition: from the objectively measurable physical variables of sound (such as pressure, intensity, frequency, and so on) to the subjective perceptions, among which the most important in our time is beyond doubt *loudness*.

Everyone knows that when the physical intensity of a sound is increased, our sensation or awareness of its loudness increases also —provided that the physical increase is not so small as to lie below our threshold of difference perceptions. However, almost nobody knows—and this includes a great many competent and honest specialists in acoustics—just *how* the two changes are correlated. The subjective impression proves to be a complex derivative of the objective physical intensity, or of the objective sound pressure, for that matter.

There is a psychological or psychodynamic principle called the Weber-Fechner law. It says that if the actual intensity of a sound is increased from I_1 to I_2, then the apparent loudness (as noted in the mind of the normal listener) will appear to rise in the ratio of the common logarithm of I_2/I_1. For example, a tenfold increase of actual sound intensity, measurable in watt-per-square-meter units, should seem to double the original loudness of sound. And a hundredfold increase in the objective intensity should seem to result in a sound three times as loud as the original one.

In keeping with the principle of this law, a unit of *loudness level* has been devised, called the *phon*. It is tied directly to the decibel: indeed, the definition of 1 phon is that it is the loudness, or subjective perception in the listener, of a standard sound whose physical intensity is just 1 dB. Similarly, a standard sound (1,000 hertz) with 40 dB intensity is assigned a loudness level of 40 phon, and so on.

The phon has been credited with two principal advantages: first, it constitutes the smallest perceptible variation in loudness level; and second, because of its logarithmic basis it can express in the small numbers of its unit the actual loudness levels around us, which vary so much when expressed in ordinary ratios.

In the multiple "decibel ladder" already presented, the corresponding loudness levels in phons would be numerically just about identical. It is true that, strictly speaking, the loudness level in phons and the intensity or power of the sound in the decibel unit will correspond only with respect to sounds of 1,000 hertz frequency. However, we have already mentioned the A scale weightings or adjustments, which take much of this into account.

The loudness level measured by 130 to 140 phons actually hurts the ears and is commonly referred to by names such as "threshold of feeling," though it might perhaps better be called the limits of possible response to auditory sensation, or even the maximum of human toleration.

Thus, the enormous dynamic range between the threshold of hearing itself (at the just-audible level of 0 dB), and the threshold or maximum of resulting sensation is compacted to within 140 dB, which shows that the logarithmic short cut saves both space and time for metrologists.

The complexity of human perception of loudness of sounds is apparent again and again. The Weber-Fechner principle calls for an apparent doubling of loudness with each 10 dB gain in the physical intensity of the sound. This is just about the way things work in the middle frequencies that humans can hear. However, outside this middle bandwidth of the sound spectrum, loudness increases at a rate more rapid than two times for each 10 dB gain. Even more precisely, if the power of a high-frequency sound is increased 10 times, its apparent loudness will more than double. To a considerable extent the same thing can be said of the effect of a tenfold increase in the power of a sound at a low frequency.

High-fidelity specialists have long known this, and they insist that if the various frequencies of a record player are in good balance when it is playing softly (low volume), it will not do simply to increase the power across the board at all frequencies by a like ratio in order to get good reproduction at a loud (high-volume) level. The only way to preserve the apparent balance is to increase the power in the middle frequencies by a *higher* ratio than in the upper and lower frequencies. Our strange ear-mind network will then receive the "mix" that conforms to proper balance—approximating that of a listener in a favored seat at the original performance!

From phon to sone—The phon is an accepted unit of perceived "loudness level" or "equivalent loudness." It is tied to the decibel (dB) of sound intensity. Thus it has both subjective (perception) aspects and objective (physical intensity) aspects. However, in use the phon has some shortcomings and confusions, for human perceptions of loudness are in many ways inconsistent and misleading.

Efforts to deal better with such problems have led to the design of a special additional unit of "loudness" as such. It is called the *sone*, which is sometimes symbolized simply by the letter *s*, just as the *phon* is sometimes indicated merely by the letter *p*.

The phon typically is given the same numerical value as the decibel (dB) rating of the sound being evaluated. The sone, on the other hand, changes in a more rapid and complex way as the physical intensity of the perceived sound waves increase or decrease.

A loudness level of 1 sone is attained when a listener with normal hearing perceives a pure 1,000-hertz tone whose intensity is 40 dB. This decibel is the "perceived loudness" decibel, sometimes designated by its initial letters: PLdB. Thus a *loudness level* of 40 phon corresponds to a *loudness* of 1 sone. However, a loudness level of 50 phon corresponds to a loudness of 2 sone; and a loudness level of 60 phon to a loudness of approximately 5 sone.

The far more rapid change in magnitude of the sone is clear: over the important range of 40 to 60 phons, the 50% gain in phons contrasts with a 500% gain in sones.

At higher loudness levels (measured in phons), the loudness in sones grows somewhat less steeply. Thus 100-phon loudness level corresponds to a loudness of slightly less than 100 sones. While the phon rating has grown 250%, from 40 to 100 p, the sone rating has shot up 10,000%, from 1 s to nearly 100 s.

The originators and advocates of the sone declare that its changes reflect far more closely than those of the phon the actual subjective loudness perceptions of normal intact human hearing, dealing with so many varied and mixed sounds.

From the "ladder"-of-loudness list earlier in this chapter we here offer a few selected situations with their corresponding ratings in phons (p) and sones (s):

SITUATION	PHONS	SONES
Typical bedroom, no TV or talk	40 p	1 s
Light freeway traffic	70 p	8 s
Heavy truck passing at 15 meters	90 p	35 s
Loud shout at 15 meters distance	100 p	95 s

Sones are seldom used at loudness levels over 100 phon.

Most of the actual sounds important to humans are mixtures of many diverse frequencies, emitted at various intensities. Our loudness sensations seldom result from a pure tone of just one frequency or pitch. Using only the phon unit can lead to confusions. Thus, two sounds of the same or closely similar frequencies, if heard together, will not gain much in phon ratings. If each alone has a 20-phon loudness level, both together will be perceived at 23-phon loudness level, not at 40-phon. Three such sounds will be heard as a 25-phon mixture, not a 60-phon blend—and so on.

Phon ratings also can mislead when we hear together two or more sounds of the same or closely similar intensities, if they differ widely in their frequencies (pitches).

Supporters of the sone say it can be used to reduce such difficulties. Intensity measurements are made on sophisticated electronic devices called "octave band analyzers." These report in sones the intensities in each of many octaves, from frequencies as low as 20 hertz to as high as 21 kilohertz. These separate octave intensities are then combined so as to fit the fact that our hearing, when dealing with mixtures of several or many sounds, responds most strongly to the most intense among them.

The highest sone rating is taken at full value. It is then increased by 30% of the sum of the lesser sone ratings in other octaves. Result: a sone total for the complex sound mixture. Such sophisticated procedures reflect the patterns and biases of human hearing reacting to complex sound mixtures.

Most commonly, sound specialists make use of the decibel and its offshoot, the phon, rather than the sone. Greater familiarity and convenience are basic factors here.

Human auditory perceptions are flawed and limited in many ways. Our hearing perceptions form only a small part of the spectrum of vibrational frequencies. Dogs, bats, dolphins and other "toothed whales" in the seas hear as sound vibrations to which we are deaf. Our "ultrasound" is basic to the lives of many fellow creatures. We must resort to meters and detectors when our ear-mind networks cannot cope.

Both the phon and the sone are relatively recent additions to the tools of acoustics and sound analyses. The phon emerged from proposals made in 1933 by J. Steudel, a German specialist in electronic acoustics. Not many years later the addition of the sone was urged by two American experts, S. S. Stevens and H. Davis. Their published data tended to show that the sone could yield measurements or estimates matching subjective loudness perceptions of listeners with normal hearing.

Other acoustic units also have been proposed. A novelty among them is the *noy*, pronounced as in the words *noise* or *annoy*. It was offered to measure or compare the noisiness or annoyance in unwanted sounds (noises) as we perceive them. It is pointed out that Noise "X" may be subjectively perceived as louder, but not necessarily more annoying than Noise "Y," which has lower physical intensity. In concept, the noy and the sone are closely related, but the former remains a mere smile-provoking outsider among the units in this field.

Human perceptions of loudness, pitch and other aspects of hearing cannot avoid ambiguities and complexities. Professor Rex Nelson, a physicist at Occidental College, Los Angeles, has noted that "After all, the subjective evaluation of loudness is very tricky and somewhat controversial."

Indeed, the author has been forced to realize that these subjective obstacles are overshadowed only by the fascinations and attractions of the subject matter itself.

46 RING OUT, WILD DECIBELS!—
SOME MEASUREMENTS RELATED
TO THE DEFENSE
OF AUDITORY INTEGRITIES

. . . full of sound and fury, signifying . . .
—SHAKESPEARE,
Macbeth

Elected Silence, sing to me
And beat upon my whorlèd ear,
Pipe me to pastures still and be
The music that I care to hear.
—GERARD MANLEY HOPKINS,
The Habit of Perfection

. . . the howling jet planes . . .
shaking the mountain
When one of them over-passes the speed of sound.
—ROBINSON JEFFERS,
Prophets (1963)

Human hearing like human seeing, or sight, is a faculty of astounding sensitivity in the frequency bandwidth for which it is adapted. The adaptability and the durability, relatively speaking, of both our hearing and our seeing systems merit both admiration and gratitude.

Yet, present-day human aims would be sadly frustrated were we obliged to depend solely or even primarily on either hearing or seeing for objective measurements of the matter waves that we call sound or of the electromagnetic waves that we call light and related radiations.

The alterations and enlargements in the science and technology of acoustics have been extraordinary, especially in the past generation or two. Acoustics has in fact undergone three great phases or episodes. At first it dealt almost entirely with subjective sound— sound as perceived by the human ear-mind network. Next, it expanded its scope and aim to include measurement of the physical sound vibrations by which those subjective human perceptions were evoked. Finally, it broke free of the human limitation in another respect: it began to deal also with vibrations at frequencies totally unhearable to humans (ultrasound, or supersonic sound) and with vibrations in media other than the atmosphere. This includes vibrations in water, in solids, even in plasmas, and so on. A good share of modern seismology deals with Earth vibrations closely akin to sound waves.

A wealth of new material has emerged from the realms of vibrations at frequencies far beyond those that we humans can hear. Dogs can respond to tones whose frequencies are as high as twice the 20,000 hertz commonly considered to be the upper limit of ordinary, unaided human hearing. For centuries it was known that bats emit audible squeaks or squeals, many of which seem to cluster around the frequency level of 5,000 hertz. Only rather recently has it been demonstrated that these are the "low tones" of bats.

For their hunting radar—a radar based solely on sound, not radio waves—bats emit brief sound pulses at frequencies about ten times as great—as much as 40,000 to 50,000 hertz. With their large and complex ears they receive back again from their insect prey the echoes of these tiny, highly directional sound pulses. Thus they are able to determine both direction and range of these insects, and to catch them while in full flight in the darkest caves or moonless nights outdoors. Also, bats are able, even with their eyes sealed shut, to fly through complicated obstacle courses, without becoming entangled in threads or wires. Again, their sound radar, or sonar, is at work, and all at frequencies that we humans can "hear" only after our instruments have picked them up and then shifted them to frequencies within our limited bandwidth of hearing.

A bat sound-pulse at 50,000 hertz represents a cluster of sound waves only a little more than ½ millimeter long. The high resolution and high directionality essential to the blind hunting by the bats can be provided only by such extremely short, high-frequency waves.

In the depths of the sea too, sound sonar is constantly at work and has been since time immemorial. Dolphins make use of extraordinary patterns of squeaks and clicks which they emit, just as a radar transmitter sends out pulses. They listen to echoes and thus, without any usable light for their eyes, locate obstacles in water. Thus too they find and pursue and capture the fish that are their prey.

Other mammals that have returned to the sea as their habitat make similar use of their own sound systems for determining direction and range. Submarine hunting by seals and sea lions would be almost impossible but for their extraordinary use of methods that antedated by eons the sonar of submarines and surface vessels.

As these lines are written, word comes from Dr. Thomas C. Poulter, shedding new light on how sea lions hear and hunt under the sea. This authority on the sound and hearing systems of living organisms has found, together with his associates at the Stanford Research Institute, that the strange stiff whiskers that figure so prominently in the appearance of a sea lion represent a sort of antenna system. Their whiskers may well be up to 10,000 times more sensitive in picking up the echoes of their sound sonar than are their small ears.

It was known that sea lions emit small click-like sounds, actually quite complex in structure, and that the echoes of these sounds provide the information they need to find and capture their prey and avoid obstacles in the dark depths. However, new insights followed the implanting of electrodes in the nerves leading from the whiskers to the sea lion brain, as well as in the nerves leading from the ear to the brain. The pattern of nerve impulses supports the idea that it is via the whiskers, rather than the ears as receptors, that these intelligent creatures acquire most of the information that makes possible their survival in the sea. The sophistication of the ancient sea lion sonar system is revealed by the finding that they are able to change the frequency patterns of their emitted clicks according to the size of the objects they are pursuing, and their distance from such objects.

Bats, sea lions, dolphins, whales and others emit their own special sound patterns in order to listen to the echoes that come back to them. The silent sea proves to be alive with strangely assorted sounds, whenever predatory life abounds in it. By tuning their emitted frequencies and altering their patterns of pulses, these sound-dependent creatures are able to avoid confusing interferences or cross-talk from other species and even from other hunting members of their own kind.

"How's that again?"—the human dilemma—Human listeners are constantly beset, however, by problems arising from the obscuring of sounds they want or need to hear by sounds that they do not want to hear but cannot avoid hearing. This frustrating sort of situation is not limited to cocktail parties or to poorly chaired meetings where vocal participation is incessant and simultaneous.

The subjective impression that our minds report is significant. It appears to us that the intensity and loudness—as well as the intelligibility—of the sounds of interest to us have been reduced by the competing sounds that we wish would go away. It does not matter whether or not we call these competing sounds "noise"—their effect on the sounds that we seek to ferret out and respond to is to make the latter appear to have lost loudness.

A rather crude figure of speech says that one sound drowns out another. Acoustic specialists prefer to speak of the "masking" effect that one sound actually shows with respect to one or more others. Though these masking effects are complex and difficult to analyze quantitatively, they cannot be disregarded.

The metrologists of acoustics sometimes prepare "masking audiograms." These are intended to indicate the amounts of masking attributable to a particular definable tone. Such audiograms are laid out in decibel units, shown as a function of the frequency of the masking tone or tones. In other words, the apparent loss of loudness because of the masking tone is expressed by the decibel, the ever-convenient summarizer of ratios.

Somewhat analogous to these graphic analyses are the loudness contour charts or diagrams. These are graphic curves showing the levels of sound pressure required at each indicated frequency in order to create in the mind of a typical listener the sensation of a selected loudness. Commonly, such a graph includes several different curves, each one applying to a different and larger subjective loudness level than the one below it.

Needless to say, these graphic summaries are all based on experiments involving listeners with normal hearing. Also, almost needless to say, they depend heavily on the decibel concept to adjust them to other procedures in acoustic measurements.

Speech, the unpowerful indispensable—Humankind is distinguished by speech, shaped by speech, sustained by speech and—it sometimes seems—obsessed by speech. Yet the physical power —the energy flow per unit time—of even notably loud speech is actually quite small. Compared with the work done when we lift even a rather light object, the work that our vocal cords do on the atmosphere is extremely small, however devastating the words thus enunciated may seem to their hearers. It has been estimated by some of the "believe it or not" kind of statistical manipulators that if every one of the billions of individuals now alive on Earth were to talk at the same time, the combined power of their entire vocalization would amount to no more than 300 kilowatt.

On the other hand, the noise power of a typical large jet engine being revved up for takeoff is so enormous that to match it, every living human, including the merest infants, would have to scream at the top of her or his voice!

We have up to now dealt by design with *sound,* both as objectively measured and as subjectively perceived. Efforts also have been made to provide particular measures and procedures to deal with the unwanted and disturbing sounds called "noise," as distinct from wanted or at least tolerable sounds to which the term "noise" does not properly apply.

Inevitably, concepts developed in connection with the labors of measuring or estimating sound have been extended also to the ever-increasing realm of noise.

Closely analogous to the sone is another unit, known as the *noy* and designed to measure a physical variable called "perceived noise level." It is symbolized by the letter L. If N represents the decibels of perceived noise, as distinct from perceived *sound,* then the relationship is:

$$N = 2^{(L-40)/10}$$

As before, a perceived noise level L of 40 dB corresponds to an N of 1 noy, and an L of 60 dB to an N of 4 noy, an L of 80 dB to an N of 16 noy, and so on.

This esoteric-looking equation is not suitable for use in efforts to answer the ancient tongue-twisting conundrum: "What kind of a noise annoys an oyster?" The answer to that early excursion into the emotional effects of noise pollution remains as before: "A noisy noise annoys an oyster." Or a human being, for that matter.

The noy unit was proposed in 1959 by K. D. Kryter in the *Journal* of the Acoustical Society of America. It was there defined in terms of the perceived, or subjective, noisiness in the frequency bandwidth between 910 and 1,090 hertz—a total width of 180 hertz, centered on the standard 1,000-hertz frequency. The object of its application was random noise whose sound pressure was 40 dB above the 0.000 2 microbar level. Since 1 microbar is equal to 0.1 pascal, the SI pressure unit, the sound pressure of the random noise in this case was 2×10^{-5} pascal.

The assumption on which the noy unit is proposed and employed is that the perceived noisiness, or noy magnitude, increases with the same ratio to the physical (objective) intensity of that noise as does the perceived loudness of non-noisy sound in relation to the physical (objective) intensity of that sound. This assumption appears acceptable to leading acoustic authorities.

The takeoff of a large jet aircraft—for example, a three- or four-motored modern transport plane—is accompanied by about 110 noys of perceived noisiness. This is numerically rather close to the 120 dB level of the objective loudness (based on pressure) of a jet takeoff at a distance of some 60 m, as given in our "loudness ladder" in the preceding chapter.

Measuring noise, an all-important effect—What, however, *is* noise, in terms of acoustic measurements? In contrast to the concept of the bel and the decibel ratio units, which originated outside the world of acoustics itself (in the realm of the electrical circuitry of telephone communication), the concept of noise, whose origin *was* within the acoustic area, has spread far beyond it. *Noise* today is a term used in optics, in radio reception, and in many other areas of our proliferating technology. It is applied freely, and with complete clarity to those who use it, in situations in which the signals that are sought and recorded at *no time* are translated into sounds.

What, to begin with, is noise in its original acoustic sense? It is simply unwanted or undesirable sound, which competes with or obscures the sought-for sound. Much noise is random and so scattered over a broad sound spectrum, from low to high frequencies, that it is called "white" noise—a term borrowed from optics, in which the spectrum of light from a star, such as the Sun, is so spread over the visible colors (frequencies) that its total effect is that of "white" light. Other noise originates in ways by no means random or frequency-spread. For example, if in the midst of an important telephone conversation, your ears are assailed by a high-pitched pure whistling note of some kind, that is none the less "noise" for the fact that it is concentrated and pure.

A reasonable if rather primitive analogy may appeal to gardeners and agronomists. Weeds have been defined as unwanted plants. They are physically first cousins to the wanted growth, whether floral or edible. But they are where they should not be, in the judgment of the grower, not in that of Mother Nature. Similarly, noise consists of signals which interfere with the signals that are sought and whose reception is the aim of the operation that the noise is, to a greater or lesser extent, polluting.

One man's meat is another man's poison. One woman's weed may be another's favorite wild flower. The background music imposed on increasing numbers of offices and institutions in the United States may be "noise" to lovers seeking to carry on a whispered conference in some corner. Their talk and that of others in the same enclosure may be considered mere "noise" by a listener trying to enjoy a favorite melody that happens at that moment to be emerging from the speakers concealed in the ceiling . . .

The extraordinary adaptability of the decibel ratio unit has been illustrated in one after another of these special acoustic units devised to deal with the increasing problems of loudness, noise, and other physical aspects of the sounds that assail us. But the trail of the decibel leads us to still more measures proposed and adopted under the pressures of our era. For example, there is the problem of the transiency, or relative brevity, of noises that, though their loudness level is subjectively insufferable and objectively dangerous to the integrity of the hearing when they are sufficiently prolonged, are ordinarily ended so quickly that their victims are not immediately aware of lasting damage. Such events include the passing of a particularly noisy truck, the takeoff of a jet plane, or that most massive and catastrophic sky-borne assault on our auditory environment, the sonic boom.

A unit of transiency or its opposite, relative duration, has been devised to deal with such noisy interludes. It is commonly called the "10-dB time" and is defined as the duration (in seconds or even minutes) of the period during which the transient noise remains within 10 dB of its maximum value. If, for example, the maximum value, in terms of pressure, were 120 dB, the 10-dB time would be the period during which the noise did not diminish below the 110-dB level.

This would seem to be a diminution of $\frac{1}{12}$, or 8.5 percent. But the ratios are quite different when we go back to the objective physical measurements, such as intensity (measured in the W/m^2 unit combination) or pressure (measured in the pascal unit). Thus, the 10-dB time is the period during which *intensity* remains at or above 10 percent of its peak value. During 10-dB time, in other words, intensity does not drop by more than 90 percent. In terms of the pressure in the noise (sound), the 10-dB time is the period during which the pressure remains at or above 32 percent of its peak level. Thus, during that 10-dB time the pressure does not drop more than some 68 percent, from its peak level.

Suppose that, using a modern "decibel meter," we find that certain intolerable events near a modern airport produce a 120-dB peak noise, and that the 10-dB time associated with a typical event of this kind is 30 seconds. If we convert the decibels into the equivalent noy units of perceived-noise level, this means that the peak is at 256 noy, and a loss of 10 dB from that peak brings us down to the 128-noy level.

Thus, the 10-dB time is also the interval during which the perceived noise, as measured by the noy unit, remains at or above 50 percent of its peak magnitude.

All this stresses again the unavoidable distinctions between (1) the objective physical variables, such as power, intensity and pressure in sound; and (2) our subjective perceptions of the resulting sounds or noises. These distinctions do complicate acoustic measurements, but also keep them attuned to urgent problems of what is sometimes rather euphemistically referred to as "the human condition" in these concluding decades of the twentieth century.

The phrase "10-dB time" is somewhat reminiscent of another

familiar Americanism: "down time," which is the elapsed time during which a piece of apparatus or equipment is out of commission. The 10-dB time, for those noise-polluting events to which it is worthwhile to apply the concept, is likely to result in "down time" for the humans whose hearings have been assailed. It may well be an interval of miserable emotional shock, of heart-pounding, nerve-tormenting terror.

The objective measuring devises used to measure loudness today do not suffer from emotional or psychosomatic ill effects. The standard sound-level meters are, or should be, capable of a reproducible objectivity, even when subjected to peak intensities. They combine four interlinked components: (1) a microphone, compliant enough to distinguish in its dynamic response between, say, 100 dB and 130 dB conditions in the air around it; (2) an electronic amplifying system; (3) a system of selectable frequency-weighting networks, capable of at least three different response patterns; and (4) an output meter, calibrated in decibel units.

The choice of networks includes the A scale, already mentioned; the B scale, which we need not analyze here; and the C scale. This last is arranged to give relatively greater weight to frequencies in the middle range (70 to 4,000 hertz) than to frequencies either above 4,000 or below 70 hertz. In general the readings when the C scale network is in action are identical with the over-all sound-pressure levels. These are considered to be unbiased in relation to frequency. That is, they neither overvalue the pressure differences in the highest audible frequencies (as does the A scale), nor do they undervalue the pressure differences in the lowest audible frequencies.

The decibel concept has become virtually inseparable from such carefully calibrated standard sound-level meters. Many users of such meters tend to forget that the decibel was born or shaped to measure differences of power and voltage in electric circuits, not primarily to deal with sound as it passes through the atmosphere.

Changing uses of the decibel unit—The decibel concept is applied now even to measurements of the ways in which levels of sound pressure attenuate, or diminish, with increasing distance from a small, pointlike sound source. With such a source we can assume that, at a reasonable distance out, the waves have nearly spherical form with the source as center. Under these conditions it is found that with each doubling of the distance from such a center, the level of measured sound pressure diminishes by 6 dB. This is, in fact, another way of saying that the measured pressure varies inversely as the square of the distance from the source. When that distance doubles, the pressure drops to $\frac{1}{4}$ of its former amount; when that distance trebles, the pressure falls to $\frac{1}{9}$; and so on.

One can fantasy that, had the decibel concept been so widespread in the seventeenth century as it has become in the latter half of the twentieth century, Isaac Newton might even have been tempted to tell his contemporaries that the magnitude of the force of gravitation diminishes by 6 dB with each doubling of the distance from the center of a massive attracting body, such as the Earth, the Moon or the Sun! Instead he made the statement in the form of the inverse-square law; and so too did Coulomb in describing how the forces between pointlike electrical charges diminish as the distances between them increase.

The bel and decibel were born because electrical, electronic and finally acoustic technologies dealt more and more with enormous amplifications (dB increases) and diminutions or attenuations (dB decreases). It is indeed the same decibel that we encounter in the specifications of high-fidelity equipment and the instrumentation that is part of virtually every scientific and technological laboratory today. In order to properly link the "different" decibels, we must recall that the original was intended to compare conveniently the ratios of two *powers,* a first power (P_1) and a final power (P_2), and for such power comparisons the decibel is defined as 10 times the common logarithm of the ratio (P_1/P_2) of those powers. This is true whether the power is consumed in an all-electric circuit, as in a long telephone line, or is the intensity (power per unit area) in sound as it moves through air or water.

On the other hand, to repeat for emphasis, when the decibel is used to compare two pressures or forces, then it is defined as 20 times the common logarithm of the ratio between them. For example, the decibel equivalent of a voltage amplification that raises an initial voltage of V_1 to another and far greater voltage V_2 is 20 times the logarithm to the base 10 of the ratio V_1/V_2. And when the level of a sound, such as the standard tone of 1,000 hertz, is greatly increased, then the initial pressure difference (P_1) and the increased pressure difference (P_2) are compared by stating that P_2 is a certain number of dB greater than P_1; with these decibels defined as 20 times the logarithm of the ratio P_2/P_1.

The decibel and the phon enjoy a common distinction: they were the only units given definitions at the first International Acoustical Congress, held in Paris in 1937, two years before the beginning of World War II.

More than one proposal has been made to rename the decibel. It has so totally displaced its parent, the bel, that its independent status seems to some specialists to call for a new name of its own. Among suggestions thus far offered is that of the Bell Telephone Laboratories: the *decilit,* being a composite of "decilogarithmic unit." Other suggestions have included *logit, decomlog, decilog,* and even *decilu,* uncomfortably similar to Desilu, name of a television producing company.

Apparently *decibel,* however, will continue to ring out in the worlds of the measurement of sound, of electronics, and of the circuitries that serve the proliferating networks of worldwide communication and even the signals sent from man-made satellites in space back to the parent planet Earth.

Measuring pitch—*Sound-intensity* level, measured in decibels, is essentially an objective physical magnitude, based clearly on the 1,000-hertz reference tone with an intensity of 10^{-12} watt per square meter, while *loudness,* measured in sones, is essentially a subjective magnitude. Similarly, sound *frequency,* measured in hertz, is an objective physical magnitude, which instruments can determine, while *pitch,* though derived from frequency, is essentially a subjective magnitude.

A unit has been devised to measure this subjective quality of pitch. It bears the rather mellifluous name *mel,* related to the Greek *melikos* or *melos* ("song"). The same syllable appears in the English word *mel*ody, signifying a pleasing pattern of successive pitches or intervals.

Paradoxically, but inevitably, both the objective units of the hertz and the decibel are required to establish the reference tone on which the mel unit is based. This is the customary pure tone of

1,000-hertz frequency at an intensity level 60 dB above the listener's threshold. Accordingly, it is a sound whose pressure differences are 1,000 times greater than those of the threshold tone, and whose intensity accordingly is 1 million times greater than that of the threshold tone.

In some determinations of the mel, 40 dB rather than 60 dB has been used, which represents a pressure 100 times greater and an intensity 10,000 times greater than the threshold tone.

The reference tone is assigned the value of 1,000 mel of pitch. Any other tone which is judged by the subject to have N times the pitch of the reference tone is regarded as having N times the 1,000 mel pitch of the reference tone. For example, a new tone, judged to have twice the pitch of the reference tone, would be assigned 2,000 mel of pitch, and so on.

Comparing frequencies of musical notes—In terms of frequency, the interval of the octave represents a doubling of frequency. Thus if a piano's middle C is tuned to 256 hertz (cycles per second), its C above middle C should be tuned to twice that, or 512 hertz. The octave interval is a pitch relationship to which our ears are so attuned that we hardly think of it as a musical interval; rather it seems a kind of identity or semiunison between the higher and lower pitch.

The octave has been converted into one of several logarithmic measures of the ratios of the frequencies provided by common intervals on the piano or other musical instruments. This measure bears the name of *millioctave*.

Other measures devised to do the same service in units of somewhat different size are called the *cent*, the *savart*, and the *modified savart*. The musical interval of the octave, as from middle C to the C above it, is a frequency ratio of 1 to 2, since the higher C has just double the frequency of the lower. This is the largest and the simplest of the frequency ratios between any two successive notes among the other ten in the complete scale (five white keys and five black) and the key note of C. In the simple key of C, the sequence is eight white keys (C,D,E,F,G,A,B, and the C above) interspersed by five black keys (which can be called C♯, D♯, F♯, G♯ and A♯). No black key lies between E and F or between B and C. Relative to the key note (middle C in this example) there are eleven frequency ratios to be compared.

When the millioctave is the unit used for such comparisons, the octave ratio of 2 becomes 1,000 millioctaves. In the modified savart unit, it becomes 300 modified savarts. In the simple savart unit, it becomes 301 savarts. In the cent unit, it becomes 1,200 cents. Thus, the octave, or any other frequency ratio or musical interval, is 1.2 times larger numerically when expressed in cents than when expressed in millioctave units. An interval in modified savart units is one quarter as large numerically as in cent units, and 0.3 times as large as in millioctave units, etcetera.

Each of these alternative units for expressing frequency ratios is based on the common logarithms to the base 10, as is also the bel and the decibel (which were not, however, intended for comparing frequencies of musical notes). The logarithm of 2 to the base 10 is 0.301 03. Accordingly, it would be true to say that the frequency ratio of any true octave interval is equivalent to the ratio represented by 3.010 3 decibel. But this would be far too coarse a unit to express the frequency ratio between, for example, the key note of C and the D just above it.

Instead, we can choose between several more convenient options. We can multiply the octave logarithm (0.301 03) by the constant 3,321.93, and thus convert it into 1,000 millioctave units. Or we can multiply it by the constant 1,000, thus converting it into 301 ordinary savart units. Or we can multiply it by the constant 996.578, thus converting it into 300 modified savart units. Finally, we can multiply it by the constant 3,986.314, thus converting it into 1,200 cent units.

Whatever constant is chosen will of course be applied to the logarithms of each and every other frequency ratio that is being compared. The constant employed determines which ratio unit results; and one can compare diverse ratios only when one unit is used consistently.

The great and amazingly versatile German scientist Hermann Helmholtz, when in the 1860s he wrote his classic, *On the Sensations of Tone as a Physiological Basis for the Theory of Music,* used logarithms, pure and simple, for comparing dozens of different musical intervals drawn from many periods and cultures. His British translator later added also ratio units in cents. For example, the "equal major third" interval, with a logarithm of 0.100 34, had 400 cents; and the "equal major seventh," with logarithm of 0.275-94, had 1,100 cents.

A later classic in the field, Alexander Wood's *The Physics of Music,* published first in 1944, includes a cogent discussion of the measurement of pitch intervals, and indicates some preference for the modified savart, calling it a unit of convenient size since one such savart "represents approximately the smallest pitch interval that the human ear can appreciate." Wood, however, noted some advantages of the cent also; it provides neatly rounded numbers of 100 cents for each semitone interval and 200 cents for each whole tone in the tempered scale. Likewise the cent had by then become the unit most widely used in the published literature on intervals as well as in "instruments designed to measure musical intonation."

The cent is used also in the following. Since the time of Johann Sebastian Bach the scales of "just intonation" have been replaced by the tempered scale, which that master supported so stoutly in his *Well-Tempered Clavichord.* By slightly modifying the frequencies of the more accurate "just intonation" scale, the tempered scale permits freedom of change from one key to another. Any note on the piano keyboard, white or black, can become the starting (lowest) tone for a scale—a tempered scale, it must be stressed. Thus the pianist or organist can transpose readily from the key of C to that of C♯ or D to D♯ or E or F, and so on, in major or minor modes.

The simplest key to use in an example is that of C major. In the following table of frequency ratios we begin with the fourth C (C_4) and end, twelve half steps later, with its octave, C_5. Standard tuning gives the tone A above C_4 a frequency of just 440 hertz. Thus C_4 itself has a frequency of 261.6 hertz and C_5 of twice that, or 523.3 hertz.

The table gives the common musical names for the indicated intervals, the decimal ratios of the frequencies forming those intervals, the values of those decimals in cent units, and the actual frequencies (in hertz) of the tones in an accurately tuned instrument.

No black key is found between the E and the F or between the

B and the C on a standard keyboard. Hence the familiar scale pattern of two successive full steps followed by a half step (E to F), then three successive full steps followed by another half step (B to C). Those eight notes, numbered 1 through 8 in the table, form the octave. In the full chromatic scale, playing white and black keys as they follow one another, there are 13 tones in the octave, hence 12 intervals with relation to the initial or lowest tone (here C_4).

Scale of Equal Temperament (from C_4 to C_5)

TONE NUMBER	SYMBOL FOR KEYS WHITE	BLACK	COMMON NAMES OF THE INTERVAL	DECIMAL RATIO	RATIO IN CENTS	FRE-QUENCY IN HERTZ
1	C		Unison	1.0	0	261.6
		C♯, D♭	Semitone or minor second	1.059	100	277.2
2	D		Whole tone or major second	1.122	200	293.7
		D♯, E♭	Minor third	1.182	300	309.3
3	E		Major third	1.260	400	329.6
4	F		Perfect fourth	1.335	500	349.2
		F♯, G♭	Augmented fourth or diminished fifth	1.414	600	370.0
5	G		Perfect fifth	1.498	700	392.0
		G♯, A♭	Minor sixth	1.587	800	415.3
6	A		Major sixth	1.682	900	440.0
		A♯, B♭	Minor seventh	1.782	1,000	466.2
7	B		Major seventh	1.888	1,100	493.9
8	C		Octave	2.0	1,200	523.3

The beauty of the logarithm-based units such as the cent is that by subtracting or adding the cent values, we can determine the musical intervals between any two tones among the 13 in the full chromatic scale. From C to E is a major third, with a ratio measured as 400 cents. From E to G♯ (400 cents to 800 cents) is also a difference of 400 cents; hence it too is a major third in the tempered scale. And from G♯ to the C above (800 cents to 1,200 cents) is again a difference of 400 cents, hence another major third. Similarly from F (500 cents) to A (900 cents) is a difference of 400 cents, and so is a major third; and from G to B, since the difference is also 400 cents, the interval is again a major third.

Before the adoption of the tempered scale, a major third in the scale of "just intonation" had to show the frequency ratio of 5 to 4, or 1.25 when expressed as a decimal. In cent units this was 386.3 cents, not 400, as in the tempered scale. The discrepancy is 13.7 cents. Thus, relative to the C below, the E in our accepted tempered scale is tuned somewhat too high. In terms of decimals, the ratio is 1.260 instead of 1.250. Similarly, a perfect fifth in the scale of just intonation had a frequency ratio of 64 to 45, or 1.422, but in the tempered scale that interval shows a frequency ratio of 1.498. The former ratio corresponds to about 609.8 cents; the latter to just 700 cents. Again the interval in the tempered scale is slightly larger than the exact ratios called for.

So attuned are our ears to the tempered scale that music lovers gifted with perfect pitch might find their favorite works sounded somewhat strange were they performed in a scale of just intonation. Indeed, a true scale of just intonation would require sixteen different intervals, rather than the twelve in the tempered chromatic scale. For example, the table preceding shows just one interval (the minor seventh) between the major sixth and the major seventh. In the scale of just intonation, however, three intervals are required in that gap: the harmonic minor seventh (968.8 cents), the grave minor seventh (996.1 cents), and the minor seventh (1,017.6 cents).

The interval between the major sixth and the major seventh in the tempered scale is the difference between 900 cents and 1,100 cents, or 200 cents. The corresponding interval in the scale of just intonation was the difference between 884.4 cents and 1,088.3 cents, or 203.9 cents. Those two interval differences are very close, but the just-intonation system required four subintervals to bridge the 203.9 cents difference, whereas the tempered system bridges the 200 cents difference with just two subintervals.

Ratio units such as cents and savarts and millioctaves are tools extremely useful and often utterly essential for analyzing the relationships between musical effects and the physical changes of vibration frequency that call them into being.

From melody to menace—The transition from measurements related to melody and harmony, back to measurements related to contemporary sound pollution and damage from intolerable noise, is unpleasant but unavoidable here.

Sound pollution, or excessive and deleterious noise, has become a local, national and international problem. This book is devoted to a broad survey of metrology, not to diagnoses, prognoses or therapeutics. Yet the most important applications of the decibel and the other units intended to measure loudness levels and perceptions are bound to lie increasingly in the areas where the sound and noise situations are such as to endanger human hearing and human health in general.

One of the most eminent authorities in the field, Dr. Samuel Rosen, professor of otology at the Mount Sinai School of Medicine, in New York, has stated that "it would seem that loud noise can increase body tensions, which can then affect the blood pressure, the functions of the heart, and the nervous system." He concludes that city dwellers "with heart disease, high blood pressure, and emotional illnesses need protection from the additional stress of noise."

Our eyelids more or less shut out disturbing light when we sleep, but, as Dr. Rosen notes, regarding the influences of excessive noise on sleepers, "The reflex effect which causes constriction of blood vessels occurs with equal intensity during sleep."

The Surgeon General of the United States not long ago estimated that between 6 and 16 million Americans were on their way to deafness from noise directly associated with their occupations.

Laboratory experiments have revealed that guinea pigs, subjected to short spells of noise that was supposedly tolerable though well above normal, actually developed swollen membranes within their ear systems, while the vital hair cells (cilia) within the inner ear showed deterioration and destruction.

The evidence along these lines already is enormous and is increasing each year. At the very least, a substantial proportion of urban dwellers stand to suffer premature hearing losses; at the worst, the incidence of deafness in virtually every degree can be expected to increase, despite the phenomenal gains in insight that have taken place in the medical specialty of otology.

The poet Robinson Jeffers once addressed his fellow Americans with the warning or incitement: "You making haste, haste on decay . . ." Almost nowhere are the evidences of haste more patent than in the realm of the exaggerated assaults on our sensitive and vulnerable hearing organs.

The velocity of sound through the atmosphere plays a particularly crucial role in the most spectacular and aggravated of all the events involved in the great complex of sound pollution, the "sonic booms." These assail broad areas on the surface of the earth when aircraft aloft "break the sound barrier"—meaning that the propulsive force exerted by their jets drives them through the air at velocities greater than that of sound itself.

Such velocity is meaningfully measured in a dimensionless number called the Mach number, for its originator, the important Austrian physicist and philosopher of science Ernst Mach (1838–1916). The Mach number is simply the ratio between the actual velocity of the plane relative to the air around it and the velocity of sound waves in that same air. At Mach 1 the plane moves *at* the velocity of sound; at Mach 1.1 it moves 10 percent faster, and so on.

This Mach number is dimensionless (just as is the decibel), because it matters not a whit what system of units is used: the Mach number will not be altered whether velocity is measured in meter per second, mile per hour, or furlong per fortnight.

There are Mach numbers for moving objects under water as well as for flying objects in the air and other gaseous media. When a plane exceeds a velocity of Mach 1, it is outstripping the sound waves that normally precede it and herald its approach to the ground and to other planes in the air. It pushes ahead of itself, by the sheer brute force of its jet thrust, a "shock wave." The effects of this shock wave spread out to the ground, at the velocity of sound itself, through each layer of air down to terra firma.

When a listener on the ground hears a plane flying at less than Mach 1 overhead, at each instant that its sound is audible he receives some sound energy which left the vicinity of the plane at a different time during its flight. Thus, the energy of sound that reaches a point on the ground is moderate and diluted, and somewhat extended in time.

However, when a plane attains velocity of greater than Mach 1 in flight, there will be points or areas on the ground whose distance from the moving plane is diminishing at just the same rate (in meter per second or any other velocity measure) as the rate with which sound travels through that intervening air.

The result is that sound energies that left the plane during an extended portion of its flight course all arrive at or near that one patch of the Earth's surface at the same instant. The consequence is a huge, sudden hammer blow of pressure there. It is much like the way the total of energies added to a hammer during its downward swing are all focused on the head of the nail that it finally strikes.

In physical terms, the typical sonic boom from a supersonic plane flying overhead brings a sudden and unprepared jump of some 1,400 to 2,000 pascal in the atmospheric pressure which had previously prevailed at that part of the surface of the ground. A split second later that abrupt increase of air pressure is followed by a pressure deficiency of like magnitude.

Thus the blast of a sonic boom can be described as a pressure difference of about 3,000 to 4,000 pascal, packed into a brief instant. That pressure difference amounts to roughly 3 or 4 percent of the standard atmospheric pressure that normally prevails in typical inhabited areas on Earth.

This pressure change is fantastically huge compared with the pressure differences involved even in noises of very high decibel ratings. No wonder that windows are broken, plaster is cracked, old structures are heavily damaged, landslides occasionally are started, and precious archaeological sites are damaged irreparably by the battering of sonic booms.

Such airborne booms are inseparable from the overland flights of supersonic transport (SST) planes. The sonic boom or blast is but one, and one of the most sensationally shocking, among the manifold forms of pollution in terms of sound and noise that have become an ever more menacing part of our world and the world it threatens to become within the next decades.

47 SMILE THE WHILE . . .

He scoffs at the uproar of the city, and hears no shouts of
a driver.
> —JOB 39:7
> (New American Bible translation)

Noise now belongs to almost all of us, or vice versa. Accordingly
its effects on humans may be suggested to some extent even without
resort to units and numerical magnitudes, or even to crescendos of
denigrating adjectives, themselves a source of noise and emotional
after-effects.

One of the most perceptive examples of the allegorical-fictional
rather than the metrological approach to this intense problem of
our time was published during the autumn of 1972. It originated in
the inspired typewriter of Art Buchwald, invaluable and seemingly
inimitable contemporary columnist and social commentator. He
has helped many to laugh, but "only when it hurts."

The heading above the following piece, when published in Art
Buchwald's column in the editorial section of the *Los Angeles
Times,* was "It's Hard to Find Out." By Mr. Buchwald's kind per-
mission we are privileged to include it here.

BY ART BUCHWALD

NEW YORK—The City Council of New York has just passed
New York's first comprehensive law to control noise. The antinoise program is expected to go into effect in the next two years.
How do New Yorkers feel about it? I went out into the streets to
find out.

The first man I spoke to was walking down the Avenue of the
Americas. "Sir, how do you feel about the new antinoise law that
was just passed?"

"What did you say?"

"I said how do you feel about Mayor Lindsay's plan to outlaw the din in New York City?"

"Is he going to outlaw gin? I'm a vodka man myself. So it won't affect me."

"Not gin—din. He wants to lower the decibel count in New York City."

"I don't know anything about decimals, but if I know Lindsay, he'll lower the decimals and up the taxes."

"Thank you very much, sir."

"My wife drinks gin. I'm not sure she's going to like it," he said, walking off.

I went over to Fifth Avenue and spoke to a lady with a shopping bag. "The Environmental Protection Administration of New York has declared war on noise pollution," I said. "How does that grab you?"

Her lips started to move, but I couldn't hear her.

"What did you say?" I shouted.

"I said I'm going to vote for Nixon and Agnew."

"No, I'm not polling you on the presidential campaign. I'm trying to find out what you think about noise."

"What's he running for?"

"He's not running for anything. Noise is an issue—not a person," I yelled.

"I wouldn't know anything about that. I live in Brooklyn."

Just then a policeman came up to me and asked, "Why are you shouting at this lady?"

"I wasn't shouting at her. I was asking her a simple question."

"Why are you shouting at me?" the policeman demanded angrily.

"I wasn't shouting. I'm sorry. I was shouting, but that was because she couldn't hear me with all that construction going on over there."

"Suppose everyone in New York shouted. What kind of city do you think we'd have?"

"That's just the point," I said. "I'm asking people what they think about the new antinoise law."

"What new antinoise law?"

"The City Council passed a new law, and as soon as the mayor signs it, you can give out summonses to people who make too much noise."

"You mean with everything else we have to do, we police are going to have to give out summonses for noise?"

"Either that or take a bribe," I said.

"Get off Fifth Avenue before I take you in," he screamed.

"You don't have to shout," I said as I headed for Eighth Avenue.

On Eighth Avenue, I walked up to a man and said, "I want to talk to you about the earsplitting noise in New York."

He threw his hands high in the air. "Take my wallet. It's in my left breast pocket."

"This is not a stickup. I'm doing a poll."

"Here's my watch. Just don't shoot me."

"Mister, put your hands down. I want to talk to you about noise."

"I got two boys," he cried. "Take the money and run."

A crowd started to gather, and I decided to get out of there.

As I walked away, one of the young men in the crowd shouted after me, "What's the matter—no guts?"

48 IT'S A DRAG!—VISCOSITY AND ITS VARIATIONS

What's done we partly may compute
But know not what's resisted.
                    —ROBERT BURNS,
                    Address to the Unco Guid

Viscosity is a physical variable that merits wider understanding than it appears to have gained among most of those whose professions or employments do not oblige them to deal with it directly.

Some motorists, if they respond to the word at all, associate it with the idea of density, as when the service station attendant asks if they wish to change to "20 weight" oil when winter comes. Others probably confuse it with the notions of stickiness or tackiness.

A modern dictionary gives some help by defining viscosity as "that property [or, as the author of the present book prefers to say, that measurable physical variable] which in a fluid resists the force tending to cause that fluid to flow." Thus, it is the effect of fluid drag, holding back the fluid, whether liquid or gas, from responding instantly and without limit to forces tending to bring about flow.

Yet even these concepts, broad as they may be, require expansion by example and analogy in order to reveal the many aspects and surprises that viscosity presents, packed away like a nest of Chinese boxes.

Viscosity is shared in widely varying degrees by substances of extraordinary differences, the thinnest gases, liquids of the utmost variety, and even amorphous noncrystalline solids, or substances that we have come to think of as solid since their viscosities are so great that they appear to be proof against flow within the time spans in which individual humans commonly survive.

Even the vast and relatively thin rock crusts that underlie the continents and those that underlie the ocean floors of our Earth are known now to respond to forces in manners that can only be comprehended in the terms of the concept of viscosity. Thus, from the thinnest stratosphere on high, through the sea waters that cover more than two thirds of the surface of the Earth, down to the crusts between human habitations and the interior mantle of the Earth—viscosities are operative and indispensable to full understanding of the responses that varied forces elicit from different kinds of matter.

Viscosity is the key concept of a lively and rapidly expanding branch of science called *rheology,* devoted to the study of the deformation and flow of matter—all matter, any matter.

Flow of any kind involves deformation. That term, in science, has no denigrative connotation. Flow of whatever sort cannot take place without changes in the alignments, spacings and interactions among the molecules and atoms composing the flowing substances. Previous states of all flowing matter are progressively deformed, or perhaps a more neutral word would be "re-formed." Such reformations are largely within the realm of the viscosity concept.

The concept of flow itself must be broadened to the utmost. It is not restricted to the movements of liquids or gases through pipes and channels, nor to the shifts and shape changes of strata and layers, as in the Earth. Every passage of a solid body through a fluid is a flow, for the fluid must flow past the body as the latter moves. Thus, viscosity is basic to the way in which the gentle rain falls from heaven, to the roaring of a hopped-up dragster down the strip, to the speeding of a jet plane through the stratosphere, and so on, world without end. Even the passage of blood through our arteries, veins and capillaries is inextricably involved with viscosity.

Throughout this broad spectrum of activities and interactions, scientific analysis of viscosity provides accepted ways to measure the drag that fluids offer by way of resistance to forces tending to deform or re-form them.

Viscosity and drag are inseparable. The greater the fluid's viscosity, the greater the measurable drag. The inverse of viscosity might be called "flowability," or even simply "fluidity." The least-viscous substance is necessarily the most fluid or most readily flowable, and vice versa. In some ways it might be more convenient, at times, to employ a measure of inverse viscosity, or flowability, than of viscosity itself. Such measures have been proposed and are sometimes used, but in the main it is viscosity that has priority, just as in electrical measurements preference is traditionally given to resistance (measured by the unit of the ohm) rather than to conductance, its inverse (measured by the unit of the mho).

Viscosity is most easily understood by discarding a number of familiar notions that simply happen to be wrong. One such notion is that a solid body moves past or through the liquid or gas layer immediately next to it. The fact is that the molecules of liquid or gas closest to the ship or plane, or closest to the walls of the pipe, tend to cling to and stay with that solid substance.

The relative motion in the fluid takes place increasingly outward from the solid surface. One learns to think of an endless series of thin layers parallel to the surface and increasingly far from it. Each of these thin fluid layers slips with respect to the layers on either side of it. It is like a deck of cards neatly stacked in a pile. You press the top card toward some direction—say, to your left. The bottom card remains just about where it was before on the table. However, each card above that bottom has moved toward the left a bit less than the one above it.

If your hand's force has pushed the top card 1 centimeter in 1 second, then a card halfway down the pile has moved $\frac{1}{2}$ cm/s, one three quarters of the way down has moved $\frac{1}{4}$ cm/s, and so on. The speed of change of position is proportional to the distance away from the static table, which represents the surface past which a fluid is moving in the typically complex fashion sometimes called "laminar flow."

Flow of this sort is smooth, regular and disciplined. It is free from the vagaries of turbulence, whirlpools and internal chaos. Yet, when velocities are attained beyond those in which laminar flows can continue, the concept of viscosity is still applicable, though in ways far more involved and complex than before.

How can viscosity be measured? What are the requirements of a unit combination or single unit suited to provide magnitudes of viscosity for all the enormous range of instances in the world of measurements?

Think of a cube-shaped volume in a motionless fluid, each edge of that cube having a unit length of 1 meter. Its volume is thus 1 cubic meter, and each of the six faces of the cube has an area of just 1 square meter, and four corners, each a 90-degree angle.

We are looking at the cube from one side. The surface below, which we call B, rests on a flat, motionless solid. Another flat solid (movable) rests on the surface above, which we call A. Thus between the 1 m² area of B and that of A there is a distance of just 1 meter of fluid.

Now, much as we did with the deck of cards, we apply a unit force, just 1 newton, to surface A, trying to push it toward the left, with respect to surface B just 1 m from it and parallel to it. The rate at which A moves leftward, as measured from B, will depend on the viscosity of the fluid in that imaginary cube. The greater the viscosity, the lower will be the speed with which surface A moves leftward with respect to surface B.

The proper unit combination for viscosity measurement— This is the basic situation, and from it we can derive the suitable or inevitable unit combination for measuring viscosity. On the one hand we have the stress—a so-called "shear stress"—exerted on the imaginary cube and deforming it at a measured rate, so the side we have been looking at, formerly a square, has now become a trapezoid, with no right angles. On the other hand, we have the rate at which this deformation is being produced.

The stress is the forcelike aspect; the deformation rate is the flowlike aspect. And the viscosity is measured by comparing the former (as numerator) to the latter (as denominator) in a fraction.

Moving step by step, how shall we measure the shear stress? We take the SI unit force (1 newton), apply it to the two unit areas (A and B), each of 1 square meter area, and separated by a distance of 1 meter measured at right angles to those surfaces. Thus the standard shear stress, in SI terms, is 1 newton per square meter (1 N/m²).

How shall we measure the resulting rate of deformation? If the speed with which A has moved leftward with respect to B is x meter per second, then in 1 second there has been a movement of x m. Or, we can with equal justice say that in y second a movement of 1 meter is completed. Once again we must include a reference to the 1 m distance between surfaces A and B, for if we were dealing with a 2 m separation, the movement of A with respect to B would be twice as great, and so on.

The result is that we measure the rate of deformation by the fraction 1/s, since the two lengths have canceled out. In this fraction, s represents the number of seconds required for a completed movement of 1 meter by A, when A is separated from B by just 1 m.

Next we form a fraction whose numerator is the measure of shear stress and whose denominator is the measure of deformation rate. The fraction at first looks like this:

$$\frac{N/m^2}{1/s}$$

When this is simplified, it becomes $N\ s/m^2$, or newton-second per square meter. From a dimensional point of view, this measure of dynamic viscosity boils down to force times time/area. Even more simply, it can be regarded as the time (s) that corresponds to unit force (1 N) divided by unit area (1 m^2).

Times, or durations, and viscosities are indeed inseparably connected. The standard methods for measuring viscosities are based on stop-watch timing of specific operations involving flow in the fluid being tested. Some viscosimeters determine the time required to let fluid fill or empty a vessel of known capacity. Others observe the time required for a ball of known size and density to fall a measured distance through the fluid, the downward force being supplied by its net weight when immersed in the fluid being tested.

In the International System (SI) the unit of dynamic viscosity, being 1 Ns/m^2, is somewhat awkward to name in full—"the newton-second per square meter." It has been given a single name, the *poiseuille,* abbreviated Pl. That name has seldom been used outside France. Rules now adopted discard it for scientific uses.

An older and far more widely used viscosity unit has a similar name, the *poise,* abbreviated P. It is based on the centimeter-gram-second (cgs) units of the dyne-second per square centimeter (dyn-s/cm^2). The magnitude of the Pl is just 10 times that of the P; hence 1 P equals 0.1 Pl.

Both the Pl and the P are named in honor of a distinguished French physician and physicist, Jean Poiseuille (1799–1869). His interest in the circulation of the blood in humans led him into a series of measurements of the flow of liquids through fine pipes (capillaries) in response to varying pressures.

The results of these historic measurements are summed up in what is now called Poiseuille's law. It is still useful in analyzing flows of all kinds, not being restricted to that of the blood, by any means. It can be simply stated as follows: If R is the radius of a tube and L its length, with P the pressure difference between the tube's two ends, while C stands for the dynamic viscosity of the fluid flowing through the tube, then the volume per second (V) of fluid that will pass through that tube is:

$$V = \frac{0.392\ 70\ R^4\ P}{C\ L}$$

The constant 0.392 70 is the ratio pi (π) divided by 8.

This law is commonly stated in another form, in which the unknown at the left is the viscosity (C) and the other variables, including the delivered volume per second (V), are known. In that form the law shows how the viscosity of a fluid can be measured by recording the volume per second of that fluid which is pushed through a tube of known dimensions by a measured pressure.

Poiseuille, with this law, revealed that, when other things remain equal, doubling of the radius of a tube allows 16 times the former per-second volume of fluid to flow through. On the other hand, halving the radius of a small tube diminishes the volume rate to one sixteenth of its former amount.

The law applies to flows whose velocities are small enough, relative to their viscosities, to allow the flow to remain smooth and "laminar," without turbulence and other obstructing agitations.

The two related units symbolized by P and P1 are unfortunately not the only measures that have been widely applied in the past to the measurement of viscosities. Almost every imaginable unit combination of force times duration divided by area has been employed, with the result that there exist elaborate tables for converting readings in one unit combination into another.

For example, there is the viscosity unit known as the kilopond second per square meter (kgf s/m²). The kilopond is the force (weight) of a mass of 1 kg under standard gravitational acceleration. One kgf s/m² equals 0.101 972 Pl.

Viscosities have also been measured in units of the poundal second per square foot (pdl s/ft²), equal to 0.671 968 Pl. As noted earlier, the poundal is the force that accelerates a *mass* of 1 pound by 1 ft/s². Further, there is the viscosity measure called the pound-force second per square foot (lbf s/ft²), equal to 0.020 885 Pl. And still another unit, called the reyn in honor of the eminent fluid dynamicist Osborne Reynolds, is based on the pound-force second per square inch, and equal to 1.457×10^{-5} Pl.

Dynamic versus kinematic viscosity—All of these, and some others not specified here, are combinations designed to measure *dynamic viscosity,* which is also called *absolute viscosity,* for it is not related to the *density* of the fluid being tested. So-called kinematic viscosity, on the other hand, is a different measure, consisting of the dynamic viscosity divided by the density of the fluid at the temperature prevailing when the test was made.

The unit combination in which kinematic viscosities are expressed is one which divides area by unit time—as, for example, square meter per second (m²/s) in the SI units, or cm²/s and ft²/s in other systems.

The designation *stokes* (abbreviated St) is given to the unit of 1 cm²/s in honor of G. G. Stokes (1819–1903), Irish physicist and mathematician, a man of many substantial scientific attainments.

It was Stokes who formulated the laws in harmony with which a small sphere falls through a viscous fluid (either a liquid or a gas). The typical pattern for such a fall is an acceleration until the body attains a limiting velocity—the maximum speed—which it does not exceed the rest of the way down. At that limit, the retarding drag of the fluid on the body just counteracts the net force with which gravity draws the body downward. That equality is reflected in steady speed, meaning zero acceleration.

Stokes was the first person to show why a mouse may fall 500 feet yet strike the ground more slowly than an elephant that has fallen 15 feet. He studied the surface behaviors of fluids also, in papers analyzing the propagation of waves over deep water. He led the way in discovering that the *group* velocity of sea swells as they move across the ocean's surface is only half as great as their *phase* velocity. Thus, as seen from above, it appears that new swells constantly arise in the rear of a train of swells, move forward through the group, then die out at the head of the column.

Stokes's great law of drag—The collective name "Stokes's law of drag" has been used at times to summarize his extraordinarily rich contributions to the world of measurements of viscosities. Stokes's best-known principle can be summarized as follows. If R is the radius of a small ball, D the difference between its density and that of the fluid in which it is allowed to drop by its own weight, g the prevailing gravitational acceleration, and C the dynamic viscosity of the fluid, then V, the limiting velocity that the ball will attain as it drops, is proportional to the fraction $R^2D \, g/C$.

Clearly, this means that if the ball's radius is doubled, its limiting velocity will be quadrupled. But does not a larger sphere have more resistance to motion through a viscous fluid than a small one? Will not its drag be greater? We must recall that doubling the radius means increasing the mass of the sphere 8 times. Meanwhile, its cross section increases only 4 times. Hence, its limiting velocity rises.

In the International System of units, the unit combination suited to measure kinematic viscosity is the square meter per second ($1 \text{ m}^2/\text{s}$). This is 10^4 (10,000) times the size of the stokes unit. It is also equal to 10.763 times the kinematic viscosity unit called the square foot per second (ft^2/s).

Both dynamic (absolute) and kinematic viscosities are presented in tables and used in solving important problems. Always it is essential to note the kind of viscosity being used. Otherwise, chaos or confusion may result.

Viscosity is an indicator of how a fluid moves in response to a measured stress. This is true even when that stress or pressure is supplied solely by the fluid's own density (mass per unit volume). Thus viscosities permit predictions of how swiftly a fluid will flow out of measured openings in containers. Just as the radius of a falling sphere is important in Stokes's law, so the radius or diameter of an opening, vent, or pipe is important in such problems of flow from containers.

Water is our most familiar fluid by far. In the case of density, water supplies the standard by which other substances are compared, the result being called their "specific gravities." What is the viscosity of pure water? The answer can be given only in terms of particular temperatures, for viscosity changes markedly as temperatures are raised or lowered.

Between the temperature at which ice melts and that at which water boils—between 0° C and 100° C—water loses more than 83 percent of its viscosity. This means that water, just before it boils away under normal atmospheric pressure, has become about 6 times as fluid, or "flowable," as it was when it first became liquid at 0° C.

Here are typical dynamic viscosities for water at various temperature levels, all stated in terms of the *poise* (P), which should be noted as 0.1 of the SI unit of the poiseuille (Pl): at 0° C, 1.8×10^{-2} P; at 4° C (the maximum-density temperature), 1.58×10^{-2} P; at 10° C, 1.3×10^{-2} P; at 20° C, 1.0×10^{-2} P; at 30° C, 8×10^{-3} P; at 95° C, 3×10^{-3} P; at 100° C, 2.8×10^{-3} P. The decline of viscosity with rise in temperature is clear.

It is common to use temperatures of either 0° C or 20° C for basic viscosity measurements. Many very-high-viscosity oils, however, often are measured at far higher temperatures than those, in order to reduce their viscosities and thus reduce the time required for testing.

At 20° C these liquids are more viscous than water by the factors shown in parentheses: ethyl alcohol (1.2 times), mercury (1.55), benzyl ether (5.33), soya bean oil (69.3), olive oil (84), light machine oil (102), heavy machine oil (233), castor oil (986), glycerine (1,490), and so on.

Liquids less viscous than water at this same temperature include: ether (0.233), chloroform (0.58), methyl alcohol (0.597), benzene (0.652), among others.

Some viscosities of great importance and interest are measured at levels in the millions of poise (P) units. Thus, at 20° C, tar or pitch shows about 3×10^8 P (about 100 billion times that of water). Glass (soda glass) when heated to 575° C shows a viscosity of about 10^{13} P, or 10^{15} (1,000 million million) times that of water at 20° C.

All these liquids and seeming-solids exhibit diminished viscosities as their temperatures are raised. Quite the reverse is the rather surprising behavior of gases, which gain viscosity as their temperature is increased. Ordinary air at 20° C has a measured dynamic viscosity of about 1.81×10^{-4} P. This means that it is about 180 times as flowable as water at the same temperature. At 0° C, however, air's viscosity is down to 1.71×10^{-4} P. That of hydrogen (H_2) gas, at 8.6×10^{-5} P, is lower still, only about half that of the air around us. (No special viscosity figures will be introduced here to indicate the effect of smog and other pollutants typical of our deteriorating ambiences.)

At 99° C, air shows about 27 percent greater viscosity than at 0° C. A notably high viscosity in a gas is that of mercury vapor at 300° C, with some 5.3×10^{-4} P, or 400 times the viscosity of 0° C air. However, the viscosity of *liquid* mercury at 20° C is about thirty times greater than that of the mercury vapor at this high temperature.

Though hydrogen gas has a markedly low viscosity, that of the gas of H_2O, or water vapor, is also quite low—only 9×10^{-5} P at 0° C, and about 47 percent above that at 100° C. Ammonia gas (NH_3) is likewise low in viscosity, with some 9.6×10^{-5} P at 0° C.

The viscosity of oxygen gas (O_2) is even higher than that of ordinary air at like temperatures—about 1.9×10^{-4} P at 0° C, which is about 9 percent more than air at that temperature.

It may seem strange that no mention has been made here of how viscosities change as pressures change. Again this physical variable holds surprises. Changes of pressure in both gases and liquids cause very little change in their viscosities, so long as temperatures are held constant.

This absence of effect is by no means self-evident. Quite the contrary. Leonhard Euler (1708–83), the Swiss mathematician, who was one of the greatest pioneers in the study of fluid dynamics, actually assumed at the outset that the fluid drag of a liquid moving under pressure through a pipe would be proportional to the pressure of the liquid on the pipe's walls.

The fact, as abundant measurements have validated, is that the drag of the walls of the pipe on the flowing fluid is virtually independent of that pressure. Likewise, the drag of the waters of the sea on the hull of a submarine submerged to a depth of 10 meter is not perceptibly less than when it has submerged to a depth of 100 meter. The drag is highly and intricately related to the velocity with which the submarine travels through the water, but not to the pressure exerted by that water!

Perhaps even more surprising is the absence of pressure effect on the viscosities of gases. For instance, when bits of dust are allowed to drop through a tube filled with air at normal pressure, they soon attain their limiting velocity and fall at a steady speed. The same kind of dust will fall with almost the same speed through a tube filled with air at a pressure only 2 percent as great as the atmospheric normal, provided that the temperature is the same. The density of the gas in the first tube is fifty times that in the second; yet the viscous effect of the gas, as a fluid, is virtually unchanged.

So long as the velocity of the solids through the fluid or the fluid past the solids remains moderate, these surprising relationships continue to be found by measurement. However, when the velocities are forced to ever higher levels, the stage is reached in which the patterns change drastically, and fluid drag begins to climb in a new and power-hungry manner.

Viscous versus inertial effects—Every movement of solids through fluids or of fluids around solids involves two types of effect: (1) the viscous effect, and (2) the inertial effect. In some situations the first dominates the second; in others the second dominates and virtually negates the first.

How can we measure, or combine measurements, so as to be able to anticipate which relationships will prevail in a particular case? The answer is provided by a famous "number," the Reynold's number, abbreviated Re. It is named for its enunciator, that same British engineer-scientist Osborne Reynolds (1842–1912). In the early 1880s he arrived at a relationship that, like the famous Mach number, is dimensionless, meaning that it remains the same no matter what unit system is used for measuring viscosity, density, length and the other physical variables.

If S is the speed of a fluid with respect to the tube through which it moves or the body that moves through it; if L is the diameter of the tube, or the diameter of the sphere moving through the fluid;

if V_d is the dynamic viscosity of the fluid; and if D is its density, then

$$Re = \frac{D \cdot L \cdot S}{V_d}$$

The fraction D/V_d is another name for the reciprocal or inverse of the kinematic viscosity of the fluid. Hence, if we represent the kinematic viscosity by V_k, the relationship can be represented with just three rather than four variables:

$$Re = \frac{L \cdot S}{V_k}$$

The results are expressed always and solely as pure numbers, or numerics. There is no unit or unit combination attached, any more than in the case of the Mach number, or a number of other dimensionless relationships of prime importance in the world of measurements.

But what is the meaning of the Re number?

When it is small, it means that the fluid viscosity is dominant, and the so-called inertial forces are relatively unimportant in determining what goes on. On the other hand, when Re is large—say, for example, in the range of 8 to 10 thousand—it means that the viscous effects are relatively unimportant or even negligible, and the inertial effects are dominant. It means that the sheer momentum of the fluid layers, as they race past the walls of the pipe or past the surface of the body speeding through the fluid, is such as to overwhelm and obscure the viscous interactions.

When Re is high, we find that uniform, smooth laminar flow no longer exists. Turbulence and cavitation agitate the fluid. A kind of chaos of motion, a madness with a method in it, has replaced the old, smooth, more predictable patterns. Above all, the differences become evident as we measure the resulting fluid drag on the body moving through the liquid or the liquid moving through tubes and passageways.

So long as Re is a low number, the fluid drag is roughly proportional to the product of three measurable variables: speed times size times dynamic viscosity of the fluid. Thus, for a body of moderate size, the drag will rise roughly in proportion to the speed of its motion through the fluid. And the power required to maintain that speed at a steady level will increase approximately as the square of the speed.

For example, if a power of X is needed to maintain a speed of Y, then a power of 4 X will be needed to raise the speed to 2 Y. This is not too difficult to demonstrate in principle. Power is the product of a force times the speed at which it is exerted. The drag or force against a body moving with a uniform speed through a fluid is always equal to the force that propels that body forward. If the drag were greater than the propulsive force, the speed of the body would reduce; if the propulsive force were greater than the drag, the speed of the body would increase. The fact that speed holds constant shows that the drag is just being countered.

Now, at a low Re number, the fluid drag increases just about in proportion to the speed through the fluid. Since this is true, and the power is the product of speed times forward-moving force, then the power must be proportional to the square of the speed, and the speed to the square root of the applied power.

Because of these relationships, a small fish swimming slowly in the water can increase its speed about 30 percent by using about 69 percent more power.

Why high speeds mean high drag—However, when the Re number becomes very high, as it does when airplanes and automobiles are forced to very high speeds, then the relationships are completely altered. The drag is no longer proportional to the product of the speed times size times dynamic viscosity (of the air). Instead it becomes proportional approximately to the *square of the speed* times the *square of the size* times the density of the fluid.

One looks in vain for any mention of viscosity in this pattern that applies to high Re numbers and to high speeds. Viscosity has ceased to matter enough even to merit a mention in this rule of thumb. Now, if one seeks to double the size of the moving object (the plane or racing car), its drag is increased four times. Likewise, if one seeks to double a previous speed, the drag is increased four times, and so on.

What is the result in terms of the power required to attain and maintain some particular speed? It means that the required power goes up as the *cube* of the speed, not as the square of the speed, as at low Re numbers. For example, to increase the speed of an automobile from 150 km/hr to 300 km/hr means overcoming an increase of four times in the fluid drag of the air against that car. Thus, the force propelling it forward must be squared; and this times the increased speed means that the applied power must be increased to 8 times what it was at the lower speed.

These relationships at high Re numbers are approximate, but they are close enough to what actual measurements reveal to validate the value of the relationship first enunciated by Osborne Reynolds.

It is perhaps strange to realize that viscosity, as a variable, makes so very little difference at speeds that are high in terms of the size of the objects involved. At large Re numbers, it is fluid density, measured by kg/m^3, rather than viscosity, that makes the great difference. Density determines how much power must be consumed in accelerating masses of the liquid or gas out of the way of the speeding plane or racing car.

There are unexpected relationships when the Re numbers are notably high as well as when they are safely low. For example, there is the so-called Paradox of d'Alembert. It tells us that if there were a totally nonviscous fluid, of whatever density, then it would be possible to push a sufficiently streamlined body through it with no fluid drag at all. Of course, there is no such fluid, but the importance of streamlining with fluids of low viscosity, such as the air, is indicated none the less.

The devices actually used to measure viscosities do so under conditions described by low Re numbers, not high ones. There have been dozens and scores of viscometers. Some of them, like the early Engler devices, gave readings in degrees, much as the Baumé and other hydrometers gave density readings in degrees that required conversion into actual densities, measurable by kg/m^3 or comparable unit combinations.

The better modern viscometers can be read directly in units such as the poise (P) or the poiseuille (Pl). A typical device, used in central Europe, is called the Höppler viscometer. Its principal part is a steeply slanted glass tube into which the fluid to be tested is placed. Small spheres of various sizes and densities are supplied. A suitable sphere is chosen and dropped through the tube, and its passage between two measured markings is timed.

By suitable choice of spheres, it is possible to make measurements in an enormous range, from about 10^{-4} P to 10^5 P, which is a ratio of 1 to 10^9. A thermometer is, of course, an indispensable part of the device. It is important that the temperature be held constant as the sphere drops, otherwise the reading cannot be relied on.

Other well-known viscometers are identified by such names as the Redwood, the Sayboldt Universal, and the Sayboldt Fural. Each operates by the timed flow of the fluid to be tested, out of or into a measured container.

The Redwood, for example, measured kinematic viscosities by correspondences such as these: 51.7 seconds means 10 centistoke (cSt); 123.1 s means 30 cSt; 283.9 s means 70 cSt. (Thus a full reading may require 4 minutes or more!) Corresponding times on the Sayboldt Universal device are given as 58.8 s, 140.9 s and 323.4 s, respectively.

The Sayboldt Fural is intended especially for work with extremely high viscosities, as heavy oils. Its measurements are made at 122° F, and a 20 s time means a 50 cSt reading; while one of 188.2 s means 400 cSt.

Conversions of kinematic viscosities, whether in stokes, centistokes, or other appropriate units, are made by multiplying these readings by the densities of the fluids at the temperatures at which the testing was done. Thus a reading in stokes times the density gives the corresponding reading in poises; a reading in centistokes times the corresponding density gives the reading in centipoises; and so on.

The dependence of viscosities on temperature is striking, and water is by no means the most vivid example of this fact. With water, each rise of 1° C is accompanied by a loss of about 3.5 percent of its kinematic viscosity. With oils, such as castor oil, the loss is far greater—about 8.4 percent per one degree C rise in temperature. And for pitch or tar in the vicinity of 20° C, each increase of one degree C is accompanied by a loss of about 30 percent in kinematic viscosity.

Since kinematic viscosity is the dynamic or absolute viscosity divided by the density, it is clear that if increased temperatures result in decreased density—that is, in expansion of any given mass of the substance—then the effect of the increase in absolute or dynamic viscosity is thereby increased. Only a substance with no coefficient of expansion would maintain an unchanging ratio between its kinematic and dynamic viscosities as its temperature was raised.

Dynamic viscosity includes an interesting relationship to another physical variable known as the modulus of elasticity. Strange as it may seem, one may deal with both elastic effects and viscous effects in the same substances.

If the viscosity is divided by the elastic modulus of a substance or a physical system, the result is a time—a certain number of seconds, few or many. This is the time known to scientists and technicians as the "relaxation time." It is an indicator of how soon the substance or physical system can be expected to adjust or respond to sudden changes in the forces, effects or changes to which it has been subjected.

The greater the viscosity, the greater the relaxation time (for any given elastic modulus), because a less flowable or deformable body will take longer before the elastic forces have their way with it. On the other hand, for any given viscosity—the greater the elastic modulus, the briefer the relaxation time.

Longest relaxation times are exhibited by physical systems characterized by high viscosity and low elastic modulus. Such a body needs a long time—and has a "hard time"—to readjust to sudden shifts of forces.

The reader is at liberty to draw useful analogies relating to the readiness (or lack of it) with which an individual adjusts to the blows of fate or the forces of circumstance. A relatively short relaxation time can be a psychological boon, if not a physical one. On the other hand, characters of low viscosity may prove altogether too compliant for their fellow humans to depend on.

VII / Nuclear Disintegrations and Some Other Pressing Problems

No one—not even the most brilliant scientist alive today —really knows where science is taking us. We are aboard a train which is gathering speed, racing down a track on which are an unknown number of switches leading to unknown destinations. No single scientist is in the cab, and there may be demons at the switch.

Most of society is in the caboose looking backward. Some passengers, fearful that they have boarded an express train to hell, want to jump off before it is too late. That option, it would appear, is no longer open, but at least the passengers can discuss matters among themselves, and keep a hand on the brake.
　　　　　—RALPH E. LAPP,
　　　　　　The New Priesthood (1965)

49 RADIOACTIVE DISRUPTIONS OF MATTER, AND THE CASE OF THE CURIE

The wrecks of matter and the crush of worlds.
—JOSEPH ADDISON,
Cato (1713)

. . . Consider Harry Truman
That innocent man sailing home from Potsdam—
rejoicing, running about the ship, telling all and sundry
That the awful power that feeds the life of the stars
had been tricked down
Into the common stews and shambles.
—ROBINSON JEFFERS,
Moments of Glory (1948)

We all know that generals today study strategy in terms of
megadeaths . . . that today's weapons aim directly at civil-
ians, and that perhaps only the military will go scot-free.
As far as I know, there is not one theologian who would
accept that a soldier may aim direct . . . at civilians. In
this situation the Christian must object, even if it costs him
his life. I should add that it would seem to me logical that
in such a war a Christian may not participate, even as a
kitchen-helper.
—DON LORENZO MILANI,
contemporary Italian theologian and moralist, in a
letter published in *Rinascita*

In the late 1960s, a patient in an American hospital was to be injected with a radioactive substance preparatory to a diagnostic test. The injection was prepared in a syringe by a student X-ray technician. The physician, a radiologist, noting that the syringe seemed unusually large, asked a student aide to verify the dosage.

The aide consulted the student technician, who told him that the entire dose was to be administered; the student aide thus informed the physician, and the injection took place.

Two months later the patient was dead, the cause—radiation poisoning. The student technician had not been aware that a millicurie (mCi) is 1,000 times greater than a microcurie (μCi). Nor did the radiologist who administered the injection recognize the thousandfold error.

Each year in the United States some eight million doses of radioactive medicines are administered. Nearly three thousand organizations and institutions hold licenses authorizing them to perform such administrations and, hence, to keep the radioactive substances on hand for the purpose.

Radioactivity and its effects on humans and other living creatures constitute the newest and most critical areas in the wide world of measurements. Fascination and alarm are inevitably intermingled in this realm. Human safety and welfare are totally at stake, and not merely because of the unquenched threat of military uses of nuclear weapons, but because radiation hazards involve us all, and there is almost literally nowhere to hide from them.

The account of the fatal overdose of radioactive medicine in an unidentified American hospital appears in a lengthy and important report to the Congress by the U.S. General Accounting Office (GAO). In sharp and specific terms it expresses disapproval of the manner in which the Atomic Energy Commission (AEC) has failed to realize its responsibilities relative to licensing, inspection and control of some 12,600 institutions to which it has issued more than 16,000 licenses for the use of radioactive materials, including the 3,000-odd permitted to make medical administrations of these substances.

The report, rendered in 1972, reveals that the AEC has enforced but laxly its own regulations regarding the handling and use of radioactive isotopes in medicine and industry. Precisely because they deal with current realities, rather than with claims or printed forms, portions of this report provide a prelude to essential concepts involved in the measurement of radioactivity and its effects on living creatures.

Among the many commercial institutions and factories licensed to use and process radioactive isotopes, a number were found by AEC inspectors to be repeatedly in violation of clear regulations governing such use and processing. Yet they were neither delicensed, fined, nor publicly named as violators.

The GAO reported one flagrant instance of a factory (unidentified) inspected 23 times between 1960 and 1971, and 18 times found to be disregarding AEC regulations. One inspection revealed unshielded radioisotopes on the factory's loading dock and in its laboratory. In four of that factory's unrestricted areas, AEC inspectors' instruments revealed excessive levels of radiation; likewise in its lobby, and even on its roof. In one instance the radiation level was found to be 22 times in excess of the AEC's "permissible" level.

Multiple warnings to discontinue these violations went unheeded. After a visit in 1970, the AEC inspector commented in his report to his headquarters that this licensee's "indifferent attitude toward his defined problems continues to be a source of amazement."

Amazing also, save perhaps to those familiar with the ambiguous role and objectives of the AEC, is the absence of effective sanctions against such repeated and indifferent violators. The GAO, exercising its "watchdog" functions, urged in its report that offenders of this kind either should lose their licenses to use radioactive substances or should be fined for violations.

"We recommend," declared the report, "that the AEC develop and apply criteria under which licenses will be suspended or revoked and under which civil penalties will be assessed." Such "enforcement actions," the report proposed, should be "sufficiently severe to provide licensees with incentives to comply with AEC's regulations."

The AEC was established in part to stimulate and in part to regulate the "peaceful" uses of the atom, as well as to provide for the nuclear weaponry of the nation. Its record reveals the basic dichotomy between those disparate and contradictory functions. Generally, regulation has been sacrificed to stimulation. No other conclusion can account for important and alarming events of the past years.

Questions of permissible, approvable or "safe" levels of exposure to the "ionizing" radiations of radioactive substances have attracted much attention since the nuclear age dawned in the terrible glare of the A-bombs exploded at Hiroshima and Nagasaki in August 1945. The subsequent nuclear testing in the atmosphere of ever more powerful fission and fusion bombs forced upon much of the world the awareness that radioactive substances were poisoning the air we breathe, the water we drink, and much of the food that we and our children consume.

Possibly no problems involving measurements have ever been argued more urgently. At length the nuclear giants of the world agreed to suspend atmospheric testing and instead to "go underground" where, supposedly, radioactive products could not leak out. Yet there remains still a vast and alarming unawareness of the demonstrable facts and powerful probabilities of radioactive contamination, especially from numerous nuclear electric-power installations to whose planning and construction the AEC remains committed.

The concluding chapters of this survey of the world of measurements seek to provide clarity and essential information for those who believe that people should remain masters of their fates.

We find once again that careful examination of the applicable units of measurement provides clues to better understanding of the physical processes and physiological consequences involved. The units devised to deal with radioactive interactions can be followed like the thread that Ariadne gave to Theseus to lead him safely out of Labyrinth, where lurked the dread Minotaur.

In our own hazardous era, the atomic nucleus is like the monster at the heart of the confusing trap. Like the Minotaur, it has already claimed its thousands and tens of thousands of sacrifices, not all of whom, by any means, were resident in or near Hiroshima and Nagasaki.

Nuclear energy, we have long been assured, has so many peaceful and beneficial applications that we should not think of it as a threat, but rather be grateful that it is about to proliferate among us and near our great population centers in the form of nuclear plants for generating electric power, and so forth.

The enormous private power industry, and some portions of the publicly owned power industry too, have mounted massive campaigns in the major media to warn that a power shortage is at hand or imminent, and that none but the blind or malicious will oppose whatever kind of generating facilities are proposed and sanctioned by the AEC as the responsible supervisory agency.

Two types of radioactive measurements—Two kinds of measurements have been found relevant and needful in the world of radioactivity, less than three quarters of a century old. One kind of measurement provides standards for assessing the self-destructiveness of a particular instance of radioactivity—that is, the extent or rate at which its atoms destroy themselves and emerge as something other than they were before. The other kind of measurement provides standards for assessing the extent or rate at which radiations or fragments from such radioactive and other high-energy processes disrupt the matter on which they impinge and through which they pass. That is to say, the extent to which they dismember normal balanced and neutral atoms of ordinary matter.

Terms like *destruction* and *disruption* are neither melodramatic nor farfetched. Radioactivity in its inception and its physical effects is a basically destructive sequence. The stable building blocks of matter either "self-destruct" or are induced by human intervention to tear themselves asunder; and the resulting radiations and explosively flying particles tear away the most accessible components of normal matter, the electrons that surround and neutralize the positively charged atomic nuclei of matter, animate or inanimate.

The A-bombs that blasted Hiroshima and Nagasaki, and the H-bombs that were developed somewhat later, made use of nuclear transformations called, respectively, "fission" and "fusion." Fission is a kind of nuclear disintegration in which myriads of heavy, complex nuclei explode, the bits and pieces including two or more lighter, less complex nuclei for each of the nuclei lost. Fusion is a violent crashing together of two or more light nuclei to form a heavy, more complex nucleus. However, not until the late 1930s was it known that fission reactions could be induced by human intervention, and not until after the fission bomb, or A-bomb, was produced in the mid-1940s was there even a possibility of inducing fusion reactions on Earth. (Fission explosions now serve as the "kindling" to initiate fusion explosions!)

Nearly two score years of study, experiment and measurement led from the first discovery of natural radioactivity to the first consummation of nuclear fission by human intervention. Obviously, radioactivity means more than nuclear bombs or nuclear reactors. This is, indeed, reflected by the concepts and units of measurement devised for this area, so replete with interest, import and menace.

Assessing extent in radioactive interactions—How extensive in fact are nuclear self-disruptions or radioactivities? Before the dawn of the nuclear era in 1945, more than three hundred possible nuclear configurations or combinations had been identified and their principal characteristics measured. Since then, the number has been enormously increased. Today the extensive tables of "isotopes" offer data on more than 1,300 different combinations of protons and neutrons that can hang together for greater or less periods of time to form a nucleus. The number is still being increased by research.

Each such combination of nuclear species is known as a *nuclide*, and no two nuclides have identical combinations of protons and neutrons. In terms of complexity and numbers of component particles, nuclide differences are enormous. A nucleus may be as simple as that of ordinary hydrogen, most abundant element in the universe. It consists solely of one proton, no neutron. A complete hydrogen atom, having one positive electrical charge thanks to the proton, holds also a single electron, the smallest known negative charge. In the classical and somewhat outdated image of a hydrogen atom, this planetary electron orbits endlessly around the proton nucleus, in which virtually all the atom's mass is concentrated.

A nucleus, on the other hand, also may be as complex and crowded as that of nobelium, one of the "transuranium" elements. One of the seven isotopes or forms of nobelium is identified by a nucleus composed of 102 protons and 155 neutrons, 257 nuclear particles in all, uneasily packed together. During the brief average lifetime of such an atom, 102 electrons, arranged in layers or shells, circumambulate this crowded core.

The mass of such a super-congested structure as this nucleus is more than 250 times that of a single hydrogen atom. Instability is the price of its complexity. A myriad of ordinary hydrogen atoms, if left in peace, persist indefinitely. A myriad of atoms of this nobelium-257, more formally symbolized by $^{257}_{102}No$, will rend themselves asunder at a rapid rate, regardless of such ambient conditions as temperature, pressure, and so on.

The rate of disintegration or decay of a nuclide is customarily given in terms of its "half life," or half period, symbolized frequently by $T_{\frac{1}{2}}$ (T sub-one-half). The half life of nobelium-257 is given as 20 second. This means that, beginning with any considerable number of such nuclides, within 20 second that number will be reduced to one half, since the other half will have disintegrated or decayed into various fragments and particles. By the end of 40 second, the remaining number will be one quarter of the starting number; by the end of 60 second (1 min) it will be down to one eighth.

Similarly, two minutes reduces the remainder to $\frac{1}{64}$, three minutes to $\frac{1}{512}$, four to $\frac{1}{4,096}$, and five to $\frac{1}{32,768}$ of the starting number. By the end of 400 second or 20 half-life periods, the number of remaining nobelium-257 nuclides will be less than one millionth of the number at the beginning.

This example happens to show a fairly short half life. Among the more than one thousand known radioactive nuclides, the half lives range all the way from about 5 billion years, or nearly 10^{17} s, in the case of uranium-238, down to fractions of a second so brief that even the most sophisticated modern procedures can barely catch the characteristics of the nuclides before they disintegrate. Among the most common radioactive fission products of nuclear reactors, the average half life is about 10^5 s, or roughly 1 day.

From the longest computed half life to the shortest determinable half life the ratio is probably about 10^{18} or 10^{19} to 1. Thus the mortality rates and the life expectancies of the many radioactive nuclides differ enormously. This extreme range of variation is a striking characteristic. Clearly it contrasts sharply with the non-radioactive nuclides, of which about 270 are identified. These are all stable or permanent. Accordingly, they have no half lives.

Half life is not the only possible measure for comparing the relative durabilities of the known nuclides. If the half life is multiplied by the factor of 1.443 the product is the mean or average lifetime of that nuclide. This is a concept more familiar to students of populations and demography. For example, in a group of one million persons, what will be the average remaining years of life? That average for one million healthy youngsters, aged two to five years, assuredly will be greater than for one million senior citizens, aged seventy-two to seventy-five years. The mean or average remaining lifetime for living creatures depends largely on the attained ages of those creatures when the actuarial estimate is made.

Age does not matter—to nuclei—Nuclear mortality, however, is utterly different from human and animal mortality. The life expectancy, or mean lifetime, for a large number of any particular nuclide is in no way affected by how "old" each individual nuclide is. The half life does not diminish, even after a large number of half life periods have elapsed. Old nuclei and atoms are no more and no less prone to self-destruct or disintegrate than are young ones. Indeed, the ordinary concepts of "old" and "young" are inapplicable here. The probabilities of disintegration within a specific period of time are determined solely by the particular configuration of protons and neutrons composing the nuclide under investigation. A nuclide that has, for example, survived during one hundred half-life periods has the same probability of future survival that it had before.

One cannot single out one particular nucleus among the myriads in any ordinary sample of that species, and predict how soon or late it will explode and become something else. But one can predict with considerable precision what proportion of any considerable number of that nuclide will survive unchanged at the end of a specific time interval, and also what is the mean or average lifetime remaining to the structures that have not yet rent themselves asunder. In the case of nobelium-257, for example, the average lifetime is 1.443 times 20 s, or nearly 28.9 s. In the case of uranium-238, on the other hand, it is about 7.2×10^9 (billion) years.

Still another index is sometimes used to describe the disintegrative proclivities of the many known nuclides. It is the reciprocal, or inverse, of the mean lifetime, and its usual name is the *radioactive decay constant* for that particular nuclide. It is commonly symbolized by Greek lambda, λ.

Here is a typical use for this constant. In the case of the nobelium-257, since its mean lifetime $(1/\lambda)$ is approximately 29 s, the radioactive decay constant is $\frac{1}{29}$, or 0.034 5, which means an average of 0.034 5 radioactive disintegrations per second *per* nuclide of the substance in question. The atomic mass of this nuclide is very nearly 257. Hence 1 mole of it has a mass of 257 gram, and by definition contains $6.022\ 2 \times 10^{23}$ individual nuclei. If a laboratory is able to produce just one millionth (10^{-6}) gram of nobelium-257, this amount contains 2.343×10^{15} nuclei. The decay constant (λ) times the number of nuclei subject to decay gives, as product, the actual number of disintegrations per second. In this case, that product is about 8×10^{13} disintegrations per second.

For uranium-238, however, the radioactive decay constant is only about 10^{-17} disintegrations per second per nucleus. One mole of this uranium isotope has a mass of 238 gram and contains $6.022\ 2 \times 10^{23}$ separate nuclei. Hence the mass of one-millionth gram contains 2.53×10^{15} nuclei. Accordingly, the average number of disintegrations per second from this sample must be only 0.025, which means an average of only one disintegration each 40 s.

If the size of the sample is increased to 1 gram, the average number of disintegrations per second is increased to about 25,000; and if the sample is further increased to 1 kilogram of uranium-238, about 25 million disintegrations per second can be expected.

From A to Z among the nuclides—Every nuclide is identified by two meaningful numbers. The larger of the two, called the A number, is the total of nuclear particles, both protons and neutrons, that it contains. The smaller, called the Z number, is the total of protons only. Since each proton is a positive charge, the Z number determines the net positive charge of the nuclide, and thus indicates how many electrons (negative charges) it can capture and retain.

The Z number is known also as the *atomic number,* whereas the A number is the *mass number.* A chemical element includes all of the nuclides that have one particular atomic, or Z, number. There may be as many as 30 or more different A numbers associated with a single Z number. Each such nuclide is called an *isotope* of that particular element. For example, platinum (Pt) with a Z number of 78 has 32 known isotopes, of which five are stable (nonradioactive) and the remainder radioactive. Mercury (Hg) with a Z number of 80 has 26 identified isotopes, seven being stable and the rest radioactive. Lead (Pb) with a Z number of 82 has 28 isotopes, four being stable, the rest radioactive. Bismuth (Bi) with a Z number of 83 has 19 isotopes, only one being stable, the rest radioactive.

Beginning with Z number of 93, every isotope of the 13 most complex elements has been artificially produced. Not one has been found to occur naturally on Earth. Also, even among the elements from hydrogen (1) to bismuth (83), two elements are included which were produced artificially in the 1930s and 1940s, because they had never been found on Earth. These are technicium (Tc) with Z number 43 and about 12 radioactive isotopes; and promethium (Pm) with Z number 61 and about 12 radioactive isotopes also. These represent "all-radioactive islands" located amidst elements which show one or more stable isotopes.

The nine radioactive elements beginning with polonium (Po) with Z number of 84, through uranium (U) with Z number of 92, each have at least one isotope found on Earth to some extent. However, these "naturally radioactive" elements have between them about 143 known isotopes, an average of nearly 16 per element; and by far the majority of these isotopes have been produced only by artificial means.

Above these lies another group, eight elements with Z numbers from 93 through 100. They have between them a total of about 98 isotopes, all radioactive, and all without exception have been produced only artificially. This is an average of about 12 isotopes per element. These nuclides range between 229 and 257 in their A or mass numbers.

The fact that this group averages 12 known isotopes per element rather than 16, as in the group just below it, suggests that the increasing crowding and complexity of the nuclei leads to ever greater instabilities, and hence less likelihood that isotopes will have lifetimes long enough to be fully identified and measured.

This choking-off effect becomes most apparent in the final group, with Z numbers of 101 and up. At the time of this writing, five such elements have been identified and produced in tiny quantities, with an average of about four isotopes per element. Keen and not always entirely amiable competition has marked the difficult researches into these most elusive and complex radioactive ultraelements. The center for these efforts in the United States has been Berkeley, where a group of specialists is headed by Albert Ghiorso; in the USSR it has been Dubna, where Georgi N. Flerov heads a team at the Joint Institute for Nuclear Research. Recently, additional valuable results have been added by a team at the Oak Ridge National Laboratory in Tennessee.

In 1964 the Dubna group reported production of an isotope of element 104 and proposed for it the name kurchatovium (Ku) for the Russian scientist Igor I. Kurchatov. Later the Berkeley group challenged this claim, stating however that they had produced another isotope of element 104 and suggested for it the name rutherfordium (Rf). A similar contest developed with regard to element 105, for which Flerov of Dubna proposed the name nielsbohrium (Ns), while Ghiorso of Berkeley offered the counterproposal hahnium (Ha).

By the end of the 1970s not only the nomenclature but the principal measured characteristics of these and even higher elements very likely will have been settled. Final choices for names of new elements are traditionally made by decisions of the nomenclature committee of the International Union of Pure and Applied Chemistry.

Whatever element 104 is finally called, it is already certain to be credited with an isotope having an A number of 257; hence with a nucleus into which are packed 104 protons plus 153 neutrons, and with a half life of 4.3 second. This and other determinations were made on the basis of the artificial production of about 3,000 atoms of the isotope, a number so tiny in atomic terms that it truly beggars description!

About 320, or 25 percent, of all the nuclides thus far identified and measured to a greater or lesser extent have been found to occur naturally on Earth. Of these, some 270, or roughly 21 percent of all known nuclides, are stable isotopes, surviving indefinitely, so far as is now known. Another 50, or about 4 percent of all known nuclides, are radioactive isotopes that have been found occurring naturally on Earth. The remainder, or about 75 percent of the grand total, became knowable and known only after they were brought into being for varying periods of time (mostly quite brief), by processes conducted in such man-made devices as nuclear reactors and high-energy accelerators.

This much is manifestly clear: radioactivity, or nuclear instability, far from being an exceptional phenomenon in the nuclear world, is widespread. It is indeed a majority occurrence if we take into account all those many hundreds of different combinations of protons and neutrons that have been induced to form a nuclear configuration lasting long enough to be measured as to its mass and also traced into its decay products, sometimes called "daughter" products. Stability and indefinitely unchanging survival are characteristic only of a minority of the nuclides now known to science.

Artificial nuclides in use—Many of the radioactive nuclides produced by human intervention have lifetimes so short that they have attained little practical usefulness, however revealing they may be for the advancement of nuclear theory. Nevertheless, a substantial number of the so-called "artificial" radionuclides have sufficient permanence to be applied to various medical or industrial uses. In a 1971 publication the U.S. Postal Service listed radionuclides that are mailable, and informed prospective shippers how such radionuclides must be packaged and in what strengths they may be sent when properly packaged.

This list includes more than 250 radionuclides, and represents more than 90 different elements or Z numbers. The individual radionuclides range all the way from H-3, a form of hydrogen with a nucleus composed of 1 proton and 2 neutrons (and hence with Z number of 1), to three isotopes of Californium (Cf), with Z number 98. Californium is an element never found in nature; it exists only as a result of human intervention. The three specified radionuclides of Californium are Cf-249, Cf-250, and Cf-252; their nuclei contain always 98 protons, and in these three isotopes there are also 151 neutrons, 152 neutrons, and 154 neutrons, respectively. Their half lives, in the same order, are 360 years, 13 years, and 2.65 years.

The Postal Service list includes also nine different radioisotopes of iodine, Z number 53; six of cesium, Z number 55; six of gold, Z number 59; seven of thorium, Z number 90; eight of uranium, Z number 92; and six of plutonium, Z number 94.

Most of the radionuclides in this list could be shipped in quantities up to 1 *millicurie* per package, provided that package was suitably strong and secure. About 30 other radionuclides could be shipped only in quantities up to 0.1 millicurie per package or 0.01 millicurie per package. These latter substances required greater caution because of the nature of their radioactive emissions. (The explanation of "curie" and "millicurie" presented later on makes the character of these precautions clear.)

The same Postal Service announcement stated that any unpackaged radioactive device, such as a clock with radioactive dial, would be accepted for shipment only if, at a distance of 4 inches from the device, it had a "dose rate" no greater than "10 millirem per hour." Nor would any package be accepted that showed on its outside surface a dose rate greater than 0.5 millirem per hour. (The rem and the millirem will be explained later.)

Radioactivity, whether in a naturally occurring substance or in one produced by human devices, is synonymous with nuclear disintegration. Such disintegration is an explosive emission of particles and of rays (electromagnetic vibrations) of various kinds and intensities.

As the extraordinary and almost incredible phenomena of radioactivity were discovered and roughly assessed in terms of energy releases, it was inevitable that some sort of unit was needed for the comparison of the relative rates (occurrences per second) of such disintegration in different radioactive substances.

The unit that emerged is called the curie (Ci).

The Curie couple and the curie unit—The curie unit originally was named to commemorate Pierre Curie (1859–1906), French chemist and codiscoverer with his wife, Marie Sklodowska Curie (1867–1934), of the radioactive elements polonium (Po) with Z number 84, and radium (Ra) with Z number 88. Today the curie unit is considered to be named also in honor of **Marie Curie**, the Polish-born chemist and physicist, who began the great work and, after her husband's untimely death from an accident, carried it forward with exemplary tenacity and courage.

The Curies launched their vast labors in a search for the active principle or component in the mixed uranium ores from which the physicist Henri Becquerel (1852–1908) had accidentally found strange rays emanating, rays which could expose well-wrapped photographic plates and ionize the air.

To the first true element isolated by the Curies they gave the name polonium, after Marie's native land. Polonium has an atomic number (Z number) of 84, and the form that the Curies first extracted corresponds to the isotope today known as Po-210, with a half life of 138 days. Next they extracted an even more powerfully radioactive element, which they called radium, because of its potent rays. It has the atomic number of 88, and the isotope that the Curies first found was the one now called Ra-226, with a half life of 1,590 years.

Radium as an early symbol—Radium in the early years of the twentieth century became a sort of symbol for radioactivity as a whole. Careful measurements of small available radium samples showed that an emanation or gas of some sort was constantly emitted from it. This emanation could be removed from the parent radium, and would continue to show its own radioactivity, though it declined rather rapidly. This proved to be the gaseous element radon, with a Z number of 86 and a mass of 222. Its half life is known now to be 3.825 days.

The curie unit, when first defined, said nothing at all about *numbers* of disintegrations or even about the disintegration process itself. The definition, as adopted at a radiography congress in Brussels in 1910, merely stated that the curie unit should be the amount of radioactivity emitted by the radon in equilibrium with 1 gram of radium.

One gram of radium contains about 2.7×10^{21} radioactive atoms, each with its nucleus. The *average life* of these radium nuclei is about 2,294 years per nucleus; this is equivalent to the 1,590-year half life already mentioned.

Each nucleus of radium, during disintegration, emits what is called an alpha particle and turns into a radon nucleus. In this transformation there has been a loss of 4 units of nuclear mass and of 2 units of electric charge. The alpha particle, obviously, has carried both away.

The myriads of radon nuclei thus produced constantly from 1 gram of radium have an average life expectancy of 5.52 days, corresponding to their half-life period of 3.825 days. Thus, the radon dwindles rapidly, so far as concerns its survival in the form of radon. Without specifying just what the radon turns into, we note that surrounding 1 gram of active radium atoms there will always be a constant quantity of radon. The radiation of this radon, not that of the parent radium, was chosen as the standard for the new curie unit.

Radon disintegrates into polonium, which has a half life of 3.05 minutes. With the emission of another alpha particle, a polonium nucleus transforms into the nucleus of a radioactive isotope of lead, with a half life of 26.8 minutes. This in turn emits a beta particle (not an alpha) and becomes a radioactive isotope of bismuth, with a half life of 19.7 minutes; and so on, on the way to the stable and nonradioactive end product of lead, Pb-206.

In the complete uranium transformation chain, radium forms the sixth link, and radon the seventh. They became the bases for the curie unit because of their availability and relative convenience in use. (The dangers of radiation poisoning were then so little realized that some physicians advised their patients to drink radium-infused waters and even administered them to their own families!)

A strange suitability—Strange as it may seem, part of the suitability of the radium-radon equilibrium base for the curie unit arose from the fact that the average life of radium nuclei is about 153,000 times that of the average life of radon nuclei. The amount of radon gas constantly available around 1 gram of radium (assuming that this gas is allowed to accumulate and disintegrate normally) proves to be about 0.66 cubic millimeter at standard atmospheric pressure.

This is, apparently, an exceedingly small amount of radon gas—known formerly as "radium emanation." However, as measurements improved and knowledge increased, it was found that this tiny volume of radon was producing about 3.61×10^{10} nuclear disintegrations per second. That is to say: 36,100 million nuclear disintegrations per second!

Later, this original radium-radon-equilibrium definition of the curie was discarded. Instead, the curie became the quantity of any radionuclide that would show this same number of nuclear disintegrations per second, regardless of how much energy was radiated away with each disintegration. Since the radon was in equilibrium with the radium, the original definition necessarily implied that 3.61×10^{10} was the number per second of the nuclear disintegrations that would occur in a gram of radium, as well as in the associated amount of radon emanation. Thus, at one stage in its evolution, the curie unit was synonymous with the amount of any radionuclide showing the same number of nuclear disintegrations per second as 1 gram of radium.

An active wish to give the curie a broader and more objective base led finally to cutting it loose from radium or any other particular radionuclide. In 1953, the International Commission on Radioactive Units meeting in Copenhagen decided that henceforth the curie was to be that amount of any radioactive substance showing 3.7×10^{10} disintegrations per second. Thus the number was rounded upward to 3.7 from the original 3.61.

The curie unit is in fact a good deal larger than is quite convenient for most practical measurements today. Consequently it is found principally in the form of its submultiples, such as the millicurie (mCi), the microcurie (μCi), or the nanocurie (nCi). Even 1 nCi represents 37 nuclear disintegrations per second.

The rutherford of radioactivity—A smaller and better-rounded unit of radioactive disintegrations was proposed in 1946, just after the dawn of our nuclear age. The unit, called the *rutherford* (Rd), was defined as one million nuclear disintegrations per second, or (a distinction with very little difference) the quantity of any radioactive nuclide that shows just that number of disintegrations per second. Thus 3.7×10^4, or 37,000, Rd equal 1 Ci, and 1 Rd equals 2.7×10^{-5} Ci.

The rutherford was named to commemorate Ernest Rutherford (1871–1937), New Zealand-born physicist, one of the two or three scientific giants who first explored, measured, and interpreted radioactivity. It was he, in fact, who first pictured the strange, small nucleus that concentrates in a tiny fraction of an atom's volume almost all of its mass and energy.

The rutherford unit was proposed by I. F. Curtis and E. V. Condon, and in 1949 won approval from the National Research Council of the United States. However, it is seldom employed in practice. The curie remains the unit of preference for radio*activity*. That italicizing is done deliberately: both curie and rutherford are, essentially, measures of *activity,* which is not synonymous with energy or power or with the effects of that radioactivity on matter other than that which has disintegrated.

Characteristics of the curie unit—In millicuries and microcuries the radiometrologists have common denominators for radioactivities of the most diverse kinds. Thus, the same unit may be applied to the activity of a single purified radionuclide or to the combined activity of a mixture of many different radionuclides. Such mixtures are the rule rather than the exception in naturally radioactive ores, such as uranium-bearing rocks. There, in a single, intermingled mass, will be found half a dozen or more radioactive "daughter products," as well as the parent uranium from which all have descended by successive radioactive transitions.

The curie, properly considered, is restricted to measurements of nuclear disintegrations only. Thus, it is not applicable to Roentgen rays (X-rays). Those are powerful electromagnetic radiations arising from energy changes produced in the shells or layers of electrons closest to the nucleus (but not actually within it) in complex atoms in the target of the X-ray tube.

The following chapter presents another unit, the *roentgen* (R), devised originally to deal with X-ray effects, but applied now also to the effects of radioactive radiations.

The curie is a physical, not a physiological measurement. Yet, it can indicate something of the degree of exposure of living tissue to radioactive substances that are placed or absorbed inadvertently into the body.

Thus, the unfortunate patient (mentioned at the beginning of the chapter) mistakenly given an injection of 1 millicurie rather than 1 microcurie of the indicated diagnostic radionuclide, acquired in her body substances undergoing an average of 37 million nuclear disintegrations per second, rather than just 37,000 such disintegrations, which correspond to the activity of a single microcurie.

In the course of the two months of subsequent illness and decline prior to her death, her body was subjected to somewhere between 15 and 25 million million (trillion) nuclear disintegrations in excess of the number that would have occurred there had the error of dosage not been made.

Radioactive isotopes of any element form the same chemical combinations that are formed by nonradioactive isotopes of the same element. For example, the human thyroid gland has an affinity for iodine and concentrates iodine compounds within itself. This concentration does not distinguish between iodine-127, the nonradioactive isotopic form found in nature, and some eighteen other isotopes of iodine, with mass numbers (A numbers) ranging from iodine-121 to iodine-139, all radioactive, and with half lives ranging from a bit less than 3 seconds to more than 17 million years. Chemically speaking, each such atom is iodine and enters into compounds with virtually the same alacrity as does normal, stable iodine-127.

Radioactive substances, accordingly, seldom dispose themselves uniformly throughout the body during the period prior to ultimate excretion in sweat, urine or feces. They tend to concentrate in and around certain preferred organs and systems of the body, according to the chemical nature of the electron patterns surrounding their unstable nuclei.

Iodine is mentioned here merely as one possible example among many. The postal shipping regulations referred to earlier include nine different isotopes of iodine, all variously radioactive: the iodines with the mass numbers 124-5-6, 129, 131-2-3-4-5. The first three of these are the iodine isotopes whose nuclei contain respectively 3, 2, and 1 neutrons fewer than the 74 neutrons found in the one stable nuclide, iodine-127. The last six of these are the isotopes whose nuclei contain respectively 2, 4, 5, 6, 7 and 8 neutrons more than the 74 nuclear neutrons that alone provide stability and indefinite endurance when combined with the 53 nuclear protons characterizing all iodine nuclides, radioactive or not.

These numerical relations regarding iodine nuclides emphasize that radioactivity is not linked solely with "too many neutrons for the number of protons in a nucleus." There may also be "too few neutrons" for the particular number of protons present.

The curie, in conclusion, deals with radioactive or nuclear-disintegrative causes rather than effects. It is linked to the careful counting of the rates—number per unit time—of nuclear disintegrations. Each such disintegration transforms one radionuclide into some other nuclide. A few varieties of such disintegrations produce a stable, nonradioactive nuclide to replace the one that has rent itself apart. Most disintegrations, however, result in a new nuclide that is also radioactive, though with different half life and consequently a different probability of disintegrating later on.

Each of the hundreds of different radioactive transitions now known is accompanied by its own characteristic type of radiation. These radiations, during the days when radioactivity was first discovered and measured, received three tentative and ambiguous labels: alpha, beta and gamma "rays." Each of these three broad categories of radioactively produced radiations plays a role in the following presentation of additional units and concepts devised to meet the needs of this acutely critical and contemporary area of the world of measurements.

50 THE ROAD TO THE ROENTGEN, UNIT OF EXPOSURE AND DOSAGE FOR IONIZING RADIATIONS

Science has . . . been charged with undermining moral-ity, but the charge is unjust. A man's ethical behavior should be based effectually on sympathy, education, and social ties and needs . . .
—ALBERT EINSTEIN

. . . the capacities of individuals who can pursue science in the only way in which true scientific knowledge can be advanced are continually thwarted by the lopsided encour-agement of research in a society which consecrates its finest gifts to gigantic preparations for destroying human life.
—LANCELOT HOGBEN,
Foreword to *Science for the Citizen* (3d ed., 1951)

Late in 1895 and early in 1896 Wilhelm Konrad Roentgen pre-sented to the world an important discovery. By means of high-voltage electric discharges in vacuum tubes he had produced a novel kind of radiation, invisible to the human eye, to which "all bodies are transparent."

Roentgen, an esteemed member of the faculty of the University of Würzburg, called his discovery X-rays, stressing the unknowns about their nature. Against his wishes they became known also as Roentgen rays.

In the second installment of his announcement, he summarized careful measurements indicating the extent to which these X-rays not only penetrated but actually disrupted and disassembled nor-mal matter. This summary announced that the X-rays discharged electrically charged bodies in air. It did not matter whether those bodies were charged negatively (meaning that they had been given an excess of electrons) or positively (meaning a deficiency of electrons). If the X-rays passed through the air around such charged bodies, those bodies lost their charge, in either case.

This meant, indeed, that the X-rays converted ordinary dry air, which is a tolerably good insulator, into a conductor. The air, traversed by the X-rays, short-circuited the charge, permitting it to leak away. That, at least, was one obvious way of visualizing the process. The common and accepted name for the process is "ionizing." The X-rays ionized the air around the charged bodies.

The innocuous verb "to ionize" masks a potent process. Every normal, nonionized atom is electrically neutral, because for each positively charged proton in its nucleus its environs include one negative charge, or electron. From the outside, the nonionized atom shows no net charge. Some substances, of the kind called metals, are composed of atoms which, in mass, allow their external electrons to stray widely. Hence, they are conductors and can discharge a charged body when an electric circuit is formed with their aid.

Air, however, is normally not such a conductor. The fact that it became so when traversed by the new X-rays meant that these rays were powerful enough to tear apart many of the atoms composing the molecules of air. An atom is ionized when one or more of its external electrons is ripped off, resulting in a so-called ion pair; one partner of such a pair is the positive part—the formerly neutral atom, now lacking one (negative) electron—and the other partner is the negative part of the pair—either the torn-off electron alone, or a formerly neutral atom to which that extra electron attaches itself for the time being. Here we shall treat the negative ions as if they were all single, free electrons.

How does ionized air discharge a body that has been, for example, positively electrified? Such a body lacks some of its normal complement of electrons. Its surface atoms contain more protons in their nuclei than electrons around those nuclei. When such a body is surrounded by air that has been ionized, the free electrons (negative ions) in the air are attracted to it. They supply the body's missing negative charges. The body becomes *neutralized*—another way to say that its former positive charge disappears.

The sensitive instruments called electroscopes, used by Roentgen and others in studying the new X-rays, showed that such discharges proceeded rather rapidly. The stronger the X-rays passing through the air near a charged body, the more rapidly its electrification was reduced.

If the body had been negatively charged—meaning that it had been given more electrons than needed by its atoms for neutrality—then the positive ions in the air would be drawn to it. Their deficiency of electrons would be supplied from the charged body. And so ionization of the air or any other normally insulating substance reduces and finally eliminates the charge on any body with which it comes into contact.

The fact that the new X-rays were "ionizing radiations" meant that they were more violent in their effects than visible light, which does not ionize. It meant, further, that the X-rays also were more violent in their effects than most of the invisible ultraviolet radiation whose frequencies are greater than those of visible light.

This comparison is important, for despite the unknown X in their name, the rays were soon identified as electromagnetic oscillations, part of the same family that includes light, radio waves, and so forth.

Roentgen's researches—Roentgen was not content merely to note that his new rays ionized air, through which ordinary visible light also passes, though without ionizing consequences. He went further and charged electrically bodies encased in solid paraffin, normally an insulating substance. Then he radiated the paraffin with the X-rays and noted that the electrification disappeared in this situation also. Obviously, the atoms of the paraffin, a hydrocarbon substance, were likewise being torn asunder by the X-rays.

At about the same time in Paris, Henri Becquerel, who had found other and even stranger radiations coming from uranium salts, was testing their effects on electrified bodies. In his 1896 paper revealing radioactivity to the world for the first time, Becquerel noted quite correctly that the natural and continual radiations from his uranium substances bore "great resemblance" to the new rays studied in Germany by Roentgen. In particular Becquerel noted that these rays ionized the air through which they passed.

Likewise, at this significant period of early 1896, at the famed Cavendish physical laboratory at Cambridge, its director J. J. Thomson (1856–1940) produced and measured the new X-rays.

He too found that they made conductors out of the air and other normally nonconducting gases through which they were passed.

Thus, from the first, both the rays of Roentgen in Germany and those of Becquerel in Paris were established as *ionizing* radiations, capable of disrupting the outer electron shells of the atoms they encountered.

When Marie and Pierre Curie conducted their heroic and difficult search for the "active principles" in the radioactivity of uranium ores, they constantly used Pierre Curie's electroscope to measure and compare the intensity of the radiations from the substances they isolated at various stages in the long processes. Ionization then had already become a measure of the relative strengths of radioactivities.

But there were important differences as well as significant likenesses between the X-ray and the radioactive radiations. This became clear when small beams of each were passed between the poles of a powerful magnet, or between two metal plates, where a substantial voltage difference (charge) was maintained, as by a battery.

Curvature as clue—The X-ray beam was not bent, either by the magnetic field or by the electric one. Being an electromagnetic oscillation, it did not consist of particles carrying electric charges, either negative or positive.

On the other hand, a beam of radiation from one of the new radioactive substances or mixtures of substances behaved in a trebly curious manner, in either the magnetic or the electric field. For example, if such a beam was shot between the poles of a magnet with the south pole piece above and the north pole piece below, then part of the radiation would proceed straight ahead, unaffected by the magnetism, just as did the X-ray beams. However, as seen by an observer looking toward the approaching beam, another part of the radiation would be curved toward his right; and a third part of the radiation would be curved, even more sharply, toward his left.

The designation alpha rays was given to the right-curving portion, beta rays to the left-curving portion, and gamma rays to the straight-on portion.

Similarly, if such a mixed beam was shot between two metal plates, the upper plate being kept charged negatively with respect to the lower one, then the gamma rays would be found to proceed straight ahead unbent, but the alpha rays would be bent upward (toward the negative plate), while the beta rays would be bent downward (toward the positive plate).

Both magnetic and electrostatic behaviors of these new rays revealed (1) that the alpha rays must consist of positively charged particles; (2) that the beta rays must consist of negatively charged particles; and (3) that the gamma rays must consist of a radiation without charge, just as did the rays that Roentgen had so recently produced in Würzburg.

Careful measurements, based on the known strengths of the magnetic fields and the electrostatic fields, as well as on the degree of resulting curvatures of the alpha and beta rays, led to unequivocal conclusions. The beta rays must be composed of streams of electrons, the least massive of known particles, each comprising one negative charge. The alpha rays, on the other hand, must be composed of some assemblages of matter or of particles whose total charge was doubly positive, and whose total mass was very nearly the same as that of an atom of helium, the element with atomic number 2 and with mass number 4.

Only one answer seemed possible to the alpha-particle problem: each alpha particle must be, in effect, like a helium nucleus or nuclide, a combination of two protons (each with one positive charge) and two neutrons, but without the two external electrons typical of every complete helium atom. This 2-proton-2-neutron alpha structure was firmly locked together, and remained so even after repeated collisions with atoms or molecules as it flew out from the exploding nucleus from which it emerged.

Describing disintegrations—Measurements extending through decades led to several simple but extraordinary conclusions regarding radioactive disintegrations. Each definable disintegration process results either in the emission of an alpha particle, with atomic mass 4, or in the emission of an electron, whose mass is less than $\frac{1}{7,000}$ that of an alpha particle. In addition, such particle ejections may be accompanied, either seemingly at the same instant or possibly a short time later, by gamma radiation, the X-raylike electromagnetic effect (substantially more powerful than X-rays, however).

Most important, in terms of our present interest, was the fact that each of the three types of radiation (alpha, beta and gamma) was an *ionizing* radiation, though to different degrees in the cases of different transitions.

Radioactive transitions from one kind of nuclide to another are typically of two kinds. Ejection of an alpha particle (helium nucleus) means that the "after" nuclide will have an atomic (Z) number 2 less than that of the "before" nuclide, while the mass (A) number of the "after" nuclide will be 4 less than that of the "before" nuclide. Thus each alpha particle ejection means that nuclear protons have diminished in number by 2, that nuclear neutrons also have diminished in number by 2, and that the nuclear mass accordingly has diminished by 4.

On the other hand, ejection of a beta particle means, more or less figuratively speaking, that a nuclear neutron has thrown out an electron and become a nuclear proton. Thus the "after" nuclide must have an atomic number 1 higher than the "before" nuclide, but the mass number remains unchanged, or virtually so. Each beta particle ejection means that the nuclear protons have increased by 1, the nuclear neutrons have decreased by 1, but the total nuclear mass has remained essentially unchanged.

Some mass has, of course, been converted into energy—the energy of the high-speed motions of the alpha and beta particles and the electromagnetic energy of the gamma radiations. This energy has resulted from the disappearances of small amounts of mass, by the cosmically fundamental Einstein relationship: $E = Mc^2$, or 9×10^{16} joule of energy per kilogram of "vanished" mass. The loss of mass in radioactive transformations of nuclides is not sufficient, relative to the total mass remaining, to prevent accurate counting of the numbers of protons and neutrons packed together in the "before" and the "after" nuclides.

Tracing chains of disintegrations—How are these two basic kinds of nuclear disintegration distributed in a typical long chain of transformations? Let us call the roll of the transformations that lead from uranium-238 to one stable isotope of lead, as mentioned before.

The mass of the lead nuclide (Pb-206), the end product of the chain, is about 13.5 percent less than the mass of the U-238 nuclide at the start of that chain. Close to two thirds of this mass loss can be accounted for by the fact that PB-206 contains 22 fewer neutrons than does U-238. The remaining part of the loss is attributable to the fact that Pb-206 contains ten fewer protons than U-238. Thus, through some fifteen separate transformations, a uranium nuclide with 92 protons and 146 neutrons, or 238 nuclear particles, has been, step-by-step and with widely varying time lags, reduced to a nuclide with 82 protons and 124 neutrons, or 206 nuclear particles in all.

During the steps in the uranium-to-lead chain, there are eight alpha particle discharges, each of which eliminates 2 protons and 2 neutrons from the remaining nucleus, or 16 protons and 16 neutrons in all. Likewise there are six beta particle discharges, each of which eliminates 1 neutron and replaces it with 1 proton. Hence the net loss of protons throughout the complete chain of transformations is 16 minus 6, or 10. And the net loss of neutrons is 16 plus 6, or 22. The grand total of nuclear mass loss is thus 10 protons plus 22 neutrons, or 32 nuclear particles, from the original uranium nuclide total of 238, the difference being lead's total of 206.

Nuclear arithmetic, once a few essentials of measurement are mastered, is really quite precise and dependable!

Not only must the mass contents of the original nuclide be entirely accounted for. There is also the matter of electric charge. Charge, like mass-energy, is conserved in all nuclear transformations. The original uranium-238 nuclide had 92 protons; hence 92 positive charges. From the nuclear site of that original supply of positive charge, a total of 8 alpha particles emerged, each with 2 protons joined to 2 neutrons. Thus, 16 protons were fired out, which would leave 76 nuclear protons. However, each of the 6 beta discharges resulted from the transformation of 1 nuclear neutron into a nuclear proton. Each such change was the loss of 1 negative charge (electron), and the subtraction of a negative charge or quantity is identical with the addition of a positive charge or quantity. Thus, in the end product (lead-206) 82 nuclear protons remain, each contributing 1 positive charge to the nucleus.

In sum, during the in-between transformations, the radioactive rays carried away from the nuclides 16 positive charges (protons) and 6 negative charges (electrons). The loss of an electron being tantamount to the addition of a nuclear proton, the 16 is reduced by 6 and becomes a net loss of only 10 nuclear protons, corresponding to the difference between the atomic numbers of uranium (92) and lead (82).

Thus, during the ages between the first disintegrations of uranium-238 atoms and the ultimate emergence of the resulting lead-206 atoms, the spaces around the nuclides all along the multistep chain have been bombarded by 14 different particles for each nuclide that made the multitransformation journey. Of these particles, 8 have been alpha units, each with 2 protons and 2 neutrons; and 6 have been electrons.

Complete nuclear accountancy—The original amount of electric charge in and around the changing nuclides has been conserved or accounted for, just as all the mass-energy has been conserved or accounted for.

Similar careful bookkeeping shows how, despite repeated violent expulsions and changes, the particles and charges are accounted for accurately in the sequence that leads from thorium-232 to lead-208.

In the case of this thorium transformation chain, the mass of the remaining nuclide is diminished by about 10.3 percent, of which again two thirds can be traced to the reduction of 142 nuclear neutrons in thorium to 126 in lead, while the remaining one third can be traced to the reduction of 90 nuclear protons in thorium to 82 in lead. This is a ten-link chain of transformations, shorter than the uranium chain.

Intermediate in number of links is the so-called actinium transformation chain, with twelve transformations. It begins as an isotope of uranium, U-235, also called actino-uranium, distinguished by 143 neutrons packed with the 92 protons in its nucleus. There is about 140 times as much uranium-238 on or near the surface of the Earth as there is uranium-235. Hence the uranium transformation chain, so-called, is more important than the so-called actinium transformation chain.

The latter chain has eleven steps, during which the 92 protons are diminished to 82 and the 143 neutrons to 125, resulting in the isotope of lead known as Pb-207. Since U-235 finishes finally as stable Pb-207, each nucleus of the latter contains 10 fewer protons and 18 fewer neutrons than the original U-235 nucleus. Of the total mass loss, about 36 percent is accounted for by reduction in the proton count and the other 64 percent by reduction in the neutron count of the surviving nuclide.

In spite of their diversities, these three basic transformation chains—the uranium, the thorium and the actinium—have much in common. Between them they include about three dozen different links or steps, each from one describable and measurable radionuclide to another, until arrival at the end of the line: one of the three isotopic forms of heavy, dull, respectable and permanent lead.

In addition to these three multilink chain transformations there are found in a state of nature a few lighter radioactive isotopes that make but one transformation and become a stable or permanent, nonradioactive substance. For example, samarium-147, with 62 protons and 85 neutrons, emits 1 alpha particle, and thus becomes an isotope of neodymium (Nd-143), with 60 protons and 83 neutrons. About 15 percent of all samarium atoms found on Earth are this naturally radioactive kind, and about 12 percent of all neodymium atoms on Earth are in the isotopic form Nd-143. Also, the common element potassium, which has two stable isotopes, K-39 and K-41, has a naturally radioactive isotope, K-40, a nuclide with 19 protons and 21 neutrons. With a half life of 1.4 billion years, these K-40 nuclides emit an electron and transform into the stable calcium nuclide, Ca-40, with 20 protons and 20 neutrons. This is a classic case of how the expulsion of a nuclear electron accompanies the shift of one positive nuclear proton into a neutral nuclear neutron.

Of all potassium atoms found on the Earth's surface, including those in living tissues, about 12 in 100,000 are the naturally radioactive K-40 type. The rest are totally stable and nonradioactive.

Finally, there is another naturally radioactive transition in which the nuclide rubidium-87, with 37 protons and 50 neutrons, emits one nuclear electron (beta particle) and thus transforms into stable strontium-87, with 38 protons and 49 neutrons. Of all rubidium atoms on the surface of Earth, nearly 28 percent are the naturally radioactive Rb-87 variety. The rest are a stable isotope known as Rb-85. Strontium-87 makes up nearly 7 percent of all strontium atoms found on the surface of the Earth. Most of the rest are in the form of another stable isotope, Sr-88.

Thus we have surveyed the principal types of natural radioactivity taking place on Earth for long ages, prior to the human intervention that reached a portentous climax in the mid-1940s. The period since then has increased immeasurably the threat to humanity's survival in the event of unchecked international conflict, and the threats to humanity's health and hereditary integrity from misuse of the so-called "peaceful atom" if radioactive pollution or poisoning results from power-generating plants now in full course of preparation.

A specific particle emerges from a nucleus with each alpha discharge—it is the helium nucleus combination, with 2 protons and 2 neutrons interlocked and moving as a unit. The particle ejected from a nucleus in a beta discharge is the common, garden-variety electron. Yet it is helpful to note that these ejected electrons did not exist within the nucleus *as electrons* before they were fired out of it. Before a beta discharge, the nucleus is a configuration only of protons and neutrons. Under certain circumstances of instability, however, a nuclear neutron will unburden itself of part of its energy. The result is the emergence of what becomes, in that act, an electron that previously had not existed as such. Since the electron is indivisible from its negative charge, the neutron from which that negative charge was removed becomes a positive proton. When one takes away a negative charge from a neutral body, what is left behind is a body with a positive charge. Otherwise the strict bookkeeping or conservation of charge could not be maintained.

Gamma rays as recoil products—But what of the gamma rays, those electromagnetic oscillations with far higher frequency and energy even than the X-rays of Roentgen? They are emitted from the nucleus by a process that could be compared somewhat freely to the recoil of a gun as it fires off a projectile. The projectiles fired from the nucleus are either the alpha particle or the beta particle. The recoil, so to speak, may leave the nucleus energized or excited. It is like a spring that has been tensed.

In a short time—very short—or perhaps in a time not quite so short, that burden of surplus nuclear energy is fired off, in the form of a photon of gamma radiation. Each radionuclide transition that is accompanied or followed by a gamma-ray discharge has its own characteristic level of gamma-ray energy.

Again the question arises: What common denominator can be found, for measurement purposes, between these three so-different yet interrelated radiation forms? That common denominator is their ability to ionize the atoms they encounter on their way from the nucleus that emitted them in the first place.

We have seen how, from the days of the first measurements Roentgen made on his mysterious new X-rays, it was demonstrable that X-rays ionized normally neutral and nonconducting substances through which they plowed their way. It is not surprising, then, that the substantially more energetic gamma rays resulting from nuclear disintegration are even more strongly ionizing than are X-rays.

Yet it is not gamma rays that provide the most powerful and spectacular ionization following radioactive disruptions. That distinction must be given to the massive alpha particles. Depending on the particular nuclear transformation that launches them, alpha particles are fired out with velocities varying between about 14 and 21 million meter per second and with energies ranging from 4 to 9 million electron volt per particle. With these nuclear projectiles, just as with bullets from a gun, the kinetic energy of the flying object varies as the *square* of its velocity.

Yet, the alpha projectiles, massive and energetic as they are, are found to have far less power to penetrate matter than the other two radiations (beta and gamma). This is an indication not that they are less energetic or effective, but on the contrary that they are so interactive, so very ionizing in their action on the atoms they encounter, that they rapidly transfer their large energies into sundering actions on those atoms.

The limited penetration of alpha particles is the result and the evidence of their ionizing power. If they did *not* interact with the atoms along the way, they could dash on over long distances, losing but little energy as they went. Measurements show that typical alpha rays penetrate not more than between 2.5 and 9 cm of ordinary air before they have run their course—that is, have lost so much of their original kinetic energy that they can no longer ionize atoms in the air.

Even a thin film of metal is sufficient to stop a beam of alpha particles. The atoms in the metal are so much more densely packed than those in air at standard pressure that the ionizing disruptions are far denser along the path.

A typical measurement of alpha ionizing power shows that the alpha rays emitted in the nuclear disintegration of the bismuth isotope Bi-214, also called "Radium C," produce about 25,000 ion pairs per centimeter of travel when they are moving between 1 and 3 cm away from the nuclei from which they were fired. This rate rises to a peak of about 60,000 ion pairs per cm of path when the distance from the origin has increased to between 6 and 7 cm.

Here is another apparent paradox. The alpha particles are more effective ionizers when they have traveled farther, yet we know that each ion pair that an alpha particle leaves in its wake has reduced its remaining kinetic energy, and hence its velocity! The solution lies in the fact that when alpha particles linger longer in the vicinity of a nearby atom—that is, when they do not dash past too quickly —they are more effective in pulling one of its external electrons away from that atom.

There is, in short, an optimum velocity for ionizing. Very soon after that optimum is attained, the alpha particle is slowed down so completely that it ceases to move with enough energy to produce any more ion pairs along its path. Thus, there is a fairly slow climb to its peak rate of ionizing, and then a rather rapid cut-off to zero in its ionizing effect.

A single alpha particle, ejected in a typical radioactive disintegration, may ionize—that is, tear apart—about a quarter of a million atoms before it ceases to be able to ionize further. The massive ionizing effectiveness of the alpha particles results from their high initial energies, their large mass, and their double positive charges (2 protons). Each alpha particle thus exerts a double pull (two plus charges) on each external electron it encounters in the environs of the atoms near which it passes.

Besides the atoms actually torn asunder, or ionized, a great many of the other atoms along the way will have been energized or "excited" by the passing alpha particles and will later radiate away their excess energy in the form of photons of light, or even of ultraviolet radiation.

Far-flying beta rays—Far more penetrating than alpha rays are the beta rays, composed of electrons ejected from disrupting nuclei. They can be detected to a distance of about 100 times that of the alpha rays in the same substance. Beta rays are ejected at extremely high velocities, ranging from more than 99 percent of c, the velocity of light itself, down to about 70 percent of c. Beta particles show far wider spread or variations of velocity than do alpha particles, for reasons that we need not explore here.

These rays of energetic electrons emerging from the disrupting nuclei produce ionizing and exciting effects in matter, and especially in living tissue, by a variety of physical processes, some direct, others indirect but none the less effective. Indeed, taking alpha and beta particle effects together, it is known that about two thirds of the total energy is scattered throughout tissue that is disrupted or ionized, by means of so-called "secondary electrons"— electrons not directly ripped off by collision with a speeding helium nucleus, but taken from their past atomic places by processes several times removed from such primary collisions.

Alpha, beta, gamma—helium nuclei, electrons, and electromagnetic photons—all share in common varied abilities to disrupt and dismember the normal structures of atoms, hence of molecules, and hence of the ultracomplex molecules—the giant biomolecules —that make possible all the incredibly subtle and sensitive processes of life and, above all, of the transmission, and the preservation or modification of the hereditary characteristics of living things. Ionizing radiations are to genetic stability in humans and animals of all kinds much as devastating earthquakes are to the integrities of communities composed of frail structures whose materials cannot be used to rebuild them once they have been disrupted.

Extent of ionization in representative substances—normal air, water, animal and human tissues of different kinds—is the basis for the measurements that encompass critical aspects of X-ray and the three separate radioactive radiation types: alpha, beta and gamma. The catch-all term "ionizing radiations" can and does include them all, even though the alpha and beta rays are floods of particles, not radiation in the narrow sense of electromagnetic oscillations.

The role of the roentgen—Ionizing, then, is the separation of an electron from the remainder of the atom in which it previously had been bound. The result is a pair: an electron lacking the rest of the atom; and an incomplete atom lacking an electron. (The cases of the tearing-away of two or more electrons from an atom can be disregarded to keep the explanation simpler.)

The character and concentration of radiation streaming through a gas, a liquid, or even a solid can be described in terms of the extent of resulting ionization. The basic unit called the roentgen (R) was devised originally to measure X-rays only, by measuring the degree of resulting ionization in dry air at standard temperature and pressure (s.t.p.). Later the roentgen was used also to measure gamma rays from radioactive substances. Both X-rays and gamma rays, being electromagnetic vibrations far more rapid than those of light, are members of the same radiation family.

The roentgen can be described as the dose of radiation which produces 2.082×10^{15} ion pairs per cubic meter of such air (or 2.082×10^9 per cubic centimeter). Thus, if electrostatic measurements in irradiated air indicate the presence of 5.5×10^{15} ion pairs per cubic meter, then the dose present is $5.5/2.082$, or 2.64 R.

Often it is more convenient to use the mass (in kg) than the volume (in m^3) of the "dosed" air. Then, 1 R is equivalent to 1.61×10^{15} ion pairs per kilogram, or 1.61×10^{12} per gram, of the standard air.

Energy is required to pull electrons away from atoms and keep the resulting ion pairs apart; since each pair has one negative ion (the electron) and one positive ion (the incomplete atom), the negatives try to rejoin the positives, and vice versa. The amount of energy is perhaps the most meaningful aspect of the roentgen unit. It corresponds to about 8.69×10^{-3} joule (0.008 69 J) per kilogram of standard air, and to 8.69×10^{-6} joule per gram. Its energy equivalent can also be stated in terms of electron volts per kilogram or gram. In fact, for each ion pair present there has been an expenditure (absorption) of about 33 eV of energy.

This absorption makes clear why X-rays and gamma rays become weaker—that is, are attenuated—as they traverse the gases, liquids, or solids that they ionize. A kilogram of water or of soft living tissue absorbs between 14.8 and 15 percent more energy than does a kilogram of standard air from equivalent X-rays or gamma rays that pass through.

Charge, measured in terms of the coulomb, can also be used to describe the roentgen unit. A dose of 1 roentgen represents about 2.58×10^{-4} coulomb of separated negative charges and a like total of separated positive charges "at large" in 1 kg of standard air. This reflects the fact that each negative ion (solitary electron) has a charge of just $1.602\,191\,7 \times 10^{-19}$ coulomb. Since there are 1.6×10^{15} such charges in 1 kg of standard air at 1 R dose (and a like number of positive charges), this is the necessary consequence.

Because the roentgen is supposed to be used only for X-ray and gamma-ray dosages, when the unit called the *rep* was defined for use with alpha, beta, and other particle radiations, the rep also was shaped in such a way that a 1 rep dose in 1 kg of standard air represented also an energy absorption of 8.69×10^{-3} joule. Thus the roentgen is to X-ray and gamma-ray dosages as the rep is to alpha, beta, and related particle radiation dosages.

From its first adoption until about 1956, the roentgen did double duty. It served not only as a unit of *exposure* doses, but also as a unit of *absorbed* doses, for example the dosage absorbed by a living creature, or by a portion of the body of such a creature. Since the late 1950s, however, increasing efforts have been made to reserve the roentgen for measurement of ionizing radiation being emitted by some source or sources, and to devise related but separate units to measure the dosage to bodies exposed to such radiation. Hence, we have now the *rad* and the *rem* units. They, as well as the rep, will be presented more fully in the following chapter.

It can be noted here, however, that 1 rad is defined not as an energy absorption of just 8.69×10^{-3} joule per kilogram, but rather as 10^{-2} joule per kilogram, or about 15 percent greater energy absorption.

In order to arrive at the total dosage acquired by a body of known mass when subjected to measured ionizing radiations, two unit combinations are used: the gram roentgen and the gram rad. The total mass of the body in gram is multiplied by the measured dose in roentgen or rad, as the case may be. The product is the "integral dose" for the body.

A dose is a total effect or sum, stated without regard to the length of time in which it was accumulated. Time rate is often important in dose measurements, however. As a rough analogy, we can compare two persons, each of whom has taken a dose of twenty tablets of some kind of medication. In one case, however, the dose may have been ingested at the rate of one per day; in the other at the rate of one every quarter hour!

Dose rate of ionizing radiation is measured in roentgen per second or rad per second, as the case may be. More gradual rates may be measured in per-day or per-year terms.

A particular and peculiar unit combination is called the roentgen-per-hour-at-one-meter (rhm). It is the intensity of a source of gamma rays, shielded according to certain specifications, which at a distance of 1 meter provides a dose rate of 1 roentgen per hour. Such a 1-rhm source will be found to exhibit about 1 curie of radioactive disintegration; which is to say 3.7×10^{10} disintegrations per second.

51 ARRIVAL AT THE REP, THE RAD, AND THE REM, PLUS CONTINUING PROBLEMS OF VITAL IMPORT TO ALL THE LIVING AND THE YET-TO-BE-BORN

Said one among them—"Surely not in vain
My substance of the common Earth was ta'en
And to this Figure moulded, to be broke,
Or trampled back to shapeless Earth again."
    —EDWARD FITZGERALD,
        Rubáiyát of Omar Khayyám (1859)

Life's norm is lost: no doubt it is put away with Plato's
Weights and measures in the deep mind of God,
To find reincarnation, after due time and their
    own deformities
Have killed the monsters: but for this moment
The monsters possess the world. Look: forty thousand
    men's labor and a navy of ships, to spring a squib
Over Bikini lagoon.
    —ROBINSON JEFFERS,
        What of It? (1948)

Considerable confusion and uncertainty have attended the interpretations of the three units mentioned in the title to this chapter. Our purpose here is to attain clarity, if not total, then at least as nearly so as the facts allow.

A few years after the first atomic bombs were unleashed on Hiroshima and Nagasaki, H. M. Parker, a specialist in radiology, proposed two new units for use in measuring and comparing radiation dosages (effects) on human beings and other living creatures.

The first of these, he suggested, should be named the Roentgen Equivalent Physical, or *rep* for short. The second was to be named the Roentgen Equivalent Man, or Mammal, and to be known as the *rem* for short.

The rep has been called a measure of ionizing radiation in human tissue, though it would seem more meaningful to consider it a guide to the damage, actual or potential, by such radiation. For a time the rep was referred to as "the tissue roentgen." Also it has sometimes been called the *parker,* for the specialist who proposed it.

As noted earlier in connection with the ionizing effects of alpha rays, the denser the substance being irradiated by a particular ionizing radiation, the greater the extent of the ionization produced within it—so long as the radiation has not been attenuated to the point where it no longer has energy sufficient to produce new ionizing disruptions.

This increase in the density of ionization (ion pairs per cm of travel) with the increase in the density of the substance traversed (measured in kg/m^3 or some equivalent unit combination) applies to living tissue no less than to inanimate matter.

Consider, for example, a uniform exposure to 1 R of ionizing radiation. According to the original definition of the roentgen, this would have to be ionizing radiation emitted by X-ray apparatus, or possibly the gamma radiations emitted by disintegrating nuclei in the course of radioactive transformations. If this 1 R of exposure operates on dry air at standard pressure, the energy absorbed in producing ion pairs per kilogram of that air is—as noted previously—0.008 69 joule.

This provides a standard that the rep unit extends by analogy to substances other than dry air, and also to dry air and other substances when the ionizing radiation is not X-ray or gamma ray (electromagnetic) in nature, but is rather the particulate, or particle-formed, radiation of alpha rays and beta rays as well.

First, let us dispose of the case of dry air subjected to alpha rays and/or gamma rays from nuclear disintegrations. Whenever such air is absorbing energy in ion-pair production at a rate of 0.008 69 joule per kilogram of mass, then that air is receiving an ionizing dose of 1 rep. It is, in other words, receiving the physical equivalent of 1 roentgen.

Now as to the situation when the substance being dosed by ioniz-ing radiation is not air but living tissues, for which the rep was conceived. The actual energy absorption in ion-pair formation will be less where the tissue is less dense, as for instance in very loose and light fatty tissue (known to be less dense than water, in which it floats). On the other hand, the energy absorption will be greater in denser tissue.

However, as a rough average, soft human tissue absorbs energy in ion-pair formation at about 15 percent higher level than normal dry air. Hence, about 0.009 9 joule per kilogram of soft tissue cor-responds to a dosage of 1 rep. This is an average only. Actual meas-urements of widely differing living tissues have shown absorption rates ranging all the way from less than 0.006 5 joule per kg, or about 34 percent below the average figure, to a few percent greater than the average figure for tissue.

When we turn to measurements of absorbed energy for dense bones of the human skeleton or large animal skeletons, the energy absorption rates rise as much as tenfold, even to 0.1 joule per kilo-gram. In simple terms this means that an exposure rate corre-sponding to 1 R when based on the behavior of ordinary air may produce a dosage effect of from 10 to 12 times as much when dense bone is being irradiated.

The important essentials of the rep unit are, first, that it is used for biological research, to measure or compare radiation dosage absorbed by living tissues; and, next, that it was designed especially to deal with the effects of the ionizing particle rays from nuclear disintegrations: the alpha and the beta rays, not the gamma rays.

A few years after the proposal that resulted in the rep unit, an additional unit, now called the rad, was proposed. In 1953 it was approved by the International Commission on Radiology, and in 1956 it was substituted for the roentgen unit for clinical work on ionizing dosages from *either* X-ray or radioactive sources. Thus the rad is officially applicable to all kinds of ionizing radiations with which humans and other living creatures are confronted or threat-ened.

The definition of the rad is relatively simple. It is the absorption of 100 erg per gram of energy in the formation of ion pairs in the substance in question. As noted before, this is equivalent to the absorption also of 0.01 joule of energy per kilogram of such substance.

The rad is adapted both for biological and physical research work. It is restricted neither to inanimate nor to animate matter as such. In general a dosage of 1 rad is very nearly equivalent to 1 roentgen (of exposure). This flows from the fact that 1 rad is found by experiment to equal between 1.05 and 1.15 roentgen of exposure, depending on the substance exposed to the ionizing dose.

The unit name *rad* had existed in relative obscurity for a generation before the present rad was launched. In 1918, the year World War I ended, a proposal was made to give the name rad to the dose of X-rays that would kill a mouse. It is hardly necessary to observe that a rat rad would not necessarily equal a mouse rad, or that a cockroach rad would be gigantic!

A unit that mentions man—The rem, or Roentgen Equivalent Man, was proposed in 1950 by H. M. Parker at the same time that he proposed the rep unit. Its aim is to take into account the specific responses of various bodily tissues and biological systems to each of the *different* ionizing radiations by which humans may be assailed. In this it goes beyond the other radiation units thus far mentioned, for it has been extended to the effects of radiations or beams that do not arise from radioactive disintegrations, but are produced in various particle accelerators and so-called "atom-smashing" devices, such as cyclotrons, synchrotrons, linear accelerators, and so forth.

Some of these devices produce high-energy beams composed of protons, others produce beams of ions formed by stripping away external electrons from various atomic nuclei. Likewise, there are nuclear reactors that produce beams or fluxes of neutrons of various velocities and energies.

Indeed, the world of scientific technology has multiplied enormously and fearsomely the varieties, the energies and the physiological dangers of ionizing "radiations." And all this has accelerated to a peak during the past generation, roughly speaking.

The intent of the rem unit, as proposed by Parker, was that it should represent the dosage, from whatever type of radiation, that produced in a particular part of the human body the same rate of energy absorption from ionization caused by X-rays from a roentgen tube operating at between 200 and 250 kilovolt of electromotive force.

Another description of the rem, which amounts to virtually the same thing though it sounds different, calls it the product of the rad (dosage measurement) times a factor or multiple known as the "relative biological effectiveness" (RBE).

The rather bland description of "certain modifying factors" has been used to describe the RBE concept, but we shall be more specific here.

The RBE of 200 to 250 kV X-rays must equal unity, since that value was chosen to define the rem. In fact all X-ray and gamma-ray radiations are assigned an RBE of just 1. So too are beta rays, those streams of electrons emitted from disintegrating atomic nuclei, provided that their energies attain 1 MeV or more. Less energetic beta rays, because they move less speedily through human tissue, are more effective ionizers than the higher-energy electrons. Hence, beta particles of just 0.1 MeV (equal to 10^5 eV) are assigned an RBE of 1.08, or 8 percent greater than the 1 MeV beta rays.

Next on our way up the RBE ladder we arrive at the chargeless neutrons, emitted in vast numbers in atomic fissioning processes—for example, those in nuclear reactors in which the fuel consists of such expensively produced radioisotopes as plutonium. Can neutrons, which have no electric charge, produce ionization in atoms and molecules that they encounter? The fact is that neutron fluxes indeed provide powerful and often catastrophical ionizing agents.

Neutrons are usually divided into two major categories, according to the velocities with which they move from their sources into and through the matter at which they are directed. Neutrons moving with kinetic energies below 1 MeV are commonly called "thermal" neutrons, since their kinetic energies are about on a par with those of molecules of matter incessantly vibrating at ordinary temperatures. Neutrons moving at or under this general level of "room temperature" kinetic energies are also often called "slow neutrons." Such neutrons are assigned RBE factors in the range from 2 to 5.

"Fast" neutrons, on the other hand, are found in the energy

range between 1 and 10 MeV, and they are assigned the RBE factor of 10. This means, roughly, that such fast neutrons produce at least double the ionization or damage to human tissue that the most energetic thermal neutrons produce. And the latter produce five times the biological damage of a corresponding X-ray or gamma-ray radiation exposure.

The proton, positively charged nuclear particle, has a mass of some 1,800 times that of an electron, and slightly less than the mass of a neutron particle. Proton radiation of 1 MeV energy is assigned an RBE of 8.5, whereas the less speedy protons with an energy of but 0.1 MeV are given an RBE rating of 10. Once again, this means that less swiftly moving particles, because they do not so rapidly leave the vicinity of the atoms they pass, are found to be more—not less—effective in ionizing.

Finally, we come to the alpha particles from radioactive disintegrations, far more massive particles than neutrons or protons, since each alpha particle is formed from two neutrons and two protons. Alpha particles with energies of 5 MeV are assigned an RBE factor of 15, whereas those with energies only one fifth as much, or 1 MeV, are assigned the high RBE factor of 20.

These RBE factors are based on measurements, not on abstractions. The underlying measurements have dealt principally with the relative ionizing effects on water of the different-named radiations. After all, water forms the bulk of human tissues, and the density of human tissue, as an over-all average for the body, does not differ much from the density of water.

The key to the RBE for each defined type of radiation can be considered to be the number of ion pairs it produces in a measured distance of travel through water.

For a level of 1 RBE, the approximate ionization rate is 100 pairs per micron, or millionth of a meter. This corresponds to 100,-000 ion pairs per mm. Thus, 2 RBE would correspond to 200,000 ion pairs per mm; and 20 RBE to 2,000,000 ion pairs per mm.

Applying the rad and the rem—How would the rad and rem units be applied in a specific instance of human exposure to mixed ionizing radiations? Assume the not impossible event that through some error of judgment or some mechanical failure a person is exposed to 0.5 rad of gamma radiation, 0.2 rad of thermal neutron radiation, and 0.15 rad of fast neutron radiation. This hypothetical case could be the result of an accident at a nuclear reactor plant.

The total dosage, if measured merely by the rad unit, would be 0.85 rad. But this would provide little indication of the actual extent of the consequent physical damage to the victim of the catastrophe. In order to come nearer to that, we must use the rem. The following tabulation shows the total damage as 3.0 rem, numerically more than 3½ times greater than suggested by 0.85 rad.

TYPE OF RADIATION	RAD X RBE FACTOR	RESULTING DOSAGE (in rem)
Gamma	0.5 × 1	0.5
Thermal neutron	0.2 × 5	1.0
Fast neutron	0.15 × 10	1.5
	Total	3.0

Another example of the use of the rem: Exposure of human eyes to 250 roentgen of gamma radiation is known to cause cataracts. This radiation would be equivalent to 250 × 1 RBE, or 250 rem. Likewise, exposure to 25 rep of fast neutrons also causes cataracts, for 25 × 10 RBE equals 250 rem.

Effects of radiation dosage are inevitably dependent on the location and extent of the exposed bodily region. Exposure of an entire body to between 300 and 600 roentgen—average 450 R—would be lethal. Specifically, half of the individuals thus exposed would be dead within a few weeks as direct consequence of the bodily changes wrought by that radiation.

Yet assurances of the following kind can be found in publications prepared and distributed by government agencies: ". . . 10,000 roentgens may be safely applied to a small part of the body in radiation therapy." That quotation comes from *Atomic Radiation,* prepared by RCA Service Company, under contract to the United States Air Force and monitored by the Aero Medical Laboratory at Wright Patterson Air Force Base, Ohio. It appears on page 17 of the third revision of that study, which was first prepared in 1957.

Beyond doubt it is true that restricted and highly localized radiation exposures can be sustained with less danger to health and life than can far smaller exposures of the entire body or major parts of it. However, careful review of events, publicized and little publicized, during the past generation is bound to produce a skeptical attitude toward such statements on ionizing-radiation dosage as "may be applied safely," or "have no discernible effect."

To be meaningful and discussable, measurements and descriptions of exposures to external ionizing radiations must include the most precise possible data on the type and energies of the radiations involved, and the location and extent of the bodily areas that received the resulting dosage.

Low-energy radiation produces its principal effects on or very near the surface—the skin and layers just below. High-energy radiation penetrates deeper and works its damage in the interior of the body.

In most instances it is insufficient to limit measurements and analyses to dosage only, whether the units used be the rep, the rad or the rem. External gamma radiation in particular penetrates to great depths in living tissue and there, far below the surface, excites atoms in complex biomolecules to radiate internally and thus gives rise to radiation damage far below the skin surfaces through which the gamma rays first penetrated.

Ultraviolet effects—The damage is not directly proportional to the total radiation energy, in joule, that falls on the body. Rather it is related to the particular type of radiation. An obvious example may make this clear. During the course of a month of normal healthy living, a person's skin may absorb a substantial amount of energy from visible light without ill effects. Yet exposure to a smaller amount of ultraviolet light on a hazy day outdoors may produce painful sunburn, blistering, and so forth.

Ultraviolet light is not uniform in its physical effect. Ultraviolet radiations of about 2,000 angstrom wavelength, corresponding to about 1.5×10^{15} hertz frequency, penetrate the skin only superficially, to depths between 0.01 and 0.1 mm. But ultraviolet radiations of about 3,400 angstrom wavelength, or 8.8×10^{14} hertz, find the outer layers of skin less opaque and penetrate to depths between 0.1 and 1.0 mm. Even these produce only relatively superficial burns, however painful they may feel for a time.

Actual ionization—the tearing-away of outer electrons from atoms—begins only with the very highest ultraviolet frequencies, and continues on through the X-ray frequencies above them, and through the gamma-ray frequencies which lie higher still.

The range of electromagnetic frequencies that have ionizing effects on atoms of matter is quite large. This range is most easily measured and compared in terms of octaves of *bandwidth*. For example, the total bandwidth of visible light is about 1 octave. This means that the highest-frequency violet light that the human eye can perceive has just about double the frequency of the lowest-frequency visible red light.

Ultraviolet radiations range from about 8×10^{14} hertz frequency all the way up to about 3×10^{16} hertz, just below the "softest," or least energetic, X-rays. This, from the lowest to the highest ultraviolet frequency, is a range of about 37 or 38 times, which is more than 5 octaves. It is only near the upper end of this range, with frequencies at the level of 10^{16} hertz or greater, that ultraviolet radiations become ionizing in their effects on the atoms they encounter.

X-rays have a range from about 3×10^{17} hertz to 10^{19} hertz. This represents a ratio from lowest to highest frequency of more than 33 times, equivalent to about 5 octaves or frequency doublings. The X-ray photons are emitted by energy changes—jumps from higher to lower energy states—of the innermost electrons of complex atoms, such as those in the metal targets of X-ray tubes.

Still more energetic are the photons emitted by jumps from higher to lower energy states of nuclei themselves. These are the gamma rays, products of nuclei that have been "excited" by the recoil kick of the expelled alpha and beta particles in the disintegration "explosions." The gamma rays thus produced are scattered throughout another broad bandwidth. It ranges from about 10^{19} to 10^{21} hertz, a ratio of 100 times from lowest to highest frequency; hence equivalent to between 6 and 7 octaves.

Nor does the ascent of the radiation-energy ladder end at 10^{21} hertz frequency. Beyond that lie the "ultra" gamma rays that are linked with the bombardment of Earth by extremely energetic particles and nuclei from the depths of space. These all, together, form the so-called "cosmic rays." Their origins are not fully understood, but they are the most energetic radiations and particles that have been detected, surpassing even the products of the most elaborate and expensive particle accelerators now operating on Earth. Cosmic gamma rays have been measured at energies up to one hundred times those of the most powerful gamma rays originating in nuclear disintegrations occurring either naturally or as the result of human intervention and arrangements.

The enormous scale of ionizing radiations—Electromagnetic radiations capable of ionizing matter have thus been measured over the enormously wide range between about 10^{16} and 10^{23} hertz—a ratio of 10 million times, equivalent to 23 octaves, or frequency doublings.

Measurements of the visible-light spectra of different substances reveal the rich detail and complexity of the emission lines by which various gases are identified. The spectra, or energy patterns, of X-ray and gamma-ray radiations are less well known, but no less complex and revealing. This part of the world of measurement is being intensively explored. It is helping to unlock secrets of the structures of the many hundreds of different nuclides that come into existence and survive for widely different periods of time.

Measurement of X-ray and gamma-ray spectra is increasing astronomical information as well as nuclear understanding. For thousands of years astronomers depended on the information that the single octave of visible light brought to the Earth. Beginning about the middle of the twentieth century a new region of astronomy began to be explored—radio astronomy, which has speedily revealed startling new aspects of our universe. During the 1960s two other regions of astronomy were added: ultraviolet and X-ray astronomy. To be complete, we doubtless should mention also the astronomical possibilities of infrared-radiation measurements.

All these areas—infrared, ultraviolet and X-ray—had been closed until it was possible to send instruments above the atmosphere on orbiting satellites. The atmosphere absorbs and screens off all or most of these radiations from bodies elsewhere in space. Conversely, with reference to X-rays and gamma rays, their ionizing action on the atoms in the molecules of the atmosphere so depletes their energy that, to a great degree, they are prevented from reaching ground-based instruments. Until quite recently, information from space on X-ray and gamma-ray activity has been largely lost to human analysis.

Already X-ray astronomy has brought great surprises and amazing insights, and more will follow during the coming decades, if sufficient funds are provided to support it properly. The time may come when there will be astronomers and astrophysicists specializing not only in X-ray analysis but even in gamma-ray analysis of data from space, utilizing measurements made from orbiting satellites or even from instruments placed upon the surface of the airless Moon itself.

All this follows from the essential fact that the ionizing action of electromagnetic radiations more energetic than the ultraviolet takes place only in encounters with matter, such as the matter composing the atmosphere. There is no ionizing in empty space, for its emptiness means precisely the absence of atoms that such radiation could tear asunder.

Ionizing action is thus a common denominator for the potencies of all kinds of high-energy rays or particle flows, whether they be electromagnetic and chargeless, like those called X-rays and gamma rays; or particulate and charged, like alpha rays, beta rays, proton beams, and ion beams; or even particulate and uncharged, like neutron fluxes and beams.

The overriding importance of ionization is especially apparent when the object of measurement is the interaction with living tissues of these high-energy rays and beams. The biomolecules, the organic structures on which life processes depend, are enormously and indeed almost incredibly complex. A moderately complex inorganic molecule may be formed as a configuration of a dozen, a score, or even two score atoms. A moderately complex biomolecule may include hundreds of thousands of separate atoms—atoms of carbon, oxygen, hydrogen, and various other elements.

The fragility and interdependence of biomolecules is as striking as their complexity. Ionizing radiations, though they may tear away but one electron from one component atom of a biomolecule, can destroy its ability to function properly and can alter its destiny in the body of which it forms one unit among myriads. All living creatures above the level of the viruses are formed from one or more cells. A human being is formed from about 10^{14} cells—that is, one hundred times a million times a million, or one hundred *trillion* separate cells. Each of these varied cells is a complex of ever-changing and ever-renewing substances made up of myriads of molecules; and the molecules themselves are composed of great numbers of atoms. Life processes, however, cannot be evaluated reasonably on the basis alone of the enormous *numbers* of these units and subunits.

These are problems not of addition but rather of *multiplication.* A very tiny amount of ionizing disruption can wreak terrible and perhaps lethal havoc in a living organism so complex as humans and the "higher" mammals. It has been estimated that a full-body dose of 1,000 roentgen, deadly to a human, actually disturbs or disrupts an average of only 1 molecule out of 10 million molecules in that doomed body.

Consider the gene, the tiny biological entity that controls the transmission and development of hereditary characteristics in all living creatures. Even the simplest cells, such as those of the bacterium *E. coli,* contain 10,000 genes. A typical gene is formed from structural units containing about 80 chemical bonds linking together atoms of carbon, oxygen, hydrogen, nitrogen and phosphorus in intricate patterns. Such units are repeated about 1,000 times in each separate gene. Thus a single gene includes roughly 80,000 chemical bonds, any one of which may be broken by ionizing radiations. And even the primitive *E. coli* cell has 800 million such vulnerable chemical bonds!

In 1972 appeared a revealing study by Dr. Theodore T. Puck of the Eleanor Roosevelt Institute for Cancer Research at Denver. Entitled *The Mammalian Cell as a Microorganism* (Holden-Day, Inc.), it is a treasury of vital information on radiation risks and damages. Dr. Puck stresses there that "Bond breakage at any part [of a gene] can lead to a new stable molecule . . . Each of these alterations would represent a change in the biological properties of the cell. Finally, since the genes replicate themselves, such changes would either destroy the replication mechanism, or cause an error to be handed down to each of the offspring, thus compounding ever more the results of the original informational alteration."

In contrast, a molecule of water has only two chemical bonds, one from the oxygen atom to each of the two hydrogen atoms. The greater part of the mass of the body of a human or other mammal consists of the water molecules it contains. Even radiative disruption of the chemical bonds in such water molecules can wreak biological havoc. Ionizing radiation breaks up water molecules into "free radicals," which interact and combine to form peroxides (H_2O_2) and other substances poisonous to living cells.

Poisonous molecules such as these circulate until taken in by a complex molecule—for example an enzyme molecule, a form of protein. Enzymes are essential catalysts that mediate chemical transformations of myriads of other molecules. One poisoned or perverted enzyme molecule may alter tens or hundreds of thousands of other molecules. Thus, a seemingly tiny radiation damage multiplies into cell injury and cell destruction. Or into the wild and unchecked cell proliferations called cancers. The evidence is ample and increasing that there literally is no "safe" minimum level of supplementary ionizing radiation to which human populations can be exposed without harm to the dosed individuals or their offspring.

Cockroaches can take it—Awareness of mortality is not the only penalty of what is often called "the human condition." We share with great numbers of our fellow-species of higher animals a vulnerability to ionizing radiations. There are, however, forms of life, notably among the millions of insect species, whose physiology enables them to survive dosages that spell swift death for us. Outstanding among these is the great insect group called the cockroaches (not "roaches," if etymology is to be kept in harmony with entomology).

Cockroaches have survived largely unchanged on Earth since about 350 million years ago. What is more, an average cockroach can take 60,000 roentgen of radiation "without batting an antenna." Authority for this assurance, which may provide more comfort for cockroach lovers than humanitarians, is William Sullivan, Jr., a research worker in the United States Department of Agriculture, and a long-time specialist on insect controls and insecticides. Possibly the survivability of the cockroach justifies a variation on an often-told observation: We don't know just what sophisticated instruments of mass destruction will be used in a third world war, but we can be certain the fourth will be fought by cockroaches!

The "safe level" fallacy—The history of nearly three quarters of a century since concentrated sources of ionizing radiations were activated in the human environment shows that dosage rates once deemed harmless or safe proved later on to be distinctly and horribly harmful. The hypothesis of a threshold level below which radiation dosage can safely be incurred is fallacious, though it emerges still in various disguises and camouflages.

Both in the United States and in the Soviet Union, the two first and principal nuclear powers, serious underestimations were made of biological damage from "negligible" radiation dosage rates. For example, a 1959 work, *Nuclear Power,* by David Voskoboinik (Moscow), reported that "at present the permissible gamma-radiation dose in constant total-body exposure is considered to be 0.05 roentgen [50 mR] per working day." This is equivalent to about 0.25 R per week, 1 R per month, and 10 or 12 R per year. It would be difficult in the mid-1970s to find reputable, knowledgeable, and independent specialists in radiomedicine who would confirm the "permissibility" of such long-run dosage rates.

A 1963 publication of the U.S. Atomic Energy Commission, *Radiation in Perspective,* by F. L. Brannigan, tells simply of the "general line" of AEC regulations for the protection of workers in the atomic energy industry and related fields. The AEC standard was that in any one year a worker was not to receive total dosage above 12 rem, which when converted is very nearly 12 roentgen of exposure. Over a worker's entire period in the industry the yearly *average* should not exceed 5 rem.

Thus a worker so employed during forty years should accumulate a total dose no greater than 200 rem. If, on the other hand, his work subjected him to 12 rem per year, then such employment would have to be terminated in about 16 years, when the total dose would reach 192 rem.

Abundant evidence now available indicates that these limits are fearsomely permissive and inadequate.

The average lethal dose, sufficient to kill most mammals, is about 400 rad. However, not long ago some investigators mistakenly believed that as much as 100 times this dose, or 40 to 50 thousand rad, was required to inhibit reproduction of individual cells in the bodies of mammals.

Using techniques that permit laboratory maintenance and reproduction of such cells outside the bodies they came from, Dr. Puck has measured survival rates of simple cells subjected to various radiation dosages. He has found that the mean lethal dose, symbolized by the ominous D°, is only some 100 rad. This is the amount of radiation that reduces to 37 percent the proportion of exposed cells that remain viable—that is, which continue both to live and self-reproduce. Some normal mammalian cells, in fact, show mean lethal doses as low as 75 rad.

The earlier excessive estimates had been biased by analogy with the considerable radiation dosages required to kill some simple nonmammalian organisms, such as protozoa (paramecium) and bacteria (notably *E. coli*).

Public bodies charged with setting "maximum permissible dose" limits for individuals and populations have been obliged repeatedly to reduce limits promulgated only short times before. Between 1952 and 1959, for example, four such significant reductions took place in the standards set by the National Committee on Radiation Protection, the International Committee on Radiation, or committees established by them in advisory capacities. The permissible average total body exposure (dose) per week in 1952 was set at 0.03 rem. In spring 1958 it was cut to 0.01 rem; in autumn 1958 to 0.003 rem; and in spring 1959 to 0.002 rem. Thus within eight years, more than 92 percent of the "permissibility" was eliminated.

In 1971 the peaceful atomic-energy programs of the United States, under standards issued by the Federal Radiation Council, were restricted only to these limits: (1) the annual average dose received by United States residents as a whole should not exceed 0.17 rad per person; and (2) no individual should receive more than 0.5 rad per year. A dose of 0.17 rad, or 170 millirad, per year corresponds very nearly to 0.003 rem per week.

Two qualified scientific investigators, John W. Gofman and Arthur R. Tamplin, were assigned in 1963 by the AEC to evaluate the hazards of atomic radiation—"to assess the cost in human disease and death for all sorts of . . . nuclear energy programs, including nuclear electricity." By late October 1969 they had concluded that if the average exposure of the United States population reached the allowed 170 millirad annual dose, "there would be, in time, an excess of 32,000 cases of fatal cancer or leukemia per year, and this would occur year after year."

Their announcement of this conclusion was followed by denials and protests from leading spokesmen for the AEC and from Representative Chet Holifield of California, chairman of the Joint Committee on Atomic Energy of the Congress. The conclusion was controverted also by spokesmen for manufacturers of nuclear-power-generating equipment and public utilities using or planning to use nuclear generation of electricity. A typical statement was that made in June 1970 by Frederick Draeger, of the Pacific Gas and Electric Company: "There is no evidence that 170 millirads is harmful, and any new plant will actually emit only an infinitesimal fraction of that amount."

At just about the same time in Britain, Dr. Alice Stewart published statistics indicating that 250 to 350 millirads of radiation received by embryos during gestation (from X-ray pictures taken of their mothers' abdominal regions) had produced increase of about 25 percent in the rate of subsequent childhood cancers and leukemias.

Such malignancies commonly become apparent about five years or more after the radiation exposures. Some forms of cancer, in fact, show themselves about ten years later; others even twenty years later.

Each year in the United States about one third of a million persons die from cancer or leukemia. The rate has been about 17 deaths per 10,000 of population. Cancer risks are usually reckoned in terms of total dosages received by people who have attained age thirty. Ordinary natural background radiation (analyzed in greater detail in following pages), plus the average dosage from medical and dental X-rays, provides mean annual dosage of about 1.7 rad per year, or 50 rad by the time age thirty is attained.

An additional 170 millirad of dosage per year could be expected to increase that average cancer-leukemia death rate by 10 percent, for the population as a whole. And for the part of the population that actually received the individual limit of 500 millirad per year, the rate per 10,000 of deaths due to malignancies would be increased by about 30 percent—that is, from 17 to 22 per 10,000.

In all these comparisons the close equivalence of the rad and the rem has been applied.

Effects on births as well as deaths—Increased radiation dosage results in increased genetic mutations, meaning here mostly the birth of defective and abnormal babies. Experts have estimated at between 10 and 100 rad the dosage of ionizing radiation that will double the ordinary spontaneous rate of genetic mutations. A central figure of 50 rad seems the most probable at present. If this is used, then the addition annually of 170 millirad of dosage would increase by about 10 percent the genetically-determined diseases that now appear. On the other hand, if just 10 rad of additional dosage doubles the present rate of genetic defects, then the addition of 170 millirad per year means a 50 percent increase in such defects.

Genetic defects and diseases frequently shorten the lives of their victims, though that is not the only harm they cause. The rate of additional deaths annually attributable to genetic defects is bound to be increased even by a supplemental dosage of 170 millirad yearly. To deaths attributable to the added cancers and leukemias must be added also deaths attributable to the added defective and abnormal babies born.

Such factors led Dr. Puck, in the work cited, to call for careful reconsideration of the current levels of maximum allowable doses with a view to again diminishing the level by an "appreciable" amount. He suggests that the factor might well be as high as 10— that is, a cut to one tenth of the preceding standard. He warns, furthermore, that a "maximum permissible dose" figure is in any case a "poor substitute for the critical information which is actually needed." This information should include numerical data on how many cancers, birth defects and other pathologies are likely to result from subjecting the population to the indicated added dose.

The 1960s and the early 1970s, Dr. Puck makes clear, have disclosed a whole new group of human diseases "apparently capable of being induced by radiation." These are by no means ailments and defects that were unknown before 1945. They have plagued humanity and caused untold grief to parents and family members as far back as we can imagine. However, until the past couple of decades their genetic origin was unknown. Now they are recognized as products of inherited damage to genes and chromosomes in the cells; the passing along of changes originally wrought by radiation on reproductive cells.

In the late 1950s mongolism (Down's syndrome) was demonstrated to be the result of certain chromosomal abnormalities. Now many other birth defects and abnormalities have been traced to similar origins. In fact, Dr. Puck estimates that between 0.5 percent and 1 percent of all live births "are accompanied by chromosomal abnormalities that may produce the most deep-seated diseases in these babies." And in nearly 70 percent of all these cases, the pathology includes mental deficiencies.

In 1969, for example, with nearly 3.6 million births in the United States, between about 18,000 and 36,000 were marked by such chromosomal defects, most of them tragic and devastating in their effects on the lives of the families involved. Such consequences cannot be assessed solely or even mainly in terms of the dollar cost of the care of the defective children.

Radiations always around or within us—Genetic defects at the level of their appearance in the millennia prior to the late 1940s are beyond doubt linked closely to the effects of the ionizing radiations constantly present in minute quantities everywhere on Earth. Nowhere on this planet is there a site wholly free from all traces of such radiation. The sources are found in the granites and basalts that make up the great crustal sheets underlying the continents and the ocean floors. They are found also in tiny but undeniable amounts in every human and animal body.

One frequently encounters references to the level of such inescapable background radiation. It is often used as a standard or measure in the effort to suggest that a supplementary radiation exposure is negligible, for its additional effects can be no greater than the natural ionizing radiation dosage already present.

A normal adult human body contains some 5 or 6 ounces of the element potassium, without which life would cease. The overwhelming proportion of these potassium atoms are either in the isotope called K-39 (about 93 percent) or that called K-41 (about 7 percent). However, about 12 potassium atoms in each 100,000 are in the radioactive isotope K-40. It has a half life of some 1.4 billion years.

In the simplest possible mode of measurement, the interior of a human body thus is the arena for about five hundred nuclear disintegrations per second by such K-40 atoms. In total dosage effect this adds up to about 0.02 rad per year (20 mrad), or just about $\frac{1}{600}$ of the "allowable" vocational dose of 12 rem per year. (The rough equivalence of the rad and rem units in this case allows them to be compared without elaborate conversion.)

Incidentally, the energies of both the beta radiation (electron) and the gamma radiation (electromagnetic) from K-40 disintegrations are about 1.5 MeV. In a lifetime of the traditional three score and ten years, an individual body would receive from internal K-40 disintegrations about 1 rad of dose.

Another constant source of natural and unavoidable bodily internal radiation is the radioactive form of carbon, C-14. Carbon atoms account for about 18 percent of the total mass of a human body. Carbon is a part of all organic compounds, and in the air we breathe. Carbon dioxide (CO_2) is seldom absent from the air, and is exhaled by every body. Indeed, we are lucky if smog-plagued urban atmospheres are not loaded also with poisonous carbon monoxide (CO)!

About 98.9 percent of the carbon atoms in our bodies are the naturally stable form, carbon-12. The remaining 1.1 percent are the radioactive form, carbon-14. A typical adult human body thus contains an average of some three thousand disintegrations per second of C-14 atoms. In terms of dosage, which is the object of attention here, this amounts to about 2 millirad per year, or about $\frac{1}{10}$ the dosage rate from the K-40 disintegrations.

Every human body contains also ultratiny amounts of radium

and other naturally radioactive substances, ingested largely in drinking water, food, and the air we breathe. This was true even before nuclear testing in the atmosphere and other, more peaceful nuclear undertakings began to play a role in raising exposure and dosage rates for groups and populations.

Sensitive Geiger counters, the great detecting devices for ionizing radiations, will click at average rates about once every two seconds in almost any typical residential location. When rainfall has brought down dissolved substances from the atmosphere, the Geiger click rate may rise to nearly once each second.

Thus, from outside as well as within our bodies, there is constant but limited background ionizing radiation. The cosmic radiations, mentioned before, account for about 0.1 millirad per day even near sea level, and more at higher altitudes, which have less atmospheric protection above them.

All in all, from within and without, an average individual in the eons before dawn of the current nuclear age sustained an exposure estimated at between 0.15 and 0.2 roentgen per year, 1.5 to 2.0 R per decade, and 10.5 to 14 R in a life span of seventy years.

This contrasts with the *additional* 200 rem accepted in the AEC guideline for the lifetime of a worker in nuclear industry, or the average of 5 rem additional dosage per year.

It all adds up—The effects of ionizing radiation on living beings is a cumulative one. False assumptions or analogies led many to suppose that the effects of small doses in time were, so to speak, thrown off; and that later additional dosage began, as it were, with an undosed body. The probability of physiological and, above all, genetic damage is one that grows greater as the total bodily dose grows greater, from whatever source or at whatever time.

Genetic defects have always occurred. Stillbirths and the live birth of monstrous or defective children are, however, increased by radiation dosage, especially by dosage concentrated in the genital and gonadal areas of male or female bodies. It was not true, before 1895, that the natural background ionizing radiation—then unknown—was without effect because unknown. It has beyond doubt always played a role in producing some, at least, of the mutations to which every living species is subject.

Such genetic jumps or changes are hazardous, whether they can be laid to radiation dosage or to chemical factors without any ionizing component. Very few "abnormalities" are advantageous. Most of them are harmful to health. Many are catastrophic.

The one defensible conclusion today is that each and every level of supplementary dosage by ionizing radiation is undesirable. The most rigorous examination by independent and unbiased experts should be given to every procedure or plan that threatens to increase ionizing dosages to individuals, to groups and to total populations.

In particular this applies to the young and to foods—milk, water, grain, fish, meat and vegetables—necessary for their sustenance and growth.

Seemingly small exposure levels of ionizing radiation in certain environments, such as the sea, become concentrated in the bodies of species living there. The predator-prey relationships may multiply such concentrations repeatedly. It is not necessary here to retell the familiar story of how catches of tuna and other fish revealed alarmingly high radioactivity after the insane epoch of atmospheric testing of the A-bombs and H-bombs.

Nuclear disintegrations have formed only the most spectacular aspects of the ionizing radiations around us. X-ray exposure also has wrought substantial harm. Millions of persons have been over-radiated before the inevitable menace of such radiation could be established and publicized. Even today many patients and doctors need information and careful deliberation before deciding whether this or that additional X-ray exposure is justified by the total situation. The decision may be positive; its benefit in diagnosis outweighs its risk. But in other instances the decision may be negative. And under no circumstances should the exposure in roentgen units be greater than is absolutely required to expose the film.

The realm of ionizing radiations, for which the tools of the roentgen, the rep, the rad and the rem were devised, requires utmost caution and skepticism as well as careful measurements. Danger must be sensed in any statement or suggestion that there is some level of supplementary radiation dosage that is truly "safe" or free from supplementary risks.

The threats to human life, present and yet to be born, do not arise from the total amounts of energy that ionizing radiations carry to human bodies, for far greater energies in heat radiations and visible light radiations can be encountered and absorbed without harm. The danger arises from the fact that the energies of ionizing radiations are packaged in such large and destructive photons or particle-bullets. The fact that they are indeed *ionizing* radiations means that they can and do disrupt the chemical bonds not only in molecules of air or water, but also in human tissues, and in the marvelously complex and vulnerable structures of the chromosomes and the genes that compose them.

52 CONCERNING UNITS, CRITICAL AND COMIC

What's in a name? that which we call a rose
By any other name would smell as sweet.
—SHAKESPEARE,
Romeo and Juliet

Difficulties aplenty beset all, professionals as well as lay persons, who deal with the units devised to measure the effects of radioactivity and other dangerous sources of ionizing radiations around us in this alarming era.

These difficulties result largely from the alien and, till so recently, utterly unfamiliar nature of the effects with which these new measurements seek to grapple. Usually the processes of measurement, to the extent that they are successful, impart a measure of understanding, if not actual mastery, of the problems that have called them forth. This is the case only to a limited degree in the area of nuclear disruptions and their consequences on living creatures.

The obscurities characteristic of these regions of the wide world of measurements have been stressed with wry humor by Sheldon Novick, editor of the useful periodical *Environment,* published by the Committee for Environmental Information, of St. Louis, Missouri.

In an editorial, he put it this way:

> Each scientific specialty has its units of measurement that are convenient to the specialists therein, but tend to be mysterious to the uninitiated.
>
> Nuclear science, which deals with the objects farthest removed from human experience, has the least comprehensible units. Our readers, perhaps with a shred of irritation, will recognize the

curie as the unit of measurement of radioactivity, and roentgens, rems, and rads as the hopelessly confusing units of radiation exposure.

How is one to make the term "picocurie" comprehensible? A curie is the radioactive equivalent of a gram of radium. The curie, being inconveniently large (through no fault of the lady for whom it is named), has been divided and sub-divided. A millicurie is one-thousandth of a curie; a microcurie is one millionth of a curie; and one-millionth of a microcurie is a picocurie. It is only when a curie has been divided into a millionth of a millionth that radiation safety experts feel comfortable with it. They allow us so many picocuries in each liter of our drinking water (a liter, as we all know, is 0.95 U.S. liquid quart).

Many of the units of nuclear science have a whimsical sound, in direct proportion to their incomprehensibility. The "barn" for instance. The barn is a measure of the likelihood that, for instance, a uranium nucleus will be split by a neutron that happens along. It [the barn] is defined as 10^{-24} square centimeters. That is a small enough area, but the name of the unit derives from the observation that, in the subnuclear world, this would be as easy to hit as, yes, a barn door.

Now 10^{-24} is a good round number, and those of us who remember our college algebra know that it is equal to a number which, if we want to impress you, we write 0.000 000 000 000 000 000 000 001.

Another vexing problem is the expression "parts per million." This is a convenient way of talking about pollutants or food additives that are present in small quantities. But how much is a part per million? My own favorite comparison is with one ounce of vermouth in 7,350 gallons of gin. The *News and Pesticide Review,* which is often at pains to explain how harmless a few parts per million of insecticide can be, offers some more similes: A part per million is like one minute in 1.9 years; one inch in 16 miles; one ounce of salt in 62,500 pounds of sugar; one ounce of sand in 31¼ tons of cement.

Another little known technical term is the boggle. This is a measure of the time a writer sits staring blankly at his typewriter trying to think up some way of explaining terms like "roentgen." It is equal to 1.5 ounces of Irish whiskey. A roentgen, by the way, is 10^{14} boggles.*

* Reprinted from *Environment,* June 1970, by permission. Copyright 1970, Committee for Environmental Information.

The instance of the barn unit, cited by the editor of *Environment,* is but one case in which whimsy and humor have attended the choice of a name for a measure of most serious import. The barn has been in use for a generation and appears likely to continue in indefinite use as an indicator of the probabilities of specific nuclear processes and interactions, such as absorption, scatter and fission, induced by the firing of particle projectiles through matter.

The physicists C. P. Baker and H. G. Holloway, working in Chicago in 1942, devised the unit and named it—or rather masked its identity, for *barn* was coined originally as a code word in those days of the clandestine preparations for the monstrous flowering that finally unfolded at Hiroshima and Nagasaki in August 1945.

The old expression "You couldn't hit the broad side of a barn!" operated in this choice. Actually the probability of bringing about a particular nuclear hit or interaction varies enormously. There are certain nuclear interactions or results measured at a level of about 10,000 barn (b); and on the other hand some electron bombardment processes that typically are measured at about 10 microbarn (equal to 10^{-11} b)! That is a probability range of 1 to 10^{15} times, or 1,000 million million.

The peculiar area of 10^{-24} square centimeter (equal to 10^{-28} m²) was not assigned to the barn on the basis of mere whim. The typical radius of an atomic nucleus is about 10^{-12} cm or 10^{-14} m. Hence the cross-sectional area of such a nucleus as target for particle bombardment can be thought of as about 3.1416 times the square of that radius. The square of 10^{-12} cm is 10^{-24} cm², or 10^{-28} m².

The barn thus corresponds to the probability of a hit or interaction on strictly geometrical grounds. Probabilities of interaction, however, do not correspond to geometrical abstractions in the nuclear and subnuclear worlds. It is as if a nucleus mysteriously expands or contracts as the particle by which it is bombarded is changed in kind (neutron, proton, or electron) and in character (high, medium, or low velocity).

Measurements made in terms of the barn unit thus indicate whether a particular interaction is observed more often or less often than corresponds to the apparent actual cross section of the nuclear target. For example, a measurement resulting in 10^3 b suggests behavior as if the nucleus had expanded to 1,000 times its apparent cross section; and a measurement resulting in 10^{-3} b, as if the nucleus had shrunk to one-thousandth of that same apparent cross section. The barn is thus an analogy in more than one sense of that word.

In the 1950s an attempt was made to add to the barn a far smaller unit of the same general character to be called the *shed*, which is definable, very roughly speaking, as a small barn. The shed was assigned a cross section of 10^{-48} cm^2, equal also to 10^{-52} m^2. The shed, however, was so excruciatingly tiny that it never caught on, and is virtually unknown except as a novelty. For example, the very smallest measurements in the barn unit, in the neighborhood of about 10^{-11} b, would correspond to 10^{13} shed, or 10 terashed (10 Ts)!

The barn is built just about right to hold its varied stores of interactions in the range between 10^{-11} b and 10^4 b.

For readers addicted to the believe-it-or-not approach, the conversion of common odds or run-of-the-mill probabilities in their barn equivalents may hold some interest. The barn is not intended for such uses, but since it does measure probabilities, it is not inconceivable that a code for 99 to 1 odds could become 10^{22} barn! A probability of 1 hit out of each 100 firings, in other words.

The extraordinarily small probabilities of producing any specific nuclear interaction are thus made more vivid. High probability in nuclear reactions is at a level of about 10,000 barn, which means in coldly statistical terms that each 10^{24} subatomic projectiles, fired through an area 1 cm^2 in cross section, will produce only 10^4 hits or successes per target nucleus or particle in the area.

This appears to be a vanishingly tiny prospect of a hit. However, if within that target area there are 10^{18} target nuclei—which is not at all unlikely—then the odds do rise to that level of 1 hit in 100 tries, or 99 to 1 against a hit. Under those circumstances, 1 out of each 100 subatomic projectiles fired through the area can be expected to produce the sought-for interaction.

The art of subnuclear and particle marksmanship, indeed, depends on firing enormous numbers of projectiles at enormous numbers of targets in order to attain sufficient numbers of hits among the myriads of inevitable misses!

Laughing with early lasers—Makeshift units rich in whimsy or humor have occasionally been devised by scientists and engineers. Thus, during the early 1960s the newly born solid-state lasers were being tested. Their unprecedentedly powerful, pure and coherent beams of red light emerged from synthetic ruby crystals in great bursts. When suitably focused, these laser beams concentrated so much radiant energy within a tiny area that they could quickly vaporize bits of metal on which they fell.

Thus emerged, for a time, the estimation of laser power in units of the *gillette,* determined by how many stacked double-edged razor blades a laser's beam would burn through. That gillette has been replaced by the more conventional SI units of the joule, to measure the energy content of separate laser bursts or pulses, and the watt, to measure the power output of continuously operating gas lasers.

Concern with questions of the nature and names of their units of measurement is common among scientists and technicians of this era. In part it is the kind of attention that craftsmen lavish on the tools they employ constantly and wish to keep as convenient and efficient as possible. In part also the concern reflects sentimentality and nostalgia.

The attachment of the name of an eminent scientist or mathematician to a unit or to a former unit combination that now becomes a single-named unit has been a fairly common occurrence during the past couple of generations. The advocacy and opposition that such proposals evoke provide interesting and revealing evidence of emotional overtones that might seem, at first, to be quite alien to science as such.

During the early 1920s, the Deutsche Physikalische Gesellschaft, roughly the counterpart of the American Institute of Physics (AIP), proposed that the name *hertz* be given to the unit of frequency, equal to 1 cycle per second. Sharp and sarcastic opposition was voiced, however, by the well-known physical chemist Hermann Walter Nernst (1864–1941).

Nernst remarked acidly, "I do not see the necessity of introducing a new name; by the same reasoning one might as well call one liter per second one 'Falstaff'!" Nevertheless the name hertz was adopted for this purpose, and today it is generally considered to be highly useful, if not indeed indispensable in many areas of measurement.

The falstaff has not, in the meantime, been given the support to which it might be entitled as a suitable unit of intake. However, other units of rate have been proposed and have received fairly substantial support. One example is the *benz,* named for the pioneer German automotive engineer and inventor Karl Benz (1844–1929). The benz, if ever adopted, would be a velocity of 1 meter per second (1 m/s). Support for the benz unit proposal has come principally, as might be expected, from German sources.

Opposition not unlike that voiced by Nernst against the hertz as a frequency unit has been directed against the unit called the *jansky,* now making its way in radio astronomy and related areas of the world of measurements. The jansky is a unit of measurement of intensities of received radiations, especially of radio waves from astronomical bodies and satellites in space.

The jansky equals 1 watt per square meter per hertz of the received radiation (1 W/m^2Hz). In general it has the same function as the so-called "flux unit" defined in a preceding chapter on radiation. The W/m^2 fraction represents a power density, as received at the antenna, or "dish." The hertz relates this power density to the frequency of the waves, thus providing the essential bandwidth component in the total measure.

The jansky is named in honor of K. G. Jansky (1905–1950), the electrical engineer who in 1931 detected and identified radio waves reaching the surface of the Earth from outer space, thus opening the door to the vast and fascinating world of radio astronomy.

It still goes on—New proposals to create eponymous units appear from time to time and elicit both opposition and support. The correspondence columns of many scientific and technical journals seldom lack letters relative to such suggestions.

As this chapter was being prepared, the correspondence columns of *Physics Today,* monthly of the American Institute of Physics, carried a proposal from H. M. Cassel of Miami Beach, Florida, that a unit of surface tension be named for Josiah Willard Gibbs (1839–1903), the American genius of thermodynamics, creator of the famous "phase rule."

The proposed *gibbs* of surface tension would equal the unit combination 1 erg per square centimeter (1 erg/cm^2). That, in turn, would inevitably correspond to 1 millijoule (10^{-3}J) per square meter in the basic units of the International System. It would also correspond to 10^{-3} newton per meter (10^{-3} N/m), for surface tension is a variable which may be measured either as an energy per unit surface area, or as a force per unit surface length.

This is not the first unit proposed to be named in honor of Gibbs. In 1951, R. B. Dean, writing in a chemical journal, urged that Gibbs's name be given to a unit of adsorption, the process whereby matter is collected or massed on surfaces. (Not to be confused with *ab*sorption, which involves the taking of matter into volumes of other matter.)

Dean's proposal was that the *gibbs* unit of adsorption should equal the concentrating of 10^{-10} mole of any substance per square centimeter of the adsorbing surface. This is equivalent to a concentration of 10^{-6} mole per square meter of such surface, in terms suitable for the International System.

It would violate the ground rules of eponymy in physical units if the name gibbs were approved for two diverse units—one to measure the variable of surface tension, the other to measure the process of adsorption.

Besides the seriousness of eponymous proposals such as these, there is a wonderful area of pseudo-eponymous spoofing, arising from the fabrication of unit names to describe and decry personal defects. This happens in almost every laboratory and research center. It is especially evident where graduate students suffer and sweat for years to attain their doctorates. For them ironical eponymy becomes an escape, if not an utter necessity for emotional survival.

The spoof-eponymous units are strictly unofficial and surreptitious, yet none the less widespread and oft-repeated for all that. At least two, and possibly more than two, prominent scientific personalities' names have been taken for such units of self-love or personal vanity. Since neither of them was called Gilligan, let us use that name as a cover for this report.

During coffee breaks and bull sessions it was told and retold how the *gilligan* was the natural unit for this vanity of vanities. But, some tellers insisted, ordinary run-of-the-mill personalities could not be rated by the whole gilligan; it was excessively large. Even a well-nourished ordinary ego would have to be measured, they suggested, in such a submultiple as the milligilligan or perhaps even the microgilligan!

Another eminent authority, thus eponymously and privately cut down to size, was a famed physical chemist. He was reputed to be an extraordinarily effective operator when it came to demanding and receiving high salaries from the institutions at which he carried on his ascending career. His name not being Dedoe, we shall assume that it was. The *dedoe,* then, became the unit of private greed or moneygrabbing, and many another scientist was declared to have reached no better than a millidedoe on this scale.

Thus, humorous units have been eponymized for measuring such human weaknesses as indolence, ability to bore, philandering, repetitiousness, obscurity of communication, forgetfulness, quick temper, and so on and on.

At Cambridge University the splendid theoretical physicist Paul Adrien Maurice Dirac (1902–1984) so impressed his students by his absence of volubility that there arose the *dirac,* unit of prevalence of silence during discourse. Just as the ohm of electrical resistance measures the opposition to current flow in response to an electromotive force, so the dirac suggested reluctance to speak unless speaking is unavoidable.

53 EPONYMOUS PROCESSIONAL—
A LAST ROUNDUP OF UNITS

Measures, not men have always been my mark . . .
   —OLIVER GOLDSMITH,
      The Goodnatured Man (1768)

   *A quantity like time or any other physical measurement
   does not exist in a completely abstract way. We find no
   sense in talking about something unless we specify how we
   measure it . . . a definition by the method of measuring a
   quantity is the one sure way of avoiding talking nonsense. . . .*
      —HERMANN BONDI,
         Relativity and Common Sense (1964)

[What Bondi here calls a "quantity" is referred to in the present
book as a "physical variable."]

Eponymous units, other than those presented in the previous pages,
are assembled here in alphabetical order. Some are obsolescent or
obsolete, yet still appear in publications that will be found in every
well-stocked library of science and technology.

   amagat. Paradoxically, there are actually two amagat units,
each named for the French scientist Emile Hilaire Amagat (1841–
1915). One is the amagat *volume* unit, the other the amagat *density* unit. The former is the volume of 1 mole of gas at standard
temperature and pressure. It is equal approximately to 0.022 413 6
cubic meter. The amagat density unit, on the other hand, is the
density of a gas, 1 mole of which occupies just 1 amagat volume
unit. The amagat density unit accordingly corresponds to 44.615-
768 mole of the gas per cubic meter.

   Bailling, Beck, and Brix scales. Various scales of liquid
density devised during the middle years of the nineteenth century

to calibrate hydrometers bearing corresponding names. Readings in each can be converted into specific gravities by means of appropriate tables.

balmer. A unit of wave number, based on the number of electromagnetic waves in 1 cm of length. Thus 1 balmer equals 1 wave per cm or 100 waves per m. Visible light lies in the range of about 13,000 to 26,000 balmer. Named for the Swiss optical scientist and mathematician J. J. Balmer (1825–98). The names *kayser* and *rydberg* are also given to this unit. The higher the number of balmer units, the higher the frequency of the waves and the shorter their wavelengths.

baud. This name has been given to two different units: (1) a unit of information, identical with the unit called the *bit* in computer practice; and (2) a unit of signaling rate in radio and telegraphic communications, equal to 1 pulse per second. The baud is named for the French inventor and communications engineer J. M. Baudot (1845–1903), who devised an early form of teletype. A signaling rate of 100 baud equals 100 pulses per second.

Beaufort scale. A sequence of wind velocities originated by a British admiral, F. Beaufort (1774–1857) and adopted for maritime use during the middle third of the nineteenth century. It rates velocities by means of 13 numbers, ranging from 0 (for wind speeds up to 1 mile per hour) to 12 (for hurricane velocities of 75 mph or more). The relationship between the Beaufort number (B) and the wind velocity in mph (V) is roughly $B = 0.66 \ V^2 \cdot V^{\frac{1}{3}}$, or $0.66 \ V^{\frac{2}{3}}$. Thus a V of 10 mph corresponds to about 3.1 B and a V of 50 mph to about 9.5 B. The Beaufort scale is widely used on weather maps and reports.

blondel. A unit of luminance, equal to $1/\pi$ nit, or 0.318 3 nit. The *apostilb* is another name for the same unit. The blondel is named for the French scientist A. Blondel (1863–1938).

Bohr magneton. The natural and smallest possible unit of magnetic moment, named for the eminent Danish atomic physicist Niels Bohr (1885–1962). It is the magnetic moment of the single spinning electron in the hydrogen atom at its lowest energy or ground state. Though important in atomic and magnetic measurements, the Bohr magneton is defined in ways that do not always agree with one another. In SI units it equals 1.165×10^{-25} weber meter. The Bohr magneton is sometimes symbolized by the Greek beta, β, and sometimes by "mu sub-B," or μ_B. Another natural magneton, somewhat less important, is the so-called *nuclear* magneton, based on the magnetic moment of the proton and, hence, with a magnitude about 1,840 times the Bohr magneton. The nuclear magneton is commonly symbolized by mu sub-N, or μ_N.

Bohr radius or length. A length equal to the first Bohr radius of the hydrogen atom—that is, to 5.3×10^{-11} m.

brewster (B). Unit of the retardation of light resulting from stress applied to the substance through which the light is transmitted. Named for the Scottish physicist David Brewster (1781–1868), discoverer of the law relating the polarizing and refracting characteristics of transparent substances. A substance is said to have 1 B of "stress optical coefficient," if when subjected to 1 pascal of stress it retards light passing through it by 1 part in 10^{12}. Similarly, 1 B may equal a retardation of 1 part in 10^7 as a result of a stress of 10^5 pascal, and so on. The retarded light passes through the substance at a right angle to the direction in which the stress is applied.

brig. A unit used to express ratios. One brig equals 10 decibel, 2.302 59 neper, or 3.321 9 octave. The brig is also known as the *dex,* or decimal exponent. Thus 10^2 has the value 2 brig; 10^3 the value 3 brig; and so on. It is named for the English mathematician Henry Briggs (1561–1630).

brinell number. A hardness number, named for the Swedish engineer J. A. Brinell (1849–1925). It depends on the extent of penetration by a specific device. If a steel sphere with a diameter of D mm is pressed by a force of F and makes an indentation with a depth H in the tested substance, then the brinell number for that substance is F/D·H times 0.318 3, which is the reciprocal of pi (3.1416). Brinell invented his hardness tester and devised its associated scales.

chad. A unit of neutron flow, or flux, named for the British physicist James Chadwick (1891–1974), discoverer of the neutron, one of the two principal nuclear particles. Two widely differing definitions have been proposed for the chad unit: (1) a flux of 10,000 neutrons per square meter per second, and (2) 10^{16} neutrons per square meter per second. Obviously, the latter is 10^{12}

times as great as the former. Nuclear reactors sometimes emit neutron fluxes about as great as the latter rate. The chad has not yet been officially adopted.

Clark degree. A unit of water hardness, named for the Scottish scientist Thomas Clark (1801–67). A measure of 5° Clark corresponds to 70 parts per million of calcium carbonate in water; and 15° Clark to 210 ppm.

dalton. A unit of atomic mass, named for the British chemist John Dalton (1766–1844). The dalton equals $\frac{1}{12}$ the mass of a neutral atom of carbon-12. Its measured value is $1.660\ 33 \times 10^{-27}$ kg. One dalton equals 1 international atomic mass unit. The mass of a hydrogen atom is 1.008 1 dalton; of a neutron, 1.009 dalton; and of a proton 1.007 6 dalton.

darcy. A unit of measurement of the mechanical or porous permeability of layers of rocks, soil or other substances; named for H. P. G. Darcy (1803–58), French hydraulic engineer. One darcy is the permeability of a substance through which a pressure gradient of 1 atmosphere per cm drives a fluid with 1 centipoise viscosity at a rate of 1 cc per sq cm of cross section. The darcy has been used especially by geologists of the petroleum industry.

darwin. A unit proposed for measuring rates of evolutionary change in living species; named to honor Charles Darwin (1809–82), British biologist and developer of evolutionary theory. If a species increases 1000-fold in 1,000 years, or diminishes in a like time to $\frac{1}{1,000}$ of its former numbers, its rate of change in either case is 1 darwin.

einstein. A unit of photoenergy, or light energy, named to honor Albert Einstein (1879–1955), outstanding theoretical physicist of the twentieth century. One einstein equals $3.989\ 9 \times 10^{-10}$ joule times the frequency in hertz of the light in question. The resulting unit is used to measure the light energy required to bring about a specific chemical change in a specified mass (1 mole) of a particular substance, such as the photographic emulsion in a light-sensitive photographic film.

eotvos (E). The eotvos, a unit to measure changes in Earth's gravitational intensity observed as a result of horizontal movements on the Earth's surface, was named for the eminent Hungarian geophysicist and educator, R. Eotvos (1848–1919). One E corresponds to a change of 10^{-7} gal or galileo of acceleration resulting from a horizontal movement of 1 meter. One gal is an acceleration unit equal to 10^{-2} m/s². Hence the E unit equals a change of 10^{-9} m/s² in acceleration, resulting from a shift of 1 m in horizontal direction on the Earth's surface.

erlang. A unit of traffic intensity in telephone systems, named for the Danish technician A. K. Erlang (1879–1924). One erlang corresponds to an hourly rate of 1 telephone call of 1 hour duration. Telephone traffic is measured by the number of calls per unit time, the average duration of the calls being measured in the same time unit.

faraday or Faraday's constant. A unit of quantity of electricity or charge, named for Michael Faraday (1791–1867), the great chemist and electrical experimenter. It is the amount of electricity that sets free, or neutralizes, 1 gram-molecular equivalent (1 mole) of any ion from an electrolytic solution. It is measured as 9,648.7 coulomb per mole. For some electrochemical measurements it is preferred to the coulomb, the SI unit of charge or electric quantity.

fermi. Unit of atomic length, equal to 10^{-15} meter, named for the Italian-American physicist Enrico Fermi (1901–54). 1 square fermi equals 10^{-30} m²; and 1 cubic fermi equals 10^{-45} m³.

finsen unit (F.U.). A unit of measurement of the intensity of ultraviolet radiation, named for the Danish scientist N. R. Finsen (1860–1904). The F.U. equals 10^5 watt per square meter, when the wavelength of the radiation is 2.967×10^{-7} meter.

fourier. A unit of thermal resistance, equal to the so-called "thermal ohm." Named for J. B. J. Fourier (1768–1830), French thermodynamicist and mathematician. One fourier corresponds to an entropy flow of 1 watt per kelvin in response to a temperature difference of 1 K.

fraunhofer. A measure of width reduction in lines of the spectrum, named for J. Fraunhofer (1787–1826), German spectroscopist. When the width of a spectrum line corresponds to a bandwidth just 10^{-6} of the wavelength of the center of that line, then that line has a 1 fraunhofer width. If its width is double, that is 2 fraunhofer, and so on.

fresnel. A seldom-used frequency unit equal to 10^{12} hertz or cycles per second; named for the French optical scientist Augustin Jean Fresnel (1788–1827).

galvat. Unit of electric current equal to 1 ampere; named for the Italian natural scientist Luigi Galvani (1737–98).

hardness scales or **numbers.** See the entries for the brinell and mohs scales. There are also hardness scales or rating routines associated with devices and systems named for Knoop, Meyer, Rockwell, Shore, and Vickers.

Hardness is in fact not a single, clearly defined physical variable, but rather a complex or combination of diverse concepts. Principal among them are these: (1) the relative ability of different substances to scratch each other; (2) the relative ability of substances to take and keep a cutting edge; (3) the relative ability of substances to resist abrasion and wear.

Some substances may rate high on one or two of the three, not on the other or others; hence "hardness" as a name means little unless the particular testing method and scale has been specified. There is no single or accepted hardness measure included in the International System. Yet hardness measurements are constantly made and are of prime importance for many industrial and technical uses.

hartley. An early unit to measure amount of information in communications systems, named for the American electrical engineer R. V. L. Hartley (1888– ?), who offered the unit in a paper presented in 1927. The hartley equals 3.219 bits, the more widely used modern unit of information. Hartley was the first to formulate the relationship, now widely recognized, that exists between the amount of information a system can transmit and the frequency bandwidth required to transmit it.

hartree. A proposed natural or atomic energy unit, named for the British physicist D. R. Hartree (1897–1958); it is equal to $4.950\ 5 \times 10^{-18}$ joule.

helmholtz. Unit of the moment of an electrically charged double layer or dipole, named for Hermann L. Helmholtz (1821–94), German physicist, physiologist and physician. One helmholtz equals 1 debye per square angstrom, or 3.335 64 coulomb per meter of length.

kapp line. A measure of magnetic induction, named by and for the British scientist Gisbert Kapp (1852–1922). Each line equals a flux of 6,000 maxwell.

kayser. Another unit of wave number equal to the balmer, already described, but named in this case for the German optical scientist J. H. G. Kayser (1853–1940).

lambert. A unit of luminance equal to 0.318 3 stilb or 0.318 3 candela per square centimeter. Named for the German optical scientist Johann Heinrich Lambert (1728–77).

langley. A unit of radiation energy density, now equal to 1 calorie per square centimeter, or 10^4 cal/m², hence equal to 41,868 joule per m². Used especially for solar radiation as it reaches the surface of the Earth. Named for Samuel Pierpont Langley (1834–1906), former director of the astrophysical laboratory at the Smithsonian Institution. The original langley unit was the "per minute" counterpart of the present unit. The change was made in 1947. The solar power reaching the Earth's surface is about 2 langleys per second, or about 83,736 watt per m².

lorentz. A unit equal to the Bohr magneton stated in wave numbers. It relates to the extent to which the frequency of light is shifted when the light source is acted on by a magnetic field. The lorentz is equal to 46.689 per tesla meter (1/Tm). It is named for the Dutch theoretical physicist H. A. Lorentz (1853–1928).

mache. An arbitrary unit of radioactive concentration, equal to 3.7×10^{-7} curie concentrated in a volume of 1 m³. Now obsolete, this unit was named for the Dutch scientist H. Mache (1876–1954).

macleod. An arbitrary unit of gas pressure, named for the British scientist Herbert MacLeod (1841–1932). The macleod single unit corresponds to a pressure of 0.1 mm of mercury or to 13.332 24 pascal in the SI pressure unit. From this level onward, pressure reductions are expressed by the sequence of negative exponents with the base value 10. Thus a pressure of $\frac{1}{100}$ of the above amount would be 2 macleod; one of $\frac{1}{1,000}$ that amount, 3 macleod; of $\frac{1}{10,000}$ that amount, 4 macleod; and so on. MacLeod invented the vacuum gauges identified with his name.

MacMichael degree. A measure of dynamic viscosity, in degrees on the scale associated with a rotating test device developed early in the twentieth century. The MacMichael-degree readings can be converted into units of the poiseuille (Pl) or the poise (P) by suitable conversion charts. It is named for R. F. Mac-Michael, American chemist and rheologist, active in the early twentieth century.

mayer. A unit for measuring heat capacities, named for J. R. Mayer (1814–78), German thermodynamicist, physiologist, and physician. The mayer equals 1 joule per ° C temperature change in 1 gram of substance.

mercalli scale. A numerical scale of relative earthquake *intensities,* named for its originator, Giuseppe Mercalli (1850–1914), Italian seismologist and priest. Unlike the Richter scale, which indicates the absolute *magnitude* of the earthquake itself at its underground source, the Mercalli scale estimates earthquake results at some specific place on the surface of Earth. An earthquake can have but one magnitude, but can have a different intensity at each of scores of places on the surface. The following indicates the dozen different Mercalli intensity indications:

1. Felt only by very few persons, under circumstances that favor their noticing it.

2. Felt by a few persons who are at rest, especially on upper stories of buildings. Some suspended objects may be noticed to swing somewhat.

3. Felt noticeably indoors, especially on upper stories, but in many cases not identified as an earthquake. The vibrations feel somewhat like those of a truck driving past.

4. Felt, if during daytime, by many persons indoors, but by few outdoors. If at night, some persons are awakened. Disturbances to doors, windows, dishes, et cetera.

5. Felt by almost all persons. If at night, many are awakened. Breakage of some windows and dishes, et cetera.

6. Felt by all persons. Many are frightened and run outdoors. Heavy furniture sometimes upset. Slight damage.

7. All persons able to do so run outdoors. Substantial damage to buildings of poor design. Quake noticeable to people riding in automobiles.

8. Substantial damage to structures of ordinary design and slight damage even to those of special design. Great damage to badly built structures. Ejection of small quantities of mud and sand at various places.

9. Extensive damage even to structures of special design. Partial collapse of extensive structures. Shifting of buildings off their foundations. Breakage of underground pipes. Cracking of ground clearly visible.

10. Destruction of most frame and masonry structures and their foundations. Bending of streetcar and railway tracks. Pronounced cracking of ground. Shifting of masses of sand and mud. Waters of rivers and ponds are slopped over their banks.

11. Few masonry structures are left standing. Broad fissures in the ground. Underground pipelines rendered entirely unusable. Great bending of rails. Slumps and landslides in soft ground.

12. "Total" damage. Formerly level ground surfaces now show tilts and waves. Distortions of both levels and lines of sight. Objects are found to have been thrown up in the air during the quake.

Mercalli scale ratings of intensities observed in three substantial New Zealand quakes are here compared with the Richter scale measures assigned to the magnitudes of the same quakes:

January 23, 1855, S. W. Wairarapa, N. Z.: 11 plus Mercalli; Richter scale, approx. 8. March 9, 1929, Buller, N. Z.: 11 Mercalli; 7 Richter scale. February 3, 1931, Hawkes Bay, N. Z.: 11 Mercalli; Richter scale about 7.75. These were reported by R. C. Hayes to the Seventh Pacific Science Congress.

Mohr cubic centimeter. A unit in saccharimetry, named for the German scientist K. F. Mohr (1806–79). Its magnitude is determined by the reciprocal of the density of water at 17.5° C;— hence it equals 1.000 13 cm³, the exact volume occupied by 1 gram of air-free water at that temperature.

mohs hardness number or scale. System devised by Friedrich Mohs (1773–1839), German mineralogist, for comparing hardness of materials. It is based on a sequence of ten carefully defined substances arranged in the order of their ability to scratch those lower in the list. From hardest to least hard they are: (10) diamond, (9) corundum, (8) topaz, (7) quartz, (6) feldspar, (5) apatite, (4) fluorite, (3) calcite, (2) gypsum, and (1) talc.

Mohs number for hardened tool steel lies between 7 and 8; for chilled copper, between 6 and 7; and for a healthy human fingernail, between 2 and 3. Mohs original sequence has been expanded by the insertion of several additional materials of graduated hardness into the gap between his 10 and 9, and others between his 8 and 7, as well as between 7 and 6. The result is a 15-step scale, with diamond at 15 and talc still at 1.

Even with these refinements, the Mohs scale is a method of approximate ranking, not of precise quantitative measurement of a physical variable.

mooney. The mooney unit is a measure of plasticity, especially of pre-vulcanized rubber or comparable synthetics, named for Melvin Mooney(1893–1968), American chemist and rheologist. It is determined by a specific procedure in which the material under test is rotated slowly while at a designated temperature; the resulting torque (twisting force) on a disk is noted in relation to the length of time that the measured twisting has been going on.

planck. Unit of "action," the physical variable that represents energy multiplied by duration (time). Named for the German thermodynamicist and physicist Max Planck (1858–1947). One planck equals 1 joule second (1 J s). It is equivalent also to 1 meter of length times a momentum of 1 meter kilogram per second.

poncelet. A unit of power, named for the French natural scientist Jean Victor Poncelet (1788–1867). It is just 100 times the magnitude of the watt, the SI power unit. The poncelet has been mistakenly stated to equal 980.665 W; the correct value is 100 W. It is a metric power unit, as is the watt.

preece. A unit of electric resistivity, named for the Welsh scientist William Henry Preece (1834–1913). It was based on the resistance, measured in megohms, of a cube with sides 10^7 meter long; 1 preece equals 10^{13} ohm meter. It is now virtually obsolete.

prout. A unit of nuclear binding energy, named for the Scottish scientist William Prout (1786–1850). It equals $\frac{1}{12}$ the binding energy of the deuteron structure and, hence, corresponds to 1.855 $\times 10^4$ electron volt of energy (18,550 eV). Binding energies of heavy nuclei are not much more than some 40 prout, but in light nuclei the level may be above this. The prout is rarely used in practice.

ray. A unit of acoustic or mechanical resistance, equal to the acoustic ohm and named for J. W. Strutt, Lord Rayleigh (1842–1919), British physicist and pioneer in acoustic measurements.

rayl. A unit of specific acoustic impedance, which is the ratio between the sound pressure and the particle velocity at a surface. It is named for the same Lord Rayleigh. The SI unit combination in which the rayl is measured is newton-second per cubic meter. The rayl is also the product of the density of a gas times the velocity of sound through it.

rayleigh (R). A measure of the luminous intensity of the night sky and of the aurora in such a sky, named after the fourth Lord Rayleigh (1875–1947), son of the foregoing Rayleigh. One rayleigh equals 10^6 photons per square centimeter and, hence, also 10^{10} photons per square meter. The ordinary night sky has a luminous intensity measured between 200 and 300 R, but auroral displays may be measured between 1,000 and 1,000,000 R. In terms of luminous power the rayleigh equals about 5.272×10^{-25} watt/ m² steradian times the frequency in hertz of the light being measured.

Richter scale. The accepted scale for comparing earthquake magnitudes, devised by C. F. Richter (1900–1985) of the California Institute of Technology, Pasadena. It is based on the measured response of a precisely defined type of seismograph instrument. Thus, Richter 0 means a response of 1 micron (equal to about 0.000 04 inch) by such an instrument, located at a distance of 100 km from the earthquake's epicenter, which is the point on the surface of the Earth directly above the earthquake's actual focus or hypocenter.

Magnitude, which is the physical variable indicated by the Richter scale, is meant to be the energy in the underground waves produced by the earthquake or explosion to which a Richter number is assigned. It is estimated that an increase of one Richter number corresponds to an increase of 250 times in the *total* seismic energy of the quake or explosion. Thus a Richter number of 3 would indicate about 15,625,000 times the *total* energy indicated by a Richter number of 0. However, when the object of attention is the energy of the triggering earth slippage or explosion that launches the total seismic energy, the ratio appears to be about 10 to 1, rather than 250 to 1. Thus an increase of one Richter number indicates an increase of about 10 times in the initial or causal energy. The Richter scale is thus approximately logarithmic in structure, and Richter 6 represents about 1 million times the initial energy of Richter 0.

In contrast, the Mercalli scale of observed earthquake intensities is approximately arithmetic, so that Mercalli 6 indicates about double the intensity of Mercalli 3, and so on.

A Richter number of 3 corresponds to about 1 million joule of seismic energy, and Richter 6 to more than 10^{13} joule of such energy. The kind of earth substance between the hypocenter of the quake and the seismograph instrument must be taken into account in determining the Richter number assigned to the event. Softer substance, of the kind called "alluvium," absorbs initial energies in frictions to a larger extent than do hard rock layers.

The basic instrument used in arriving at the Richter number assigned to an earthquake is the Wood-Anderson torsion seismograph with 0.8 second period and a magnification factor of 2,800.

The zero of the Richter scale by no means indicates zero energy or even negligible energy. Very small tremors may be assigned negative Richter scale numbers, such as -0.5 or even -1.0 Richter. Upwards of a million separate earthquakes are recorded around the world each year, their absolute magnitudes differing widely on the Richter scale. The Richter scale ratings assigned by various seismological centers and authorities do not always precisely agree, but they seldom deviate by more than 0.3 or 0.4 Richter, which is rather close correspondence in view of the many factors that must be taken into account.

The highest recorded Richter magnitudes thus far are somewhat below 9. Following are several notably large earthquakes of recent years with their Richter scale ratings: November 25, 1941, in western Portugal, 8.3 Richter; August 15, 1950, Assam, India, 8.5 Richter; March 4, 1952, Tokaichi, Japan, 8.6 Richter; December 4, 1957, Altai, Outer Mongolia, 8.6 Richter; May 22, 1960, Lebu, Chile (chief shock), 8.9 Richter.

It is estimated that the famous Lisbon, Portugal, earthquake of 1755 had a Richter magnitude of about 8. The San Francisco earthquake of April 18, 1906, is rated at Richter 8.25, below two other earthquakes in that same year—January 31 at Colombia, Ecuador, Richter 8.6; and August 18, at Valparaiso, Chile, Richter 8.6 also.

rockwell number. An index of hardness, depending on the depth of penetration achieved by a hard steel or diamond cone under a measured force. Named for Stanley P. Rockwell, American engineer active in the 1920s. His penetrometers were made in different sizes, and the depths by which their hard cones penetrated tested materials were indicated by a number, called the rockwell number.

rydberg. A unit of wave number, named after the Swedish spectroscopist Janne Rydberg (1845–1919). Identical with the Balmer unit, already defined here. Violet light, with a wavelength of 4×10^{-6} m, has a rydberg (wave number) of 2,500.

rydberg (R). A very different unit, being the natural-energy unit corresponding to the energy required to ionize (take away the electron from) an atom of hydrogen. One rydberg of energy equals about 13.6 electron volts (eV) of energy. This unit too is named for Janne Rydberg.

sabin. A unit of acoustic absorption, named for Wallace C. Sabine (1868–1919), American acoustician. The sabin, also known as the "open-window" unit and the "total-absorption" unit, equals the sound absorption of a surface 1 ft^2 in area with a reverberation coefficient of 1. The absorption of 1 m^2 of the same kind of material equals 10.76 sabin. In a closed hall, 1 square foot of open window has exactly 1 sabin of sound-removing or absorbing effect, hence the "open window" designation.

scheiner degrees or numbers. A system for indicating relative speeds of photographic emulsions, named for Julius Scheiner (1858–1913), German astrophysicist. Two different Scheiner scales have been used, one the B.S.I., the other the American. Relative speeds of 1:4:8:16:64:128 are indicated, respectively, in the B.S.I. by 17°, 23°, 29°, 35°, and 38°; and in the American version by 12°, 18°, 24°, 30°, and 33°. Thus a degree difference of 38 degrees minus 17 degrees, or 18 degrees, corresponded to the ratio change from 1 to 128 in the B.S.I. scheme; whereas a 21-degree difference in the American scheme corresponded to that same ratio of 1 to 128 in relative speeds of the emulsions.

siegbahn. Unit for measuring wavelengths of X-rays, named for Karl Manne Georg Siegbahn (1886-1978), Swedish physicist and X-ray specialist. The siegbahn, also known as the X-unit, was very nearly 10^{-13} m in length. It was based on a definition that assigned 3,029.04 siegbahn units to the grating spacing of calcite crystals at $18°$ C, using the so-called "first order" reflection of the X-rays under test.

störmer. A unit of momentum for analyzing the behaviors of charged particles from space as they approach the Earth's magnetic field; named for Fredrik C. M. Störmer (1874–1957), Norwegian geophysicist and mathematician, a specialist in the study of auroral and polar discharges and cosmic-ray effects. His unit of momentum is the one at which a particle can circle around the equator near the surface of the Earth without being intercepted. The störmer unit is calculated from data that includes the mass of the particle, its electric charge, and its velocity as it approaches the Earth from space.

svedberg (S). A unit for measuring the rate of sedimentation of molecules of organic compounds when they are centrifuged (spun rapidly). Named for Theodor Svedberg (1884–1971), Swedish physical chemist. The S works out actually to be a unit of time: the very short time of 10^{-13} second. Sedimentation rates produced in centrifuges may measure from less than 1 S to several hundred S. Svedberg created the ultracentrifuge as a marvelous tool for analyzing the compositions and characteristics of colloidal mixtures, of both organic and inorganic substances.

talbot. A unit of luminous energy, named for the British inventor and pioneer of photography William Henry Fox Talbot (1800–77). One talbot equals 1 joule of energy of light with a luminous efficiency of 1 lumen per watt. Hence 1 talbot equals 1 lumen-second. It is sometimes called the *lumerg*. The talbot per second is identical with the lumen unit. A more exactly analogous name for it might be the *lumjoule,* rather than the lumerg; but the former has not been used.

troland. A unit of illumination of the retina of the human eye, named for Leonard Thompson Troland (1889–1932), engineer, psychologist and optical inventor. One troland is the visual stimulation experienced by a normal observer, when the entrance pupil of his eye has an opening of 1 square millimeter area, and the eye is observing a surface whose illumination is at a level of 1 international candle per square meter. The names *photon* and *luxon* have at times been used for the troland unit. The troland, in any case, is a physiological and psychophysical unit, not solely a physical one.

violle. A unit of luminous intensity, named for the French optical scientist Jules Violle (1841–1923). It is defined as the luminous intensity actually emitted by 1 cm^2 of incandescent platinum, at its "freezing" temperature under standard atmospheric pressure. By test, 1 violle equals about 20.2 candela, the SI unit of luminous intensity. The definition of the candela makes it the luminous intensity of just $\frac{1}{60}$ cm^2 of area of a perfect blackbody or "full" radiator at 2,045 K, the platinum "freezing" point. It might seem that the luminous intensity of 1 cm^2 of such platinum should be about 60 candela, rather than 20.2 as mentioned above. The difference is due to the fact that platinum is not a perfect blackbody radiator; hence it emits just a bit more than $\frac{1}{3}$ of the light intensity that such a radiator would theoretically emit. The violle unit is based on measured intensity, not on theoretical blackbody light intensity.

young. A unit suggested to measure the so-called Y stimulus in the trichromatic, or triple-primary-color, system of approximating actual colors. It is named for Thomas Young (1773–1829), many-sided British scientist and scholar, one of the pioneers of modern theories of color vision. This is the same amazing Dr. Young for whom Young's modulus is named. Among his other achievements he became a great linguist and cryptographer.

54 TIME-HONORED QUESTIONS, REFERRED TO THE READER FOR FURTHER CONSIDERATION

Mark this . . . It is the measure you give
that will be measured out to you . . .
—MARK'S VERSION OF THE GOOD NEWS
(Mark 4:24, translated by H. J. Schonfield)

I know I have the best of time and space
and was never measured and never will be measured.
—WALT WHITMAN,
Song of Myself

. . . the true beginning of our end.
—SHAKESPEARE,
A Midsummer Night's Dream

Who has cupped in his hands the water of the sea,
    and marked off the heavens with a span?
Who has held in a measure the dust of the earth,
    weighed the mountains in scales
    and the hills in a balance? . . .
Behold the nations count as a drop of the bucket,
    as dust in the scales:
    the coastlands weigh no more than powder . . .
Before him all the nations are as nought,
    as nothing and void he accounts them.
—Isaiah 40:12, 15, 17
(New American Bible trans.)

Then the Lord addressed Job out of the storm and said:
". . . I will question you, and you tell me the answers!

Where were you when I founded the earth?
   Tell me, if you have understanding.
Who determined its size; do you know?
   Who stretched out the measuring line for it?
Into what were its pedestals sunk, and who laid the cornerstone,
While the morning stars sang in chorus
   and all the sons of God shouted for joy?

      • • •

". . . have you seen the gates of darkness?
Have you comprehended the breadth of the earth?
   Tell me, if you know all:
Which is the way to the dwelling place of light,
   and where is the abode of darkness,
That you may take them to their boundaries
   and set them upon their homeward paths?

      • • •

"Have you fitted a curb to the Pleiades,[1]
   or loosened the bonds of Orion?
Can you bring forth the Mazzaroth in their season,[2]
   or guide the Bear with its train?
Do you know the ordinances of the heavens;
   can you put into effect their plan
      on the earth?

      • • •

"Can you send forth the lightnings on their way,
   or will they say to you, 'Here we are'?
Who counts the clouds in his wisdom?
   Or who tilts the water jars of heaven
So that the dust of earth is fused into a mass
   and its clods made solid?"

                  —Job 38:1–38, *passim*
                  (New American Bible trans.)

[1] Also translatable as "Can you bind up the Pleiades in a cluster," or "Can you bind up the cluster of the Pleiades . . ."
[2] ". . . bring forth the Mazzaroth . . ." means to produce or order the signs of the zodiac at their proper times.

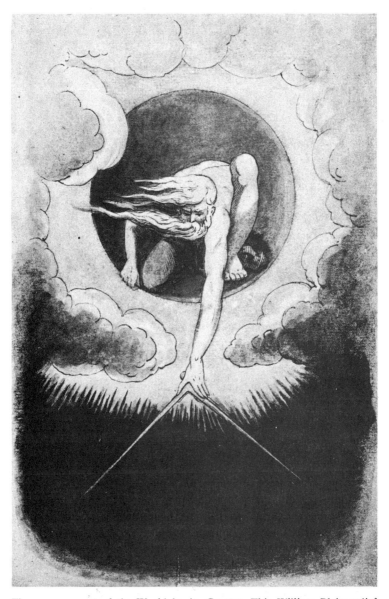

The measurement of the World by its Creator. This William Blake relief etching with watercolors is commonly known as "The Ancient of Days." It forms the frontispiece to Blake's 1794 work *Europe, a Prophecy*. In the present book, for obvious reasons, it is given the opposite position. (COURTESY OF THE TRUSTEES OF THE BRITISH MUSEUM)

Guide to Multiple and Submultiple Prefixes

The 14 presently approved prefixes are listed in alphabetical order, each followed by its abbreviation, and also by an indicator of its effect in increasing or decreasing the magnitude of the unit to which it is prefixed. Thus 10^3 means "times 1,000" for the prefix *kilo-*, while 10^{-3} means "times $1/1,000$" for the prefix *milli-*.

atto- (a) 10^{-18} · centi- (c) 10^{-2} · deci- (d) 10^{-1} · deka- (da) 10 · femto- (f) 10^{-15} · giga- (G) 10^9 · hecto- (h) 10^2 · kilo- (k) 10^3 · mega- (M) 10^6 · micro- (μ—Greek *mu*) 10^{-6} · milli- (m) 10^{-3} · nano- (n) 10^{-9} · pico- (p) 10^{-12} · tera- (T) 10^{12} ·

(The multiple prefixes giga-, mega-, and tera- are capitalized when abbreviated—as G, M, and T, respectively. Other prefixes retain lower-case when abbreviated.)

Symbols for Important Single Units of Measurement

An asterisk (*) indicates that the following unit is not within the International System (SI), but is still in fairly common use. Capital letters are used in abbreviations of unit names of eponymous origin.

A (ampere of electric current) ·*Å (angstrom of wavelength) · At (ampere-turn of magnetomotive force) · *atm (standard atmosphere, a pressure unit) · *°C (degree Celsius of temperature) · cd (candela of luminous intensity) · *dB (decibel, see *Index*) · *eV (electron volt, an energy unit) · F (farad of electric capacitance) · *°F (degree Fahrenheit of temperature) · g (gram of mass) · *gal (galileo or gal of acceleration) · *G or Gs (gauss of magnetic flux density) · *Gb or Gi (gilbert of magnetomotive force) · H (henry of inductance) · Hz (hertz of frequency) · J (joule of energy, work, or quantity of heat) · K (kelvin of temperature) · kg (kilogram of mass) · lm (lumen of luminous flux) · lx (lux of illumination) · m (meter of length) · nt (nit of luminance) · *Mx (maxwell of magnetic flux) · Ω (ohm of electrical resistance) · Pa (pascal of pressure) · rad (radian of plane angle) · s (second of time) · T (tesla of magnetic flux density) · V (volt of potential difference, electromotive force, or voltage) · W (watt of power) · Wb (weber of magnetic flux) ·

Symbols for Important Compound Units

A/m (ampere per meter of magnetic field strength) · cd/m² (candela per square meter of luminance—equal to the *nit* unit) · J/K (joule per kelvin of entropy) · J/kg·K (joule per kilogram kelvin of specific heat) · kg/m³ (kilogram per cubic meter of density) · m/s (meter per second of velocity or speed) · m/s² (meter per second squared of acceleration) · m²/s (square meter per second of kinematic viscosity) · N/m² (newton per square meter of pressure—equal to the *pascal* unit) · N·s/m² (newton second per square meter of dynamic viscosity) · rad/s (radian per second of angular velocity) · rad/s² (radian per second squared of angular acceleration) · V/m (volt per meter of electric field strength) · W/m·K (watt per meter kelvin of thermal conductivity) · W/sr (watt per steradian of radiant intensity) ·

INDEX

abnormalities related to ionizing radiations, 684–85

A-bombs, 223, 638, 688; measuring energies of, 223

absolute temperature scales, 295, 324, 392, 395

absolute viscosity, 621

absolute zero (of temperature), 290, 295, 323, 324–25, 329–30, 394

acceleration, 198, 201–05, 217, 228, 621, 702; constant or uniform, 244–45; related to power, 244–245; *see also* gravitational acceleration

accelerators, 232, 234–36, 644, 671

acoustical absorption, 710; a. impedance, 708; a. measurements, 576–92

acre, 76–77, 79

actinium transformation chain, 659–660

action, 177–79, 194, 196, 707

admittance, 504

adsorption, 696

AEC (Atomic Energy Commission), 634–37, 681, 687

air: conditioning of, 353, 355, 360; ionization of, 652–53; resistance of, 203, 245; specific gravity of, 513; viscosity of, 624, 627; weight of, 512–13, 527–28

alcohol, 176, 307, 311, 340, 536, 557–558, 623; alcoholic content, 554–557, 559–60; alcoholometers, alcoholometry, 307, 546, 559, 573

alcohol-water mixtures, 308, 311, 565

ale, 547, 550

alloys, 187, 199–200, 453–54, 473, 487

alpha particles or rays, 647, 651, 655–657, 669, 673, 678; energies of, 662, 673; penetrating and ionizing power of, 662–64

altimetry and altitude, 204, 207, 304, 531–33

aluminum, 330, 453, 536, 540–41

American Institute of Physics (AIP), 591, 694

ammonia, 155, 359, 441–42, 624

Amontons, Guillaume, 287–89, 295–296, 323; Law of, 288–89

Ampère, André Marie, 429, 433

ampere (unit of electric current), 24–25, 172, 240, 427, 438, 491, 703; absolute, 431; international, 430

ampere-hour, 456–57

ampere-turn (unit of magnetic force), 477, 483–84, 499

amplification, 602

amu (atomic mass unit), 208–10, 212

angstrom, 187, 192, 401–02, 405, 703

angular measurement, 103, 114–15, 124, 138, 148, 150, 153, 169, 314; degree, 100, 103–04, 154; minute, 101, 104, 108–09, 114; radian, 25, 100; second, 101, 110, 114, 146, 148, 150

antimony, 331

A-number (nuclear), 641, 643, 657

apostilb, 274, 699

arc (electric), 404, 449

arc of meridian, 113, 123

are, 78–79, 122–23

area, 24, 26, 39, 47, 60, 65, 74–81, 90, 113, 122–23, 125

area-expansion coefficient, 388–89

Aristotle, Aristotelian concepts, 217, 511, 517–18, 527–28, 546

"artificial" radionuclides, 644

Aston, Allen V., 143

Astronomy (Hoyle), 102

AT (symbol for atomic time), 165–166

atmosphere (as unit of pressure), 199, 288, 304–06, 413–15, 534, 701; atmospheric pressure, 199, 312, 515, 531–34, 611; *see also* standard atmospheric pressure

atmosphere (of Earth), 193, 397, 403, 412, 521; mass of, 530–31

atomic bombs, 222

atomic clocks, 133, 136–37, 144, 152, 155–67, 169, 173–76

Atomic Energy Commission, see AEC

atomic heat, 339, 342

atomic length, 702

A CATALOG OF SELECTED
DOVER BOOKS
IN ALL FIELDS OF INTEREST

A CATALOG OF SELECTED DOVER
BOOKS IN ALL FIELDS OF INTEREST

CONCERNING THE SPIRITUAL IN ART, Wassily Kandinsky. Pioneering work by father of abstract art. Thoughts on color theory, nature of art. Analysis of earlier masters. 12 illustrations. 80pp. of text. 5⅜ x 8½. 23411-8 Pa. $3.95

ANIMALS: 1,419 Copyright-Free Illustrations of Mammals, Birds, Fish, Insects, etc., Jim Harter (ed.). Clear wood engravings present, in extremely lifelike poses, over 1,000 species of animals. One of the most extensive pictorial sourcebooks of its kind. Captions. Index. 284pp. 9 x 12. 23766-4 Pa. $12.95

CELTIC ART: The Methods of Construction, George Bain. Simple geometric techniques for making Celtic interlacements, spirals, Kells-type initials, animals, humans, etc. Over 500 illustrations. 160pp. 9 x 12. (USO) 22923-8 Pa. $9.95

AN ATLAS OF ANATOMY FOR ARTISTS, Fritz Schider. Most thorough reference work on art anatomy in the world. Hundreds of illustrations, including selections from works by Vesalius, Leonardo, Goya, Ingres, Michelangelo, others. 593 illustrations. 192pp. 7⅛ x 10¼. 20241-0 Pa. $9 95

CELTIC HAND STROKE-BY-STROKE (Irish Half-Uncial from "The Book of Kells"): An Arthur Baker Calligraphy Manual, Arthur Baker. Complete guide to creating each letter of the alphabet in distinctive Celtic manner. Covers hand position, strokes, pens, inks, paper, more. Illustrated. 48pp. 8¼ x 11. 24336-2 Pa. $3.95

EASY ORIGAMI, John Montroll. Charming collection of 32 projects (hat, cup, pelican, piano, swan, many more) specially designed for the novice origami hobbyist. Clearly illustrated easy-to-follow instructions insure that even beginning papercrafters will achieve successful results. 48pp. 8¼ x 11. 27298-2 Pa. $2.95

THE COMPLETE BOOK OF BIRDHOUSE CONSTRUCTION FOR WOODWORKERS, Scott D. Campbell. Detailed instructions, illustrations, tables. Also data on bird habitat and instinct patterns. Bibliography. 3 tables. 63 illustrations in 15 figures. 48pp. 5¼ x 8½. 24407-5 Pa. $2.50

BLOOMINGDALE'S ILLUSTRATED 1886 CATALOG: Fashions, Dry Goods and Housewares, Bloomingdale Brothers. Famed merchants' extremely rare catalog depicting about 1,700 products: clothing, housewares, firearms, dry goods, jewelry, more. Invaluable for dating, identifying vintage items. Also, copyright-free graphics for artists, designers. Co-published with Henry Ford Museum & Greenfield Village. 160pp. 8¼ x 11. 25780-0 Pa. $9.95

HISTORIC COSTUME IN PICTURES, Braun & Schneider. Over 1,450 costumed figures in clearly detailed engravings—from dawn of civilization to end of 19th century. Captions. Many folk costumes. 256pp. 8⅜ x 11¾. 23150-X Pa. $12.95

STICKLEY CRAFTSMAN FURNITURE CATALOGS, Gustav Stickley and L. & J. G. Stickley. Beautiful, functional furniture in two authentic catalogs from 1910. 594 illustrations, including 277 photos, show settles, rockers, armchairs, reclining chairs, bookcases, desks, tables. 183pp. 6½ x 9¼. 23838-5 Pa. $9.95

AMERICAN LOCOMOTIVES IN HISTORIC PHOTOGRAPHS: 1858 to 1949, Ron Ziel (ed.). A rare collection of 126 meticulously detailed official photographs, called "builder portraits," of American locomotives that majestically chronicle the rise of steam locomotive power in America. Introduction. Detailed captions. xi + 129pp. 9 x 12. 27393-8 Pa. $12.95

AMERICA'S LIGHTHOUSES: An Illustrated History, Francis Ross Holland, Jr. Delightfully written, profusely illustrated fact-filled survey of over 200 American lighthouses since 1716. History, anecdotes, technological advances, more. 240pp. 8 x 10¾. 25576-X Pa. $12.95

TOWARDS A NEW ARCHITECTURE, Le Corbusier. Pioneering manifesto by founder of "International School." Technical and aesthetic theories, views of industry, economics, relation of form to function, "mass-production split" and much more. Profusely illustrated. 320pp. 6⅛ x 9¼. (USO) 25023-7 Pa. $9.95

HOW THE OTHER HALF LIVES, Jacob Riis. Famous journalistic record, exposing poverty and degradation of New York slums around 1900, by major social reformer. 100 striking and influential photographs. 233pp. 10 x 7⅞. 22012-5 Pa. $10.95

FRUIT KEY AND TWIG KEY TO TREES AND SHRUBS, William M. Harlow. One of the handiest and most widely used identification aids. Fruit key covers 120 deciduous and evergreen species; twig key 160 deciduous species. Easily used. Over 300 photographs. 126pp. 5⅜ x 8½. 20511-8 Pa. $3.95

COMMON BIRD SONGS, Dr. Donald J. Borror. Songs of 60 most common U.S. birds: robins, sparrows, cardinals, bluejays, finches, more–arranged in order of increasing complexity. Up to 9 variations of songs of each species. Cassette and manual 99911-4 $8.95

ORCHIDS AS HOUSE PLANTS, Rebecca Tyson Northen. Grow cattleyas and many other kinds of orchids–in a window, in a case, or under artificial light. 63 illustrations. 148pp. 5⅜ x 8½. 23261-1 Pa. $4.95

MONSTER MAZES, Dave Phillips. Masterful mazes at four levels of difficulty. Avoid deadly perils and evil creatures to find magical treasures. Solutions for all 32 exciting illustrated puzzles. 48pp. 8¼ x 11. 26005-4 Pa. $2.95

MOZART'S DON GIOVANNI (DOVER OPERA LIBRETTO SERIES), Wolfgang Amadeus Mozart. Introduced and translated by Ellen H. Bleiler. Standard Italian libretto, with complete English translation. Convenient and thoroughly portable–an ideal companion for reading along with a recording or the performance itself. Introduction. List of characters. Plot summary. 121pp. 5¼ x 8½. 24944-1 Pa. $2.95

TECHNICAL MANUAL AND DICTIONARY OF CLASSICAL BALLET, Gail Grant. Defines, explains, comments on steps, movements, poses and concepts. 15-page pictorial section. Basic book for student, viewer. 127pp. 5⅜ x 8½. 21843-0 Pa. $4.95

CATALOG OF DOVER BOOKS

BRASS INSTRUMENTS: Their History and Development, Anthony Baines. Authoritative, updated survey of the evolution of trumpets, trombones, bugles, cornets, French horns, tubas and other brass wind instruments. Over 140 illustrations and 48 music examples. Corrected and updated by author. New preface. Bibliography. 320pp. 5⅜ x 8½. 27574-4 Pa. $9.95

HOLLYWOOD GLAMOR PORTRAITS, John Kobal (ed.). 145 photos from 1926-49. Harlow, Gable, Bogart, Bacall; 94 stars in all. Full background on photographers, technical aspects. 160pp. 8⅜ x 11¼. 23352-9 Pa. $11.95

MAX AND MORITZ, Wilhelm Busch. Great humor classic in both German and English. Also 10 other works: "Cat and Mouse," "Plisch and Plumm," etc. 216pp. 5⅜ x 8½. 20181-3 Pa. $6.95

THE RAVEN AND OTHER FAVORITE POEMS, Edgar Allan Poe. Over 40 of the author's most memorable poems: "The Bells," "Ulalume," "Israfel," "To Helen," "The Conqueror Worm," "Eldorado," "Annabel Lee," many more. Alphabetic lists of titles and first lines. 64pp. $5\frac{3}{16}$ x 8¼. 26685-0 Pa. $1.00

PERSONAL MEMOIRS OF U. S. GRANT, Ulysses Simpson Grant. Intelligent, deeply moving firsthand account of Civil War campaigns, considered by many the finest military memoirs ever written. Includes letters, historic photographs, maps and more. 528pp. 6⅛ x 9¼. 28587-1 Pa. $11.95

AMULETS AND SUPERSTITIONS, E. A. Wallis Budge. Comprehensive discourse on origin, powers of amulets in many ancient cultures: Arab, Persian Babylonian, Assyrian, Egyptian, Gnostic, Hebrew, Phoenician, Syriac, etc. Covers cross, swastika, crucifix, seals, rings, stones, etc. 584pp. 5⅜ x 8½. 23573-4 Pa. $12.95

RUSSIAN STORIES/PYCCKNE PACCKA3bl: A Dual-Language Book, edited by Gleb Struve. Twelve tales by such masters as Chekhov, Tolstoy, Dostoevsky, Pushkin, others. Excellent word-for-word English translations on facing pages, plus teaching and study aids, Russian/English vocabulary, biographical/critical introductions, more. 416pp. 5⅜ x 8½. 26244-8 Pa. $8.95

PHILADELPHIA THEN AND NOW: 60 Sites Photographed in the Past and Present, Kenneth Finkel and Susan Oyama. Rare photographs of City Hall, Logan Square, Independence Hall, Betsy Ross House, other landmarks juxtaposed with contemporary views. Captures changing face of historic city. Introduction. Captions. 128pp. 8¼ x 11. 25790-8 Pa. $9.95

AIA ARCHITECTURAL GUIDE TO NASSAU AND SUFFOLK COUNTIES, LONG ISLAND, The American Institute of Architects, Long Island Chapter, and the Society for the Preservation of Long Island Antiquities. Comprehensive, well-researched and generously illustrated volume brings to life over three centuries of Long Island's great architectural heritage. More than 240 photographs with authoritative, extensively detailed captions. 176pp. 8¼ x 11. 26946-9 Pa. $14.95

NORTH AMERICAN INDIAN LIFE: Customs and Traditions of 23 Tribes, Elsie Clews Parsons (ed.). 27 fictionalized essays by noted anthropologists examine religion, customs, government, additional facets of life among the Winnebago, Crow, Zuni, Eskimo, other tribes. 480pp. 6⅛ x 9¼. 27377-6 Pa. $10.95

FRANK LLOYD WRIGHT'S HOLLYHOCK HOUSE, Donald Hoffmann. Lavishly illustrated, carefully documented study of one of Wright's most controversial residential designs. Over 120 photographs, floor plans, elevations, etc. Detailed perceptive text by noted Wright scholar. Index. 128pp. 9¼ x 10¾. 27133-1 Pa. $11.95

THE MALE AND FEMALE FIGURE IN MOTION: 60 Classic Photographic Sequences, Eadweard Muybridge. 60 true-action photographs of men and women walking, running, climbing, bending, turning, etc., reproduced from rare 19th-century masterpiece. vi + 121pp. 9 x 12. 24745-7 Pa. $10.95

1001 QUESTIONS ANSWERED ABOUT THE SEASHORE, N. J. Berrill and Jacquelyn Berrill. Queries answered about dolphins, sea snails, sponges, starfish, fishes, shore birds, many others. Covers appearance, breeding, growth, feeding, much more. 305pp. 5¼ x 8¼. 23366-9 Pa. $8.95

GUIDE TO OWL WATCHING IN NORTH AMERICA, Donald S. Heintzelman. Superb guide offers complete data and descriptions of 19 species: barn owl, screech owl, snowy owl, many more. Expert coverage of owl-watching equipment, conservation, migrations and invasions, etc. Guide to observing sites. 84 illustrations. xiii + 193pp. 5⅜ x 8½. 27344-X Pa. $8.95

MEDICINAL AND OTHER USES OF NORTH AMERICAN PLANTS: A Historical Survey with Special Reference to the Eastern Indian Tribes, Charlotte Erichsen-Brown. Chronological historical citations document 500 years of usage of plants, trees, shrubs native to eastern Canada, northeastern U.S. Also complete identifying information. 343 illustrations. 544pp. 6½ x 9¼. 25951-X Pa. $12.95

STORYBOOK MAZES, Dave Phillips. 23 stories and mazes on two-page spreads: Wizard of Oz, Treasure Island, Robin Hood, etc. Solutions. 64pp. 8¼ x 11. 23628-5 Pa. $2.95

NEGRO FOLK MUSIC, U.S.A., Harold Courlander. Noted folklorist's scholarly yet readable analysis of rich and varied musical tradition. Includes authentic versions of over 40 folk songs. Valuable bibliography and discography. xi + 324pp. 5⅜ x 8½. 27350-4 Pa. $7.95

MOVIE-STAR PORTRAITS OF THE FORTIES, John Kobal (ed.). 163 glamor, studio photos of 106 stars of the 1940s: Rita Hayworth, Ava Gardner, Marlon Brando, Clark Gable, many more. 176pp. 8⅜ x 11¼. 23546-7 Pa. $12.95

BENCHLEY LOST AND FOUND, Robert Benchley. Finest humor from early 30s, about pet peeves, child psychologists, post office and others. Mostly unavailable elsewhere. 73 illustrations by Peter Arno and others. 183pp. 5⅜ x 8½. 22410-4 Pa. $6.95

YEKL and THE IMPORTED BRIDEGROOM AND OTHER STORIES OF YIDDISH NEW YORK, Abraham Cahan. Film Hester Street based on Yekl (1896). Novel, other stories among first about Jewish immigrants on N.Y.'s East Side. 240pp. 5⅜ x 8½. 22427-9 Pa. $6.95

SELECTED POEMS, Walt Whitman. Generous sampling from *Leaves of Grass.* Twenty-four poems include "I Hear America Singing," "Song of the Open Road," "I Sing the Body Electric," "When Lilacs Last in the Dooryard Bloom'd," "O Captain! My Captain!"–all reprinted from an authoritative edition. Lists of titles and first lines. 128pp. 5$\frac{3}{16}$ x 8¼. 26878-0 Pa. $1.00

THE BEST TALES OF HOFFMANN, E. T. A. Hoffmann. 10 of Hoffmann's most important stories: "Nutcracker and the King of Mice," "The Golden Flowerpot," etc. 458pp. 5⅜ x 8½. 21793-0 Pa. $9.95

FROM FETISH TO GOD IN ANCIENT EGYPT, E. A. Wallis Budge. Rich detailed survey of Egyptian conception of "God" and gods, magic, cult of animals, Osiris, more. Also, superb English translations of hymns and legends. 240 illustrations. 545pp. 5⅜ x 8½. 25803-3 Pa. $11.95

FRENCH STORIES/CONTES FRANÇAIS: A Dual-Language Book, Wallace Fowlie. Ten stories by French masters, Voltaire to Camus: "Micromegas" by Voltaire; "The Atheist's Mass" by Balzac; "Minuet" by de Maupassant; "The Guest" by Camus, six more. Excellent English translations on facing pages. Also French-English vocabulary list, exercises, more. 352pp. 5⅜ x 8½. 26443-2 Pa. $8.95

CHICAGO AT THE TURN OF THE CENTURY IN PHOTOGRAPHS: 122 Historic Views from the Collections of the Chicago Historical Society, Larry A. Viskochil. Rare large-format prints offer detailed views of City Hall, State Street, the Loop, Hull House, Union Station, many other landmarks, circa 1904-1913. Introduction. Captions. Maps. 144pp. 9⅜ x 12¼. 24656-6 Pa. $12.95

OLD BROOKLYN IN EARLY PHOTOGRAPHS, 1865-1929, William Lee Younger. Luna Park, Gravesend race track, construction of Grand Army Plaza, moving of Hotel Brighton, etc. 157 previously unpublished photographs. 165pp. 8⅞ x 11¾. 23587-4 Pa. $13.95

THE MYTHS OF THE NORTH AMERICAN INDIANS, Lewis Spence. Rich anthology of the myths and legends of the Algonquins, Iroquois, Pawnees and Sioux, prefaced by an extensive historical and ethnological commentary. 36 illustrations. 480pp. 5⅜ x 8½. 25967-6 Pa. $8.95

AN ENCYCLOPEDIA OF BATTLES: Accounts of Over 1,560 Battles from 1479 B.C. to the Present, David Eggenberger. Essential details of every major battle in recorded history from the first battle of Megiddo in 1479 B.C. to Grenada in 1984. List of Battle Maps. New Appendix covering the years 1967-1984. Index. 99 illustrations. 544pp. 6½ x 9¼. 24913-1 Pa. $14.95

SAILING ALONE AROUND THE WORLD, Captain Joshua Slocum. First man to sail around the world, alone, in small boat. One of great feats of seamanship told in delightful manner. 67 illustrations. 294pp. 5⅜ x 8½. 20326-3 Pa. $5.95

ANARCHISM AND OTHER ESSAYS, Emma Goldman. Powerful, penetrating, prophetic essays on direct action, role of minorities, prison reform, puritan hypocrisy, violence, etc. 271pp. 5⅜ x 8½. 22484-8 Pa. $6.95

MYTHS OF THE HINDUS AND BUDDHISTS, Ananda K. Coomaraswamy and Sister Nivedita. Great stories of the epics; deeds of Krishna, Shiva, taken from puranas, Vedas, folk tales; etc. 32 illustrations. 400pp. 5⅜ x 8½. 21759-0 Pa. $10.95

BEYOND PSYCHOLOGY, Otto Rank. Fear of death, desire of immortality, nature of sexuality, social organization, creativity, according to Rankian system. 291pp. 5⅜ x 8½. 20485-5 Pa. $8.95

A THEOLOGICO-POLITICAL TREATISE, Benedict Spinoza. Also contains unfinished Political Treatise. Great classic on religious liberty, theory of government on common consent. R. Elwes translation. Total of 421pp. 5⅜ x 8½. 20249-6 Pa. $9.95

MY BONDAGE AND MY FREEDOM, Frederick Douglass. Born a slave, Douglass became outspoken force in antislavery movement. The best of Douglass' autobiographies. Graphic description of slave life. 464pp. 5⅜ x 8½. 22457-0 Pa. $8.95

FOLLOWING THE EQUATOR: A Journey Around the World, Mark Twain. Fascinating humorous account of 1897 voyage to Hawaii, Australia, India, New Zealand, etc. Ironic, bemused reports on peoples, customs, climate, flora and fauna, politics, much more. 197 illustrations. 720pp. 5⅜ x 8½. 26113-1 Pa. $15.95

THE PEOPLE CALLED SHAKERS, Edward D. Andrews. Definitive study of Shakers: origins, beliefs, practices, dances, social organization, furniture and crafts, etc. 33 illustrations. 351pp. 5⅜ x 8½. 21081-2 Pa. $8.95

THE MYTHS OF GREECE AND ROME, H. A. Guerber. A classic of mythology, generously illustrated, long prized for its simple, graphic, accurate retelling of the principal myths of Greece and Rome, and for its commentary on their origins and significance. With 64 illustrations by Michelangelo, Raphael, Titian, Rubens, Canova, Bernini and others. 480pp. 5⅜ x 8½. 27584-1 Pa. $9.95

PSYCHOLOGY OF MUSIC, Carl E. Seashore. Classic work discusses music as a medium from psychological viewpoint. Clear treatment of physical acoustics, auditory apparatus, sound perception, development of musical skills, nature of musical feeling, host of other topics. 88 figures. 408pp. 5⅜ x 8½. 21851-1 Pa. $10.95

THE PHILOSOPHY OF HISTORY, Georg W. Hegel. Great classic of Western thought develops concept that history is not chance but rational process, the evolution of freedom. 457pp. 5⅜ x 8½. 20112-0 Pa. $9.95

THE BOOK OF TEA, Kakuzo Okakura. Minor classic of the Orient: entertaining, charming explanation, interpretation of traditional Japanese culture in terms of tea ceremony. 94pp. 5⅜ x 8½. 20070-1 Pa. $3.95

LIFE IN ANCIENT EGYPT, Adolf Erman. Fullest, most thorough, detailed older account with much not in more recent books, domestic life, religion, magic, medicine, commerce, much more. Many illustrations reproduce tomb paintings, carvings, hieroglyphs, etc. 597pp. 5⅜ x 8½. 22632-8 Pa. $11.95

SUNDIALS, Their Theory and Construction, Albert Waugh. Far and away the best, most thorough coverage of ideas, mathematics concerned, types, construction, adjusting anywhere. Simple, nontechnical treatment allows even children to build several of these dials. Over 100 illustrations. 230pp. 5⅜ x 8½. 22947-5 Pa. $7.95

DYNAMICS OF FLUIDS IN POROUS MEDIA, Jacob Bear. For advanced students of ground water hydrology, soil mechanics and physics, drainage and irrigation engineering, and more. 335 illustrations. Exercises, with answers. 784pp. 6⅛ x 9¼. 65675-6 Pa. $19.95

SONGS OF EXPERIENCE: Facsimile Reproduction with 26 Plates in Full Color, William Blake. 26 full-color plates from a rare 1826 edition. Includes "The Tyger," "London," "Holy Thursday," and other poems. Printed text of poems. 48pp. 5¼ x 7. 24636-1 Pa. $4.95

OLD-TIME VIGNETTES IN FULL COLOR, Carol Belanger Grafton (ed.). Over 390 charming, often sentimental illustrations, selected from archives of Victorian graphics—pretty women posing, children playing, food, flowers, kittens and puppies, smiling cherubs, birds and butterflies, much more. All copyright-free. 48pp. 9¼ x 12¼. 27269-9 Pa. $5.95

PERSPECTIVE FOR ARTISTS, Rex Vicat Cole. Depth, perspective of sky and sea, shadows, much more, not usually covered. 391 diagrams, 81 reproductions of drawings and paintings. 279pp. 5⅜ x 8½. 22487-2 Pa. $6.95

DRAWING THE LIVING FIGURE, Joseph Sheppard. Innovative approach to artistic anatomy focuses on specifics of surface anatomy, rather than muscles and bones. Over 170 drawings of live models in front, back and side views, and in widely varying poses. Accompanying diagrams. 177 illustrations. Introduction. Index. 144pp. 8⅜ x11¼. 26723-7 Pa. $8.95

GOTHIC AND OLD ENGLISH ALPHABETS: 100 Complete Fonts, Dan X. Solo. Add power, elegance to posters, signs, other graphics with 100 stunning copyright-free alphabets: Blackstone, Dolbey, Germania, 97 more–including many lower-case, numerals, punctuation marks. 104pp. 8⅛ x 11. 24695-7 Pa. $8.95

HOW TO DO BEADWORK, Mary White. Fundamental book on craft from simple projects to five-bead chains and woven works. 106 illustrations. 142pp. 5⅜ x 8. 20697-1 Pa. $4.95

THE BOOK OF WOOD CARVING, Charles Marshall Sayers. Finest book for beginners discusses fundamentals and offers 34 designs. "Absolutely first rate . . . well thought out and well executed."–E. J. Tangerman. 118pp. 7¾ x 10⅝. 23654-4 Pa. $6.95

ILLUSTRATED CATALOG OF CIVIL WAR MILITARY GOODS: Union Army Weapons, Insignia, Uniform Accessories, and Other Equipment, Schuyler, Hartley, and Graham. Rare, profusely illustrated 1846 catalog includes Union Army uniform and dress regulations, arms and ammunition, coats, insignia, flags, swords, rifles, etc. 226 illustrations. 160pp. 9 x 12. 24939-5 Pa. $10.95

WOMEN'S FASHIONS OF THE EARLY 1900s: An Unabridged Republication of "New York Fashions, 1909," National Cloak & Suit Co. Rare catalog of mail-order fashions documents women's and children's clothing styles shortly after the turn of the century. Captions offer full descriptions, prices. Invaluable resource for fashion, costume historians. Approximately 725 illustrations. 128pp. 8⅜ x 11¼. 27276-1 Pa. $11.95

THE 1912 AND 1915 GUSTAV STICKLEY FURNITURE CATALOGS, Gustav Stickley. With over 200 detailed illustrations and descriptions, these two catalogs are essential reading and reference materials and identification guides for Stickley furniture. Captions cite materials, dimensions and prices. 112pp. 6½ x 9¼. 26676-1 Pa. $9.95

EARLY AMERICAN LOCOMOTIVES, John H. White, Jr. Finest locomotive engravings from early 19th century: historical (1804–74), main-line (after 1870), special, foreign, etc. 147 plates. 142pp. 11⅜ x 8¼. 22772-3 Pa. $10.95

THE TALL SHIPS OF TODAY IN PHOTOGRAPHS, Frank O. Braynard. Lavishly illustrated tribute to nearly 100 majestic contemporary sailing vessels: Amerigo Vespucci, Clearwater, Constitution, Eagle, Mayflower, Sea Cloud, Victory, many more. Authoritative captions provide statistics, background on each ship. 190 black-and-white photographs and illustrations. Introduction. 128pp. 8⅞ x 11¾. 27163-3 Pa. $13.95

EARLY NINETEENTH-CENTURY CRAFTS AND TRADES, Peter Stockham (ed.). Extremely rare 1807 volume describes to youngsters the crafts and trades of the day: brickmaker, weaver, dressmaker, bookbinder, ropemaker, saddler, many more. Quaint prose, charming illustrations for each craft. 20 black-and-white line illustrations. 192pp. 4⅝ x 6. 27293-1 Pa. $4.95

VICTORIAN FASHIONS AND COSTUMES FROM HARPER'S BAZAR, 1867–1898, Stella Blum (ed.). Day costumes, evening wear, sports clothes, shoes, hats, other accessories in over 1,000 detailed engravings. 320pp. 9⅜ x 12¼. 22990-4 Pa. $14.95

GUSTAV STICKLEY, THE CRAFTSMAN, Mary Ann Smith. Superb study surveys broad scope of Stickley's achievement, especially in architecture. Design philosophy, rise and fall of the Craftsman empire, descriptions and floor plans for many Craftsman houses, more. 86 black-and-white halftones. 31 line illustrations. Introduction 208pp. 6½ x 9¼. 27210-9 Pa. $9.95

THE LONG ISLAND RAIL ROAD IN EARLY PHOTOGRAPHS, Ron Ziel. Over 220 rare photos, informative text document origin (1844) and development of rail service on Long Island. Vintage views of early trains, locomotives, stations, passengers, crews, much more. Captions. 8⅞ x 11¾. 26301-0 Pa. $13.95

THE BOOK OF OLD SHIPS: From Egyptian Galleys to Clipper Ships, Henry B. Culver. Superb, authoritative history of sailing vessels, with 80 magnificent line illustrations. Galley, bark, caravel, longship, whaler, many more. Detailed, informative text on each vessel by noted naval historian. Introduction. 256pp. 5⅜ x 8½. 27332-6 Pa. $7.95

TEN BOOKS ON ARCHITECTURE, Vitruvius. The most important book ever written on architecture. Early Roman aesthetics, technology, classical orders, site selection, all other aspects. Morgan translation. 331pp. 5⅜ x 8½. 20645-9 Pa. $8.95

THE HUMAN FIGURE IN MOTION, Eadweard Muybridge. More than 4,500 stopped-action photos, in action series, showing undraped men, women, children jumping, lying down, throwing, sitting, wrestling, carrying, etc. 390pp. 7⅞ x 10⅝. 20204-6 Clothbd. $25.95

TREES OF THE EASTERN AND CENTRAL UNITED STATES AND CANADA, William M. Harlow. Best one-volume guide to 140 trees. Full descriptions, woodlore, range, etc. Over 600 illustrations. Handy size. 288pp. 4½ x 6⅜. 20395-6 Pa. $5.95

SONGS OF WESTERN BIRDS, Dr. Donald J. Borror. Complete song and call repertoire of 60 western species, including flycatchers, juncoes, cactus wrens, many more–includes fully illustrated booklet. Cassette and manual 99913-0 $8.95

GROWING AND USING HERBS AND SPICES, Milo Miloradovich. Versatile handbook provides all the information needed for cultivation and use of all the herbs and spices available in North America. 4 illustrations. Index. Glossary. 236pp. 5⅜ x 8½. 25058-X Pa. $6.95

BIG BOOK OF MAZES AND LABYRINTHS, Walter Shepherd. 50 mazes and labyrinths in all–classical, solid, ripple, and more–in one great volume. Perfect inexpensive puzzler for clever youngsters. Full solutions. 112pp. 8¼ x 11. 22951-3 Pa. $4.95

PIANO TUNING, J. Cree Fischer. Clearest, best book for beginner, amateur. Simple repairs, raising dropped notes, tuning by easy method of flattened fifths. No previous skills needed. 4 illustrations. 201pp. 5⅜ x 8½. 23267-0 Pa. $6.95

A SOURCE BOOK IN THEATRICAL HISTORY, A. M. Nagler. Contemporary observers on acting, directing, make-up, costuming, stage props, machinery, scene design, from Ancient Greece to Chekhov. 611pp. 5⅜ x 8½. 20515-0 Pa. $12.95

THE COMPLETE NONSENSE OF EDWARD LEAR, Edward Lear. All nonsense limericks, zany alphabets, Owl and Pussycat, songs, nonsense botany, etc., illustrated by Lear. Total of 320pp. 5⅜ x 8½. (USO) 20167-8 Pa. $6.95

VICTORIAN PARLOUR POETRY: An Annotated Anthology, Michael R. Turner. 117 gems by Longfellow, Tennyson, Browning, many lesser-known poets. "The Village Blacksmith," "Curfew Must Not Ring Tonight," "Only a Baby Small," dozens more, often difficult to find elsewhere. Index of poets, titles, first lines. xxiii + 325pp. 5⅝ x 8¼. 27044-0 Pa. $8.95

DUBLINERS, James Joyce. Fifteen stories offer vivid, tightly focused observations of the lives of Dublin's poorer classes. At least one, "The Dead," is considered a masterpiece. Reprinted complete and unabridged from standard edition. 160pp. 5³⁄₁₆ x 8¼. 26870-5 Pa. $1.00

THE HAUNTED MONASTERY and THE CHINESE MAZE MURDERS, Robert van Gulik. Two full novels by van Gulik, set in 7th-century China, continue adventures of Judge Dee and his companions. An evil Taoist monastery, seemingly supernatural events; overgrown topiary maze hides strange crimes. 27 illustrations. 328pp. 5⅜ x 8½. 23502-5 Pa. $8.95

THE BOOK OF THE SACRED MAGIC OF ABRAMELIN THE MAGE, translated by S. MacGregor Mathers. Medieval manuscript of ceremonial magic. Basic document in Aleister Crowley, Golden Dawn groups. 268pp. 5⅜ x 8½. 23211-5 Pa. $8.95

NEW RUSSIAN-ENGLISH AND ENGLISH-RUSSIAN DICTIONARY, M. A. O'Brien. This is a remarkably handy Russian dictionary, containing a surprising amount of information, including over 70,000 entries. 366pp. 4½ x 6⅛. 20208-9 Pa. $9.95

HISTORIC HOMES OF THE AMERICAN PRESIDENTS, Second, Revised Edition, Irvin Haas. A traveler's guide to American Presidential homes, most open to the public, depicting and describing homes occupied by every American President from George Washington to George Bush. With visiting hours, admission charges, travel routes. 175 photographs. Index. 160pp. 8¼ x 11. 26751-2 Pa. $11.95

NEW YORK IN THE FORTIES, Andreas Feininger. 162 brilliant photographs by the well-known photographer, formerly with *Life* magazine. Commuters, shoppers, Times Square at night, much else from city at its peak. Captions by John von Hartz. 181pp. 9¼ x 10¾. 23585-8 Pa. $12.95

INDIAN SIGN LANGUAGE, William Tomkins. Over 525 signs developed by Sioux and other tribes. Written instructions and diagrams. Also 290 pictographs. 111pp. 6⅛ x 9¼. 22029-X Pa. $3.95

ANATOMY: A Complete Guide for Artists, Joseph Sheppard. A master of figure drawing shows artists how to render human anatomy convincingly. Over 460 illustrations. 224pp. 8⅜ x 11¼. 27279-6 Pa. $10.95

MEDIEVAL CALLIGRAPHY: Its History and Technique, Marc Drogin. Spirited history, comprehensive instruction manual covers 13 styles (ca. 4th century thru 15th). Excellent photographs; directions for duplicating medieval techniques with modern tools. 224pp. 8⅜ x 11¼. 26142-5 Pa. $11.95

DRIED FLOWERS: How to Prepare Them, Sarah Whitlock and Martha Rankin. Complete instructions on how to use silica gel, meal and borax, perlite aggregate, sand and borax, glycerine and water to create attractive permanent flower arrangements. 12 illustrations. 32pp. 5⅜ x 8½. 21802-3 Pa. $1.00

EASY-TO-MAKE BIRD FEEDERS FOR WOODWORKERS, Scott D. Campbell. Detailed, simple-to-use guide for designing, constructing, caring for and using feeders. Text, illustrations for 12 classic and contemporary designs. 96pp. 5⅜ x 8½. 25847-5 Pa. $2.95

SCOTTISH WONDER TALES FROM MYTH AND LEGEND, Donald A. Mackenzie. 16 lively tales tell of giants rumbling down mountainsides, of a magic wand that turns stone pillars into warriors, of gods and goddesses, evil hags, powerful forces and more. 240pp. 5⅜ x 8½. 29677-6 Pa. $6.95

THE HISTORY OF UNDERCLOTHES, C. Willett Cunnington and Phyllis Cunnington. Fascinating, well-documented survey covering six centuries of English undergarments, enhanced with over 100 illustrations: 12th-century laced-up bodice, footed long drawers (1795), 19th-century bustles, 19th-century corsets for men, Victorian "bust improvers," much more. 272pp. 5⅜ x 8¼. 27124-2 Pa. $9.95

ARTS AND CRAFTS FURNITURE: The Complete Brooks Catalog of 1912, Brooks Manufacturing Co. Photos and detailed descriptions of more than 150 now very collectible furniture designs from the Arts and Crafts movement depict davenports, settees, buffets, desks, tables, chairs, bedsteads, dressers and more, all built of solid, quarter-sawed oak. Invaluable for students and enthusiasts of antiques, Americana and the decorative arts. 80pp. 6½ x 9¼. 27471-3 Pa. $7.95

HOW WE INVENTED THE AIRPLANE: An Illustrated History, Orville Wright. Fascinating firsthand account covers early experiments, construction of planes and motors, first flights, much more. Introduction and commentary by Fred C. Kelly. 76 photographs. 96pp. 8¼ x 11. 25662-6 Pa. $8.95

THE ARTS OF THE SAILOR: Knotting, Splicing and Ropework, Hervey Garrett Smith. Indispensable shipboard reference covers tools, basic knots and useful hitches; handsewing and canvas work, more. Over 100 illustrations. Delightful reading for sea lovers. 256pp. 5⅜ x 8½. 26440-8 Pa. $7.95

FRANK LLOYD WRIGHT'S FALLINGWATER: The House and Its History, Second, Revised Edition, Donald Hoffmann. A total revision–both in text and illustrations–of the standard document on Fallingwater, the boldest, most personal architectural statement of Wright's mature years, updated with valuable new material from the recently opened Frank Lloyd Wright Archives. "Fascinating"–*The New York Times*. 116 illustrations. 128pp. 9¼ x 10¾. 27430-6 Pa. $11.95

AUTOBIOGRAPHY: The Story of My Experiments with Truth, Mohandas K. Gandhi. Boyhood, legal studies, purification, the growth of the Satyagraha (nonviolent protest) movement. Critical, inspiring work of the man responsible for the freedom of India. 480pp. 5⅜ x 8½. (USO) 24593-4 Pa. $8.95

CELTIC MYTHS AND LEGENDS, T. W. Rolleston. Masterful retelling of Irish and Welsh stories and tales. Cuchulain, King Arthur, Deirdre, the Grail, many more. First paperback edition. 58 full-page illustrations. 512pp. 5⅜ x 8½. 26507-2 Pa. $9.95

THE PRINCIPLES OF PSYCHOLOGY, William James. Famous long course complete, unabridged. Stream of thought, time perception, memory, experimental methods; great work decades ahead of its time. 94 figures. 1,391pp. 5⅜ x 8½. 2-vol. set.
Vol. I: 20381-6 Pa. $12.95
Vol. II: 20382-4 Pa. $12.95

THE WORLD AS WILL AND REPRESENTATION, Arthur Schopenhauer. Definitive English translation of Schopenhauer's life work, correcting more than 1,000 errors, omissions in earlier translations. Translated by E. F. J. Payne. Total of 1,269pp. 5⅜ x 8½. 2-vol. set. Vol. 1: 21761-2 Pa. $11.95
Vol. 2: 21762-0 Pa. $11.95

MAGIC AND MYSTERY IN TIBET, Madame Alexandra David-Neel. Experiences among lamas, magicians, sages, sorcerers, Bonpa wizards. A true psychic discovery. 32 illustrations. 321pp. 5⅜ x 8½. (USO) 22682-4 Pa. $8.95

THE EGYPTIAN BOOK OF THE DEAD, E. A. Wallis Budge. Complete reproduction of Ani's papyrus, finest ever found. Full hieroglyphic text, interlinear transliteration, word-for-word translation, smooth translation. 533pp. 6½ x 9¼.
21866-X Pa. $10.95

MATHEMATICS FOR THE NONMATHEMATICIAN, Morris Kline. Detailed, college-level treatment of mathematics in cultural and historical context, with numerous exercises. Recommended Reading Lists. Tables. Numerous figures. 641pp. 5⅜ x 8½.
24823-2 Pa. $11.95

THEORY OF WING SECTIONS: Including a Summary of Airfoil Data, Ira H. Abbott and A. E. von Doenhoff. Concise compilation of subsonic aerodynamic characteristics of NACA wing sections, plus description of theory. 350pp. of tables. 693pp. 5⅜ x 8½. 60586-8 Pa. $14.95

THE RIME OF THE ANCIENT MARINER, Gustave Doré, S. T. Coleridge. Doré's finest work; 34 plates capture moods, subtleties of poem. Flawless full-size reproductions printed on facing pages with authoritative text of poem. "Beautiful. Simply beautiful."–*Publisher's Weekly.* 77pp. 9¼ x 12. 22305-1 Pa. $6.95

NORTH AMERICAN INDIAN DESIGNS FOR ARTISTS AND CRAFTSPEOPLE, Eva Wilson. Over 360 authentic copyright-free designs adapted from Navajo blankets, Hopi pottery, Sioux buffalo hides, more. Geometrics, symbolic figures, plant and animal motifs, etc. 128pp. 8⅜ x 11. (EUK) 25341-4 Pa. $8.95

SCULPTURE: Principles and Practice, Louis Slobodkin. Step-by-step approach to clay, plaster, metals, stone; classical and modern. 253 drawings, photos. 255pp. 8⅛ x 11.
22960-2 Pa. $10.95

PHOTOGRAPHIC SKETCHBOOK OF THE CIVIL WAR, Alexander Gardner. 100 photos taken on field during the Civil War. Famous shots of Manassas Harper's Ferry, Lincoln, Richmond, slave pens, etc. 244pp. 10⅝ x 8¼.   22731-6 Pa. $9.95

FIVE ACRES AND INDEPENDENCE, Maurice G. Kains. Great back-to-the-land classic explains basics of self-sufficient farming. The one book to get. 95 illustrations. 397pp. 5⅜ x 8½.   20974-1 Pa. $7.95

SONGS OF EASTERN BIRDS, Dr. Donald J. Borror. Songs and calls of 60 species most common to eastern U.S.: warblers, woodpeckers, flycatchers, thrushes, larks, many more in high-quality recording.   Cassette and manual 99912-2 $8.95

A MODERN HERBAL, Margaret Grieve. Much the fullest, most exact, most useful compilation of herbal material. Gigantic alphabetical encyclopedia, from aconite to zedoary, gives botanical information, medical properties, folklore, economic uses, much else. Indispensable to serious reader. 161 illustrations. 888pp. 6½ x 9¼. 2-vol. set. (USO)   Vol. I: 22798-7 Pa. $9.95
Vol. II: 22799-5 Pa. $9.95

HIDDEN TREASURE MAZE BOOK, Dave Phillips. Solve 34 challenging mazes accompanied by heroic tales of adventure. Evil dragons, people-eating plants, blood-thirsty giants, many more dangerous adversaries lurk at every twist and turn. 34 mazes, stories, solutions. 48pp. 8¼ x 11.   24566-7 Pa. $2.95

LETTERS OF W. A. MOZART, Wolfgang A. Mozart. Remarkable letters show bawdy wit, humor, imagination, musical insights, contemporary musical world; includes some letters from Leopold Mozart. 276pp. 5⅜ x 8½.   22859-2 Pa. $7.95

BASIC PRINCIPLES OF CLASSICAL BALLET, Agrippina Vaganova. Great Russian theoretician, teacher explains methods for teaching classical ballet. 118 illustrations. 175pp. 5⅜ x 8½.   22036-2 Pa. $5.95

THE JUMPING FROG, Mark Twain. Revenge edition. The original story of The Celebrated Jumping Frog of Calaveras County, a hapless French translation, and Twain's hilarious "retranslation" from the French. 12 illustrations. 66pp. 5⅜ x 8½.   22686-7 Pa. $3.95

BEST REMEMBERED POEMS, Martin Gardner (ed.). The 126 poems in this superb collection of 19th- and 20th-century British and American verse range from Shelley's "To a Skylark" to the impassioned "Renascence" of Edna St. Vincent Millay and to Edward Lear's whimsical "The Owl and the Pussycat." 224pp. 5⅜ x 8½.   27165-X Pa. $4.95

COMPLETE SONNETS, William Shakespeare. Over 150 exquisite poems deal with love, friendship, the tyranny of time, beauty's evanescence, death and other themes in language of remarkable power, precision and beauty. Glossary of archaic terms. 80pp. 5³⁄₁₆ x 8¼.   26686-9 Pa. $1.00

BODIES IN A BOOKSHOP, R. T. Campbell. Challenging mystery of blackmail and murder with ingenious plot and superbly drawn characters. In the best tradition of British suspense fiction. 192pp. 5⅜ x 8½.   24720-1 Pa. $6.95

THE WIT AND HUMOR OF OSCAR WILDE, Alvin Redman (ed.). More than 1,000 ripostes, paradoxes, wisecracks: Work is the curse of the drinking classes; I can resist everything except temptation; etc. 258pp. 5⅜ x 8½. 20602-5 Pa. $5.95

SHAKESPEARE LEXICON AND QUOTATION DICTIONARY, Alexander Schmidt. Full definitions, locations, shades of meaning in every word in plays and poems. More than 50,000 exact quotations. 1,485pp. 6½ x 9¼. 2-vol. set.
Vol. 1: 22726-X Pa. $16.95
Vol. 2: 22727-8 Pa. $16.95

SELECTED POEMS, Emily Dickinson. Over 100 best-known, best-loved poems by one of America's foremost poets, reprinted from authoritative early editions. No comparable edition at this price. Index of first lines. 64pp. 5³⁄₁₆ x 8¼.
26466-1 Pa. $1.00

CELEBRATED CASES OF JUDGE DEE (DEE GOONG AN), translated by Robert van Gulik. Authentic 18th-century Chinese detective novel; Dee and associates solve three interlocked cases. Led to van Gulik's own stories with same characters. Extensive introduction. 9 illustrations. 237pp. 5⅜ x 8½. 23337-5 Pa. $6.95

THE MALLEUS MALEFICARUM OF KRAMER AND SPRENGER, translated by Montague Summers. Full text of most important witchhunter's "bible," used by both Catholics and Protestants. 278pp. 6⅝ x 10. 22802-9 Pa. $12.95

SPANISH STORIES/CUENTOS ESPAÑOLES: A Dual-Language Book, Angel Flores (ed.). Unique format offers 13 great stories in Spanish by Cervantes, Borges, others. Faithful English translations on facing pages. 352pp. 5⅜ x 8½.
25399-6 Pa. $8.95

THE CHICAGO WORLD'S FAIR OF 1893: A Photographic Record, Stanley Appelbaum (ed.). 128 rare photos show 200 buildings, Beaux-Arts architecture, Midway, original Ferris Wheel, Edison's kinetoscope, more. Architectural emphasis; full text. 116pp. 8¼ x 11. 23990-X Pa. $9.95

OLD QUEENS, N.Y., IN EARLY PHOTOGRAPHS, Vincent F. Seyfried and William Asadorian. Over 160 rare photographs of Maspeth, Jamaica, Jackson Heights, and other areas. Vintage views of DeWitt Clinton mansion, 1939 World's Fair and more. Captions. 192pp. 8⅞ x 11. 26358-4 Pa. $12.95

CAPTURED BY THE INDIANS: 15 Firsthand Accounts, 1750-1870, Frederick Drimmer. Astounding true historical accounts of grisly torture, bloody conflicts, relentless pursuits, miraculous escapes and more, by people who lived to tell the tale. 384pp. 5⅜ x 8½. 24901-8 Pa. $8.95

THE WORLD'S GREAT SPEECHES, Lewis Copeland and Lawrence W. Lamm (eds.). Vast collection of 278 speeches of Greeks to 1970. Powerful and effective models; unique look at history. 842pp. 5⅜ x 8½. 20468-5 Pa. $14.95

THE BOOK OF THE SWORD, Sir Richard F. Burton. Great Victorian scholar/adventurer's eloquent, erudite history of the "queen of weapons"–from prehistory to early Roman Empire. Evolution and development of early swords, variations (sabre, broadsword, cutlass, scimitar, etc.), much more. 336pp. 6⅛ x 9¼.
25434-8 Pa. $9.95

THE INFLUENCE OF SEA POWER UPON HISTORY, 1660–1783, A. T. Mahan. Influential classic of naval history and tactics still used as text in war colleges. First paperback edition. 4 maps. 24 battle plans. 640pp. 5⅜ x 8½. 25509-3 Pa. $12.95

THE STORY OF THE TITANIC AS TOLD BY ITS SURVIVORS, Jack Winocour (ed.). What it was really like. Panic, despair, shocking inefficiency, and a little heroism. More thrilling than any fictional account. 26 illustrations. 320pp. 5⅜ x 8½. 20610-6 Pa. $8.95

FAIRY AND FOLK TALES OF THE IRISH PEASANTRY, William Butler Yeats (ed.). Treasury of 64 tales from the twilight world of Celtic myth and legend: "The Soul Cages," "The Kildare Pooka," "King O'Toole and his Goose," many more. Introduction and Notes by W. B. Yeats. 352pp. 5⅜ x 8½. 26941-8 Pa. $8.95

BUDDHIST MAHAYANA TEXTS, E. B. Cowell and Others (eds.). Superb, accurate translations of basic documents in Mahayana Buddhism, highly important in history of religions. The Buddha-karita of Asvaghosha, Larger Sukhavativyuha, more. 448pp. 5⅜ x 8½. 25552-2 Pa. $9.95

ONE TWO THREE . . . INFINITY: Facts and Speculations of Science, George Gamow. Great physicist's fascinating, readable overview of contemporary science: number theory, relativity, fourth dimension, entropy, genes, atomic structure, much more. 128 illustrations. Index. 352pp. 5⅜ x 8½. 25664-2 Pa. $8.95

ENGINEERING IN HISTORY, Richard Shelton Kirby, et al. Broad, nontechnical survey of history's major technological advances: birth of Greek science, industrial revolution, electricity and applied science, 20th-century automation, much more. 181 illustrations. ". . . excellent . . ."–*Isis*. Bibliography. vii + 530pp. 5⅜ x 8¼. 26412-2 Pa. $14.95

DALÍ ON MODERN ART: The Cuckolds of Antiquated Modern Art, Salvador Dalí. Influential painter skewers modern art and its practitioners. Outrageous evaluations of Picasso, Cézanne, Turner, more. 15 renderings of paintings discussed. 44 calligraphic decorations by Dalí. 96pp. 5⅜ x 8½. (USO) 29220-7 Pa. $4.95

ANTIQUE PLAYING CARDS: A Pictorial History, Henry René D'Allemagne. Over 900 elaborate, decorative images from rare playing cards (14th–20th centuries): Bacchus, death, dancing dogs, hunting scenes, royal coats of arms, players cheating, much more. 96pp. 9¼ x 12¼. 29265-7 Pa. $11.95

MAKING FURNITURE MASTERPIECES: 30 Projects with Measured Drawings, Franklin H. Gottshall. Step-by-step instructions, illustrations for constructing handsome, useful pieces, among them a Sheraton desk, Chippendale chair, Spanish desk, Queen Anne table and a William and Mary dressing mirror. 224pp. 8⅛ x 11¼. 29338-6 Pa. $13.95

THE FOSSIL BOOK: A Record of Prehistoric Life, Patricia V. Rich et al. Profusely illustrated definitive guide covers everything from single-celled organisms and dinosaurs to birds and mammals and the interplay between climate and man. Over 1,500 illustrations. 760pp. 7½ x 10⅛. 29371-8 Pa. $29.95

Prices subject to change without notice.

Available at your book dealer or write for free catalog to Dept. GI, Dover Publications, Inc., 31 East 2nd St., Mineola, N.Y. 11501. Dover publishes more than 500 books each year on science, elementary and advanced mathematics, biology, music, art, literary history, social sciences and other areas.